COMPUTATIONAL MATHEMATICS AND ANALYSIS SERIES

COMPUTATIONAL MATHEMATICS: THEORY, METHODS AND APPLICATIONS

COMPUTATIONAL MATHEMATICS AND ANALYSIS SERIES

Additional books in this series can be found on Nova's website
under the Series tab.

Additional E-books in this series can be found on Nova's website
under the E-book tab.

COMPUTATIONAL MATHEMATICS AND ANALYSIS SERIES

COMPUTATIONAL MATHEMATICS: THEORY, METHODS AND APPLICATIONS

PETER G. CHARETON
EDITOR

Nova Science Publishers, Inc.
New York

For permission to use material from this book please contact us:
Telephone 631-231-7269; Fax 631-231-8175
Web Site: http://www.novapublishers.com

LIBRARY OF CONGRESS CATALOGING-IN-PUBLICATION DATA

Computational mathematics : theory, methods and applications / editor, Peter G. Chareton.
p. cm.
Includes index.
ISBN 978-1-60876-271-2 (hardcover)
1. Numerical analysis--Data processing. I. Chareton, Peter G.
QA297.C636 2009
518.0285--dc22 2010042682

Published by Nova Science Publishers, Inc. † New York

CONTENTS

PREFACE

Computational mathematics involves mathematical research in areas of science where computing plays a central and essential role, emphasizing algorithms, numerical methods, and symbolic methods. Computation in the research is prominent. Computational mathematics emerged as a distinct part of applied mathematics by early 1950s. This new and important book gathers the latest research from around the globe in the study of this dynamic field and highlights such topics as: coherence-homotopies of higher order, Vandermonde Systems: theory and application, numerical conformal mappings for waveguides, computational study of 3D affine transformation, commutativity formulas for fundamental group entropy, the completion of fuzzy metric spaces, and others

Chapter 1 - Analytical and numerical spectral methods are developed for the linear (normal-mode) instability study of steady solutions to the nonlinear barotropic vorticity equation (BVE) governing the motion of a two-dimensional ideal incompressible fluid on a rotating sphere. The four types of BVE solutions known up to now are considered, namely, the zonal (one-dimensional) flows in the form of a Legendre-polynomial (LP) of degree n, and such non-zonal (two-dimensional) flows as Rossby-Haurwitz (RH) waves, Wu-Verkley (WV) waves and modons. A unified approach to the normal-mode instability study of these solutions is suggested. A conservation law for disturbances of each steady solution is derived and then used to obtain a necessary condition for its normal-mode (exponential) instability. According to these conditions, Fjörtoft's [1] average spectral number of the amplitude of any unstable mode must be equal to a special value. In the case of the LP flows or RH waves, this value is related only with the basic flow degree. Unlike these results, the above-mentioned value for the WV waves and modons depends both on the basic flow degree and on the spectral distribution of the mode energy in the inner and outer regions of the basic flow. Peculiarities of the instability conditions for different types of modons are also discussed. The new instability conditions specify the spectral structure of growing normal-mode disturbances localizing them in the phase space. Note that for the LP flows, the new condition complements the well-known Rayleigh-Kuo and Fjörtoft conditions related to the zonal flow profile. As to more complicated two-dimensional steady flows (RH waves, WV waves and modons), the instability conditions obtained here have no analogues. The maximum growth rate of unstable modes is estimated too, and the orthogonality of any unstable, decaying and non-stationary mode to the basic flow is shown in the energy inner product. The numerical spectral method for the normal mode instability study consists of three parts: the calculation of the elements of stability matrix, the solution of the complete eigenvalue problem, and the construction of unstable normal modes on the sphere. The spectral method uses the spherical harmonics as the basic functions for representing solutions and disturbances by their Fourier series on the sphere. Some analytical and numerical examples are given too.

It should be stressed that the analytical instability results obtained here can also be applied for testing the accuracy of computational programs and algorithms used for the numerical stability study. Also note that Fjörtoft's spectral number appearing both in the instability conditions and in the maximum growth rate estimates is considered as the key parameter in the linear instability problem of ideal flows on a sphere.

Chapter 2 - Leonhard Euler's influence on mathematics was enormous. He wrote an impressive quantity of papers that contained innumerable new results and his innovative techniques and procedures led a profound change in the structure of mathematics and in its basic principles. In this article, after a brief description of Euler's life, I will discuss some aspects of his works to pure and mixed mathematics. In particular, I will deal with his contributions to the rise of the concept of functions, to development of the theories of series and differential equations. I will also mention Euler's study on the calculus of variations and will, finally, highlight the importance of his studies in mechanics, especially the point-mass mechanics, rigid body dynamics, and fluid dynamics.

Chapter 3 - This chapter puts into practice a geometrical model of political economics studies developed by the authors, which comprises the consideration of a problem of political competition with two opposed parties. They will attempt to capture the greatest number of voters of a discrete population of elements. It is supposed that these parties can modify their policies to a certain degree. The authors' purpose is to determine the optimun position or positions for the party in terms of guaranteeing the maximum number of voters.

On one hand, an algorithm is implemented that tries to solve this problem in a certain case of Spanish politics, simulated in some data partially based on the survey of public opinion and Fiscal Politics Study n° 2615 (July 2005) of the CIS (Sociological Investigations Center of Spain). On the other hand, the possible Nash equilibrium positions are studied under these certain limits. Due to the discrete characteristics of the problem, computational geometry techniques are applied.

Chapter 4 - The notion of an inverse system is well known and widely used in mathematics.The category of inverse systems pro-HTop is the fundamental tool in establishing of proper homotopy and shape theory. Using homotopies of higher orders, the authors will present notions of coherent map, coherent homotopy and coherent inverse system, corresponding categories and their relations.

This manuscript is organized as follows:

- Introduction
- Coherent systems and coherent maps
- Level coherent category
- Coherent shift and coherent category
- Relations of coherent categories
- Appendix: Strict ordering vs ordering for directed sets

In Section 1 is presented the construction of the category pro-HTop with objects arbitrary inverse systems (usually, by HTop is denoted the category of topological spaces and homotopy classes), and of the coherent category CPHTop with objects strictly commutative inverse systems and coherent maps of order ∞.

The inverse systems could be strictly commutative or commutative up to homotopy. The authors introduce the notion of coherent inverse system, the system which is commutative up to homotopy, and homotopies are connected with homotopies of higher and higher order (coherence of order ∞).

The notion of a coherent map of order ∞ between such system is introduced and the notion of level homotopy of two coherent maps

This allows in Section 2. to deal with the level coherent category Coh^{A} over a directed set A.

Explicitly is constructed a level category pro-H^2Top^N of arbitrary inverse sequences and strong fundamental sequences (maps between sequences of order of coherence 2)

It is shown that the subcategory pro-H^2Top^N having as objects coherent inverse sequences is isomorphic to Coh^{N} – level coherent category over natural numbers.

As an application, the following theorem is proven. If in a strictly commutative inverse sequence the spaces are replaced by homotopy equivalent spaces, one obtains an inverse sequence, which is commutative only up to homotopy. The two inverse sequences are obviously isomorphic in the category pro-HTop.It is proven that two inverse sequences are isomorphic in pro-H^2Top^N, and consequently in Coh^{N}. Moreover, they are isomorphic in the category Coh, constructed in Section 3.

In Section 3 by introducing the notion of coherent shift it is defined the notion of homotopy between two coherent maps of coherent inverse systems and is formed the category Coh of coherent inverse systems and coherent maps defined using homotopies of all orders (i.e. the order of coherence is ∞).

In Section 4, different types of coherent categories are considered: Coh - category of coherent inverse systems and coherent maps, CPHTop - category of strictly commutative inverse systems and coherent maps. It is showt that CPHTop is a full subcategory of the category Coh [13].

In the Appendix it is shown that in the construction of coherent category CPHTop of commutative inverse systems the authors can use the strict ordering $<$ for directed sets as well as more usual ordering $\leq$. The resulting coherent categories are isomorphic.

Chapter 5 - The authors investigate the application of a meshless method for the stable and accurate solution of inverse problems associated with two-dimensional Helmholtz-type equations in domains with smooth and piecewise smooth boundaries, as well as in the presence of boundary singularities. More precisely, the governing equation and available boundary conditions are discretized by the method of fundamental solutions (MFS). It is well known that inverse problems are ill-posed and, consequently, their associated MFS system of linear algebraic equations is ill-conditioned. Therefore, the stability of the solutions to the aforementioned inverse problems is a key issue and this is usually taken into account by employing a regularization method. These difficulties are overcome in the case of the first type of inverse problems investigated herein, i.e. inverse problems for Helmholtz-type equations in two-dimensional domains withoutboundary singularities, by considering the Tikhonov regularization method (TRM).

The existence of boundary singularities affects adversely the accuracy and convergence of standard numerical methods. Solutions to such problems and/or their corresponding derivatives may have unbounded values in the vicinity of the singularity. In this situation, these difficulties are overcome by combining the TRM with the subtraction from the original

MFS solution of the corresponding singular solutions/eigenfunctions, without an appreciable increase in the computational effort and at the same time keeping the same MFS discretization. Examples for both the Helmholtz and the modified Helmholtz equations, as well as the types of inverse problems considered, are presented and carefully investigated.

Chapter 6 - The Vandermonde matrix, defined by $\widetilde{V}_n(i,j) = x_i^{j-1}, x_j \in \mathbb{C},$ is ubiquitous in mathematics and engineering. Its use includes, for example, polynomial interpolation, coding theory and signal processing, where the matrix for the discrete Fourier transform is a Vandermonde matrix. There is an extensive literature on numerically solving systems of linear equations when the matrix is given by a Vandermonde matrix. The objective of this chapter is twofold: to gain structural properties and explicit formulas useful to further analytic works in many problems directly connected to the inversion of the Vandermonde matrix and to the least-squares problems; to gain, as a consequence, fast and efficient algorithms.

Chapter 7 - In this chapter the authors consider different semilocal convergence results applied to Newton's method to numerically solve nonlinear equations. Let $\{x_n\}_n^{\infty} = 0$ be the Newton sequence to approximate the root of x^* a nonlinear equation ($f(x)$=0). A semilocal convergence result imposes conditions on the starting point (x_0) in order to guarantee the convergence of the Newton sequence to the root x^*. Here the authors consider the well known Kepler's equation

$$E = e\sin E + M,$$

as a function test in order to compare some well known semilocal convergence results. So the authors consider Kantorovich's theorem, Smale's α-theory and some of their variants. These kind of results are not only convergence theorems. In fact, they can be seen as existence and uniqueness results. So the authors generalize the work of Zhen Shi-Ming to obtain the values of the eccentricity e for which Kepler's equation has a solution just by applying one of the aforementioned semilocal convergence results. In addition, the authors compare the regions of existence and uniqueness of solution given by these theorems.

Chapter 8 - The discrete maximum principle (DMP) is an important qualitative property of various discretized elliptic equations. Conditions that ensure the DMP have drawn much attention, including geometric properties for FEM discretizations. This chapter starts with a brief summary of some background on the DMP, including the algebraic case, and nonobtuseness or acuteness type conditions for FEM. When lower order terms are included in the operator, the DMP can be ensured for sufficiently fine mesh, under uniform acuteness or strict non-narrowness in the case of simplicial or rectangular FEM meshes, respectively. (Similar conditions also appear for prismatic FEM.)

The authors' main interest is formed by nonlinear elliptic systems under standard linear or bilinear FEM discretizations. The authors first present their previous results on systems with second and zeroth order terms, then extend them to the case involving first order terms. The presentation includes a detailed exposition of the required theory, which needs a generalization of the usual underlying algebraic DMP and some Hilbert space background. The geometric properties of the FEM mesh are also discussed. In many applications the DMP implies (or reduces to) a natural requirement of nonnegativity for the approximations of the corresponding nonnegative physical quantities. Such applications are given to reaction-diffusion processes and diffusion-dominated transport systems, respectively.

Chapter 9 - Acoustic or electro-magnetic scattering in a waveguide with varying direction and cross-section can, if the variations takes place in only one dimension at a time be re-formulated as a two-dimensional scattering problem. By using the so-called Building Block Method, it is possible to construct the scattering properties of a combination of scatterers when the properties of each scatterer are known. Hence, variations in the waveguide geometry or in the boundary conditions can be treated one at a time.

The authors consider in this work acoustic scattering, but the same techniques can be used for both electro-magnetic and some quantum scattering problems.

By suppressing the time dependence and by using the Building Block Method, the problem takes the form of the Helmholtz equation in a waveguide of infinite length and with smoothly varying geometry and boundary conditions. A conformal mapping is used to transform the problem into a corresponding problem in a straight horizontal channel, and by expanding the field in Fourier trigonometric series, the problem can be reformulated as an infinite-dimensional ordinary differential equation. From this, numerically solvable differential equations for the reflection and transmission operators are derived.

To be applicable in the Building Block Method, the numerical conformal mapping must be constructed such that the direction of the boundary curve can be controlled. At the channel ends, it is an indispensable requirement, that the two boundary curves are (at least) asymptotically parallel and straight. Furthermore, to achieve bounded operators in the differential equations, the boundary curves must satisfy different regularity conditions, depending on the properties of the boundary.

Several methods to accomplish such conformal mappings are presented. The Schwartz-Christoffel mapping, which is a natural starting point and for which also efficient numerical software exists, can be modified in different ways to round the polygon corners, and the authors show algorithms by which the parameter problem can be solved after such modifications. It is also possible to use the unmodified Schwartz-Christoffel mapping for regions with smooth boundary, by constructing an appropriate outer polygon to the considered region. Finally, the authors show how a so-called zipper algorithm can be used for waveguides.

Chapter 10 - According to the results of the authors' computational study, up to now the most effective method to compute the parameters of 3D affine transformation model is based on a symbolic-numeric algorithm. This algorithm computes the parameters of the 3-point problem employing analytical expression developed by computer algebra (Dixon resultant or reduced Groebner basis), then using these values as initial values for a Newton - Raphson method with Krylov iteration to solve the N-point problem of the determined model developed from the overdetermined model via computer algebra. This method is fast, robust and has very low complexity. Criteria for selecting an appropriate triplet from data points is also given.

However, for relatively small system, $N < 200$ data points, the solution of the original overdetermined system via Newton – Raphson method with deflation can be more efficient.

Although, in case of the Helmert transformation (7 parameters), the general Procrustes method is very efficient, its application for 3D affine transformation needs a time consuming iteration process if the scale factors are strongly different, therefore up to now it seems not a good choice.

Chapter 11 - In computers the discrete plane/space is used since this world is digital. The discrete space is usually defined by a grid. There are three regular grids in two dimensions:

the square grid, the hexagonal grid and the triangular grid. The square grid is the most used due to the simplicity of the Cartesian coordinate frame. There are two types of neighbors defined naturally: the city-block and the chessboard neighborhoods. The hexagonal grid is the simplest one, since there is only one natural neighborhood among the hexagons of the grid. The triangular grid is a little bit sophisticated (there are three types of natural neighborhoods), however it has some nice and interesting properties. In the digital space the usage of the usual (Euclidean) distance may lead to some strange phenomena. Instead, path-based, so-called digital distances can be used based on the neighborhood structure of the grid. Digital distances are frequently used in computerized applications of geometry, e.g., in image processing, in computer graphics. There are two main approaches to define digital distances: distances based on neighborhood sequences in which the used types of neighbors is varied along a path; and weighted distances, where various types of steps have various weights (lengths). In the present chapter distances based on neighborhood sequences on the triangular grid are detailed. First, an effective coordinate system is presented to the grid. By this system the grid can handle as easily, as the square grid by the Cartesian frame. After the definitions of neighborhood sequences, paths and distances, some results are detailed: a greedy algorithm that provides a shortest path, formula to compute the distance from a point to another point defined by a given neighborhood sequence. Interesting properties of these distances, such as non-metrical distances are shown (triangular inequality and symmetry may be violated). A necessary and sufficient condition to define metrical distances is proved. Some details on digital circles based on these distances are also presented. Finally some further directions of research and open problems close the chapter.

Chapter 12 - The concept of Σ-products was introduced and the normality has been investigated since 1959. This article is a survey for this study in a half century. There seems to be a certain stream in its long history. The purpose of this article is to show the vague stream as visible as possible. The contents consist of the following four chapters, each of which has several sections. The authors start from the beginning of the study and end with the most recent results which will appear in the near future.

Chapter 13 - The problem of the completion of fuzzy metric spaces is one that has received major attention by researchers in the area of fuzzy topology. In fact, several authors have studied such a problem for different notions of metric fuzziness. In this article the authors discuss the fuzzy metric completion for fuzzy metric spaces in the sense of Kramosil and Michalek. This notion of metric fuzziness has an evident appeal due to its close relationship with probabilistic metric spaces. In particular, it is well known that Sherwood's construction of the completion of a Menger space endowed with a continuous t-norm, can be extended in a natural way to the fuzzy setting to deduce that every fuzzy metric space (in the sense of Kramosil and Michalek) has a completion which is unique up to isometry. The authors apply this construction to the study of the problem of the completion of strong fuzzy metric spaces, non-Archimedean fuzzy metric spaces, fuzzy metric groups and intuitionistic fuzzy metric spaces. The completion of fuzzy metric spaces (in the sense of George and Veeramani) is also discussed.

Chapter 14 - Homotopy theory has been sporadically applied to economic theory mainly in order to simplify the aggregation of preferences among agents (decision makers) in social choice and to design stable algorithms in computable general equilibrium models. These applications, while dealing with relevant issues, do not consider explicitly the influence of information asymmetries on the behaviour of agents, which constitutes a leading argument in current economic theory.

The present paper aims to fill the existing gap and illustrate the main consequences derived from applying homotopy theory to an economic system where agents are asymmetrically informed. Indeed, the authors show that, when information asymmetries among agents are explicitly considered, a homotopic approach can be used as a destabilizing device in economic equilibrium theory.

The authors use homotopy techniques to illustrate how the information sets determining the choices of agents can be modified to induce any a priori assigned economic equilibrium. More precisely, the authors investigate the conditions under which a homotopy can be defined such that a predetermined choice is imposed on an economic agent. In this way, choices and, consequently, equilibria are proved to be perfectly manipulable when such conditions apply.

Besides its already important economic applications, the authors' model displays immediate extensions in management and decision theory.

Chapter 15 – Presents a biographical article about Minggen Cui, who is an outstanding pioneer in the field of reproducing kernel theory.

In: Computational Mathematics
Editor: Peter G. Chareton, pp.1-34

ISBN:978-1-60876-271-2

Chapter 1

ANALYTICAL AND NUMERICAL METHODS IN THE LINEAR STABILITY STUDY OF IDEAL FLOWS ON A SPHERE

Yuri N. Skiba*

Centro de Ciencias de la Atmósfera, Universidad Nacional Autónoma de México, Ciudad Universitaria, México, D.F., 04510, Mexico

ABSTRACT

Analytical and numerical spectral methods are developed for the linear (normal-mode) instability study of steady solutions to the nonlinear barotropic vorticity equation (BVE) governing the motion of a two-dimensional ideal incompressible fluid on a rotating sphere. The four types of BVE solutions known up to now are considered, namely, the zonal (one-dimensional) flows in the form of a Legendre-polynomial (LP) of degree n, and such non-zonal (two-dimensional) flows as Rossby-Haurwitz (RH) waves, Wu-Verkley (WV) waves and modons. A unified approach to the normal-mode instability study of these solutions is suggested. A conservation law for disturbances of each steady solution is derived and then used to obtain a necessary condition for its normal-mode (exponential) instability. According to these conditions, Fjörtoft's [1] average spectral number of the amplitude of any unstable mode must be equal to a special value. In the case of the LP flows or RH waves, this value is related only with the basic flow degree. Unlike these results, the above-mentioned value for the WV waves and modons depends both on the basic flow degree and on the spectral distribution of the mode energy in the inner and outer regions of the basic flow. Peculiarities of the instability conditions for different types of modons are also discussed. The new instability conditions specify the spectral structure of growing normal-mode disturbances localizing them in the phase space. Note that for the LP flows, the new condition complements the well-known Rayleigh-Kuo and Fjörtoft conditions related to the zonal flow profile. As to more complicated two-dimensional steady flows (RH waves, WV waves and modons), the instability conditions obtained here have no analogues. The maximum growth rate of unstable modes is estimated too, and the orthogonality of any

* Tel. : (52-55) 5622-4247, Fax: (52-55) 5622-4090, 5616-0789, E-mail: skiba@servidor.unam.mx

unstable, decaying and non-stationary mode to the basic flow is shown in the energy inner product. The numerical spectral method for the normal mode instability study consists of the three part: the calculation of the elements of stability matrix, the solution of the complete eigenvalue problem, and the construction of unstable normal modes on the sphere. The spectral method uses the spherical harmonics as the basic functions for representing solutions and disturbances by their Fourier series on the sphere. Some analytical and numerical examples are given too.

It should be stressed that the analytical instability results obtained here can also be applied for testing the accuracy of computational programs and algorithms used for the numerical stability study. Also note that Fjörtoft's spectral number appearing both in the instability conditions and in the maximum growth rate estimates is considered as the key parameter in the linear instability problem of ideal flows on a sphere.

Keywords: Ideal incompressible fluid, vorticity equation solutions, linear instability

1. INTRODUCTION

In hydrodynamics, the vorticity equation

$$\Delta\psi_t + J(\psi, \Delta\psi + 2\mu) = 0 \tag{1}$$

governs the motion of an ideal incompressible fluid on a rotating unit sphere S. In equation (1) written in the non-dimensional form, $\psi(t,x) \equiv \psi(t,\lambda,\mu)$ is the stream function, μ is the sine of latitude, λ is the longitude, $\Delta\psi$ and $\Omega = \Delta\psi + 2\mu$ are the relative and absolute vorticity, respectively, Δ is the Laplace operator on S and

$$J(\psi, f) = (\vec{k} \times \nabla\psi) \cdot \nabla f = \psi_\lambda f_\mu - \psi_\mu f_\lambda \tag{2}$$

is the Jacobian ($\vec{k}$ is the unit normal to S). Hereinafter, f_t, f_λ and f_μ denote correspondingly the derivatives of $f(t,\lambda,\mu)$ with respect to t, λ and μ.

Since equation (1) captures many features of the large-scale dynamics of the barotropic atmosphere, it is also used in meteorology under the name of barotropic vorticity equation (BVE). Thus, the stability of exact solutions to this equation is not only interesting hydrodynamic problem, but also important problem of atmospheric dynamics providing insight into deeper understanding of the low-frequency atmosphere variability and climate predictability [2-6]. The four classes of BVE solutions have been known by now: the simple zonal flows $\psi(\mu)$ depending only on μ, and the Rossby-Haurwitz (RH) waves [7-9], Wu-Verkley (WV) waves [10] and modons [11-16] whose streamfunction $\psi(t,\lambda,\mu)$ depends on all time-space variables.

The linear stability of shear flows has been studied since the classical work by Lord Rayleigh [17] who obtained the necessary condition for instability. According to this condition, the velocity $U(y)$ of an unstable shear flow must have an inflection point at some point y_i, that is, $U_{yy}(y_i)=0$. This condition was later improved by Fjörtoft's [18] who proved that $U_{yy}(U-U_i)$ must be negative in some point of any unstable flow ($U_i \equiv U(y_i)$). In the case of zonal flows on a sphere, the instability condition by Rayleigh is known as the Rayleigh-Kuo condition [19]. Since the Legendre polynomial (LP) flows form an orthogonal basis in the set of zonal flows, their instability is of special interest [20-22].

Despite of many works devoted to the instability of BVE solutions (see [23, 24]), this problem is still far from its complete resolution even for the shear (or zonal) flows. Indeed, the instability criterion given in [25] can be used only for very simple plane parallel flows, while Tung's instability criterion [26] requires finding a special function that makes its application not easier than that of the famous theorems by Liapunov [27] and Arnold [28]. As a result, for half a century, the classical conditions by Rayleigh-Kuo and Fjörtoft have been the only simple and constructive methods for checking the linear stability of zonal flows [24]. Both conditions are only the necessary instability conditions, and, for example on a sphere, any sufficiently strong LP flow of degree $n \geq 3$ satisfies them. Thus, the effectiveness of these conditions can be quite scanty. Moreover, in the case of an unstable flow, they provide no information on the growth rate and time-space structure of unstable disturbances. Some limits on the growth rate of unstable modes is set by Howard's semicircle theorem [29, 30]. A necessary condition for the instability of LP flow of degree n as well as an estimate of the maximum growth rate of unstable modes were obtained in [31]. This condition complements those by Rayleigh-Kuo and Fjörtoft. During the last decades, new important results on the spectrum of linearized operator and stability of ideal plane (mainly shear) flows have been obtained by Friedlander, Latushkin, Lin, Vishik, Yudovich, Howard, Strauss, Marchioro, Pulvirenti, Belenkaya, Stanislavova, Shvydkoy, Chen, Bärmann, Gierling and Rebhan [32-50]. It should be noted that the linear instability of ideal flows is generally a more difficult problem than that of viscous flows. Indeed, unlike a viscous fluid, the spectrum of linearized operator for an ideal fluid may have a continuous part and finite accumulation points [22, 38, 44-46, 51-54]. The spectrum of linearized operator is studied in [35, 38, 39, 45, 46], the stability of eigenfunctions of the Laplacian in the 2D Euler equation is discussed in [24] in the case of a flat 2D torus, and some classes of unstable flows are obtained in [39, 41, 47, 48] (see also the classical works [49, 50]).

The stability of the Rossby waves on the β-plane has been studied in [55-57] and on a sphere - in [3, 31, 58-60]. The instability of the WV waves and modons has been analyzed numerically in [4, 13, 16].

On a sphere, the first instability conditions for the RH waves, WV waves and modons were developed just recently [31, 61]. In the present work, we give a unified approach to the normal mode instability study of such stationary solutions to equation (1) as the LP flow, RH wave, WV wave, monopole, dipole and quadrupole modons. A conservation law for infinitesimal perturbations to either solution is derived and used to obtain a necessary condition for its exponential (normal mode) instability. The condition imposes a restriction on the mean spectral number χ of the amplitude of each unstable mode (here χ is the square of spectral number by Fjörtoft [1]), that is, on the spectral distribution of the mode energy. More precisely, $\chi = n(n+1)$ for unstable modes of any LP flow or RH wave of degree n, and $\chi^{-1} = \delta\chi_\sigma^{-1} + (1-\delta)\chi_\alpha^{-1}$ for unstable modes of a WV wave or any modon, except for the modon with uniform absolute vorticity whose unstable modes must satisfy the condition $\chi = \chi_\sigma$. Here δ is the fraction of the mode enstrophy corresponding to the outer region of the solution (see (41)), $\chi_\alpha = \alpha(\alpha+1)$, $\chi_\sigma = \sigma(\sigma+1)$, and α and σ are the degrees of the spherical harmonics representing the solution in its inner and outer regions, respectively. Thus, the instability conditions for the LP flows, the RH waves and the modons with uniform absolute vorticity depend only on the solution degree. Unlike this, the instability conditions for WV waves and all other types of modons depend not only on the solution degree, but also on the spectral distribution of the mode energy in the inner and outer regions of the solution.

The new instability conditions specify the spectral structure of growing disturbances localizing them in the phase space. They are especially useful for testing the computational programs and algorithms designed for the linear instability study [2, 16, 22, 52-54]. In the case of a Legendre-polynomial flow, the new condition complements the well-known conditions by Rayleigh-Kuo and Fjörtoft [17-19].

The maximum bounds of the growth rate of unstable modes are also estimated, and the orthogonality of the amplitude of each unstable, decaying, or non-stationary mode to the basic solution is shown in the energy inner product. It should be noted that both the instability conditions and the estimates of the maximum growth rate of unstable modes use the mean spectral number by Fjörtoft [1]. In this connection we can say that it is the parameter of paramount importance in the linear instability problem of ideal flows on a sphere.

2. Hilbert Spaces and Geometric Structure of Smooth Functions on a Sphere

We note that if $\psi(t,\lambda,\mu)$ is a solution to (1) then $\psi(t,\lambda,\mu)+Const$ is the solution, too. Therefore, without loss of generality, we will consider in this work only the functions which are orthogonal to any constant function on the sphere. Let $C_0^\infty(S)$ denote the set of infinitely differential functions $f(x)$ such that

$$\int_S f(x)\,dx = 0 \tag{3}$$

In this work we will use the three inner products of these functions defined as

$$\langle f,h\rangle_k = \int_S (-\Delta)^k f(x)\overline{h(x)}\,dx\,,\ (k=0,1,2) \tag{4}$$

and the corresponding norms

$$\|f\|_k = \langle f,f\rangle_k^{1/2}\,,\ (k=0,1,2) \tag{5}$$

where $\overline{f}$ is the complex conjugate of *f*. The three Hilbert (Sobolev) spaces obtained by closing the set $C_0^\infty(S)$ in the norms (5) we denote as $W_2^k(S)$. Hereinafter, the L_2-inner product and norm will be denoted as $\langle\cdot,\cdot\rangle \equiv \langle\cdot,\cdot\rangle_0$ and $\|\cdot\| \equiv \|\cdot\|_0$, respectively. The space

$$W_2^0(S) \equiv L_2(S) = H_1 \oplus H_2 \oplus \ldots \oplus H_n \oplus \ldots, \tag{6}$$

is the orthogonal sum of subspaces

$$H_n = \{\, p(x):\ -\Delta p = n(n+1)p \,\} \tag{7}$$

of homogeneous spherical polynomials of degree *n* [62]. The subspace H_n is of dimension $2n+1$, and each its polynomial is the eigenfunction of spherical Laplace operator $-\Delta$ corresponding to the eigenvalue

$$\chi_n = n(n+1) \tag{8}$$

The $2n+1$ spherical harmonics $Y_n^m(x)$ of degree n and zonal number m ($-n \le m \le n$) form an orthonormal basis in H_n [63]:

$$\langle Y_n^k(x), Y_n^m(x) \rangle = \delta_{km} \tag{9}$$

where δ_{km} is the Kronecker delta. As a result, all the spherical harmonics $Y_n^m(x)$ ($n = 1, 2, 3, \ldots;$ $-n \le m \le n$) form an orthonormal basis in $L_2(S)$: $\langle Y_l^k(x), Y_n^m(x) \rangle = \delta_{km}\delta_{ln}$

Due to (6), any function $f(x)$ of $L_2(S)$ is represented by its Fourier series $f(x) = \sum_{n=1}^{\infty} f_n(x)$ where

$$f_n(x) = (2n+1)(f * P_n)(x) = \sum_{m=-n}^{n} f_n^m Y_n^m(x) \tag{10}$$

is the homogeneous spherical polynomial of degree n from H_n, and represents the orthogonal projection of $f(x)$ on H_n; the operation $*$ is the convolution of $f(x)$ with the

Legendre polynomial $P_n(\mu)$ of degree *n*, and $f_n^m = \left\langle f(x), Y_n^m(x) \right\rangle$, $(-n \le m \le n)$ is the Fourier coefficient of $f(x)$ [62-64].

Since $\|\nabla f\| = \|f\|_1$ [53], the Poincaré-Steklov inequality on a sphere is $\|f\| \le 2^{-1/2} \|f\|_1$

In order to estimate the distribution of the kinetic energy between different scales of an ideal incompressible two-dimensional flow, Fjörtoft [1] introduced the mean spectral number

$$\rho(\psi) = \|\psi\|_2 / \|\psi\|_1 \tag{11}$$

of the stream function $\psi(t,x)$. This is a square root of the ration of the enstrophy

$$\tfrac{1}{2}\|\psi\|_2^2 \equiv \tfrac{1}{2}\|\Delta\psi\|^2 \tag{12}$$

of flow to its kinetic energy

$$\tfrac{1}{2}\|\psi\|_1^2 \equiv \tfrac{1}{2}\|\nabla\psi\|^2 \tag{13}$$

Hereinafter, we will call $\|\cdot\|_1$ the energy norm, $\|\cdot\|_2$ - the enstrophy norm, $\langle\cdot,\cdot\rangle$ - L_2- inner product, and $\langle\cdot,\cdot\rangle_1$ - the energy (or W_2^1-) inner product.

3. Integral Formulas Related to the Jacobian

It is well known [62] that

$$\int_S J(\psi, f)dS = 0 \tag{14}$$

Let now $G = \{x \in S : \mu \in (\mu_a, 1]\}$ be a part of *S* bounded by a latitudinal circle $\mu = \mu_a$ $(-1 < \mu_a < 1)$. If the circle is an isoline of $\psi(x)$ then

$$\psi_\lambda(\lambda, \mu_a) = 0 \tag{15}$$

Since $J(\psi, f) = (\psi_\lambda f)_\mu - (\psi_\mu f)_\lambda$, we obtain

$$\int_G J(\psi, f)dS = 0 \tag{16}$$

Obviously, equation (16) is also valid for any periodic channel $G = \{(\lambda, \mu) \in S : \mu \in [\mu_a, \mu_b]\}$ on S provided that

$$\psi_\lambda(\lambda, \mu_a) = \psi_\lambda(\lambda, \mu_b) = 0 \tag{17}$$

Since $J(\psi, f)h = J(\psi, fh) - J(\psi, h)f$, equation (16) leads to

$$\int_G J(\psi, f) h dS = -\int_G J(\psi, h) f dS \tag{18}$$

which is valid for any complex differentiable functions ψ, h and f in G provided that ψ satisfies (15) or (17). The substitution $h = f$ in (18) gives

$$\int_G J(\psi, f) f dS = 0 \tag{19}$$

Let ψ be a real function and $f = f_r + \boldsymbol{i} f_i$. Then (18) leads to $\int_G J(\psi, f)\bar{f} dS = -\int_G J(f, \psi)\bar{f} dS = \int_G J(f, \bar{f})\psi dS$.

Since $J(f, \bar{f}) = -2\boldsymbol{i}J(f_r, f_i)$, we obtain

$$\operatorname{Re} \int_G J(\psi, f)\bar{f} dS = 0 \tag{20}$$

In particular, if $G = S$, then (19) and (20) lead to

$$\langle J(\psi, f), \bar{f} \rangle = 0, \ \operatorname{Re}\langle J(\psi, f), f \rangle = 0 \tag{21}$$

Also, we will use the following assertion [65]:

Lemma 1

Let $\psi(x)$, $f(x)$ and $h(x)$ be sufficiently smooth complex functions on the sphere S, and let $F(\psi)$ be a continuously differentiable function. Then

$$\langle J(\psi, f), h \rangle = -\langle J(f, \psi), h \rangle = \langle J(f, \bar{h}), \bar{\psi} \rangle = -\langle J(\psi, \bar{h}), \bar{f} \rangle, \tag{22}$$

$$\left\langle J(\psi,h),\overline{F(\psi)}\right\rangle = 0, \tag{23}$$

$$\mathrm{Re}\left\langle J(\psi,\mu),\psi\right\rangle = 0, \tag{24}$$

$$\mathrm{Re}\left\langle J(\psi,\Delta\psi),\mu\right\rangle = 0, \ \mathrm{Re}\left\langle J(\psi,\mu),\Delta\psi\right\rangle = 0 \tag{25}$$

If, $G = \{x \in S : \mu \in (\mu_a, 1]\}$ and $\psi(x)$ satisfies (15), or if $G = \{(\lambda,\mu) \in S : \mu \in [\mu_a, \mu_b]\}$ is a periodic channel and $\psi(x)$ satisfies (17) then

$$\int_G J(\psi,f)h\,dS = -\int_G J(f,\psi)h\,dS = \int_G J(f,h)\psi\,dS = -\int_G J(\psi,h)f\,dS \tag{26}$$

4. Steady BVE Solutions on a Rotating Sphere

In this work, we will consider the exponential instability of the four types of real steady solutions to (1):

(1) The LP flow

$$\psi(\mu) = aP_n(\mu) \tag{27}$$

(2) The RH wave

$$\psi(\lambda,\mu) = -\omega\mu + \psi_n(\lambda,\mu) \equiv -\omega\mu + \sum_{m=-n}^{n} \psi_n^m Y_n^m(\lambda,\mu) \tag{28}$$

where $\psi_n \in H_n$, $n \geq 2$, $\omega = 2/(\chi_n - 2)$, χ_n is given by (8), ψ_n^m is arbitrary for $m > 0$, and $\psi_n^{-m} = (-1)^m \overline{\psi}_n^m$ for a real wave [66-68]. Each RH wave represents a super-rotation flow $\psi(\mu) = -\omega\mu$ perturbed by its $2n+1$ neutral modes of eigensubspace H_n corresponding to an isolated eigenvalue with finite multiplicity $2n+1$. Besides, the isolated eigenvalues have the finite accumulation point $\nu = \omega$ [54].

(3) The WV wave

$$\psi(\lambda,\mu) = \begin{cases} X_i(\lambda,\mu) - \omega_i\mu + d_i, & \text{in } S_{in} \\ X_o(\lambda,\mu) - \omega_o\mu + d_o, & \text{in } S_{out} \end{cases} \tag{29}$$

where $S_{in} = \{(\lambda,\mu) \in S : \mu \in (-\mu_0, \mu_0)\}$ and $S_{out} = S \setminus S_{in}$, $0 < \mu_0 < 1$, and ω_i, ω_o, d_i and d_o are certain constants [10]. We will refer to S_{in} and S_{out} as to the "inner" and "outer" regions of the WV wave on the sphere, respectively. The wave (29) is antisymmetric about the equator ($\mu = 0$), and is a particular form of the general solution given by Wu [69]. Both X_i and X_o are eigenfunctions of the Laplace operator $-\Delta$ on S corresponding to the eigenvalues

$$\chi_\alpha = \alpha(\alpha+1) \text{ and } \chi_\sigma = \sigma(\sigma+1) \tag{30}$$

with real numbers α and σ, and $\omega_i = 2/(\chi_\alpha - 2)$, $\omega_o = 2/(\chi_\sigma - 2)$ for the steady WV wave. Note that, by construction,

$$\psi_\lambda(\lambda, \mu_0) = 0 \text{, and } \psi_\lambda(\lambda, -\mu_0) = 0 \tag{31}$$

(4) Modon by Verkley [12-14] and Neven [15, 16]

$$\psi(\lambda,\mu) = \tilde{\psi}(\lambda',\mu') = \begin{cases} X_i(\lambda',\mu') - \omega_i\mu + d_i, & \text{in } S_{in} \\ X_o(\lambda',\mu') - \omega_o\mu + d_o, & \text{in } S_{out} \end{cases} \tag{32}$$

where (λ',μ') is a new system of coordinates on the sphere whose north pole has coordinates (λ_0,μ_0) in the system (λ,μ), $S_{in} = \{(\lambda',\mu') \in S : \mu' > \mu_a\}$ and $S_{out} = \{(\lambda',\mu') \in S : \mu' < \mu_a\}$ are the inner and outer regions of the modon separated by the circle $\mu' = \mu_a$, and ω_i, ω_o, d_i and d_o are some constants. The modon centre moves along the latitudinal circle $\mu = \mu_0$ with a constant velocity *C*. It is shown in [60] that $\psi_{\lambda'}(\lambda', \mu_a) = C\sqrt{1-\mu_a^2}\sqrt{1-\mu_0^2}\sin\lambda'$ for Verkley's modon, and hence,

$$\psi_{\lambda'}(\lambda', \mu_a) = 0 \tag{33}$$

for a steady dipole modon (*C*=0), or monopole modon ($1-\mu_0^2 = 0$), i.e., the boundary $\mu' = \mu_a$ between S_{in} and S_{out} is a streamline for such modons. Using formulas (19), (20), (34), (35), (42) and (43) of work [15], one can make sure that condition (33) is valid for any steady quadrupole modon as well.

5. Conservation Law for Perturbations to LP Flows and RH Waves

Let ψ be some LP flow (27) and let $\tilde{\psi}$ be another solution to BVE (1). Considering $\psi' = \tilde{\psi} - \psi$ as a perturbation of the LP flow and using equation $\Delta\psi = -\chi_n \psi$, one can rewrite (1) as

$$\Delta\psi'_t + J(\psi + \psi', \Delta\psi' + \chi_n \psi') + 2J(\psi', \mu) = 0 \tag{34}$$

For a steady RH wave (28), (34) is reduced to

$$\Delta\psi'_t + J(\psi + \psi', \Delta\psi' + \chi_n \psi') = 0 \tag{35}$$

because $\Delta\psi + 2\mu = -\chi_n \psi$.

Thus, the linearized equation

$$\Delta\psi'_t + J(\psi, \Delta\psi') + J(\psi', \Delta\psi + 2\mu) = 0 \tag{36}$$

governing infinitesimal perturbations $\psi'(\lambda, \mu, t)$ of a steady solution $\psi(\lambda, \mu)$ accept the form

$$\Delta\psi'_t + J(\psi, \Delta\psi' + \chi_n \psi') + 2J(\psi', \mu) = 0 \tag{37}$$

for the LP flow (27), and the form

$$\Delta\psi'_t + J(\psi, \Delta\psi' + \chi_n \psi') = 0 \tag{38}$$

for the steady RH wave (28).

Taking the L_2-inner product of equation (37) with $\Delta\psi' + \chi_n \psi'$ and using (21), (24) and (25) we obtain a conservation law

$$\left[\eta(t) - \chi_n K(t)\right]_t = 0 \tag{39}$$

for disturbances to the LP flow, that is,

$$\eta_t(t) = \chi_n K_t(t) \tag{40}$$

where $\chi_n = n(n+1)$,

$$K(t) = \tfrac{1}{2}\|\psi'(t)\|_1^2 \tag{41}$$

is the perturbation energy, and

$$\eta(t) = \tfrac{1}{2}\|\psi'(t)\|_2^2 \tag{42}$$

is the perturbation enstrophy. Taking the L_2-inner product of (38) with $\Delta\psi' + \chi_n\psi'$ and using (21) we obtain that infinitesimal perturbations to the RH wave obey the same law (39).

Note that arbitrary real (not only infinitesimal) perturbations to the LP flow or RH wave also obey the law (39), since solution $\tilde{\psi} = \psi + \psi'$ is real, and the real part of L_2-inner product of $J(\tilde{\psi}, \Delta\psi' + \chi_n\psi')$ with $\Delta\psi' + \chi_n\psi'$ is equal to zero again due to (21). We recapitulate this result as

Theorem 1

Any infinitesimal complex perturbation (as well as arbitrary real perturbation) to the LP flow (27) or RH wave (28) obeys the conservation law (39), and its energy (41) and enstrophy (42) decrease, remain constant or increase simultaneously according to (40).

The law (39) was first obtained in [21] for infinitesimal perturbations to the LP flow. The Theorem 1 generalizes this law to arbitrary perturbations of LP flow and RH wave. Thus, although the energy norm $\|\cdot\|_1$ is generally weaker than the enstrophy norm $\|\cdot\|_2$, both norms give the same information on the linear and nonlinear stability of the LP flows and RH waves. It is interesting to note that Lin [43] constructed in a special domain on a plane a steady flow which is nonlinearly and linearly stable in the enstrophy norm but linearly unstable in the energy norm.

6. Conservation Law for Infinitesimal Perturbations to WV Waves and Modons

Let now $\psi(x)$ be a stationary WV wave (29), or modon (32). We now show that an infinitesimal perturbation $\psi'(x,t)$ of either solution is governed by equation

$$\Delta\psi'_t + J(\psi, \Delta\psi' + q\psi') = 0 \tag{43}$$

where

$$q(x) = \begin{cases} \chi_\alpha, & \text{if } x \in S_{in} \\ \chi_\sigma, & \text{if } x \in S_{out} \end{cases} \tag{44}$$

Indeed,

$$\Delta\psi = \begin{cases} -\chi_\alpha\psi + (2-\chi_\alpha)\omega_i\mu + \chi_\alpha d_i, & \text{in } S_{in} \\ -\chi_\sigma\psi + (2-\chi_\sigma)\omega_o\mu + \chi_\sigma d_o, & \text{in } S_{out} \end{cases} \tag{45}$$

Taking into account that $(2-\chi_\alpha)\,\omega_i = (2-\chi_\sigma)\,\omega_o = -2$ for either solution, we obtain

$$\Delta\psi + 2\mu = \begin{cases} -\chi_\alpha\,[\psi - d_i], & \text{in } S_{in} \\ -\chi_\sigma\,[\psi - d_o], & \text{in } S_{out} \end{cases} \tag{46}$$

Substituting (46) in (36), we obtain (43) where $q(x)$ is the step function (44) with a discontinuity of the first kind on the boundary of regions S_{in} and S_{out}.

Let us multiply (43) by $\overline{\Delta\psi' + q\psi'}$, then integrate the result first over inner region S_{in} and then over outer region S_{out} and take the real parts of the two equations obtained. Due to (31) and (33), (20) is valid both for $G = S_{in}$ and for $G = S_{out}$ giving

$$\chi_\alpha^{-1}\eta_t^{(i)} = -\mathrm{Re}\int_{S_{in}} \Delta\psi_t' \cdot \overline{\psi'}\, dS, \quad \chi_\sigma^{-1}\eta_t^{(o)} = -\mathrm{Re}\int_{S_{out}} \Delta\psi_t' \cdot \overline{\psi'}\, dS \tag{47}$$

where $\eta^{(i)}(t) = \frac{1}{2}\int_{S_{in}} |\Delta\psi'|^2\, dS$, and $\eta^{(o)}(t) = \frac{1}{2}\int_{S_{out}} |\Delta\psi'|^2\, dS$ are the parts of perturbation enstrophy (42) concentrated in S_{in} and S_{out}, respectively. The summation of equations (47) gives a conservation law for disturbances to a WV wave or modon:

Theorem 2

Any infinitesimal perturbation to steady WV wave (29), or modon (32) evolves so that

$$\left[\chi_\alpha^{-1}\,\eta^{(i)} + \chi_\sigma^{-1}\,\eta^{(o)} - K\right]_t = 0 \tag{48}$$

where $K(t)$ is the perturbation energy (41).

Note that conservation laws (39) and (48) are related with the basic flow only by means of degrees of the spherical harmonics forming this flow (degree n in (50), and degrees α and σ in (60)).

7. Unified Conservation Law for Disturbances of BE Solutions

We now combine (39) and (48) into one law. Denoting by

$$\chi(t) = \rho^2(t) = \eta(t) / K(t) \tag{49}$$

the square of Fjörtoft's spectral number (11) of disturbance ψ', and by

$$\delta(t) = \eta^{(o)}(t) / \eta(t) \tag{50}$$

and $1-\delta(t) = \eta^{(i)}(t) / \eta(t)$ the portions of perturbation enstrophy corresponding to the regions S_{out} and S_{in}, respectively ($0 \le \delta \le 1$), we can rewrite (48) as $\left\{[\chi(t)(\delta\, \chi_\sigma^{-1} + (1-\delta)\chi_\alpha^{-1}) - 1]\, K(t)\right\}_t = 0$.

Thus, the two laws (39) and (48) can be unified into one equation:

$$\left\{[p(t)-1]\, K(t)\right\}_t = 0 \tag{51}$$

where $p(t)$ is the spectral characteristic of disturbance $\psi'(\lambda, \mu, t)$, besides,

$$p(t) = \chi(t)\, \chi_n^{-1} = \chi(t)[n(n+1)]^{-1} \tag{52}$$

for LP flow (27) or steady RH wave (28), and

$$p(t) = \chi(t)\, [\delta\chi_\sigma^{-1} + (1-\delta)\chi_\alpha^{-1}] \tag{53}$$

for steady WV wave (29) or modon (32).

Due to (51), all the disturbances $\psi'(\lambda, \mu, t)$ can be divided into three sets:

$$\begin{aligned} M_+ &= \{\psi' : p(t) > 1\} \\ M_0 &= \{\psi' : p(t) = 1\} \\ M_- &= \{\psi' : p(t) < 1\} \end{aligned} \tag{54}$$

The set M_0 is a hypersurface separating the set M_- of large-scale disturbances and the set M_+ of small-scale disturbances. By (52) and (53), the spectral number $\chi(t)$ of each disturbance of M_0 is

$$\chi(t)=\begin{cases} \chi_n \quad , & \text{for the LP flow and RH wave} \\ \left\{ \delta\chi_\sigma^{-1}+(1-\delta)\chi_\alpha^{-1}\right\}^{-1}, & \text{for the WV wave and modon} \end{cases} \tag{55}$$

Note that in the case of LP flows and RH waves, M_0 contains the subspace H_n of homogeneous spherical polynomials of degree *n*, besides, any perturbation of H_n is neutral [66]. Also, Theorem 1 asserts that sets (54) are invariant sets for arbitrary (not only infinitesimal) perturbations to the LP flow (27) and RH wave (28).

8. Instability Conditions for LP Flows, RH Waves, WV Waves and Modons

The existence of continuous spectrum and finite accumulation points makes increased demands to the accuracy of numerical algorithms used to construct normal modes, especially in the case of such not so smooth solutions as WV waves and modons [16, 52-54]. In this connection, analytical instability results are of great importance for checking numerical algorithms and computational programs. We now obtain one of such analytical results, namely, a necessary condition for the normal mode (exponential) instability of solutions. Let ψ be one of the four steady BVE solutions: a LP flow, RH wave, WV wave, or modon. A normal mode of ψ can be written as

$$\psi'(t,\lambda,\mu)=\Psi(\lambda,\mu)\exp\{\nu t\}, \tag{56}$$

where

$$\Psi(\lambda,\mu)=R(\mu)\exp\{\,im\lambda\} \tag{57}$$

for any LP flow or monopole modon. Here *i* is the imaginary unit, $\nu=\nu_r+i\nu_i$, and Ψ is the mode amplitude. The real part ν_r of ν determines the growth (or decay) rate of the mode amplitude, whereas its imaginary part ν_i characterizes the mode frequency. Thus, a mode is unstable if $\nu_r>0$, decaying if $\nu_r<0$, and neutral if $\nu_r=0$. The mode energy and enstrophy are

$$K(t)=K_\Psi\exp(2\nu_r t) \text{ and } \eta(t)=\eta_\Psi\exp(2\nu_r t) \tag{58}$$

where

$$K_\Psi = \tfrac{1}{2}\sum_{n=1}^{\infty}\chi_n \sum_{m=-n}^{n}\left|\Psi_n^m\right|^2 \text{ and } \eta_\Psi = \tfrac{1}{2}\sum_{n=1}^{\infty}\chi_n^2 \sum_{m=-n}^{n}\left|\Psi_n^m\right|^2 \tag{59}$$

are respectively the total energy and enstrophy of amplitude $\Psi(\lambda,\mu)$ to mode (56).

Due to (58) and (59), the spectral number $\chi(t)$ of a mode is time-independent and coincides with the spectral number χ_Ψ of its amplitude:

$$\chi(t) \equiv \eta(t)/K(t) = \eta_\Psi / K_\Psi = \chi_\Psi \tag{60}$$

The mode values δ and p (defined by (50) and (52) or (53)) are also time-independent, and (51) and (58) lead to $\nu_r(p-1)K_\Psi = 0$.

For a growing mode $\nu_r > 0$, and hence, the last equation is fulfilled only if

$$p = 1 \tag{61}$$

By condition (61), exponential growth of energy of a mode is possible only in the set M_0 where $p=1$. This fact guarantees the fulfillment of conservation law (51) under exponential growth [70].

Taking into account (52), (53) and (60) we obtain necessary conditions for the normal mode (exponential) instability:

Theorem 3

Let $n \ge 1$, and $0 < |m| < n$. The spectral number (60) of amplitude $\Psi(\lambda,\mu)$ to any unstable mode (56),(57) of the LP flow (27) must satisfy the condition

$$\chi_\Psi = \chi_n = n(n+1) \tag{62}$$

In Theorem 3, we consider only the case $0 < |m| < n$, because any mode with $|m| \ge n$ and $m = 0$ is stable [22, 65]. Thus, for the LP flows, condition (62) complements the conditions by Rayleigh-Kuo and Fjörtoft related to the structure (profile) of the basic flow velocity (or enstrophy). It shows that the spectral number (geometric structure) of unstable disturbances strictly depends on the degree of polynomial flow. Besides, the greater is the basic flow degree *n*, the smaller is the geometric scale of unstable modes.

Example 1

A Super-Rotation Flow

$$\psi(\mu) = aP_1(\mu) = a\mu\,, \tag{63}$$

(or LP flow of degree one), is exponentially stable and all its modes are neutral [65]. Indeed, the Rayleigh-Kuo condition is not fulfilled, since the absolute vorticity derivative $\Omega_\mu = \Delta\psi_\mu + 2 = 2(\omega+1)$ does not change its sign in interval $-1 < \mu < 1$ [19]. This result immediately follows from Theorem 3 as well, because we must examine only modes with $|m| = 1$. Let $m = 1$. Then $\chi_\Psi > 2$ for any disturbance formed by spherical harmonic $Y_n^1(\lambda,\mu)$ ($n \geq 1$), while any disturbance proportional to $Y_1^1(\lambda,\mu)$ is neutral. In fact, flow (63) is Liapunov stable [60], and hence, linearly stable. Moreover, the energy and enstrophy of arbitrary perturbation to any flow of subspace H_1 is invariant [66].

Example 2

Let us Consider Some LP Flow of Degree Two

$$\psi(\mu) = aP_2(\mu) = \frac{a}{2}(3\mu^2 - 1)\,, a > \sqrt{2}/9 \tag{64}$$

Since $\Omega_\mu(\mu) = (\Delta\psi + 2\mu)_\mu = 2(1 - 9a\mu)$ changes its sign at $\mu_0 = 1/9a$, the Rayleigh-Kuo condition is satisfied, and flow (64) may be unstable. According to Fjörtoft's condition [18], flow (64) may be unstable as well, since flow velocity $u(\mu) = -\sqrt{1-\mu^2}\,\psi_\mu = -3a\mu\sqrt{1-\mu^2}$ is a decreasing function in $(-1/\sqrt{2}, 1/\sqrt{2})$, and hence,

$$F(\mu) \equiv \Omega_\mu(\mu)[u(\mu) - u(\mu_0)] > 0 \text{ for each } \mu \in (1/9a, 1/\sqrt{2}) \tag{65}$$

Thus, Fjörtoft's instability condition is also satisfied. However, flow (64) is linearly stable for any amplitude a, since new condition (62) is not satisfied. Indeed, by Theorem 3, we must analyze only modes with $|m| = 1$ or $|m| = 2$. But any disturbance formed by a single harmonic $Y_1^m(\lambda,\mu)$ (or $Y_2^m(\lambda,\mu)$) is invariant and neutral. Thus, for each m (equal to ± 1 or ± 2), (62) is not fulfilled since $\chi_\Psi > 6$ for any disturbance formed by the harmonics $Y_n^m(\lambda,\mu)$ with $n \geq 3$. We see that unlike (62), the conditions by Rayleigh-Kuo and Fjörtoft are useless for flow (64). Note that exponential stability of $P_2(\mu)$ was first numerically shown in [26].

Example 3

In fact, flow (64) is linearly Liapunov stable [60], and hence, exponentially and algebraically stable. We now show that it is Liapunov stable. Indeed, the projection a disturbance $\psi'(t)$ onto subspace H_1 is invariable, and any perturbation of subspace H_2 is stable, since $J(\psi+\psi',\Delta\psi'+\chi_2\psi')\equiv 0$ in (34), and $\psi'(t)$ evolves according to the linear equation $\psi'_t + C_n\psi'_\lambda = 0$ conserving its form, and hence, its energy and enstrophy. Then the subspace of neutral perturbations $H_1 \oplus H_2$ can be taken as the 0-class of factor space $C_0^\infty(S)/H_1 \oplus H_2$ of perturbations to flow (64), besides, two elements g and h belong to the same class only if $g-h\in H_1\oplus H_2$, and for any ψ', the corresponding class is $\{\psi'+h:\ h\in H_1\oplus H_2\}$. We now can consider only the representative perturbations ψ' orthogonal to $H_1\oplus H_2$. Then $V(t)\equiv\eta(t)-\chi_n K(t)=\sum_{n=3}^{\infty}\chi_n(\chi_n-\chi_2)\sum_{m=-n}^{n}\left|\psi_n'^m(t)\right|^2$ is the Liapunov function since it is a positive definite and $(1-\chi_2/\chi_3)\|\psi'\|_2^2 \le V(t) \le \|\psi'\|_2^2$ where $\|\psi'\|_2=\left(\sum_{n=3}^{\infty}\chi_n^2\sum_{m=-n}^{n}\left|\psi_n'^m(t)\right|^2\right)^{1/2}$ is the factor norm. Thus, (39) implies the Liapunov stability of flow (64).

Theorem 4

Let ψ be a steady RH wave (39), WV wave (40) or modon (43). Then the spectral number (85) of amplitude $\Psi(\lambda,\mu)$ to any unstable mode (79) of ψ must satisfy the condition

$$\chi_\Psi=\begin{cases} n(n+1) & \text{, if } \psi \text{ is the RH wave} \\ [\delta\chi_\sigma^{-1}+(1-\delta)\chi_\alpha^{-1}]^{-1}, & \text{if } \psi \text{ is the WV wave, or modon} \end{cases} \tag{66}$$

Note that any mode of a monopole modon with $|m|(|m|+1)>\max\{\chi_\alpha,|\chi_\sigma|\}$ is neutral [60].

9. Peculiarities of Instability Conditions for Wv Waves and Modons

We now consider peculiarities of the instability condition (66) for the WV waves and three types of modons.

1. Non local BVE solutions. Let ψ be a WV wave, or a modon by Verkley [13] or Neven [15]. Then $\chi_\sigma>0$ and $\chi_\alpha>0$, and due to (66), $\chi_\Psi^{-1}=\delta\chi_\sigma^{-1}+(1-\delta)\chi_\alpha^{-1}$ is a linear combination of χ_σ^{-1} and χ_α^{-1}. Thus, for each such a solution, the spectral number χ_Ψ of amplitude $\Psi(\lambda,\mu)$ of a growing mode is

always between the numbers χ_σ and χ_α. In particular case when all the energy of unstable mode is concentrated only in the inner region S_{in} (or only in the outer region S_{out}) then $\delta = 0$ ($\delta = 1$) and $\chi_\Psi = \chi_\alpha$ ($\chi_\Psi = \chi_\sigma$).

2. Modons with uniform absolute vorticity. Let ψ be a modon with uniform absolute vorticity in the region S_{in} [14]. According to the theorem from Appendix B of [14], the vorticity $\Delta\Psi(\lambda,\mu)$ of amplitude of any unstable mode is zero in S_{in}, that is, $\delta = 1$, and due to (66), a normal mode (56) of such a modon may be unstable only if

$$\chi_\Psi = \chi_\sigma \tag{67}$$

Thus, the instability condition for such special modon depends only on the degree of spherical harmonic that forms the modon in its outer region S_{out}.

3. Isolated modons. Let ψ be a localized modon [12] of complex degree $\sigma = -0.5 + ik$ with $\chi_\sigma = -(k^2 + 0.25) < 0$ and $\chi_\alpha = \alpha(\alpha+1) > 0$. By Theorem 4, mode (56) of this modon may be unstable only if $\chi_\Psi^{-1} = \delta\chi_\sigma^{-1} + (1-\delta)\chi_\alpha^{-1}$. Since χ_Ψ^{-1} is positive, we obtain a restriction on the part of perturbation enstrophy δ that an unstable mode can have in the region S_{out}:

$$0 \le \delta < \delta_{cr} = |\chi_\sigma|(\chi_\alpha + |\chi_\sigma|)^{-1} = (0.25 + k^2)(\chi_\alpha + 0.25 + k^2)^{-1} < 1 \tag{68}$$

As α grows or/and k decreases, the critical value δ_{cr} decreases monotonically reducing the interval $0 \le \delta < \delta_{cr}$ of potential instability. Thus, unstable disturbances of any isolated modon with rather large α and small k are mainly localized in S_{in}. In particular, if $\delta = 1$ (that is, the mode amplitude satisfies the condition $\Delta\Psi = 0$ in S_{in}) then the mode is neutral. Note that unlike this, a mode of any modon of type 2 with $\delta = 1$ satisfies the necessary instability condition (67). Further, it follows from (66) that $\chi_\Psi = \chi_\alpha\left(1 - \delta/\delta_{cr}\right)^{-1} \ge \chi_\alpha$ for unstable modes of an isolated modon, besides, the minimal value $\chi_\psi = \chi_\alpha$ corresponds to the case $\delta = 0$ when $\Delta\Psi = 0$ in the region S_{out}.

10. Estimates of the Maximum Growth Rate of Unstable Modes

We now estimate the maximum growth rate of unstable modes for the solutions under consideration. First, assume that the basic flow ψ is the LP flow (27) or RH wave (28). Substituting (56) into (37) and (38) we obtain

$$\nu\Delta\Psi + J(\psi, D_n\Psi) + 2J(\Psi, \mu) = 0 \tag{69}$$

$$\nu\Delta\Psi + J(\psi, D_n\Psi) = 0 \tag{70}$$

where

$$D_n = \Delta + \chi_n \tag{71}$$

is the orthogonal projector onto the subspace orthogonal to H_n. Thus the amplitude Ψ can be represented as the orthogonal sum $\Psi = D_n\Psi + \Psi_n$ where Ψ_n is the projection of Ψ onto H_n.

Taking the real part of the L_2-inner product of equation (69) (or equation (70)) with $\Delta\Psi$ and using (21) and (25), we get in both cases the same result:

$$\nu_r \|\Delta\Psi\|^2 = -\chi_n \operatorname{Re}\langle J(\psi,\Psi), \Delta\Psi\rangle \tag{72}$$

Thus,

$$\operatorname{Re}\langle J(\psi,\Psi), \Delta\Psi\rangle < 0 \tag{73}$$

is the necessary and sufficient condition for the mode instability. In particular, it immediately follows from (73) and (25) that the super-rotation flow (that is, LP flow of degree one) is linearly stable. Also, for arbitrary flow ψ on *S*, every zonal disturbance $\Psi(\mu)$ as well as every disturbance $\Psi \in H_k$ ($k = 1, 2, \ldots$) is stable, due to (73) and (22). Unfortunately, this condition is not easy to apply for arbitrary flow. Estimating $J(\psi, \Psi)$ with the L_2-norm and using the formula $\vec{u} = \vec{k} \times \nabla\psi$ for the solution velocity ($\vec{k}$ is the unit normal to the sphere *S*), we obtain $\|J(\psi,\Psi)\| \le \max_S |\nabla\psi| \cdot \|\nabla\Psi\| \le \max_S |\vec{u}| \cdot \|\nabla\Psi\| = C \cdot \|\Psi\|_1$ where

$$C = \max_S |\nabla\psi| = \max_S |\vec{u}| \tag{74}$$

is the maximum velocity of basic flow on *S*. The application of Schwartz inequality to inner product in (72) leads to $|\nu_r| \le C\chi_n \dfrac{\|\Psi\|_1}{\|\Delta\Psi\|} = C\chi_n \dfrac{\|\Psi\|_1}{\|\Psi\|_2} = C\chi_n\,\chi_\Psi^{-1/2} = C\,\chi_n^{1/2}$.

In deriving this estimate we used formulas (60), (59) and the fact that $\chi_\Psi = \chi_n$ for unstable modes (Theorem 3 and 4). Thus, we proved:

Theorem 5

The maximum growth rate of an unstable mode of the LP flow (27) or RH wave (28) is limited:

$$|\nu_r| \le \sqrt{n(n+1)}\,\max_S |\vec{u}| \tag{75}$$

It should be noted that number *n* in (75) characterizes not only the degree of basic flow, but also Fjörtoft's spectral number of unstable disturbance. We know that the geometric scale of unstable modes increases with the basic flow scale. Due to (75), the growth rate decreases with velocity $|\vec{u}|$ and number *n*. Thus, for the same maximum velocity, the larger is the scale of basic flow the smaller is the growth rate of its unstable modes.

Let now ψ be a steady WV wave (29) or modon (32). Substituting mode (56) into (43) we obtain

$$\nu\Delta\Psi + J(\psi, \Delta\Psi + q\Psi) = 0 \tag{76}$$

The real part of L_2-inner product of (76) with $\Delta\Psi$ gives

$$\nu_r \|\Delta\Psi\|^2 = -\operatorname{Re}\langle J(\psi, q\Psi), \Delta\Psi\rangle \tag{77}$$

We have used here formula (21) again. With (74) the Jacobian can be estimated as $\|J(\psi, q\Psi)\| \le \max_S |\nabla\psi| \|q\nabla\Psi\| \le C\,\max\{\chi_\alpha, |\chi_\sigma|\}\|\Psi\|_1$.

If we apply Schwartz's inequality to the inner product in (77) and use the last estimate and formula $\|\Psi\|_1/\|\Delta\Psi\| = \|\Psi\|_1/\|\Psi\|_2 = \chi_\Psi^{-1/2}$ we obtain $|\nu_r| \le C\,\max\{\chi_\alpha, |\chi_\sigma|\}\,\chi_\Psi^{-1/2}$ where χ_Ψ is the spectral number (60) of mode amplitude. Then condition (66) of Theorem 4 leads to

Theorem 6

The maximum growth rate of an unstable mode (56) of steady WV wave (29) or modon (32) is limited:

$$|\nu_r| \le \max_S |\vec{u}| \max\{\chi_\alpha, |\chi_\sigma|\} [\delta\chi_\sigma^{-1} + (1-\delta)\chi_\alpha^{-1}]^{1/2} \tag{78}$$

Thus, the growth (decay) rate of modes depends on the velocity maximum (74), the degrees α and σ of basic solution and the part δ of mode enstrophy concentrated in the region S_{out}. In particular, $|\nu_r| \le C \max\{\chi_\alpha, |\chi_\sigma|\} \chi_\sigma^{-1/2}$ for any modon with uniform absolute vorticity (see the second type of modons in section 9), and $|\nu_r| \le C \max\{\chi_\alpha, |\chi_\sigma|\} \chi_\alpha^{-1/2} (1-\delta/\delta_{cr})^{1/2}$.for any isolated modon of the third type (see section 9). Note that estimate (75) for the RH wave (28) is obtained from (78) as the particular case when $\chi_\sigma = \chi_\alpha = \chi_n$.

11. Orthogonality of Unstable Modes to the Basic Flow (BVE Solution)

The amplitudes of unstable modes presented in Fig.1b and Fig.2b are formed by the spherical harmonics orthogonal to the harmonics $Y_3^1(\lambda,\mu)$ and $Y_5^4(\lambda,\mu)$ which form the basic flows.

We now prove the W_2^1-orthogonality of any non-neutral or non-stationary mode to the basic flow.

Theorem 7

Let ψ be the LP flow (27), RH wave (28), WV wave (29) or modon (32). Then the amplitude Ψ of each unstable, decaying, or non-stationary mode is orthogonal to the basic flow ψ in the W_2^1-inner product.

Proof. Taking the L_2-inner product of each of the equations (69), (70) or (76) with the basic flow ψ and using formulas (19) and (21) we obtain the same result:

$$\nu\langle\Delta\Psi, \psi\rangle = \nu\langle\Psi, \Delta\psi\rangle = 0 \tag{79}$$

where Ψ is the mode amplitude. Thus, by (4), $\langle \Psi, \psi \rangle_1 \equiv -\langle \Delta\Psi, \psi \rangle = 0$ for any non-neutral ($\nu_r \neq 0$) or non-stationary ($\nu_i \neq 0$) mode. The theorem is proved.

The W_2^1-orthogonality means that the mode amplitude velocity $\vec{U} = \vec{k} \times \nabla\Psi$ is orthogonal to the solution velocity $\vec{u} = \vec{k} \times \nabla \psi$:

$$\left(\vec{U}, \vec{u}\right) \equiv \int_S \vec{U} \cdot \vec{u}\, ds = \int_S \nabla\Psi \cdot \nabla \psi\, ds = -\langle \Delta\Psi, \psi \rangle = 0 .$$

Corollary 1

Let ψ be the LP flow (27) or RH wave (28). Then the amplitude Ψ of each unstable, decaying, or non-stationary mode is both L_2-orthogonal and W_2^2-orthogonal to flow ψ:

$$\nu\langle \Psi, \psi \rangle = 0\,,\, \nu\langle \Delta\Psi, \Delta\psi \rangle = 0 \tag{80}$$

Indeed, let ψ be a LP flow (27). By (79), we get $\nu\langle \Psi, \psi \rangle = -\chi_n^{-1}\nu\langle \Psi, \Delta\psi \rangle = 0$. Obviously, the second equation (80) is also valid.

Let now $\psi = -\omega\mu + \psi_n$ be a RH wave (28) where $\psi_n \in H_n$. Then

$$\langle J(\psi, D_n\Psi), \psi_n \rangle = -\omega \langle J(\mu, D_n\Psi), \psi_n \rangle = \omega \langle (D_n\Psi)_\lambda, \psi_n \rangle = 0 \tag{81}$$

because $(D_n\Psi)_\lambda$ is orthogonal to H_n (see (71)). Taking the L_2-inner product of Eq.(70) with ψ_n and using (104), we get $\nu \langle \Delta\Psi, \psi_n \rangle = 0$, or, by the symmetry of Laplace operator,

$$\nu \langle \Psi, \psi_n \rangle = 0 \tag{82}$$

Since

$$\Delta\psi = -2\psi + (2 - \chi_n)\psi_n, \tag{83}$$

the L_2-orthogonality of Ψ to ψ follows from (79) and (82):

$$\nu\langle \Psi, \psi \rangle = 0 \tag{84}$$

The second equation (80) follows from (82)-(84).

Corollary 2

The amplitude Ψ of each unstable, decaying, or non-stationary mode of the RH wave is also orthogonal to the subspace H_1 of spherical polynomials of degree one.

It immediately follows from (28), (82) and (84) that amplitude Ψ is orthogonal to a super-rotation flow $\nu\langle\Psi,\mu\rangle = 0$, and due to (10), to the subspace H_1 of spherical polynomials of degree one.

12. Numerical Experiments

The theorems 3-7 can be used for controlling the precision of numerical instability results.

Experiment 1

Let us consider two zonal LP flows:

$$\psi(\mu) = 0.03P_4(\mu) \tag{85}$$

$$\psi(\mu) = -0.03P_5(\mu) \tag{86}$$

In this case, the instability condition by Rayleigh-Kuo [19] as well as the spherical analogue of Fjörtoft's condition [18] are satisfied.

We can use equation

$$K_t = -\int_S \sqrt{1-\mu^2}(uv)\frac{dU}{d\mu}dS - \int_S \frac{\mu}{\sqrt{1-\mu^2}}(uv)UdS \tag{87}$$

that describes the evolution of perturbation kinetic energy (41) of a disturbance $\psi'(\lambda,\mu,t)$ to the zonal flow on sphere S where $U = -\sqrt{1-\mu^2}\,\psi_\mu$, $u = -\sqrt{1-\mu^2}\,\psi'_\mu$, $v = \frac{1}{\sqrt{1-\mu^2}}\psi'_\lambda$ are the velocity components of the basic flow and disturbance, respectively. Usually, the maximum values of the amplitude of unstable mode are localized in the regions related with characteristic features of the basic velocity $U(\mu)$. Unlike the equation for perturbation energy on the β-plane [71], its spherical analogue (87) contains one more integral of the product of basic velocity U with uv. Whereas the first integral dominates principally at the sides of jets (where U_μ is large) and far off the polar regions (where $\sqrt{1-\mu^2}$ is not small), the second integral can be significant in the central parts of strong jets (where U is large), especially when such jets are located in the polar regions (where $\mu / \sqrt{1-\mu^2}$ is also large). By (87), the sign of K_t depends on the signs of the products $(uv)U_\mu$ and $(uv)U$ in various regions of the sphere S. In the case when the first integral is dominant, one can say that the growth of the perturbation energy takes place in the regions where the inclination of closed isolines of the streamfunction of disturbance is opposite to the inclination of the profile of basic flow velocity $U(\mu)$, that is in the regions where product $(uv)U_\mu$ is negative [71].

The series for disturbances were truncated by the same number ($N = 21$). The profiles of velocity $U(\mu)$ of the basic flows (85) and (86) are shown in Fig.1a and Fig.1c, while isolines of the real part of the amplitudes of their most unstable modes are presented in Fig.1b and Fig. 1.d, respectively. As it is seen from these figures, the basic flows contain subtropical jets, and the zonal wavenumbers of the both unstable modes are $m = 2$. It is easy to show [65] that if the streamfunction of the basic flow is anti-symmetric (or the flow velocity is symmetric) about equator, then all the normal modes of such flow are divided in two groups: symmetric or anti-symmetric about equator. One can see that the mode amplitude (Fig.1b) has no symmetry for flow (85) and is symmetric about equator (Fig.1d) for flow (86).

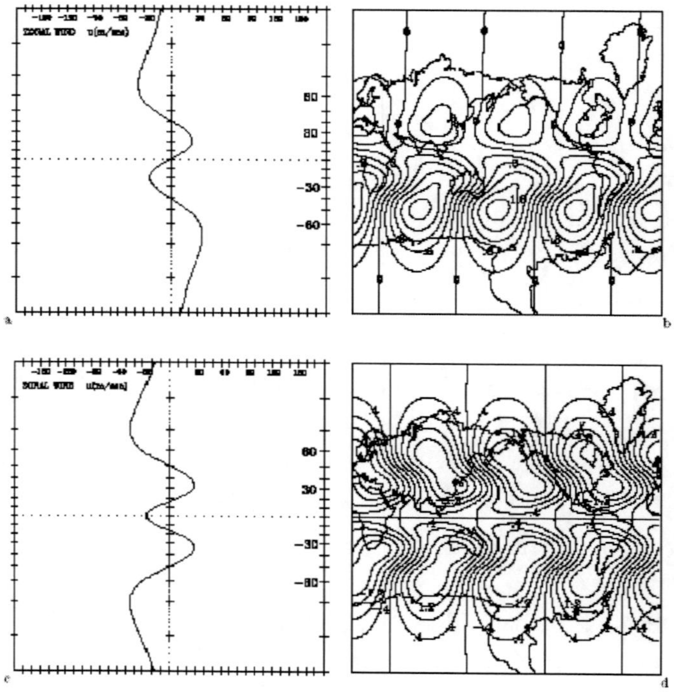

Figure 1. The profile of velocity $U(\mu)$ of zonal flows (85) and (86) (figures (a) and (c), respectively), and isolines of the real parts of amplitudes of their most unstable modes (figures (b) and (d)).

The maximum values of the mode amplitude are located in a neighborhood of subtropical jets, so the first integral is dominant in the formula (87). The inclination of closed isolines of the streamfunction of disturbance is opposite to the inclination of the profile of basic flow velocity $U(\mu)$ so that the negative product $(uv)U_\mu$ generates the instability. The orthogonality conditions established in section 11 are evidently fulfilled, and the spectral numbers χ_Ψ of mode amplitudes are 19.99 and 30.0 for the flows (85) and (86), respectively, that is, the calculated values of χ_Ψ coincide with the theoretical values $\chi_\Psi = n(n+1)$ given by theorem 3. The e-folding time and period of the mode shown in Fig.1b are 1.5 days and 4.9 days, and these values are respectively 0.67 days and 39.67 days for the mode presented in Fig. 1d.

Experiment 2

Let us consider a steady non-zonal RH wave

$$\psi(\lambda,\mu) = -\omega\mu + aP_3^2(\mu)\cos 2\lambda \qquad (88)$$

where $\omega = 2/(\chi_3 - 2) = 0.2$ [9]. In order to show numerically the existence of a critical amplitude for the RH-wave instability, the modes have been calculated for different amplitude values a of the RH wave: $a = 0.013$, $a = 0.014$; $a = 0.015$, $a = 0.023$ [54]. The basic flows for different a are presented in Fig.2a-d, while the real parts of amplitudes $\Psi_r(x)$ of the most unstable modes are given in Fig.3a-d.

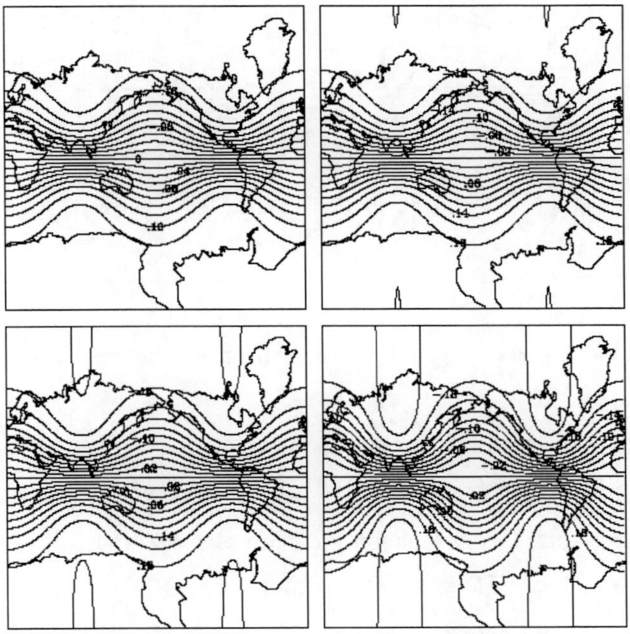

Figure 2. The RH wave $\psi(\lambda,\mu) = -0.2\mu + aP_3^2(\mu)\cos 2\lambda$ for $a = 0.013$ (a), $a = 0.014$ (b); $a = 0.015$ (c) and $a = 0.023$ (d).

The mean spectral number χ_Ψ is equal to 12.0 for all unstable modes, and hence, the instability condition (68) is fulfilled. It should be noted that the growth rate of the unstable mode increases with amplitude a, in accordance with estimate (75). The number of unstable modes also increases with a. At the same time, the period of unstable mode is changed slowly with a. The most unstable modes shown in Fig.3b-d are anti-symmetric about equator. Indeed, since wave (88) is anti-symmetric about equator, each mode of this wave is either symmetric or anti-symmetric about equator [65].

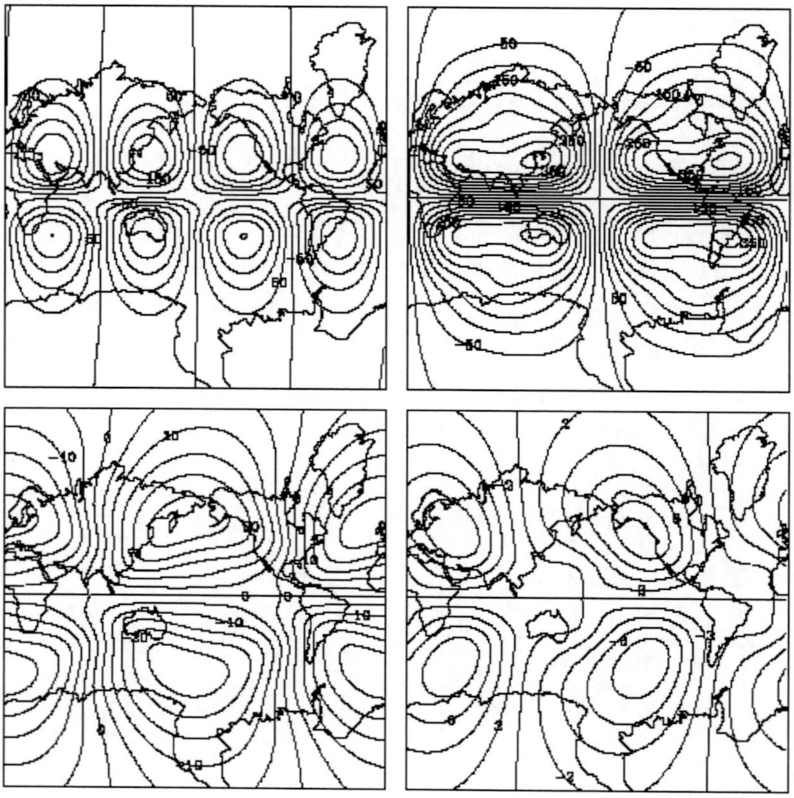

Figure 3. Isolines of the real parts of amplitudes of the most unstable modes of the four RH waves shown in Figure 2.

All the modes are divided into three sets (54), and due to (62), the amplitude $\Psi(\lambda,\mu)$ of any unstable mode must belong to the set M_0 being a hypersurface (of zero measure) in the whole space of disturbances. Any mode of the main sets M_- and M_+ is neutral. Moreover, in the case of LP flow (27) or RH wave (28), disturbances of the subspace H_n of homogeneous spherical polynomials of degree n are also neutral ($H_n \in M_0$, see Example 3). Thus, exponential growth of disturbances to LP flow and RH wave is possible only in the set M_0 / H_n being the complement of H_n to M_0.

Experiment 3

Consider now a steady and anti-symmetric about equator WV wave (29) defined by

$$\psi(\lambda,\mu)=\begin{cases} X_1(\lambda,\mu)-\omega_o\mu+d_o, & \text{in } S_1 \\ X_i(\lambda,\mu)-\omega_i\mu+d_i, & \text{in } S_{in} \\ X_2(\lambda,\mu)-\omega_o\mu+d_o, & \text{in } S_2 \end{cases} \tag{89}$$

where $S_{in}=[-\pi,\pi]\times[-\mu_0,\mu_0]$, $S_{out}=S_1\cup S_2$, $S_1=[-\pi,\pi]\times[\mu_0,1]$, $S_2=[-\pi,\pi]\times[-1,-\mu_0]$, $0<\mu_0<1$, and η_i, η_o, d_i and d_o are certain constants [10]. Since wave (52) is stationary, $\omega_i=2/(\chi_\alpha-2)$, $\omega_o=2/(\chi_\sigma-2)$ with χ_α and χ_σ defined by (30), $X_1(\lambda,\mu)=A_1P_\sigma^0(\mu)+B_1P_\sigma^m(\mu)\cos m\lambda$, $X_i(\lambda,\mu)=A_iT_\alpha^0(\mu)+B_iT_\alpha^m(\mu)\cos m\lambda$, $X_2(\lambda,\mu)=-A_1P_\sigma^0(-\mu)-B_1P_\sigma^m(-\mu)\cos m\lambda$, $T_\alpha^m(\mu)=P_\alpha^m(\mu)-P_\alpha^m(-\mu)$, $\mu_0=\sin\varphi_0$, $\varphi_0=29.99^o$, m = 2, ,α=5.7701 and $\sigma=4.5419$. The *WV* wave is shown in Fig. 4a is anti-symmetric about equator, and wherefore the real parts of amplitudes of the first most unstable modes to this flow presented in Fig.4b-d are anti-symmetric or symmetric about equator. They were calculated with truncation number $N=64$. The instability condition (66) is fulfilled with a small error. It is explained not only by the errors committed in the course of calculating the spectral number χ_Ψ, but also by the errors related with truncation number N. Indeed, this field is not as smooth as a RH wave and, due to Gibbs effect [72], the series of spherical harmonics for the WV wave converges slowly [16, 52-54].

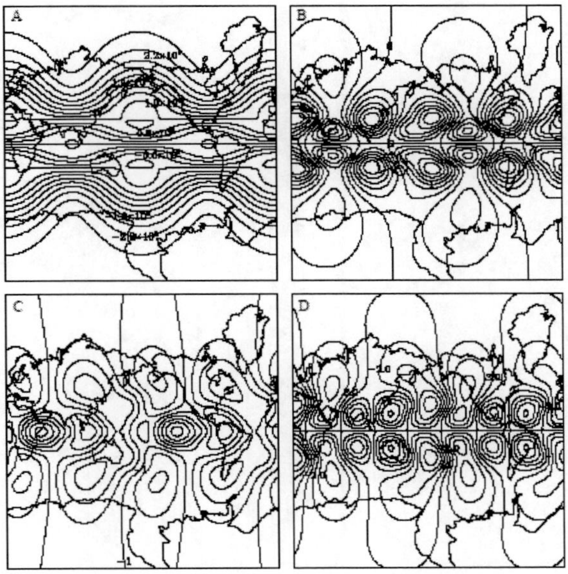

Figure 4. Anti-symmetric WV wave (a), and isolines of the real parts of amplitudes of its three most unstable modes (b-d).

Experiment 4

Consider now a steady modon by Verkley (1990) with uniform absolute vorticity in the inner region S_{in} [14] whose structure is described by formula (32). The *WV* wave is shown in Fig. 5a, while the real parts of amplitudes of the first most unstable modes to this flow are presented in Fig.5b-d. The modes were calculated with truncation number $N = 64$. Since the modon has no symmetry, the modes are asymmetric too. The instability condition (66) is fulfilled only approximately, besides the difference between the theoretical and calculated values of χ_{Ψ} is larger than in the case of the WV wave and is explained by the same reasons that were noted in experiment 4 [16]. Nevertheless, these errors slowly decrease as the truncation numbers for the basic flow (WV wave or modon) and disturbances increase.

In full accordance with the instability condition obtained by Verkley (1990), the unstable perturbations are located out of the inner region of modon (see Fig. 5b-d). Like in the case of WV wave (experiment 4), the growth rate of the three most unstable modes are close to each other (about 7 days). The fist mode is stationary, while the periods of the second and third modes are 25.37 and 10.89 days, respectively.

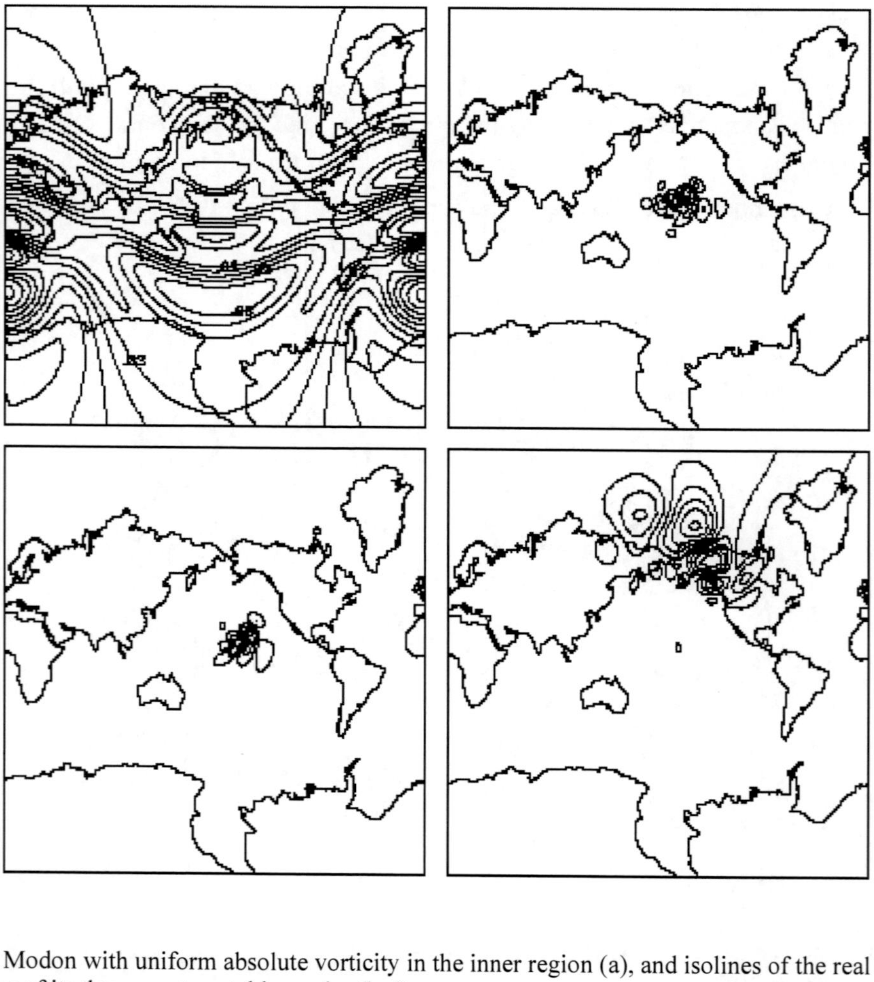

Figure 5. Modon with uniform absolute vorticity in the inner region (a), and isolines of the real parts of amplitudes of its three most unstable modes (b-d).

13. CONCLUSIONS

In this work, a unified approach to the study of exponential instability of exact solutions to the vorticity equation (1) on a sphere has been suggested. The four types of the steady solutions known up to now have been analyzed: the LP flows, RH waves, WV waves and modons. The instability of these solutions is of great importance in hydrodynamics and meteorology, in particular, for deeper understanding the low-frequency atmospheric variability and climate predictability.

The basic integral properties of the Jacobian on a sphere are given in Lemma 1. The conservation laws for disturbances of the flows are derived in Theorems 1 and 2. Each of the laws depends on the basic flow only through the degrees of spherical harmonics forming this flow. Unlike a viscous fluid, the spectrum of linearized operator for an ideal fluid may have a continuous part and finite accumulation points. For example, each RH wave represents a super-rotation flow $\psi(\mu) = -\omega\mu$ perturbed by its $2n+1$ neutral modes of eigensubspace H_n corresponding to an isolated eigenvalue with finite multiplicity $2n+1$, besides, the isolated eigenvalues have the finite accumulation point $\nu = \omega$.

The conservation laws are then used to obtain necessary conditions for the exponential instability of the LP flows, RH waves, WV waves and modons. These conditions impose a strict restriction on the spectral energy distribution of each unstable mode requiring the average spectral number χ_Ψ of its amplitude Ψ to be equal to a special number. Here χ_Ψ is the square of the famous spectral number introduced by Fjörtoft [1]. More precisely, $\chi_\Psi = n(n+1)$ for every unstable mode of the LP flow or RH wave of degree n (Theorem 3), and $\chi_\Psi = [\delta\chi_\sigma^{-1} + (1-\delta)\chi_\alpha^{-1}]^{-1}$ for every unstable mode of the WV wave or modon (Theorem 4). Here δ is the portion of the mode enstrophy concentrated in the outer region S_{out} of the solution ($0 \le \delta \le 1$), $\chi_\alpha = \alpha(\alpha+1)$, $\chi_\sigma = \sigma(\sigma+1)$, and α and σ are the degrees of spherical harmonics representing the solution in its inner and outer regions S_{in} and S_{out}. In particular, $\chi_\Psi = \chi_\sigma$ for unstable modes of the modon with uniform vorticity in the inner region, and χ_Ψ is always between the numbers χ_α and χ_σ for unstable modes of any WV wave or non-localized modon. Each unstable mode of a localized modon, in addition to $\chi_\Psi^{-1} = \delta\chi_\sigma^{-1} + (1-\delta)\chi_\alpha^{-1}$, must also satisfy the condition (68). Besides, the spectral number of any unstable mode must be limited from below: $\chi_\Psi = \chi_\alpha(1-\delta/\delta_{cr})^{-1} \ge \chi_\alpha$. It follows from (68) that as α grows or/and k decreases, the critical value δ_{cr} decreases monotonically reducing the interval $0 \le \delta < \delta_{cr}$ of potential instability. Thus, unstable disturbances of any isolated modon of rather large degree α and small degree k are mainly localized inside the inner region S_{in}.

The new instability conditions establish strict relations between the spectral number (geometric structure) of growing disturbances and the basic flow scale. In other words, they localize the unstable disturbances in the phase space and are useful in interpreting the spectral structure of growing atmospheric perturbations. For the LP flow, the new condition complements the well-known Rayleigh-Kuo and Fjörtoft conditions in the sense that while the latter are related to the basic flow profile, the former characterizes the spectral structure of unstable modes. In fact, the average geometric scale of unstable modes is proportional to the scale of LP flow (or is inversely proportional to the LP flow degree). The last assertion is true for unstable modes of the RH wave too. Note that the zonal wavenumber *m* of each unstable mode (56) of the LP flow (27) must be small enough: $0 < |m| < n$. In the case of the LP flows and RH waves, an unstable mode must belong to the complement M_0 / H_n of H_n to set M_0 (M_0 is defined by (54), and H_n is the subspace of homogeneous spherical polynomials of degree *n*). It is also proved that any LP flow of degree one or two is Liapunov stable, and hence, is exponentially and algebraically stable as well (Examples 1 and 3).

The truncation of series used in the numerical stability study with the spectral method gives rise to the spectral approximation problem [52, 53]. The new instability conditions provide exact value of the spectral number for each unstable mode, and thus allow testing the accuracy of numerical stability study algorithm and detecting possible errors in the computer programs. In particular, Theorems 3 and 4 show that in the linear stability study of a stationary flow on a sphere, Fourier series for disturbances must be truncated higher than that for the basic flow [52, 53]. For example, if we truncate the disturbances of stationary RH wave (28) by a number $N \leq n$ then, due to Theorem 4, none of the modes will be unstable. The control of the quality of calculations is especially important in the case of modons. Indeed, the derivatives of the modon vorticity are not continuous on a sphere at the boundary between S_{in} and S_{out}, and as a result, Fourier series for the modon and its perturbation converge slowly nearby this boundary (the Gibbs phenomenon). As a result, condition (66) for unstable modes of a modon is fulfilled only approximately, besides, the difference between the theoretical and numerical values of spectral number χ_Ψ decreases slowly as the resolution (series truncation number) increases. Thus, a high resolution is required to be confident in the numerical modon stability results [16]. The same is true for the WV wave.

We have estimated the upper bounds of the growth rate for unstable modes of the solutions too (Theorems 5 and 6), and showed the orthogonality of the amplitude of any unstable, decaying or non-stationary mode to the basic solution in the W_2^1-inner product (Theorems 7). In the case of LP flows and RH waves, the mode amplitude is also orthogonal to the basic flow in the L_2- and W_2^2-inner products (Corollary 1).

The analytical instability results obtained here can also be applied for testing the accuracy of computational programs and algorithms used for the numerical stability study. Note that Fjörtoft's spectral number $\rho = \sqrt{\chi_\Psi}$ appears not only in the instability conditions but also in the maximum growth rate estimates, and hence, is the key parameter in the linear instability problem of ideal flows on a sphere.

ACKNOWLEDGMENTS

The figures kindly put at my disposal by Dr. Ismael Pérez García are very much appreciated. This work was supported by the grants IN105608 (PAPIIT, UNAM, Mexico), 46265-A1 and FOSEMARNATd-2004-01-160 (CONACyT, Mexico), and 14539 (SNI, Mexico).

REFERENCES

[1] R. Fjörtoft, *On the changes in the spectral distribution of the kinetic energy for two-dimensional nondivergent flow,* Tellus 5 (1953) 225-230.

[2] A.J. Simmons, J.M. Wallace, G.W. *Branstator, Barotropic wave propagation and instability, and atmospheric teleconnection patterns,* J. Atmos. Sci. 40 (1983): 1363-1392.

[3] R.J. Haarsma, J.D. Opsteegh, *Barotropic instability of planetary-scale flows*, J. Atmos. Sci. 45 (1988) 2789-2810.

[4] P. Wu, *Nonlinear resonance and instability of planetary waves and low-frequency variability in the atmosphere*, J. Atmos. Sci., 50 (1993) 3590-3607.

[5] Yu.N. Skiba, *On the long-time behavior of solutions to the barotropic atmosphere model, Geophys.* Astrophys. Fluid Dynamics 78 (1994) 143-167.

[6] Yu.N. Skiba, 1997. *On dimension of attractive sets of viscous fluids on a sphere under quasi-periodic forcing,* Geophys. Astrophys. Fluid Dynamics 85 (1997) 233-242.

[7] C.-C. Rossby, *Relation between variations in the intensity of the zonal circulation of the atmosphere and the displacements of the semi-permanent centers of action,* J. Marine Res. 2 (1939) 38-55.

[8] B. Haurwitz, *The motion of atmospheric disturbances on the spherical earth,* J. Marine Research. 3 (1940) 254-267.

[9] Ph.D. Thompson, *A generalized class of exact time-dependent solutions of the vorticity equation for nondivergent barotropic flow,* Mon. Wea. Rev. 110 (1982) 1321-1324.

[10] P. Wu, W. T. M. Verkley, *Non-linear structures with multivalued* (q, ψ) *relationships – exact solutions of the barotropic vorticity equation on a sphere,* Geophys. Astrophys. Fluid Dynamics 69 (1993) 77-94.

[11] J.J. Tribbia, *Modons in spherical geometry,* Geophys. Astrophys. Fluid Dynamics 30 (1984) 131-168.

[12] W.T.M. Verkley, *The construction of barotropic modons on a sphere,* J. Atmos. Sci. 41 (1984) 2492-2504.

[13] W.T.M. Verkley, *Stationary barotropic modons in westerly background flows,* J. Atmos. Sci. 44 (1987) 2383-2398.

[14] W.T.M. Verkley, *Modons with uniform absolute vorticity,* J. Atmos. Sci. 47 (1990) 727-745.

[15] E.C. Neven, *Quadrupole modons on a sphere*, Geophys. Astrophys. Fluid Dynamics 65 (1992) 105-126.

[16] E.C. Neven, *Linear stability of modons on a sphere*. J. Atmos. Sci. 58 (2001) 2280-2305.

[17] L. Rayleigh, *On the stability or instability of certain fluid motions,* Proc. London Math. Soc., s1-11 (1879), 57-72 L. Rayleigh, *On the stability or instability of certain fluid motions,* Proc. London Math. Soc.

[18] R. Fjörtoft, *Application of integral theorems in deriving criteria of stability of laminar flow and for baroclinic circular vortex,* Geofys. Publ. Norske Vid.-Akad.Oslo, 17 (1950), 1–52.

[19] H.-L. Kuo, *Dynamic instability of two-dimensional non-divergent flow in a barotropic atmosphere*, J. Meteorology, 6 (1949) 105-122.

[20] P.G. Baines, *The stability of planetary waves on a sphere*. J. Fluid Mech. 73 (1976) 193-213.

[21] A.B. Karunin, *On Rossby waves in barotropic atmosphere in the presence of zonal flow,* Izv. Atmos.Ocean. Physics 6 (1970) 1091-1100.

[22] Yu.N. Skiba, J. Adem, *On the linear stability study of zonal incompressible flows on a sphere,* Numer. Meth. Part. Differ. Equations 14 (1998) 649-665.

[23] A.E. Gill, *The stability of planetary waves on an infinite beta-plane,* Geophys. Fluid Dyn. 6 (1974) 29-47.

[24] C. Marchioro, M. Pulvirenti, *"Mathematical Theory of Incompressible Nonviscous Fluids,"* Springer-Verlag, New York, 1994.

[25] M.N. Rosenbluth, A. Simon, *Necessary and sufficient condition for the stability of plane parallel inviscid flow*. Physics of Fluids, 7 (1964) 557-558.

[26] K.K. Tung, *Barotropic instability of zonal flows*, J. Atmos. Sci. 38 (1981) 308-321.

[27] A.M. Liapunov, *Stability of Motion*. Academic Press, New York, 1966.

[28] V.I. Arnold, *Conditions for nonlinear stability of stationary plane curvilinear flows of an ideal fluid,* Sov. Math. Doklady 6 (1965) 331-334.

[29] L.N. Howard, *Note on a paper of John W. Miles*, J. Fluid Mech., 10 (1961), 509-512.

[30] J. Thuburn, P.H. Haynes, P.H., *Bounds on the growth rate and phase velocity of instabilities in non-divergent barotropic flow on a sphere*: A semicircle theorem, Q.J.R. Meteorol. Soc. 122 (1996) 779-787.

[31] Yu.N. Skiba, *On the normal mode instability of harmonic waves on a sphere,* Geophys. Astrophys. Fluid Dynamics 92 (2000) 115-127.

[32] P. Chen, *The barotropic normal modes in certain shear flows and the travelling waves in the atmosphere*, J. Atmos. Sci. 50 (1993) 2054-2064.

[33] F. Bärmann, J. Gierling, E. Rebhan, *Shear flow instabilities of the jet stream and Hopf-bifurcation to periodic solutions,* Beitr. Phys. Atmosph., 70 (1997) 117-130.

[34] S. Friedlander, M.M. Vishik, *Instability criteria for the flow of an inviscid incompressible fluid,* Phys. Rev. Letts. 66 (1991), 2204–2206.

[35] S. Friedlander, M. Vishik, V. Yudovich, *Unstable eigenvalues associated with inviscid fluid flows,* J. Math. Fluid Mech. 2 (2000), 365-380.

[36] S. Friedlander, W. Strauss, M.M. Vishik, *Nonlinear instability in an ideal fluid,* Ann. Inst. H. Poincaré, Anal. Nonlinéaire 14 (1997), 187–209.

[37] S. Friedlander, W. Strauss, M. M. Vishik, *Robustness of instability for the two dimensional Euler equations*, SIAM J. Math. Anal. 30 (1999), 1343–1355.

[38] L. Belenkaya, S. Friedlander, V. Yudovich, *The unstable spectrum of oscillating shear flows*, SIAM J. Appl. Math. 59 (1999), 1701-1715.

[39] S. Friedlander, L. N. Howard, *Instability in parallel flow revisited,* Stud. Appl. Math. 101 (1998), 1–21.

[40] M. Visik, S. Friedlander, *Nonlinear instability in two dimensional ideal fluids: The case of a dominant eigenvalue,* Commun. Math. Phys. 243 (2003), 261-273.
[41] Z. Lin, *Instability of some ideal plane flows*, SIAM J. Math. Anal. 35 (2003), 318-356.
[42] Z. Lin, *Nonlinear instability of ideal plane flows*, Int. Math. Res. 41 (2004), 2147-2178.
[43] Z. Lin, *Some stability and instability criteria for ideal plane flows*, Commun. Math. Phys. 246 (2004), 87-112.
[44] Y. Latushkin, M. Vishik, *Linear stability in an ideal incompressible fluid,* Commun. Math. Phys. 233 (2003), 439-461.
[45] Y. Latushkin, Y. C. Li, M. Stanislavova, *The spectrum of a linearized 2D Euler operator*, Stud. Appl. Math. 112 (2004), 259-270.
[46] R. Shvydkoy, Y. Latushkin, *Essential spectrum of the linearized 2D Euler equation and Lyapunov-Oseledets exponents.* J. Math. Fluid Mech. 7 (2005), 164-178.
[47] C. Marchioro, M. Pulvirenti, *Some considerations on the nonlinear stability of stationary planar Euler flows.*Comm. Math. Phys. 100 (1985) 343-354.
[48] C. Marchioro, M. Pulvirenti, O*n nonlinear stability of stationary Euler flows on a rotating sphere,* J. Math Anal. Appl. 129 (1988) 24-36.
[49] W. Tollmien, *Ein allgemeiness kriterium der instabilität laminarer geichwindigkeitsverteilungen,* Nachr. Ges. Wiss. Göttingen Math. Phys., Klasse NF 1, 50, 1935, 79-114.
[50] C.C. Lin, *The Theory of Hydrodynamic Stability*, Cambridge University Press, Cambridge, 1955.
[51] L.A. Dikii, *"Hydrodynamic Stability and Atmosphere Dynamics,"* Gidrometeoizdat, Leningrad, 1976 (in Russian).
[52] Yu.N. Skiba, *Spectral approximation in the numerical stability study of non-divergent viscous flows on a sphere*, Numer. Meth. Part. Differ. Equations 14 (1998) 143-157.
[53] Yu.N. Skiba, *On the spectral problem in the linear stability study of flows on a sphere,* J. Math. Analys. Appl. 270 (2002) 165-180.
[54] Yu.N. Skiba, I. Peres-Garcia, *On the structure and growth rate of unstable modes to the Rossby-Haurwitz wave,* Numer. Meth. Part. Diff. Equations 21 (2005) 368-386.
[55] E.N. Lorenz, *Barotropic instability of Rossby wave motion,* J. Atmos. Sci. 29 (1972) 258-264.
[56] B.J. Hoskins, A. Hollingsworth, *On the simplest example of the barotropic instability of Rossby wave motion*, J. Atmos. Sci. 30 (1973) 150-153.
[57] J.L. Anderson, *The instability of finite amplitude Rossby waves on the infinite* β *-plane,* Geophys. Astrophys. Fluid Dynamics 63 (1992) 1-27.
[58] B.J. Hoskins, *Stability of the Rossby-Haurwitz wave,* Quart. J. Roy. Meteorol. Soc. 99 (1973) 723-745.
[59] Yu.N. Skiba, *Liapunov instability of the Rossby-Haurwitz waves and dipole modons,* Sov. J. Numer. Analys. Math. Modelling 6 (1991) 515-534.
[60] Yu.N. Skiba, *Dynamics of perturbations of the Rossby-Haurwitz wave and the Verkley modon*, Atmósfera 6 (1993) 87-125.
[61] Yu.N. Skiba, A.Y. Strelkov, *On the normal mode instability of modons and Wu-Verkley waves,* Geophys. Astrophys. Fluid Dynamics 93 (2000) 39-54.
[62] S. Helgason, *Groups and Geometric Analysis, Integral Geometry, Invariant Differential Operators and Spherical Functions*. Academic Press, Orlando, 1984.

[63] R.D. Richtmyer, *Principles of Advanced Mathematical Physics*, Vol.2. Springer-Verlag, New York, 1981.

[64] A.D. Gadzhiev, *On differential properties of the symbol of multi-dimensional singular integral operator*, Matem. Sbornik 114 (156) (1981), 483-510.

[65] Yu.N. Skiba, *Mathematical Problems of the Dynamics of Viscous Barotropic Fluid on a Rotating Sphere*. Dept. Numerical Mathematics, USSR Academy of Sciences, Moscow, 1989, 178 pp. (Russian) (English transl.: Indian Inst. Tropical Meteorology, Pune, 211 pp., 1990).

[66] Yu.N. Skiba, *Instability of the Rossby-Haurwitz wave in invariant sets of perturbations*, J. Math. Analys. Appl. 290 (2004) 686-701.

[67] C.-G. Rossby, *Relation between variations in the intensity of the zonal circulation of the atmosphere and the displacements of the semi-permanent centers of action*, J. Marine Res. 2 (1939) 38-55.

[68] B. Haurwitz, *The motion of atmospheric disturbances on the spherical earth*, J Marine Res 3 (1940), 254-267.

[69] P. Wu, *On nonlinear structures and persistent anomalies in the atmosphere.* Ph.D Thesis, Imperial Colledge, London, 168 pp., 1992.

[70] P. Ripa, *Positive, negative and zero wave energy and the flow stability problem in the Eulerian and Lagrangian-Eulerian descriptions*, Pure and Applied Geophysics 133 (1990), 713-732.

[71] J. Pedlosky, *Geophysical Fluid Dynamics*. Springer-Verlag, New York, 1979.

[72] H.F. Davis, *Fourier Series and Orthogonal Functions*, Allyn and Bacon, 1963.

REVIEWED BY

Dr. Denis M. Filatov, Centro de Investigacion en Computacion (CIC), Instituto Politecnico Nacional (IPN), C.P. 07738, Mexico, D.F., MEXICO Email: denisfilatov @gmail.com

In: Computational Mathematics
Editor: Peter G. Chareton, pp.35-61
ISBN:978-1-60876-271-2

Chapter 2

PURE AND MIXED MATHEMATICS IN THE WORK OF LEONHARD EULER

Giovanni Ferraro[*]
University of Molise, Campobasso, Italy

ABSTRACT

Leonhard Euler's influence on mathematics was enormous. He wrote an impressive quantity of papers that contained innumerable new results and his innovative techniques and procedures led a profound change in the structure of mathematics and in its basic principles. In this article, after a brief description of Euler's life, I will discuss some aspects of his works to pure and mixed mathematics. In particular, I will deal with his contributions to the rise of the concept of functions, to development of the theories of series and differential equations. I will also mention Euler's study on the calculus of variations and will, finally, highlight the importance of his studies in mechanics, especially the point-mass mechanics, rigid body dynamics, and fluid dynamics.

1. INTRODUCTION

Leonhard Euler was born on April 15, 1707 in Basel but lived his childhood at Riehen, near Basel, where his family moved in 1708. His father, Paul, was a Protestant minister; he had studied theology at the University of Basel, where he had also attended Jacob Bernoulli's lectures. His mother, Margaret Brucker, was of a "distinguished family whose name was well recognized in the republic of letters of which there were several scientists who shared the same name" (Fuss [1783]). In 1720, Euler was sent to the University of Basel, where he showed a clear predilection for mathematics. He was a student of Johann Bernoulli, who provided him with a private lesson once a week to help clarify the problems which occurred during his lectures and his studies (Fuss [1783]).

[*] giovanni.ferraro@unimol.it

In 1723, Euler received his Master of Arts degree with a thesis in which he drew a comparison between the Newtonian and Cartesian philosophies. Since his father wished an ecclesiastical career for him, Leonhard was obliged to study theology and oriental languages for a while; but soon Paul Euler was "convinced that his son was born to replace Johann Bernoulli and not to be the pastor of Reihen" (see Condorcet [1783]); thus Euler could continue his mathematical education and, in the years that followed, he probably read works by Galileo, Descartes, van Schooten, Wallis, Newton, Varignon, Jacob Bernoulli, Hermann, Taylor (see Calinger [1996]).

In 1726, Euler completed his studies at the University of Basel with a Ph.D. dissertation, entitled *De Sono*, on the nature and propagation of sound. At that time he had already a paper in print, a short article on isochronous curves in a resisting medium and, in the following year, he published another article on reciprocal trajectories and also entered the *Paris Academy Prize* competition with a paper on the best arrangement of masts on a ship. Pierre Bouguer (1698-1758), an expert on mathematics relating to ships, won the prize, but the Paris Academy judged Euler's work worthy of an *accessit*.

In 1726 Euler was offered a position at the new Academy of St. Petersburg, which had been erected by Catherine I on the basis of a project of her husband Peter the Great. Euler immediately accepted the position, which would involve him in teaching applications of mathematics and mechanics to physiology; however, he delayed his departure probably because he hoped to be appointed at University of Basel. This did not occur and, on 5 April 1727, Euler left Swiss for Russia. He arrived in St Petersburg on 17 May 1727 and, through the requests of Daniel Bernoulli and Jakob Hermann, Euler was appointed to the mathematical-physical division of the Academy rather than to the physiology post he had originally been offered.

After the death of Catherine I, St. Petersburg Academy had political and financial difficulties. According to Fuss [1783], it was looked upon as an Academy that annually cost considerable amounts of money without seeming to offer any applicable utility. The Academicians felt the necessity to accept the consequences of this reality and Euler decided to serve as a medical lieutenant in the Russian navy from 1727 to 1730. Euler's situation improved when two Academicians, Hermann and Bulffinger, left their positions to return to their countries. Euler received the position of professor of Physics in 1730 and finally, in 1733, was appointed to the senior chair of mathematics, which was previously held by Daniel Bernoulli. On 7 January 1734, Euler married Katharina Gsell, the daughter of a Swiss painter who taught at the St Petersburg Gymnasium. The marriage was apparently happy. They had thirteen children, but eight died at an early age and only two sons survived him[1]. In 1735, Euler was appointed director of the St Petersburg Academy's geography section; thus, he was involved in cartography and helped the astronomer and geographer Joseph Nicholas Delisle (1688-1768) prepare a map of the Russian Empire, which was printed in 1745. By the 1730s Euler had eyesight problems and in the following years he lost his right eye (see Calinger [1996]).

In the 1740s Euler had a very high reputation due to the many and brilliant successes of his works. However, after the death of the Empress Anna, the state of Russian academy was

[1] The eldest son, Johann Albrecht Euler (1734 - 1800), followed his father's footsteps; he was appointed to the chair of physics at the Academy in St Petersburg in 1766 and become its secretary in 1769. Instead, Christoph Euler (1743-1808) had a military career.

become precarious and political turmoil made the position of foreigners particularly difficult. So, in 1741, Euler accepted the offer of Frederick II, the King of Prussia, to move to Berlin, where he arrived on 25 July.

In 1744, a new Academy of Science was founded in Berlin and the French scientist Maupertuis was its first president. Euler, who was appointed director of mathematics, intensely worked for the new Academy: he oversaw various financial matters and dealt with many practical problems, such the selection of the personnel and the supervision of the observatory and the botanical gardens; he managed the publication of various calendars and geographical maps, the sale of which was a source of income for the Academy. He also supervised the work on pumps and pipes of the hydraulic system at Sans Souci, the royal summer residence, and served as an advisor to the government on state lotteries, annuities, pensions, insurance, and artillery (see Youschkevitch [B]).

Euler was very religious and fulfilled with the greatest detail all the duties of a Christian. He dislikes Wolff's philosophy since he believed the monad theory led ultimately to atheism. He felt it his duty to made a stand in the defence of religion and wrote ''Gegen die Einwurfe der Freygeister'', a tract against atheists which was published in Berlin in 1747.

In Germany, Euler was, first, on good terms with Frederick II of Prussia, but soon their relationship became bad. The king did not like Euler and had a clear preference for the Illuminists, in particular, for d'Alembert, who Euler had bitterly argued with (see Ferraro [2008b]). After the death of Maupertius, d'Alembert was also offered to become the new president of Berlin Academy, a position at which Euler aimed, too.

The Swiss mathematician became unsatisfied of his situation in Germany and tried to return to St. Petersburg, but succeeded in doing this only in 1766. In Russia, Euler was warmly welcome and could work in the best way possible. He also had several disciples (Johann Albrecht Euler, Georg Wolfgang Krafft (1701-1754), Anders Johann Lexell (1740-1784), Nicolaus Fuss (1755-1826), Stepan Rumovsky (1734-1812), and others) who helped him in preparing his papers. So, in the years that followed, Euler was able to write an enormous number of works (almost half his total papers). This occurs despite some unfortunate accidents. Indeed, after his return in Russia, he fell gravely ill; he recovered but became almost entirely blind (Fuss [1783]). In 1771, his home was destroyed by fire, however he saved his mathematical manuscripts and the Empress Catherine II helped him with a present of 6000 rubles. In the same year, though a cataract operation restored his sight for a few days, Euler became totally blind. On 10 November 1773 Euler's wife, Katharina, died and, in 1776, married her half-sister Salome Abigail Gsell.

Euler died on 18 September 1783, but St. Petersburg Academy continued to publish his manuscripts for more than thirty years.

It is not possible to give a complete picture of Euler's contributions in few pages; consequently, I will dwell upon only on some of these contributions: the rise of the concept of functions, ordinary and partial differential equation, the theory of function of two variables, infinite series, the calculus of variation, mechanics of a point-mass and rigid body, fluid dynamics, and some examples of Euler's use of differential equation in mixed mathematics.

2. The Rise of the Concept of Functions

In his first researches Euler was influenced by Johann Bernoulli and dealt with typical problems of the Bernoullian school, such as isochronous curves (cf. Euler [1726]), tautochrone curves (Euler [1727a]), reciprocal trajectories (cf. Euler [1727b] and [1727c]). In various cases Johann Bernoulli himself[2] suggested Euler the topic of his research. For instance, in his [1728], Euler stated: "The Celebrated Johann Bernoulli proposed this question[3] to me and urged me to write up my solution and to investigate these three kinds of surfaces which lead to solutions that are integrable equations. I wanted to include the solutions to these questions because they followed so easily from what I had done earlier" (Euler [1728, §. 2]).

In the Leibnizian and Bernoullian conception, analysis was not an autonomous and self-founding mathematical discipline; rather it was an instrument for solving geometrical problems: it investigated the relations between geometrical quantities by symbolic representations, which were termed analytical expressions. Euler's first papers was influenced by this conception; however, he almost immediately shifted attention towards the analytical instruments and attempted to improve them since he thought that analysis facilitated the understating and solution of geometrical and physical problems. He soon began to develop the crucial analytical instrument of his analysis: the concept of a function. The term "function" had initially been used by Gottfried W. Leibniz to denote a line that performs a special duty in a given figure or a part of a straight line that is cut off by straight lines drawn solely by means of a fixed point and points of a given curve (Ferraro [2000a]). Then, Johann Bernoulli gave the name "function" to quantities "somehow formed from indeterminates and constants" [1849-1863, 3:150]) and, in 1718, defined a function of a variable quantity as "a quantity composed in whatever way of that variable quantity and constants" (Bernoulli [1742, 241]). However, in Johann Bernoulli's work the use of the term "function" did not play a central role. Instead, Euler transformed it the fundamental concept of analysis. The evolution toward the transformation of a geometrical calculus based on the notion of a curve to a merely analytical calculus based on the notion a function lasted ten or more years and can be considered ended when Euler wrote the *Introductio in analysin infinitorum* (published in 1748). In this book, Euler gave his famous definition of a function ("A function of a variable quantity is an analytical expression composed in whatever way of that variable and numbers or constant quantities" [1748, 1, §. 4]). Later, Euler attempted to improve and develop this notion of a function mainly to adapt it the need of mixed mathematics (in particular, he introduced a distinction between continuous and discontinuous functions and investigated other type of function, even if he did not considered them as a function in the proper sense of the word) [4].

---

[2] In [1727b, 408] Euler stated that Johan Bernoulli was "the most renowned of masters" and that "not only was my teacher, greatly fostering my inquiries into such matters, but also looked after me as a patron".

[3] The problem of finding "the shortest line between two points on a surface".

[4] On Euler's concept of a function, see Ferraro [2000a]. Here, I limit myself to observing that the importance of the analytical expression in the concept of a function was due to the fact that only the relationships that were analytically expressed by means of certain determined analytical expressions were accepted as functions. More precisely, an explicit function was given by one analytical expression constructed from variables in a finite number of steps using exponential, logarithmic, and trigonometric functions, algebraic operations and

3. Ordinary Differential Equations

Ordinary differential equations were one of the main fields of Euler's studies. In this article, I limit to deal with Euler's contributions on equations with constant coefficients and Riccati equations.

In 1735, on Daniel Bernoulli's request (Fuss [1843, vol. 2: 422]), Euler investigated the equation

$$k^4 \frac{d^4 y}{dx^4} = y$$

and obtained the solution the form of a power series (Euler, [1734-35c]). Later, in his [1739], Euler realized that the trigonometric functions could the solution of the differential equations with constant coefficients. In order to solve the equation

$$2ad^2 s + s\frac{dt^2}{b} + \frac{\text{adt}^2}{g} sin\frac{t}{a} = 0 . \tag{1}$$

Euler first showed that $s = C\, cos\frac{t}{\sqrt{2ab}}$, where C is constant, was an integral of the equation

$$2ad^2 s + s\frac{dt^2}{b} = 0 . \tag{2}$$

Then he considered $s = u\, cos\frac{t}{\sqrt{2ab}}$, where u is a variable quantity, substituted this expression into equation (2) and obtained $u = D + \frac{C\, sin\frac{t}{\sqrt{2ab}}}{cos\frac{t}{\sqrt{2ab}}} - \frac{a^2 b\, sin\frac{t}{a}}{g(a-2b)cos\frac{t}{\sqrt{2ab}}}$, where D is an arbitrary constant. Hence,

$$s = D\, cos\frac{t}{\sqrt{2ab}} + C\, sin\frac{t}{\sqrt{2ab}} - \frac{a^2 b\, sin\frac{t}{a}}{g(a-2b)} . \tag{3}$$

Soon Euler found the general method for integrating higher-order differential equations with constant coefficients. On September 15, 1739, Euler communicated it to Johann Bernoulli and then published it in his [1743b] where he even applied it to differential

composition of functions. A function could also be given in an implicit form *f(x,y)*=0, where *f* is an analytical expression in the above sense.

equations of infinite order. In the letter to J. Bernoulli, Euler stated that his method extends to all equations of the form

$$y + a\frac{dy}{dx} + b\frac{d^2y}{dx^2} + c\frac{d^3y}{dx^3} + d\frac{d^4y}{dx^4} + e\frac{d^5y}{dx^5} + etc. = 0 .$$

To find the integral of this equation he considered the equation

$$1 - ap + bp^2 - cp^3 + dp^4 - ep^5 + etc. = 0$$

and observed that this expression can be resolved into factors of the type 1-αp or 1-αp +βp^2. The factor 1-αp gives the integral $Ce^{x/\alpha}$ and the factor 1-αp+βp^2 gives the integral

$$e^{-\alpha x/2\beta}\left(C\,sin\frac{x\sqrt{4\beta-\alpha^2}}{2\beta} + D\,cos\frac{x\sqrt{4\beta-\alpha^2}}{2\beta}\right). \tag{4}$$

Euler played much attention to Riccati equations. For example, in his [1732-33], Euler considered the equation

$$ax^n dx = dy + y^2 dx,$$

replaced y by dt/tdx, and obtained $ax^n tdx^2 = d^2t$. Then, he set

$$t = 1 + \sum_{k=1}^{\infty} a_n x^{k(n+2)},$$

where the unknown coefficients a_n of the series had to determined by using the method of indeterminate coefficients. He [1732-33, §. 36] found

$$t = 1 + \frac{ax^{n+2}}{(n+1)(n+2)} + \frac{a^2x^{2n+4}}{(n+1)(n+2)(2n+3)(2n+4)} + \frac{a^3x^{3n+6}}{(n+1)(n+2)(2n+3)(2n+4)(3n+5)(3n+6)} + \dots$$

After several complicate calculations, Euler showed that, for $n<-2$ and $n>0$,

$$t = K\int_0^{\infty} \frac{e^{\frac{2}{n+2}\sqrt{\frac{bax^nz^2}{1+bz^2}}} + e^{\frac{-2}{n+2}\sqrt{\frac{bax^nz^2}{1+bz^2}}}}{2(1+bz^2)^{\frac{n+1}{n+2}}},$$

where b and K are constants and e is the Neper number.

In his [1760-61], Euler considered the general Riccati equation[5]

$$dy+Pydx+Qy^2dx+Rdx=0, \tag{5}$$

where P, Q, and R are functions of x, and proved that if one solution v of this equation is known, then equation (5) can be reduced to the linear equation of the first order

$$dz-(P+2Qv)zdx-Qdx=0$$

by replacing y by $v+1/z$; so the general solution can be found by two quadratures. Then, Euler showed that, if two particular solutions of (5) are known, the general solution can be effected by a single quadrature (Euler [1760-61, §. 32]).

In [1762-63, §. 1], Euler faced the problem of finding the values of the exponent m such that the Riccati equation

$$dy\pm ay^2dx=ac^2x^{2n-2}dx \tag{6}$$

had algebraic solutions. To solve this problem, Euler set

$$y = P = cx^{n-1} + \frac{dz}{azdx}$$

and obtained

$$\frac{d^2z}{dx^2} + \frac{2acx^{n-1}dz}{dx} + (n-1)acx^{n-2}z = 0\,.$$

Then, he set

$$z = Ax^{\frac{-n+1}{2}} + Bx^{\frac{-3n+1}{2}} + Cx^{\frac{-5n+1}{2}} + Bx^{\frac{-7n+1}{2}} + \ldots$$

and derived

$$z = x^{\frac{-n+1}{2}} + \frac{(n^2-1)}{8n}\frac{x^{\frac{-3n+1}{2}}}{ac} + \frac{(n^2-1)(9n^2-1)}{8n\cdot 16n}\frac{x^{\frac{-5n+1}{2}}}{a^2c^2} + \frac{(n^2-1)(9n^2-1)(25n^2-1)}{8n\cdot 16n\cdot 24n}\frac{x^{\frac{-7n+1}{2}}}{a^3c^3} + \ldots$$

In this way the solution y to equation (6) was expressed as the ratio of two series and, if

$$(2i+1)^2n^2-1=0,$$

[5] Following the previous use Euler termed the equations of the type axndx=dy+y2dx as Riccati equations. In [1760-61] Euler gave no name to the equation dy+Pydx+Qy2dx+Rdx=0.

the series were finite and the solution was algebraic, a result already known to Daniel Bernoulli (Euler [1762-63, §§. 1-4]). Since one solution is known, Euler was able to show that the complete (general) integral of (6) is

$$y = cx^{n-1} + \frac{dz}{azdx} + \frac{e^{-\frac{2acx^n}{n}}}{z^2 \int \frac{1}{z^2} e^{\frac{-2acx^n}{n}} adx}.$$

He set

$$y = Q = -cx^{n-1} + \frac{du}{audx}$$

and found another particular solution of (6) :

$$u = x^{\frac{-n+1}{2}} - \frac{(n^2-1)}{8n} \frac{x^{\frac{-3n+1}{2}}}{ac} + \frac{(n^2-1)(9n^2-1)}{8n \cdot 16n} \frac{x^{\frac{-5n+1}{2}}}{a^2c^2} - \frac{(n^2-1)(9n^2-1)(25n^2-1)}{8n \cdot 16n \cdot 24n} \frac{x^{\frac{-7n+1}{2}}}{a^3c^3} + \ldots$$

By appropriate transformations, Euler [1762-63, §. 5] expressed the general integral in the form

$$Ce^{-\frac{2acx^n}{n}} = \frac{(P-y)z}{(Q-y)u}.$$

4. Differentiation and Integration of Functions of Two Variables

The investigation of functions of two variables arose in the context of the investigation of families of curves *y=y(x,a)* dependent on a parameter *a*. While studying these families of curves, Leibniz and Nikolaus Bernoulli discovered some crucial theorems of the theory of functions of two variables, such as the interchangeability theorem for differentiation and integration and equality of mixed second-order differentials (Engelsman, [1984, 45 and 100-106]). Euler also studied families of curves in the 1730s. In "De differentiatione" [1984], he gave a his own demonstration of the theorem of mixed differentials which he later formulated in this way:

THEOREM. If $dV=Pdx+Qdy$ then the differential of P for variable y and constant x and the differential of Q for variable x and constant y are equal (see Euler [1755c, §. 226]).

The proof runs as follows. Consider a function V of the variables x and y and put $A=V(x,y)$, $B=V(x+dx,y)$, $C=V(x,y+dy)$, $D=V(x+dx,y+dy)$. Take the differential of V, holding x constant: this produces $C–A=Qdy$. If in $C–A$ we put $x+dx$ in place of x, it produces $D–B$, the differential of which (namely, the differential of Qdy for variable x) is

$D-B-C+A$.

Now, if $x+dx$ is put into A in place of x, then B is produced, and then the differential of A, taking x to be the variable, is $B-A(=Pdx)$. Putting $y+dy$ in place of y in B-A we obtain the differential of Pdx for variable y:

$D-B-C+A$,

Since this differential is equal to the differential found in the previous operations, the theorem is proved (cf. Euler [1755c, §§. 226-228]).

Subsequently, Euler posed $dP=rdy$ (constant x) and $dQ=qdx$ (constant y), and observed that

$dPdx=rdxdy$ and $dQdx=qdxdy$.

Since the mixed differentials are equal, he had $r=q$. At this point, Euler [1755c, §§. 231-232] decided to introduce a symbolism to indicate the functions r and q in a convenient and unambiguous way. He denoted r by means of the symbol $\left(\frac{dP}{dy}\right)$, and q by $\left(\frac{dP}{dx}\right)$. Therefore, the condition that linked the finite quantities P and Q could be expressed as

$$\left(\frac{dP}{dy}\right)=\left(\frac{dP}{dx}\right).$$

In his [1984], he also proved the theorem on homogenous functions:

THEOREM. If $P(x,a)$ is a homogenous functions of degree n in x and a, and $dp=Qdx+Rda$, then $nP=Qx+Ra$.

From this theorem he easily derived the interchangeability of differentiation and integration

$$\frac{\partial}{\partial a}\int_{x_0}^{x} y(x,a)dx=\int_{x_0}^{x}\frac{\partial}{\partial a}y(x,a)dx.$$

Using the modern symbol of partial derivatives, Euler's derivation (cf. [1984, 206-209]) can be summarized as follows. Given $P(x,a)$ and put

$$Q=\frac{\partial P}{\partial x},\quad R=\frac{\partial P}{\partial a},\quad M=\frac{\partial R}{\partial x}=\frac{\partial^2 P}{\partial x\partial a}=\frac{\partial Q}{\partial a},$$

one has

$$\frac{\partial}{\partial a}\int Qdx=\frac{\partial P}{\partial a}=R=\int\frac{\partial M}{\partial x}da=\int\frac{\partial^2 Q}{\partial x\partial a}dx=\int\frac{\partial Q}{\partial a}dx.$$

It was only in his *De formulis integralibus duplicatis* [1769] that Euler introduced the notion of double integration. In this paper, Euler observed that that if one had to solve questions concerning the volume of the surface of a given body, then it occurred double integration, which denoted by the symbol

$$\iint Zdxdy.$$

The integral $\iint Z(x,y)dxdy$ was defined as a formula such that if it was twice differentiated, first with respect to x, second with respect to y, it gave $Z(x,y)dxdy$ as a differential[6]. For example, if $Z=a$ then

$$\iint adxdy=axy+X(x)+Y(y),$$

where $X(x)$ is a function of x alone and Y is a function of y alone (Euler [1769, 73]).

In [1769] Euler also stated that

$$\iint Zdxdy=\int dy\int Zdy=\int dx\int Zdy \tag{7}$$

For instance, if Z= $\iint \frac{dxdy}{x^2+y^2}$, integrating with respect to y, he obtained:

$$\iint \frac{dxdy}{x^2+y^2}=\int \frac{dx}{x} arctan \frac{y}{x}+X(x)$$

He developed $arctan \frac{y}{x}$ into series and integrated with respect to x:

$$\iint \frac{dxdy}{x^2+y^2}=X(x)+Y(y)-\frac{y}{x}+\frac{y^3}{9x^3}-\frac{y^5}{25x^5}+\dots \tag{8}$$

He then showed that formula (8) could be obtained integrating, first, with respect to x, then with respect to y:

$$\iint \frac{dxdy}{x^2+y^2}=\int \frac{dy}{y}\arctan\frac{x}{y}+f(y)=\int \frac{1}{y}\left[\frac{\pi}{2}-\arctan\frac{y}{x}\right]dy+f(y)$$

$$=X(x)+Y(y)-\frac{y}{x}+\frac{y^3}{9x^3}-\frac{y^5}{25x^5}+\dots$$

---

[6] Similarly, in [1768-1770, 1: §§. 1-11], Euler had defined an integral as an antidifferential, namely the integral $\int f(x)dx$ meant when $\int f(x)dx$ were differentiated, gave $f(x)dx$ the function $f(x)$ as a differential.

In [1769], Euler also provided a geometrical interpretation of the double integral as a volume just as the definite integral of a positive one-variable function could be interpreted as an area.

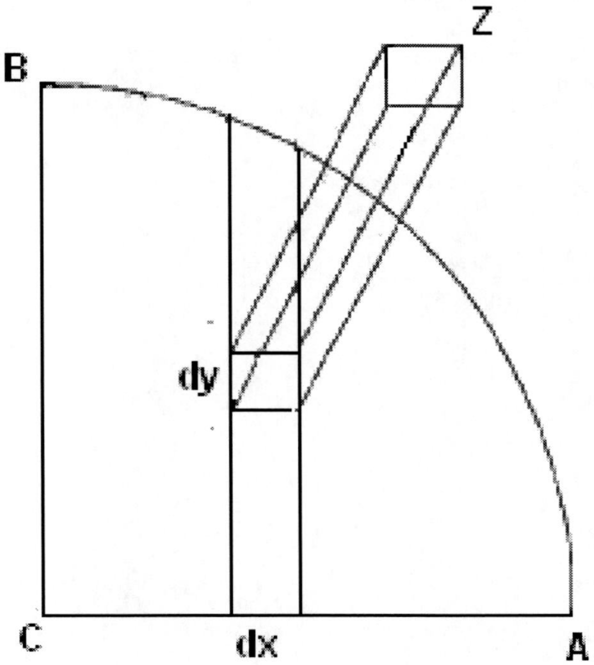

Figure 1. Euler's diagram illustrating the notion of the double integral.

5. Partial Differential Equation

In *De infinitis curvis* [1734-35b] Euler dealt with the so-called modular equations, namely differential equations of the type $dy=P(x,a)dx+Q(x,a)da$, where P and Q are algebraic functions, a is a parameter of the curve $y(x,a)$, and da is the differential of a. The investigation of these equations led Euler to the study of some partial differential equations, such as $\frac{\partial y}{\partial a}=-\frac{x}{a}\frac{\partial y}{\partial x}$ and $\frac{\partial y}{\partial a}=-\frac{x}{a}\frac{\partial y}{\partial x}+\frac{n}{a}y$. Nevertheless, he did not realize the importance of these equations in the investigation of natural phenomena. It was d'Alembert who first applied them to mechanics. For instance, d'Alembert described the motion of a stretched elastic string by means of the equation $\frac{\partial^2 z}{\partial x^2}=\frac{\partial^2 z}{\partial t^2}$ (see d'Alembert [1747]). When Euler read d'Alembert's research, he recognized the role that partial differential equations could play in applied mathematics and, in the years that followed, published several results on differential equations and their use in mechanics (see below, §. 9).

Now I will give some examples of Euler's treatment of partial differential equations. First, in order to solve the first-order differential equation

$$\frac{\partial z}{\partial x}x+\frac{\partial z}{\partial y}y=0\,, \tag{9}$$

Euler sought the total differential of a function *z(x,y)* such that

$$dz=\frac{\partial z}{\partial x}dx+\frac{\partial z}{\partial y}dy\,. \tag{10}$$

He replaced (9) into (10) and obtained

$$dz=\frac{\partial z}{\partial x}\left(dx-\frac{xdy}{y}\right)=\frac{\partial z}{\partial x}y\left(\frac{dx}{y}-\frac{xdy}{y^2}\right)=\frac{\partial z}{\partial x}yd\left(\frac{x}{y}\right).$$

According to Euler,

$$\frac{\partial z}{\partial x}y=\frac{dz}{d(x/y)}$$

was a function of $\frac{x}{y}$[7]. Therefore, he set $\frac{\partial z}{\partial x}y=f'\left(\frac{x}{y}\right)$. It follows that

$$z=f\left(\frac{x}{y}\right),$$

where *f* an arbitrary function (Euler [1768-70, vol. 3, §.138]).

Second, to solve the equation

$$\frac{\partial^2 z}{\partial x^2}=P(x,y)$$

Euler integrated it twice with respect to *x*. He first obtained $\frac{\partial z}{\partial x}=\int P(x,y)dx+f(y)$ and then

$$z=\int dx\int P(x,y)dx+xf(y)+F(y)\,,$$

where *f(y)* and *F(y)* could also be interpreted as discontinuous functions of *y* (cf. [1768-70, 3: §. 245-248]).

Third, to solve the wave equation

[7] Euler used an implicit principle: if the total differential *dz* can be written in the form ϕdu, then ϕ is a function of *u*.

$$\frac{\partial^2 z}{\partial y^2} = a^2 \frac{\partial^2 z}{\partial x^2},$$

Euler changed the variables $t=x+ay$, $u=x-ay$ and obtain $\frac{\partial^2 z}{\partial t \partial u} = 0$. He then integrated with respect to t so to have

$$\frac{\partial z}{\partial u} = h(u).$$

Hence, $z=\int h(u)du+f(t)=F(u)+f(t)$ and

$z=f(x+ay)+F(x-ay)$,

where f and F could be continuous or discontinuous functions (Euler [1768-70, 3:296]).

6. Infinite Polynomials and Series

One of the most famous Eulerian results was the solution of the Basel problem, namely the problem of finding the sum of the infinite series $\sum \frac{1}{n^2}$. In his [1734-35a]), Euler considered the power expansion of sin x

$$\frac{x}{1} - \frac{x^3}{3!} + \frac{x^5}{5!} - \frac{x^7}{7!} \pm \cdots$$

Since the equation sin x=0 has the infinitely many roots $0, \pm\pi, \pm 2\pi, \pm 3\pi, \ldots$, the infinite equation

$$\frac{\sin x}{x} = 1 - \frac{x^2}{3!} + \frac{x^4}{5!} - \frac{x^6}{7!} \pm \cdots = 0 \quad (11)$$

has the roots $\pm\pi, \pm 2\pi, \pm 3\pi, \ldots$ It is well-known that if a polynomial of even degree 2n has the 2n distinct roots $\pm a_1, \pm a_2, \ldots, \pm a_n$ $(a_i \neq 0)$, then one can write:

$$b_0 - b_1 x^2 + b_2 x^4 + \cdots (-1)^n b_n x^{2n} = b_0 \left(1 - \frac{x^2}{a_1^2}\right)\left(1 - \frac{x^2}{a_2^2}\right) \cdots \left(1 - \frac{x^2}{a_n^2}\right) \quad (12)$$

and

$$b_1 = b_0\left(\frac{1}{a_1^2}+\frac{1}{a_2^2}+\cdots+\frac{1}{a_n^2}\right) \tag{13}$$

Euler applied (12) and (13) to the infinite equation (11) and derived

$$\frac{1}{3!}=\frac{1}{\pi^2}+\frac{1}{4\pi^2}+\frac{1}{9\pi^2}+\cdots$$

and

$$\frac{\pi^2}{3!}=1+\frac{1}{4}+\frac{1}{9}+\cdots$$

In his [1734-35a, §.18], Euler also proved

$$\sum\frac{1}{n^4}=\frac{\pi^4}{90},$$

$$\sum\frac{1}{n^6}=\frac{\pi^6}{945},$$

$$\sum\frac{1}{n^8}=\frac{\pi^8}{9450},$$

$$\sum\frac{1}{n^{10}}=\frac{\pi^{10}}{93555},$$

$$\sum\frac{1}{n^{12}}=\frac{691\pi^{12}}{638512875}.$$

In his [1737], Euler proved the famous relation relating the sum of the reciprocals of the powers of the positive integers with an infinite product extended over the primes, which, in modern notation, can be written as:

$$\zeta(s)=\sum\frac{1}{n^s}=\prod(1-p^{-s})^{-1} \tag{14}$$

(the sum is over all natural numbers n while the product is over all prime numbers). Euler formulated (14) as follows.

THEOREM. If we use the series of prime numbers to form the expression

$$\frac{2^n}{(2^n-1)}\cdot\frac{3^n}{(3^n-1)}\cdot\frac{5^n}{(5^n-1)}\cdot\frac{7^n}{(7^n-1)}\cdot\frac{11^n}{(11^n-1)}\cdot etc.,$$

then its value is equal to the sum of this series

$$1+\frac{1}{2^n}+\frac{1}{3^n}+\frac{1}{4^n}+\frac{1}{5^n}+\frac{1}{6^n}+etc..$$

Proof. If we set $x=1+\frac{1}{2^n}+\frac{1}{3^n}+\frac{1}{4^n}+\frac{1}{5^n}+\frac{1}{6^n}+\ldots$ and multiply this equation by $\frac{1}{2^n}$, we have

$$\frac{1}{2^n}x=\frac{1}{2^n}+\frac{1}{4^n}+\frac{1}{6^n}+\ldots.$$

Hence

$$x\frac{2^n-1}{2^n}=1+\frac{1}{3^n}+\frac{1}{5^n}+\frac{1}{7^n}+\frac{1}{9^n}+\ldots. \tag{15}$$

If we multiply (15) by $\frac{1}{3^n}$, we have

$$x\frac{2^n-1}{2^n}\frac{1}{3^n}=\frac{1}{3^n}+\frac{1}{9^n}+\frac{1}{15^n}+\frac{1}{21^n}+\ldots.$$

Hence

$$\frac{2^n-1}{2^n}\cdot\frac{3^n-1}{3^n}x=1+\frac{1}{5^n}+\frac{1}{7^n}+\ldots.$$

If we continue the same procedure, we obtain

$$x\frac{(2^n-1)(3^n-1)(5^n-1)(7^n-1)\ldots}{2^n\cdot 3^n\cdot 5^n\cdot 7^n\cdot\ldots}=1.$$

Therefore

$$x = \frac{2^n}{(2^n-1)} \cdot \frac{3^n}{(3^n-1)} \cdot \frac{5^n}{(5^n-1)} \cdot \frac{7^n}{(7^n-1)} \cdot \frac{11^n}{(11^n-1)} \cdot \ldots \text{ Q.E.D.}$$

In the *Introductio* [1748, 1:122-124], Euler derived power series of the elementary functions. An example of his method is the expansion of the exponential function. Euler considered the equality $a^{\omega}=1+\psi$, where ω and ψ are infinitesimal, and assumed that ψ was equal to kω and that $a^{\omega}=1+k\omega$. Then he put $i=x/\omega$, where x is a finite number, and observed that

$$a^x = a^{i\omega} = (1+k\omega)^i = \sum_{r=0}^{\infty} \binom{i}{r} (k\omega)^r = \sum_{r=0}^{\infty} \binom{i}{r} \left(\frac{kx}{i}\right)^r .$$

Euler asserted that $\frac{i-1}{i}=1$, $\frac{i-2}{i}=1$, $\frac{i-3}{i}=1$, … for an infinitely large number i; therefore

$$a^x = \sum_{r=0}^{\infty} \frac{1}{r!}(kx)^r . \qquad (16)$$

By putting $(1+k\omega)^i=1+x$, Euler obtained $i\omega = \frac{1}{k}(1+x)^{1/i} - \frac{1}{k}$ and

$$\log_a(1+x) = \frac{i}{k}((1+x)^{1/i} - 1).$$

For $k=1$, he had $log(1+x)=i((1+x)^{1/i}-1)$. By applying (16) he derived

$$log(1+x) = x - \frac{x^2}{2} + \frac{x^3}{3} - \frac{x^4}{4} + \ldots \text{ [1748, 1:125-126].} \qquad (17)$$

In chapter 8 of the *Introductio*, Euler showed that

$$(\cos z + \sqrt{-1}\sin z)^n = \cos nz + \sqrt{-1}\sin nz .$$

Hence

$$\cos nz = \frac{(\cos z + \sqrt{-1}\sin z)^n + (\cos z + \sqrt{-1}\sin z)^n}{2} .$$

Another application of the binomial theorem allowed him to obtain

$$\cos nz = \cos^n z - \frac{n(n-1)}{1\cdot 2}\cos^{n-2} z \sin^2 z + \frac{n(n-1)(n-2)(n-3)}{1\cdot 2\cdot 3\cdot 4}\cos^{n-4} z \sin^4 z + \dots .$$

Then Euler put $nz=v$, where z was an infinitesimal, n an infinitely large number and v a finite number. In this case $\sin z=z$ and $\cos z=1$, hence

$$\cos v = 1 - \frac{1}{1\cdot 2}v^2 + \frac{1}{1\cdot 2\cdot 3\cdot 4}v^4 + \dots$$ [1748, 1: §§. 132-133].

7. Mechanics

In 1736, Euler published *Mechanica sive motus scientia analytice exposita*, a work that was originated by the Leibniz's programme to reformulate Newton's *Principia* in terms of the Leibnizian calculus (see Guicciardini [1999, 249]). In the preface, Euler distinguished mechanics from statics: he stated that statics investigated the comparison and equilibrium of forces, while mechanics was the science of motion. Euler considered mechanics as a rational science, whose fundamental laws are necessary truth that can be demonstrated mathematically. On his opinion, the first foundations of Mechanics were due to Galileo and, after the discovery of infinitesimal analysis, mechanics had increased and advanced enormously; however, he was not satisfied with the treatises till then published on mechanics. He explained that some mechanical works "have been undertaken by authors who do not have a thorough grasp of analysis; others have been fortified by exceedingly intricate and elaborate old-fashioned demonstrations; and yet others indeed with derivations from obscure principles". In his opinion, "what distracts the reader the most, is the fact that everything is carried out synthetically, with the demonstrations presented in the manner of the old geometry, and the analysis hidden, and recognition of which is given only at the end of the work. Hermann's work is not a great deal different also, from the manner of the composition of Newton's *Principia Mathematica Philosophiae*, from which the science of motion has benefited the most. But what pertains to all the works composed without analysis, is particularly true for mechanics. In fact, the reader, even though he is persuaded about the truth of the things that are demonstrated, nonetheless cannot understand them clearly and distinctly. So he is hardly able to solve with his own strengths the same problems, when they are changed just a little, if he does not inspect them with the help of analysis and if he does not develop the propositions into the analytical methods. This is exactly what happened to me, when I began to study in detail Newton's *Principia* and Hermann's *Phoronomia*. In fact, even though I thought that I could understand the solution to numerous problems well enough, I could not solve problems that were slightly different. Therefore I strove, as much as I could, to get at the analysis behind those synthetic methods in order, for my purposes, to deal with those propositions in terms of analysis. Thanks to this procedure I perceived a remarkable improvement of my understanding. Thus I have endeavoured or a long time now, to use the old synthetic method to elicit the same propositions that are more readily handled by my own analytical method, and so by working with this latter method I have gained a perceptible increase in my. Then in like manner also, everything regarding the writings about this science that I have pursued, is scattered everywhere, whereas I have set out my own method in a plain and well-ordered manner, and with everything arranged in a suitable order. Being engaged in this business, not only have I fallen upon many questions not to be found in previous tracts, to which I have been happy to provide solutions: but also I have increased our knowledge of the science by providing it with many unusual methods, by which it must be admitted that both mechanics and analysis are evidently augmented more than a just a little" (Euler [1736, 1:38-39]).

In *Mechanica*, Euler started with the concepts of inertia and force. He defines the inertia as the faculty of a body to maintain its state of rest or of continuing in its present state of motion in a straight line [1736, I, §. 74] and a force as "an action on a free body that either leads to the motion of the body at rest, or changes the motion of that body" [1736, I, §. 99]. Then Euler states that the force of inertia of any body is proportional to the quantity of matter, upon which it depends ([1736, I, §. 142]). In proposition 20 of the first book, he demonstrated the so-called Newton's second law of motion[8]: "The motion of the point in a direction in agreement with the direction of the force, the increment of the speed will be as the force taken with the element of time, and divided by the quantity of matter of the point is composed " ([1736, I, §. 154]), namely $a=F/m$. From this equation Euler was able to derive all differential equation necessary to describe the motion of a point-mass.

In the following years, Euler continued his mechanical study. In *Découverte d'un nouveau principe de Mecanique* [1750], Euler, for the first time, expressed the second law of motion in the form $F_x=ma_x$, $F_y=ma_y$, $F_z=ma_z$; more precisely, he wrote the equations:

$$2Mddx=Pdt^2,\ 2Mddy=Qdt^2,\ 2Mddz=Rdt^2, \qquad (18)$$

where M is the mass and P, Q, and R the components of the force on the axis (the coefficient 2 depended on the unity of measure). According to Euler, a physical continuum could be subdivided in elementary particles and one could apply differential equations (18) to these elementary particles (so the mass M could also be an infinitesimal quantity). According to Euler, this constituted a new and fundamental principle of mechanics from which any other principles or law of mechanics could be derived; in other words, mechanical problems way could be formulated in a general, analytical by means of an appropriate application of (18).

In 1765, Euler published *Theoria motus corporum solidorum seu rigidorum* [1765] a wide treatise where he introduced the concept of moment of inertia[9] of a rigid body and decomposed the motion into two elements: the rectilinear motion of the center of mass and the rotational motion about the center of mass[10]. In *Nova methodus motum corporum*

[8] At the beginning of Book I of his *Principia*, Newton formulated the second law of motion as follows: "the change of motion is proportional to the motive force impressed, and it takes place along the right line in which that force is impressed" [Newton 1687:114]. In modern terms, this definition corresponds to $F=\Delta(mv)$, where mv is the motion (momentum). The expression "change of motion" (*mutatio motus*) is not univocal in Newton and, elsewhere, Newton states that a centripetal force is proportional to the motion that it generates it in a given time [Newton 1687: 99]. This sounds as $F=ma$ (see Maltese [2002]). Some mathematicians use Newton's law in Cartesian form; however, it was Euler who based the mechanics of rigid bodies and fluid mechanics on this principle.

[9] See Problem 86. Being given a solid body actuated by a given angular velocity about some axis passing through its center of inertia, to find the elementary forces which must act on the elements of the solid in order that the axis of rotation and the angular velocity should undergo given variations in the time dt .

[10] See Problem 88. If a solid body, turning about an axis passing through its center of inertia with angular velocity w, is acted upon by some forces, to find the variation of the axis of rotation and the angular velocity at the end of a time dt .

solidorum rigidorum determinandi [1775] Euler completed the construction of the general equations of dynamics. Indeed, he formulated a system of six equations determining the motion of any body, which (except for an additional coefficient) he wrote in this way:

$$\int dM \frac{d^2x}{dt^2} = P,$$

$$\int dM \frac{d^2y}{dt^2} = Q,$$

$$\int dM \frac{d^2z}{dt^2} = P,$$

$$\int zdM \frac{d^2y}{dt^2} - \int ydM \frac{d^2z}{dt^2} = S,$$

$$\int xdM \frac{d^2z}{dt^2} - \int zdM \frac{d^2y}{dt^2} = T,$$

$$\int ydM \frac{d^2x}{dt^2} - \int xdM \frac{d^2y}{dt^2} = U.$$

These equations cover both the principles of linear momentum and of moment of momentum. Truesdell proposed to name them as Euler's fundamental laws of mechanics.

Euler also contributed to the development of the fluid mechanics. In *Principes géněraux de l'état d'équilibre des fluides* [1755a], he tackled the following general problem: "The forces which act on all the elements of the fluid being given, together with the relation which exists at each point between the density and the elasticity of the fluid, find the pressures that there must be, at all points of the fluid mass, in order that it may remain in equilibrium" [1755a, §. 21],

To solve this question, Euler [1755a, §§. 22-30] considered the mass of the fluid as composed by three-dimensional infinitesimal parallelepipeds. If the components of forces acting on an elementary parallelepiped, with one corner at the point Z of coordinates x, y, z and with dimensions dx, dy, dz, are R, Q, P and the body density is q, then the element of volume $dxdydz$ is subject to a force which has the components

$$Pqdxdydz, \quad Qqdxdydz, \quad Rqdxdydz.$$

Euler denoted the unknown pressure at the point Z by p, put $dp=Ldx+Mdy+Ndz$ and derived the general equilibrium conditions

$$L=Pq, \quad M=Qq, \quad N=Rq,$$

If p is a function of q, at each point, the equation

$$dp=q\,(Pdx + Qdy + Rdz)$$

shows that $Pdx + Qdy + Rdz$ is the total differential of dp/q.

In another paper [1755b], Euler assumed that the original state of the fluid, namely the configuration of particles and their velocities, is known at a given instant as well as the forces acting on it. Euler denoted the components of the force acting on the fluid by R, Q, P. They are known functions of z, y, x and t. The density q, the pressure p and the components w, v, u of the velocity of the fluid element that is at point Z are at time t unknown quantities. Euler supposed that, during the time dt, the element of fluid at the point Z is carried to a point Z' of coordinates $x+udt$, $y+vdt$, $z+wdt$, and the element of fluid at z, of coordinates $x+dt$, $y+dt$, $z+dt$, is carried to the point z', and after some calculations, derived the general equations:

$$P - \frac{1}{q}\frac{\partial p}{\partial x} = \frac{\partial u}{\partial t} + u\frac{\partial u}{\partial x} + v\frac{\partial u}{\partial y} + w\frac{\partial u}{\partial z}$$

$$Q - \frac{1}{q}\frac{\partial p}{\partial y} = \frac{\partial v}{\partial t} + u\frac{\partial v}{\partial x} + v\frac{\partial v}{\partial y} + w\frac{\partial v}{\partial z}$$

$$R - \frac{1}{q}\frac{\partial p}{\partial z} = \frac{\partial w}{\partial t} + u\frac{\partial w}{\partial x} + v\frac{\partial w}{\partial y} + w\frac{\partial w}{\partial z}$$

$$\frac{\partial q}{\partial t} + \frac{\partial qu}{\partial x} + \frac{\partial qv}{\partial y} + \frac{\partial qw}{\partial z} = 0.$$

8. The Calculus of Variations and the Principle of the Least Action

In 1744 Euler published a fundamental book on the calculus of variation, entitled *Methodus inveniendi lineas curvas maximi minimive proprietate gaudentes, sive solutio problematis isoperimetrici latissimo sensu accepti*. Euler's main contribution to the calculus of variation is the so-called Euler-Lagrange equation, which provides a solution to the problem of extremising an integral expression of the type:

$$J = \int_{x_1}^{x_2} F(x, y, y', \dots, y^{(n)})dx$$

where F is a given function. Euler showed this problem the solution to this problem had to satisfy the differential equation:

$$\frac{\partial F}{\partial y} - \frac{d}{dx}\frac{\partial F}{\partial y'} + \frac{d^2}{dx^2}\frac{\partial F}{\partial y''} - \dots = 0 \qquad (19)$$

In [1744] Euler approached the question in a geometrical way, however his reasoning was very general and did not depended on the particular geometrical representation. He also called for the development of a simple method or an algorithm to obtain variational equations. This algorithm was developed by Lagrange, who recognized the dual usage of the symbol dy

in Euler [1744]; indeed, it denoted both the differential *dy* of *y* with respect to *x* and the variation of the curve *y(x)* (on Euler's calculus of variation, see Fraser [1994]).

Euler applied the calculus of variations to solve several problems. In particular, in Appendix I of [1744], entitled *De curvis elasticis* [1744, 245-310], Euler dealt with elastic curves. He tackled the following problem:

PROBLEM. Consider two points *A* and *B* and the curves of given length such that their extreme points are *A* and *B* and their slopes in *A* and *B* are given, find the curve such that the value of the integral

$$\int \frac{ds}{R^2},$$

where *R* is the radius of curvature of the curve, is a minimum.

Euler showed that the solution was given by a curve *y=y(z)* such that

$$\frac{dy}{dz} = \frac{\alpha + \beta z + \gamma z^2}{\sqrt{a^4 - (\alpha + \beta z + \gamma z^2)^2}}$$

(a, α, β, γ are constants).

In Appendix II of [1744], entitled *De motu projectorum in medio non resistente per methodum maximorum ac minimorum determinando* [1744, 311-320], Euler faced the so-called principle of least action. This principle states that in all natural phenomena a quantity called 'action' tends to be minimized. Pierre-Louis Moreau de Maupertuis (1698-1759) wrote about this principle in 1744 and 1746[11]. According to Maupertuis, action could be expressed mathematically as the product of the mass of the body involved, the distance it had traveled and the velocity at which it was traveling. Maupertuis's formulation of the principle is however rather aprioristic and metaphysical way. Euler stated: "Since all effects of nature obey some maximum or minimum law, we cannot deny that the trajectories described by projectiles under the influence of some forces, will follow a property of maximum or minimum. It seems easier to define a priori by using metaphysical principles what this property actually is. However, with the necessary application it is possible to determine these curves by direct methods and then we can decide what is a maximum or a minimum" [1744, 311]. He formulated the principle in these terms. Let the mass of a moving particle be *M*, and let its speed be *v* while being moved over an infinitesimal distance *ds*. The particle will have a momentum *Mv* that, when multiplied by the distance *ds*, gives *Mvds*. Euler asserted that the true trajectory of the moving particle is the trajectory to be described (from among all possible trajectories connecting the same endpoints) that minimizes

[11] In 1751 Maupertuis's priority was disputed by Samuel König, who attributed it to Leibniz in 1707. König was not able to prove his claim (he exhibited a copy of a 1707 letter from Leibniz to Jacob Hermann with the principle, but not the original one). A bitter polemics followed and König was prosecuted for forgery. In this polemics Euler defended Maupertuis. Note that in the 17th century Pierre de Fermat had already stated the principle of least time (or Fermat's principle): "light travels between two given points along the path of shortest time".

$$\int Mvds$$

or, provided that M is constant, $\int vds$ [1744, 311-312].

9. Other Results in Mixed Mathematics

To conclude this paper, I discuss Euler's treatment of some mechanical problems. First, let us consider the problem of the oscillations of a flexible hanging chain, a classical problem of oscillating theory. This problem was first investigated by Daniel Bernoulli in his [1732-33]. In his [1781a], Euler derived the equation of motion. His procedure can be summarized as follows. Consider a uniform heavy flexible chain of length L, fixed at the upper end and free at the lower end. Let ρ be the line density of the chain and let T be the tension at height x above the lowest point. When the chain is slightly disturbed from its position of equilibrium in a vertical plane, it undergoes small oscillations. The horizontal component of the tension is $T\frac{dy}{dx}$, where y *is* the horizontal displacement of the chain at time t. If we apply Newton's second law to an infinitesimal element dx of the chain, we obtain

$$\rho dx\frac{d^2y}{dx^2}=d\left(T\frac{dy}{dx}\right). \tag{20}$$

If we suppose that the tension T is due entirely to weight of the chain below a given point x, then $T=g\rho x$, where g is the gravitational constant. By replacing this expression of T into (20), we have

$$dx\frac{d^2y}{dx^2}=gd\left(x\frac{dy}{dx}\right)$$

and, by dividing by dx,

$$\frac{d^2y}{dx^2}=g\frac{d}{dx}\left(x\frac{dy}{dx}\right),$$

which, in modern form, can be written as:

$$\frac{\partial^2y}{\partial x^2}=g\frac{\partial}{\partial x}\left(x\frac{\partial y}{\partial x}\right). \tag{21}$$

Euler assumed that the oscillation y was essentially sinusoidal with angular frequency $\omega = \sqrt{\frac{g}{f}}$ (f is the length of the simple equivalent pendulum) and wrote

$$y = Av\,sin\left(\xi + t\sqrt{\frac{g}{f}}\right), \tag{22}$$

where A and ξ are constants and $v = \Phi\left(\frac{x}{f}\right)$ is an appropriate function of the only variable x. Substituting (22) into (21), he found that $\Phi\left(\frac{x}{f}\right)$ is a solution of the ordinary differential equation

$$\frac{d}{dx}\left(x\frac{dy}{dx}\right) + \frac{v}{f} = 0 \tag{23}$$

Euler solved this equation using series and obtained

$$v = 1 - u + \frac{u^2}{1\cdot 4} - \frac{u^3}{1\cdot 4\cdot 9} + \frac{u^4}{1\cdot 4\cdot 9\cdot 16} - \ldots,$$

where $u = \frac{x}{f}$[12].

Later Euler dealt with the series solution of equation (23) in [1768-80, 2: §. 977] and in [1781b].

Today we recognize (173) as a Bessel function $J_0\left(2\sqrt{\frac{x}{n}}\right)$ of order zero and argument $2\sqrt{\frac{x}{n}}n$; however, I emphasize that for Bernoulli and Euler (173) was merely a tool for obtaining an approximate solution to a problem of physics.

Euler also investigated the problem of the vibrations of a stretched membrane. In *De motu vibratorio tympanorum* [1764], Euler derived an equation equivalent to

$$\frac{1}{c^2}\frac{\partial^2 z}{\partial t^2} = \frac{\partial^2 z}{\partial r^2} + \frac{1}{r}\frac{\partial z}{\partial r} + \frac{1}{r^2}\frac{\partial^2 z}{\partial \varphi^2} \tag{24}$$

---

12 The solution was already known to Daniel Bernoulli.

where z is the transverse displacement at time t at the point whose polar coordinates are (r, φ) and c is an appropriate constant. He assumed that the solutions had the form

$$z = u(r)\sin(\omega t + A)\sin(\kappa\varphi + B),$$

where ω, A, κ, B are constants. By replacing $u(r)\sin(\omega t+A)\sin(\beta\varphi+B)$ in (24), he derived the equation

$$\frac{d^2u}{dr^2}+\frac{1}{r}\frac{du}{dr}+\alpha^2-\frac{\kappa^2}{r^2}=0, \tag{25}$$

where $\alpha=\frac{\omega^2}{c^2}$. Euler assumed the existence of a power series solution of this equation and obtained[13].

$$u(r)=\sum_{n=0}^{\infty}\frac{(-1)^n}{n!(\kappa+1)_n}\left(\frac{\alpha r}{2}\right)^{\kappa+2n}.$$

REFERENCES

Alembert, J. le Rond d' [1743] *Traitè de dynamique*, Paris, David.

Alembert, J. le Rond d' [1747] Recherches sur la courbe que forme une corde tenduë mise en vibration, *Mémoires des l'Académie des Sciences de Berlin*, vol. 3 , 214-219.

Bernoulli, D. [1732–33] Theoremata de oscillationibus corporum filo flexili connexorum et catenae verticaliter suspensae, *Commentarii Academiae Scientiarum Petropolitanae,* 6 (1732–1733), published in 1738, pp. 108–122.

Bernoulli, Johann [1742] Remarques sur ce qu'on a donné quisqu'ici de la solutions de problêmes sur les isoperimétres; in *Opera omnia*, Lausannae et Genevae, Marci-Michaelis Bousquet et Sociorum, 1742.

Condorcet, M. J. A. N. C., Marquis de [1783] Eloge de M. Euler, *Histoire de l'Academie Royale des Sciences* 1783 (printed 1786), Paris 37-68.

Calinger, R. [1996], Leonhard Euler: The first St Petersburg years (1727-1741), *Historia Mathematica* 23 (1996), 121-166.

Euler, L. [1726] Constructio linearum isochronarum in medio quocunque resistente'', *Acta Eruditorum*, 1726, pp. 361-363 in *Leonhardi Euleri opera omnia*, Basel, Birkhäuser (afterwards: *Opera Omnia*): Ser. 2, Vol. 6, pp. 1 – 3.

[1727a] Dissertatio de novo quodam curvarum tautochronarum genere *Commentarii academiae scientiarum Petropolitanae* 2 (published 1729), pp. 126-138 in *Opera Omnia*: Ser. 2, Vol. 6, pp. 4 - 14 .

[13] See Euler [1764, 344–359]. Today equation (25) is called Bessel's equation, and the solution $u(r)$ is the Bessel function $J_\beta(\alpha r)$.

[1727b] Methodus inveniendi traiectorias reciprocas algebraicas, *Acta Eruditorum* 1727, 1727, pp. 408-412 in *Opera Omnia*: Ser. 1, Vol. 27, pp. 1 – 5.

[1727c] Problematis traiectoriarum reciprocarum solutio, *Commentarii academiae scientiarum Petropolitanae* 2, 1729, pp. 90-111.

1728] De linea brevissima in superficie quacunque duo quaelibet puncta iungente, *Commentarii academiae scientiarum Petropolitanae* 3, 1732, pp. 110-124 in *Opera Omnia*: Ser. 1, Vol. 25, pp. 1 – 12.

[1734-35a] De summis serierum reciprocarum, *Commentarii Academiae Scientiarum Petropolitanae,* 7 (1734–1735), published in 1740, pp. 123–134; in *Opera Omnia*, ser. 1, vol. 14, pp. 73–86.

[1734-35b] De infinitis curvis eiusdem generis seu methodis inveniendi aequationes pro infinitis curvis eiusdem generis, *Commentarii Academiae Scientiarum Petropolitanae.,* vol. 7, pp. 174-189 and 180-183 (there is a typographical error in page numbers of *Commentarii*) or *Opera Omnia*, ser. 1, vol. 22, 36-56.

[1734-35c] De minimis oscillationibus corporum tam rigidorum quam flexibilium methodus nova et facilis, *Commentarii Academiae Scientiarum Petropolitanae,* 7, published in 1740, in *Opera Omnia*, ser. 2, vol. 10, 17-34.

[1736] *Mechanica sive motus scientia analytice exposita. Auctore Leonhardo Eulero academiae imper. scientiarum membro et matheseos sublimioris professore* (2 vols.). Petropoli. Ex typographia academiae scientarum. 1736.

[1737] Variae observations circa series infinitas, *Commentarii academiae scientiarum Petropolitanae* 9, 1744, p. 160-188. Reprinted in *Opera Omnia,* ser. 1, vol. 14, p. 216-244.

[1739] De novo genere oscillationum, *Commentarii Academiae Scientiarum Petropolitanae*, vol. 11, 128-149, in *Opera*, ser. 2, vol. 10, 78-97.

[1743b] De integratione aequationum differentialium altiorum graduum, *Miscellanea Berolinensia*, 7, 193-242, in *Opera*, ser. 1, vol. 22, 108-149.

[1744] *Methodus inveniendi lineas curvas maximi minimive proprietate gaudentes, sive solutio problematis isoperimetrici latissimo sensu accepti*, Lausannae et Genevae, M. M. Bousquet et Soc. in *Opera Omnia*: Series 1, Volume 24

[1748] *Introductio in analysin infinitorum*, Lausannae, M. M. Bousquet et Soc., in *Opera*, ser. 1, vols. 8-9.

[1750] Decouverte d'un nouveau principe de Mecanique, Mémoires de l'académie des sciences de Berlin, vol. 6, published in 1752, pp. 185-217, Opera Omnia: Ser. 2, Volume 5, pp. 81 - 108

[1755a] Principes généraux de l'état d'équilibre des fluides, *Mémoires de l'académie des sciences de Berlin* vol. 11, 1757, pp. 217-273 in *Opera Omnia*, ser. 2, vol. 12, pp. 2 - 53

[1755b] Principes généraux du mouvement des fluides, *Mémoires de l'académie des sciences de Berlin* vol. 11, 1757, pp. 274-315 in *Opera Omnia*, ser. 2, vol. 12, pp. 54 – 91.

[1755c] *Institutiones calculi differentialis cum eius usu in analisi finitorum ac doctrina serierum*, Berlin, Academia Imperialis Scientiarum.

[1760-61] De integratione aequationum differentialium *Novi Commentarii Academiae Scientiarum Petropolitanae,* vol. 8, 3-63, in *Opera Omnia*, ser. 1, vol. 22, 334-394.

[1762-63] De resolutione æquationis $dy \underset{\cdot}{+} ayydx \underset{\cdot}{=} bx^m dx$, *Novi Commentarii Academiae Scientiarum Petropolitanae*, vol. 9, 154-169, in *Opera Omnia*, ser. 1, vol. 22, 403-420.

[1764] De motu vibratorio tympanorum, *Novi Commentarii Academiae Scientiarum Petropolitanae,* 10, published in 1766, pp. 243–260; in *Opera Omnia*, ser. 2, vol. 10, pp. 344–359.

[1765] *Theoria motus corporum solidorum seu rigidorum ex primis nostrae cognitionis principiis stabilita et ad omnes motus, qui in hujusmodi corpora cadere possunt, accommodata*. Rostochii et Gryphiswaldiae litteris et impensis A. F. Röse. 1765.

[1768-70] *Institutionum calculi integralis,* Petropoli, Impensis Academiae Imperialis Scientiarum, in *Opera Omnia*, ser. 1, vols. 11-13.

[1769] De formulis integralibus duplicatis, *Novi Commentarii academiae scientiarum Petropolitanae* 14, 1770, pp. 72-103 in *Opera Omnia*, ser. 1, vol. 17, pp. 289 – 315.

[1775] Nova methodus motum corporum solidorum rigidorum determinandi *Novi Commentarii academiae scientiarum Petropolitanae* 20, published in 1776, pp. 208-238, in *Opera Omnia*, ser. 2, vol. 9, pp. 99 – 125.

[1781a] De oscillationibus minimis funis libere suspensi, *Acta Academiae Scientiarum Petropolitanae* (1781), pt. 1, pp. 157–177, in *Opera Omnia*, ser. 2, vol. 11, pp. 307–323.

[1984] De differentiatione functionum duas pluresve variabiles quantitates involventium, in Engelsman [1984, 205-213].

Engelsman, S. B. [1984], *Families of curves and origins of partial differentiation*, Amsterdam, North-Holland, 1984.

Ferraro, G. [1998] Some aspects of Euler's series theory. Inexplicable functions and the Euler–Maclaurin summation formula, *Historia Mathematica*, 25 pp. 290–317.

[2000a] Functions, functional relations and the laws of continuity in Euler, *Historia Mathematica*, 27, pp. 107–132.

[2000b] The value of an infinite sum. Some observations on the Eulerian theory of series, *Sciences et techniques en perspective*, 4, pp. 73–113.

[2002] Convergence and formal manipulation of series from the origins of calculus to about 1730, *Annals of Science*, 59, pp. 179–199.

[2008a] *The rise and development of the theory of series up to the early 1820s*, New York, Springer, 2008.

[2008b] D'Alembert visto da Eulero, *Bollettino di Storia delle Scienze Matematiche*, 28, pp. 257-275.

Fraser, C. G. [1994] The origins of Euler's variational calculus, *Archive for History of Exact Sciences,*. 47, pp. 103-141.

Fuss, Nicolas [1783], Éloge de monsieur Leonard Euler lu a l'académie Impériale des sciences,dans son assemblée du 23 octobre 1783. Avec une liste complète des ouvrages de M. Euler. St. Pétersbourg 1783.

Fuss, P.H. [1843] *Correspondance mathématique et physique de quelque célèbres géomètres du XVIIIème siècle*, St.Pétersbourg, Académie impériale des sciences, 1843.

Guicciardini, N. [1999] *Reading the Principia. The Debate on Newton's Mathematical Methods for Natural Philosophy from 1687 to 1736*, Cambridge: Cambridge University Press, 1999.

Leibniz, G. W. [1849-1863] *Leibnizens mathematische Scriften*, edited by C. I. Gerhardt, Berlin and Halle, 1849-1863.

Maltese, Giulio [2002] On the Changing Fortune of the Newtonian Tradition in Mechanics, in kim Williams (editor) *Two Cultures Essays in Honour of David Speiser*, Birkhäuser Basel, pp. 97-113.

Newton, I. [1687] Principi matematici di filosofia naturale. Trans. by A. Pala. Torino: UTET, 1977.

Youschkevitch, A. P. [B], Biography in *Dictionary of Scientific Biography* (New York 1970-1990).

In: Computational Mathematics
Editor: Peter G. Chareton, pp.63-83

ISBN:978-1-60876-271-2

Chapter 3

APPLICATIONS OF COMPUTATIONAL GEOMETRY TO PROBLEMS OF POLITICAL COMPETITION

***M. Dolores López*[*,1], *Javier Rodrigo*[†,2] *and Sagrario Lantarón*[1]**

[1]Department of Applied Mathematical and Computer Science, Higher Technical School of Civil Engineering, Polytechnic University of Madrid, s/n. 28040 Madrid, Spain

[2]Department of Applied Mathematics, Higher Technical School of Engineering, University Pontificia Comillas of Madrid, Alberto Aguilera 23, 28015 Madrid, Spain

ABSTRACT

This chapter puts into practice a geometrical model of political economics studies developed by the authors, which comprises the consideration of a problem of political competition with two opposed parties. They will attempt to capture the greatest number of voters of a discrete population of elements. It is supposed that these parties can modify their policies to a certain degree. Our purpose is to determine the optimun position or positions for the party in terms of guaranteeing the maximum number of voters.

On one hand, an algorithm is implemented that tries to solve this problem in a certain case of Spanish politics, simulated in some data partially based on the survey of public opinion and Fiscal Politics Study n° 2615 (July 2005) of the CIS (Sociological Investigations Center of Spain). On the other hand, the possible Nash equilibrium positions are studied under these certain limits. Due to the discrete characteristics of the problem, computational geometry techniques are applied.

Keywords: computational geometry, location, game theory, political competition, equilibrium, search algorithms

* ma08@caminos.upm.es

† jrodrigo@upcomillas.es

1. INTRODUCTION

This work approaches the resolution of a political economics problem with tools of computational geometry. The points of a plane, which we will call "plane of policies," represent the different political options about two different topics. We assume that the distance between the points will give an idea of the affinity of the policies related to such topics. Furthermore, if a set of points in the plane is given, its Voronoi diagram divides the plane into regions: each point is associated with the region of the points of the plane that are closer to it than to the others of the set (Okabe et al., 2000).

Similar models of location of points have been studied in different fields such as industrial organization, image treatment, movement of robots, etc. Most of them consider the population as a continuum (Ahn et al., 2001). In this work, the new scope consists of working with a discrete population and in the application of techniques and results of computational geometry adapted to the problem.

Section 2 solves a problem of political competition with two opposed political parties. They will attempt to capture the greatest number of voters of a discrete population of elements. It is supposed that one of the parties can modify its policy to a certain degree. Our purpose is to determine the optimum position or positions for the party in terms of guaranteeing the maximum number of voters. An efficient algorithm of search for these optimum positions is put forward and it is implemented in a practical case taken from politics in Spain, analysing with it the importance of the studies previous to the electoral campaigns, which can be done through the opinion surveys.

Section 3 deals with a generalisation of the previous model, permitting both parties to change their central positions to a certain extent, which is marked by circular neighborhoods centred on the initial positions of the candidates. In this model, the possible Nash equilibrium positions are studied under these considerations. Therefore, it turns out to be a model that adjusts itself to the reality of the countries, where the political parties change their proposal according to certain criteria and, furthermore, interesting results are obtained; in many cases, there are infinite equilibrium positions.

2. GEOMETRICAL SEARCH FOR OPTIMUM POSITIONS IN THE GAME WITH RESTRICTIONS: USE OF THE OPINION SURVEYS

Many statistical works exist for the study of the intention of the citizens' vote. Along that line, we can highlight the opinion surveys in political and fiscal topics related to taxes, considerations that the society receives for the payment of taxes, the operation of public services, fiscal fraud, the politicians' valuation, etc.

However, there is a void regarding quantitative surveys in relation to these topics. In our opinion, this is a consequence of the almost total ignorance of these themes on the part of most citizens, due mainly to the scant information that the politicians reveal on questions of valuation.

In this section we want to show, through a concrete example, the influence that some information could have in the adoption of political strategies for the elections, and how more information and greater economic and political culture of the citizenship would have notorious effects on the vote decision and the electoral results.

In order to carry out the development of a concrete study, we use as a basis a theoretical model given by the following statement. This statement is established in works of political competition with some variation:

The points of a plane, which we will call plane of policies, represent the different political options about two different topics. We assume that the distance between the points will give an idea of the affinity of the policies related to such topics. Let p and q be two political parties located in the points X_1 and X_2, and the voter population represented by a finite set of types H=$\{p_1, \ldots, p_n\} \subset R^2$ (Roemer, 2001; Abellanas et al., 2006). We consider that each player captures those points that are closer to him than to the other one. To count the points each player gets, we trace the perpendicular bisector of the two positions of the players. Then, each one will get the points located in the half-plane the player belongs to. The winner will be the player who gets more points (Serra and Revelle, 1994; Smid, 1997; Aurenhammer and Klein, 2000; Okabe et al., 2000; Roemer, 2001).

In politics, a slight variation in the programs of the parties will be admitted in order to obtain a greater number of votes. We admit that only one party, for example the party p, relaxes its position—that is to say, it can move in a certain neighbourhood. This is the disk with center X_1 and radius r. We look for the best situation for p within this neighbourhood, the one that approaches it with the greatest number of voters.

In this work we develop an algorithm of search for this best position. It is created from geometric ideas, and with computational geometry techniques adapted to the problem (Abellanas et al., 2006, Preparata and Shamos, 1985).

2.1. The Opinion Surveys

There are many surveys and statistical studies of a great number of topics that entities like the CIS (Center of Sociological Investigations), the INE (National Institute of Statistic) or the CEACS (Center of Advanced Studies in Social Sciences), among others, have carried out for many years in Spain

As an example of surveys and information that one can have nowadays on the contents we are interested in, we consider the opinion and fiscal policy survey nº 2615 of the CIS. With its results, on one hand, we will simulate the remaining information that is necessary for the development of the work that we present; on the other hand, we will show the necessities of another kind of study and the lack of citizens' information and preparation in certain areas.

2.1.1. Public Opinion and Politics Fiscal Survey Nº 2615 of the CIS

Technical data:

- Place of distribution: Spain
- Universe notes: Population over 18 years
- Size of file: Designed, 2500 respondents. Featured, 2483 respondents
- Geography terms: 167 municipalities and 47 provinces
- Methodology notes: Random selection of sample by random routes
- Data were collected by questionnaires in a face-to-face interview

- For a confidence level of 95.5%, MoE (Margin of error) ±2.0% for the sample set (we suppose simple random sampling)
- Dates of production: From 1 to 13 of July, 2005
- It includes questions like these:

Opinion related to the demand degree of the citizens to the different public services; considerations that the citizen receives for the payment of taxes; evaluation of the quantity of resources that the State dedicates to the diverse public services; opinion about the necessity of the taxes increase for the improvement of the services; valuation of the quantity of taxes that the Spaniards pay; justice in the distribution of the payment of taxes.

All of these questions are answered through the following possibilities: nothing, little, quite or a lot.

2.2. The Algorithm and the Simulation

2.2.1. A Graphic Approximation of the Algorithm

The algorithm that we use for the resolution of the problem presented in the introduction has been created starting from the following graphic ideas:

1. An optimum situation for p, is always found on the boundary of the neighbourhood, and it is in the arc of the circumference located between the two points of tangent lines from X_2 (the positionn of q) to the circumference (visible part of the neighbourhood of p from X_2). See Figure 1.

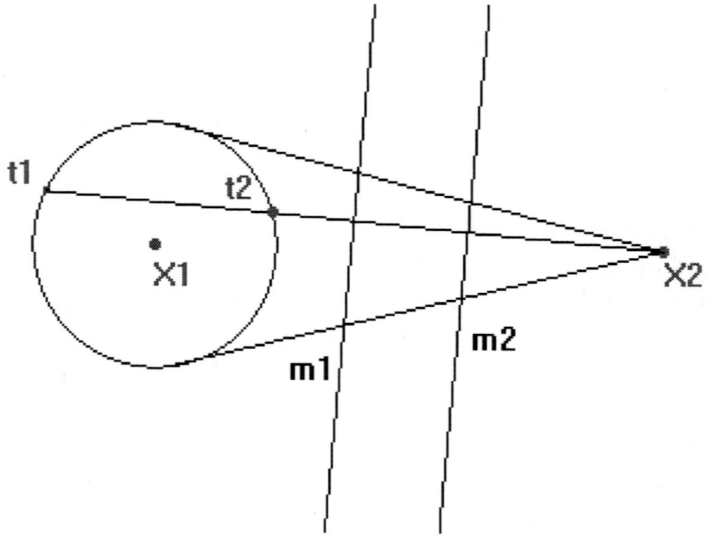

Figure 1. Zone where the optimal situation for p is located.

2. It is based on the localization of intersections of circumferences. On the one hand, we have the circumference centred in X_1 that indicates the neighbourhood of flexibility for the first political party; on the other hand, for each voter: p_i, we trace the

circumference centred in this point that goes through X_2. In the intersection zone the first party captures p_i, since in any point t_1 of this zone, the distance between p_i and t_1 is shorter than the distance between p_i and X_2 (Figure 2).

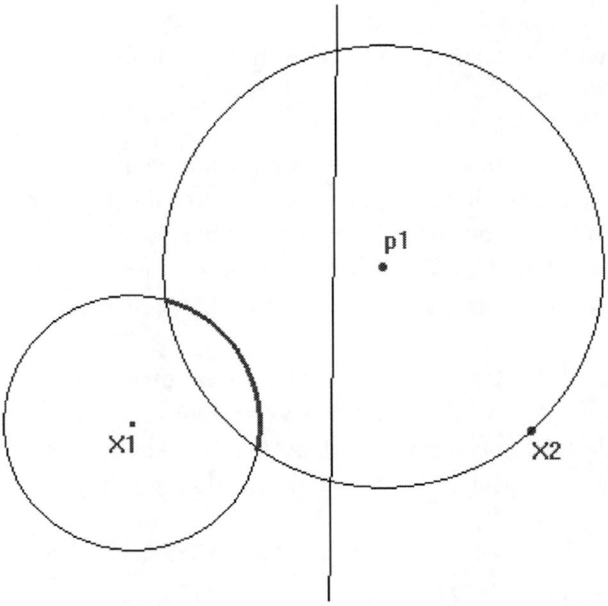

Figure 2. Captation zone for p of the point p_1.

The best area for p inside the neighbourhood will be in the arc of maximum intersection of the previous intersection zones inside the visible part (Abellanas et al, 2005).

The development of the algorithm is included in Appendix A. The program is freely available asking authors.

2.3. Simulation with an Example of the National Politics (Spain)

Following the presented competition model, the idea is to simulate some data partially based on the results obtained in the survey of the CIS nº 2615 that allow us to develop the search for the best policies of a party in relation to two specific topics and to evaluate which would be the benefits of having this information previously. This survey serves us as a statistic and numeric guide for the generation of the data: the number of interviewed or the answers to questions such as the evaluation of the quantity of resources that the State dedicates to certain services, allow us to work with data fitting reality, for the simulation of the problem at national level.

We propose to add previous survey questions like:

1. Choose the two services that you consider high-priority of the following list: Education, Public Works, Defence, Health, Housing, Justice, Work and Social Matters, Transport and Communications, Environment.

2. Knowing that the current Government dedicated the following percentages from the total expense to these services in 2005, tell me: what percentage would you dedicate to the same ones?
3. Would it affect to your vote decision to know in advance how much money the political parties would dedicate to each of the services? If so, what margin of difference with the quantity that you want to dedicate would you admit to the party that you would vote?

Questions like these allow us to choose two important topics for the citizens with quantitative information of their opinions and with effects about their possible vote decision.

We choose as policies to evaluate the investment in education and health and we generate the answers to the questions 1, 2 and 3 randomly using the real percentages of the answers in the CIS survey previously mentioned. The policies of the parties faced have been taken as:

- q the PP (Partido Popular) Spanish party's mean investment in education and the mean investment in health during its 8 years of government.
- p PSOE (Partido Socialista Obrero Español) Spanish party's mean investment in education and the mean investment in health during its 2 years of government.

These quantities have been extracted from the Consolidated General Government Budget (1997–2006): policies of expense (Chapters 1 to 8).

According to these values, we think about the possibility of studying how would the voter be affected by the expense policies' variation in education and health in the sense of finding, inside a margin, the appropriate policy to capture the biggest number of followers.

2.3.1. Algorithm Implementation

The algorithm has been implemented in the computer using the C programming language. The C is a powerful programming language, of general purpose, that allows solving problems of any scientific area in an appropriate time of execution.

Input:

- The two parties' location.
- Radius of the neighbourhood of political flexibility for a party.
- Voters location p_i, i=1,...,2276.

Output:

- Number of voters that each party always captures in any position of its neighbourhood (points p_i for which there does not exist intersection between the circumferences of the second step of section "a graphic approximation to the algorithm", and the circle centered in p_i contains the circle centered in X_I).
- Number of voters that the party that changes its politics captures by locating in the optimal region (points p_i for which there exists common intersection between the circumferences of the second step of section "a graphic approximation to the algorithm").

➢ Optimal region where the party should be located in order to capture the biggest number of voters.

2.3.2. Results

Following the ideas of the previous section, we consider

➢ The political plane settles down with the percentages of the expense dedicated to education and health, of the Consolidated General Government Budget for the expense policies (26 elements).
➢ The policy followed by the first party (PSOE) and the second party (PP) have been determined with the mean percentage dedicated to these two policies calculated from the total mean expense during the two years of PSOE party's government (years 2005, 2006) and the same for PP party (years 1997 to 2004): X_1=(0.6,1.4), X_2 =(1.6,8.9).
➢ The political flexibility neighbourhood is enlarging in the diverse studies.
➢ The voters and their preferences have been generated randomly by using the results of the CIS survey mentioned (concretely with the answers to the question related to the quantity of resources that the State dedicates to the diverse public services) that became under the government of the PSOE. Then:
  ➢ Voters that prefer more investment in education or health than the one invested by the current government.
  ➢ Voters that agree with the investment in some of the elements
  ➢ Voters that prefer a smaller investment in education or health.

A graphic representation of the problem is shown in Figure 3.

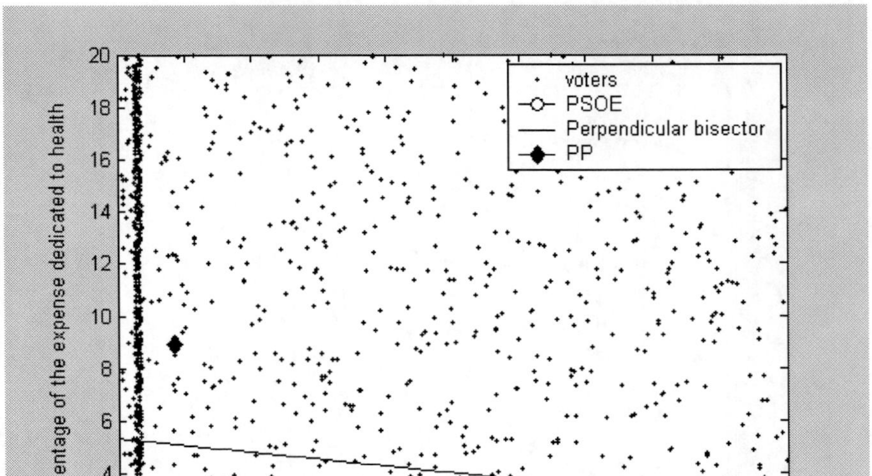

Figure 3. Graphic representation of the problem.

- For the study of the variation of the results in the capture of voters, flexibility in the politics of investment of one of the parties (in the example of Figure 4, of 0.8% for the party p) will be allowed.
-

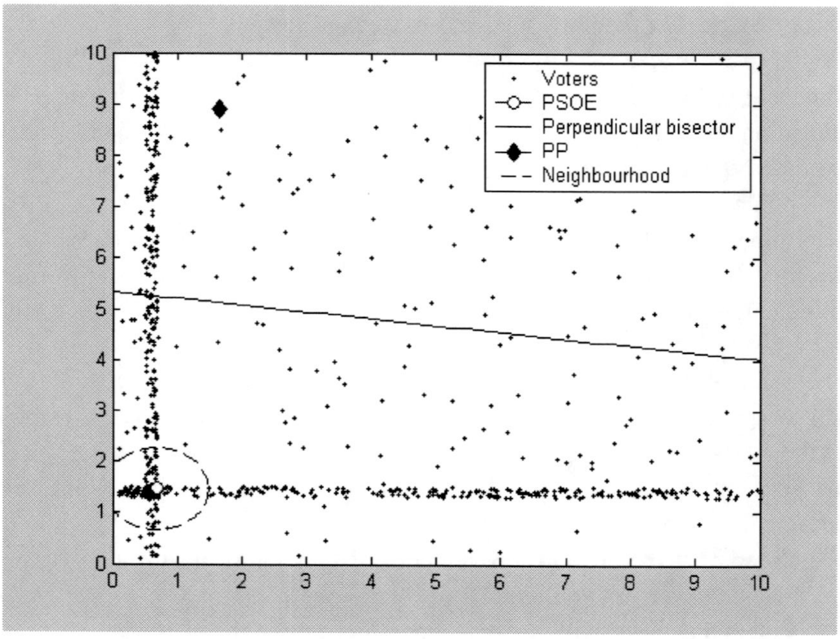

Figure 4. Flexibility in expense policy for the first party

The execution of the algorithm finds voters that even with this flexibility would not be captured by the party and voters that would be (Figure 5).

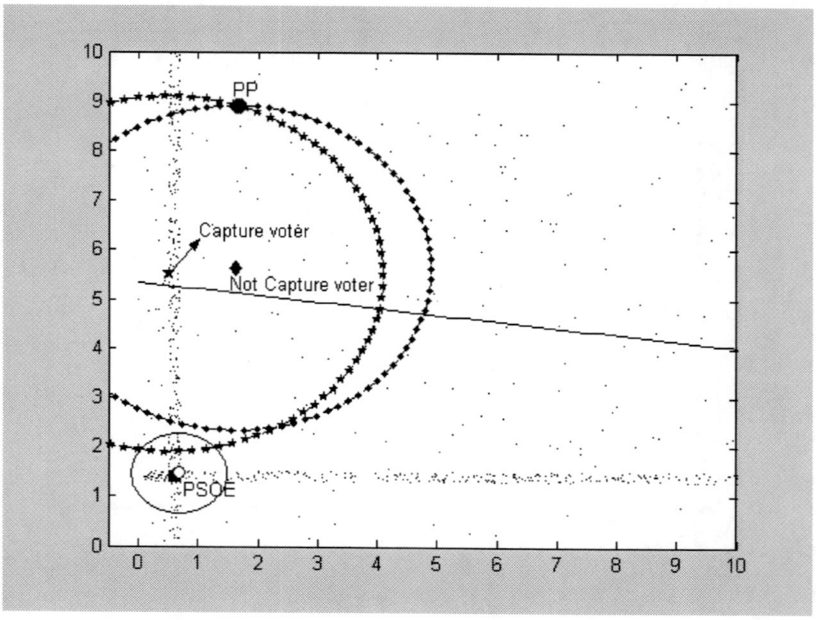

Figures 5. Study of the captures of voters with political flexibility.

The results obtained for this example are:

In the generated situation, the vote intention reflected for these two parameters would give the victory to the PSOE that would get 1277 voters against the 999 of the PP.

Study 1: Flexibility for the first party: winner party

- Political flexibility from 0.8% to the PSOE is allowed. This makes it increase its gain up to 1312 voters. This optimal situation is possible if it locates in the circumference arc defined by the points (x_1,x_2)=(1.32,1.73) and (x_1,x_2)=(1.37,1.59). That supposes an expense in education, x_1, between 1.32% and 1.37% and in health, x_2, between 1.59% and 1.73%, with $(x_1-0.6)^2+(x_2-1.4)^2=(0.8)^2$.

Study 2: Flexibility for the second party: loser party

- Political flexibility from 0.6% to the PP is allowed. This makes it increase its gain up to 1078 voters (it continues losing).
- Flexibility of 0.8%. Its gain increases up to 1138 voters. A technical tie takes place. This optimal situation is possible if it locates in the circumference arc defined by the points (x_1,x_2)=(2.25,8.44) and (x_1,x_2)=(2.28,8.48). That supposes an expense in education, x_1, between 2.25% and 2.28% and in health, x_2, between 8.44% and 8.48%, with $(x_1-1.66)^2+(x_2-8.91)^2=(0.8)^2$.
- Flexibility of 0.9%. Its gain increases up to 1166 voters. It becomes winner party. This optimal situation is possible if it locates in the circumference arc delimited by the points (x_1,x_2)=(2.312,8.350) and (x_1,x_2)=(2.316,8.355), or in the circumference arc defined by the points (x_1,x_2)=(2.332,8.377) and (x_1,x_2)=(2.339,8.386), with $(x_1-1.66)^2+(x_2-8.91)^2=(0.9)^2$.

2.4. Conclusions

An efficient algorithm has been implemented created from geometric techniques for the localization of the optimal position for a player in a competition game between two players in which one of them is allowed to move inside some margins in order to increase its gain. This algorithm has been programmed in C language and it has been executed on a problem of political competition in the Spanish environment. This situation has been simulated partly from real data collected from an opinion survey carried out by the CIS. The results obtained after the study can be summarized in the following ones:

- There are very few quantitative data about political opinions of the citizens.
- The political parties are reluctant to commit themselves in questions relative to valuation, not existing quantitative commitments in their electoral programs, for example in the line of public expense and investments.
- To keep in mind these elements and to spread them for a greater citizens' economic and political culture can be decisive for the vote decisions and for the electoral results.

- The previous knowledge on the part of the political parties of the opinions of the citizens in decisive topics for them that can be measured in a quantitatively way, could help the parties to choose an optimal strategy that assures them the bigger caption of voters. This idea is shown in the example developed inside the budgetary policy of Spain in the last years:

If we suppose the expense in education and health high-priority for the citizens (something that is supported partially in the answers of the CIS nº 2616 survey), then the variation of the investment of the losing party of the last elections (PP) of 0.9% could be decisive in the results of it.

We have not tried to do an overall analysis of items that may influence in the citizen's vote, due to the extent of them. We propose to restrict the study to those important elements for the voters in a specific period according to the results that the surveys yield. Anyway, we could enlarge the study including more than two relevant topics. The generalization of the algorithm would be based on the intersection of higher dimension neighbourhoods.

This example can be developed in numerous competition studies within the economic framework and it allows finding, in an efficient way, optimal positions for the search for the victory.

3. Geometrical Study of Equilibrium Positions in the Game with Restrictions

We propose the Nash equilibrium study in a variation of the political competition game with neighbourhood restrictions, as proposed by the authors in section 2. The variation of this section is focused on the consideration of neighbourhoods for both parties who try to win the maximum number of the voters. Thus, the movements of the opposing parties, p and q, are restricted by two separate circular neighbourhoods, $B = B(X_1, r)$, $B' = B(X_2, r')$, which guarantees that both parties cannot adopt the same politics, thus avoiding undesired equilibrium positions of the form (t, t), and also ensuring that the parties do not deviate too much from their central ideological positions X_1, X_2, resulting in a more "partisan" outlining of the game than the Downs game considered in Roemer (2001), or in other models like those presented, for example, in Lillo et al. (2007), where the parties are only interested in obtaining office.

The Nash equilibrium has been studied as a general model of competition. It was first stated by John Forbes Nash in his dissertation, *Non-Cooperative Games* (Nash, 1951), as a way to obtain an optimum strategy for games with two or more players. Plott (1997), Kramer (1973), McKelvey (1976) and others have demonstrated that pure-strategy Nash equilibrium generally do not exist when the competition takes place in a space of more than one dimension.

Various approaches to search for a resolution to this situation have been reported in the literature, among them mixed-strategy equilibria, uncovered sets, probabilistic voting, and valence criteria. See for example Laver and Shepsle, 1996; McKelvey, 1976; McKelvey,

1986; Enelow and Hinich, 1982; Londregan and Roemer, 1993; Ansolabehere and Snyder, 2000; Banks and Sundaram, 1993; Hinich and Mueger, 1995.

This section presents a discrete model of competition and it develops geometric techniques to search for equilibrium positions.

A simplification of the model has already been dealt with in previous works (Abellanas et al. 2006), and is now adapted to better reflect the reality of many countries. Thus, both competing parties that will represent the majority parties of a country are allowed to alter political positions on two items especially relevant for the citizens, in order to obtain more followers.This variability will be shown in a certain neighbourhood represented by a circle centered on the initial politics of each party, and in the radius, the flexibility each party can allow for itself in the position it offers to the voters.

Subsection 3.1 introduces the problem and develops the mathematical model, with emphasis on its geometric analysis. In subsection 3.2, necessary and sufficient conditions are presented for the equilibrium position to exist. Besides it is proved the non-unicity of the equilibrium positions when there are.

3.1. The Model

This subsection generalizes the study of Nash equilibrium in a political-competition game that can be interpreted as a discrete version of the Downs game (Downs, 1957; Roemer, 2001). This game is defined on a model like the one introduced in section 2 but with some variations.

Recall that in the model, the players are two parties denoted by p and q. The localizations they adopt in the policy plane are denoted by t_1, t_2, determined by the policies they offer. All the political positions appearing in the voter population are represented by a finite set of types H=$\{p_1, ..., p_n\} \subset R^2$.

Voter preferences over the issue space are Euclidean, so the payoff functions in the presented game are given by:

$$\Pi^1(t_1,t_2) = \text{number of points } p_i \text{ such that } d(p_i,t_1) \le d(p_i,t_2)$$
$$\Pi^2(t_1,t_2) = \text{number of points } p_i \text{ such that } d(p_i,t_1) > d(p_i,t_2) = n - \Pi^1(t_1,t_2)$$
$$\text{if } t_1 \neq t_2$$

where $d(t, p_i)$ is the Euclidean distance between policy t and position p_i.

In the case where t_1=t_2, the previous equation becomes: $\Pi^1(t_1,t_2) = \Pi^2(t_1,t_2) = \frac{n}{2}$, and each party takes half of the voters.

Both parties, p and q, will be allowed to vary their position within an neighbourhood, that is: the movements of the opposing parties with initial positions X_1, X_2 are restricted by two separate circular neighbourhoods $B = B(X_1, r)$, $B' = B(X_2, r')$. One way of guaranteeing this restriction is by assigning null gains to each of them if they are outside of their neighbourhoods, that is, define $\Pi^1(t_1,t_2) = 0$ if $t_1 \notin B$, $\Pi^2(t_1,t_2) = 0$ if $t_2 \notin B'$, with the gains previously defined if $t_1 \in B$, $t_2 \in B'$.

With this model, in section 2, the optimum positions (areas of maximum gain) for one party within its neighbourhood are determined for each position of the other party (see Figures 1 and 2).

The following section establishes the necessary or sufficient conditions for equilibrium in the proposed game.

3.2. Equilibrium with Restrictions

We start by looking at a proposition that is practically the translation of the definition of the Nash equilibrium for the gains of the proposed game.

3.2.1. Existence Conditions

Proposition 1: the equilibrium positions in the proposed game will be the positions (t_1, t_2) such that t_1 is within the area of B of maximum point gain (Abellanas et al., 2006) when q is at t_2, t_2 is within the area of B' of maximum point gain when p is on t_1.

Proof:

When q is located at t_2, which is within the area of maximum vote gain when p is at t_1, we have that if q moves to t, then $\Pi^2(t_1,t) \leq \Pi^2(t_1,t_2)$, and the same will happen with p #

Now we provide a sufficient condition so that the positions (t_1, t_2) defined in the previous proposition exist:

Proposition 2: Let p_1, ..., p_n be the positions in the plane of the n voters. We call $p^1_{i_1}, \ldots, p^1_{i_k}$ the points of the previous set, such that $d\left(p^1_{i_j}, B\right) \leq d\left(p^1_{i_j}, B'\right)$, $p^2_{i_{k+1}}, \ldots, p^2_{i_n}$ are the points which make $d\left(p^2_{i_j}, B\right) > d\left(p^2_{i_j}, B'\right)$ true, and we call $B^1_j = C\left(p^1_{i_j}, d\left(p^1_{i_j}, B'\right)\right) \cap B$, $B^2_j = C\left(p^2_{i_j}, d\left(p^2_{i_j}, B\right)\right) \cap B'$, where $C(x, R)$ is the circle with center x and radius R.

Then, if $\bigcap_{j=1}^{k} B^1_j \neq \phi$ and $\bigcap_{j=k+1}^{n} B^2_j \neq \phi$, any position (t_1, t_2) with $t_1 \in \bigcap_{j=1}^{k} B^1_j$, $t_2 \in \bigcap_{j=k+1}^{n} B^2_j$ is in equilibrium.

Proof:

In these positions, p beats the points $p^1_{i_1}$, $j = 1, \ldots, k$, wherever q is within its neighbourhood, and q beats the points $p^1_{i_1}$, $j = k+1, \ldots, n$, wherever p is. Then, if p moves, it cannot gain more than the k voters it has when q is in position t_2, and we have the same situation for q; then, the position (t_1, t_2) is in equilibrium. #

Remarks:

1. The previous result guarantees for the established cases the existence of infinite equilibrium positions. Said result contrasts with the unique equilibrium positions,

when they exist, in the usual outlining of the games without restrictions like Downs game.

2. The given condition is not necessary for the existence of equilibrium, that is, there may be equilibrium positions without this condition. The possible cases are shown in subsection 3.2.2
3. To verify the condition established in the proposition 2, we make use of the geometric construction known as arrangement of circles (de Berg et al., 1997). This arrangement can be made through a randomized incremental algorithm with an expected running time $O(n^2)$. (Edelsbrunner et al., 1992; Shadir and Agarwal, 1995). When the circles that are part of the arrangement have a non-empty intersection, this condition will occur. When not, we obtain information about the area of maximum securing of voters for each party, wherever the other may be, by attaching a label to each cell of the arrangement stating the number of circles containing it. This can be done through a standard sweeping algorithm (Bentley and Ottmann, 1979), with a running time $O(n^2 logn)$.
4. As in the case of equidistance between the positions of both parties, the voter is assigned to the first one (see section 2), and to guarantee that the position is in equilibrium, t_2 should be inside $\bigcap_{j=k+1}^{n} B_j^2$

The following result presents a necessary condition in terms of the gains so a certain position may be in equilibrium. Since the gains are considered supplementary, only the gain condition for the first party is considered.

Proposition 3: if (t_1, t_2) is an equilibrium position, then $l \leq \Pi^1(t_1, t_2) \leq n - l'$, where l is the maximum intersection in B of $C(p_i, d(p_i, B'))$, and l' is the maximum intersection in B' of $C(p_i, d(p_i, B))$, $i = 1, \ldots, n$.

Proof:

Let's suppose that there is an equilibrium position (t_1, t_2) with $\Pi^1(t_1, t_2) < l$, then by situating p at a position t in the maximum intersection of the sets B_j^1, it ensures a gain $\Pi^1(t, t_2) \geq l$. This contradicts the fact that (t_1, t_2) is an equilibrium position.

If $\Pi^1(t_1, t_2) > n - l'$ then $\Pi^2(t_1, t_2) < l'$ (the gains are supplementary); therefore, q can ensure a gain of l' in a position t' in the maximum intersection of the sets B_j^2, thus improving its gain, which contradicts the fact that (t_1, t_2) is in equilibrium. #

Remark: In the case where sufficient conditions established in proposition 2 are fulfilled, we can verify that $l + l' = n$, for which in the equilibrium positions (t_1, t_2) we have that $\Pi^1(t_1, t_2) = l$.

Now we see a sufficient condition of non equilibrium, that is, another necessary equilibrium condition.

Proposition 4: If for all $t \in B$ there exist $\frac{n}{2}+1$ points between $p_1,\dots,p_n$, $p_{j_1},\dots,p_{j_{\frac{n}{2}+1}}$, such that $\bigcap_{i=1}^{\frac{n}{2}+1} C(p_{j_i}, d(p_{j_i},t)) \cap B' \neq \phi$, and we have the same pattern for all $t' \in B'$, then there are no equilibrium positions in the proposed game.

Proof:

If there were an equilibrium position (t,t'), then $\prod^1(t,t') \leq \frac{n}{2}$ or $\prod^2(t,t') \leq \frac{n}{2}$, as the gains are supplementary. If, for example, $\prod^2(t,t') \leq \frac{n}{2}$, then for the position t of the first party in B, there are $p_{j_1},\dots,p_{j_{\frac{n}{2}+1}}$ such that $\bigcap_{i=1}^{\frac{n}{2}+1} C(p_{j_i}, d(p_{j_i},t)) \cap B' \neq \phi$, then situating the second party in that intersection there is a gain of at least $\frac{n}{2}+1$ points ($p_{j_1},\dots,p_{j_{\frac{n}{2}+1}}$), and therefore gain increases, which is a contradiction since the position was in equilibrium. #

This condition is not very practical since we have to see if it occurs for all the points in the neighbourhoods B and B', and, therefore, we have to check it for an infinite number of points. Now we see a more effective variation of this condition since it provides regions of B (and of B') that cannot be part of an equilibrium position, which limits the areas in which to search for equilibrium positions.

Proposition 5: If $t \in B$, and if there are $n-l+1$ points of the set of voters $p_{j_1},\dots,p_{j_{n-l+1}}$, such that $\bigcap_{i=1}^{n-l+1} C(p_{j_i}, d(p_{j_i},t)) \cap B' \neq \phi$, then no point $t_1 \in \bigcap_{i=1}^{n-l+1} \mathring{C}(p_{j_i}, d(p_{j_i},t))^c \cap B$ can be in an equilibrium position (t_1,t_2) ($\mathring{C}$ is the interior of C, and A^c the supplementary set of A; l follows the definition of the proposition 3).

Proof:

If there is an equilibrium position (t_1,t_2) with $t_1 \in \bigcap_{i=1}^{n-l+1} \mathring{C}(p_{j_i}, d(p_{j_i},t))^c \cap B$, then we have $\prod^2(t_1,t_2) \leq n-l$ as a consequence of proposition 3.

But as t_1 is in $\bigcap_{i=1}^{n-l+1} \mathring{C}(p_{j_i}, d(p_{j_i},t))^c$, it occurs that $d(p_{j_i},t_1) \geq d(p_{j_i},t)$, therefore $C(p_{j_i}, d(p_{j_i},t_1)) \supset C(p_{j_i}, d(p_{j_i},t))$, so $\bigcap_{i=1}^{n-l+1} C(p_{j_i}, d(p_{j_i},t_1)) \cap B' \supset \bigcap_{i=1}^{n-l+1} C(p_{j_i}, d(p_{j_i},t)) \cap B' \neq \phi$, and then situating the second party on $\bigcap_{i=1}^{n-l+1} C(p_{j_i}, d(p_{j_i},t_1)) \cap B'$ it guarantees at least $n-l+1$ voters, and it improves its gain which is a contradiction since (t_1,t_2) was in equilibrium. #

To finish this subsection, we see the non-unicity of equilibrium positions when it exists.

Proposition 6: If there is an equilibrium position (t_1, t_2), then it is not unique.

Proof:

Supposing that p_{j_1}, ..., p_{j_k} are the voters of t_1, and $p_{j_{k+1}}$, ..., p_{j_n} are the voters of t_2, then, since t_2 is inside $C(p_{j_i}, d(p_{j_i}, t_1))$ for $i = k+1, ..., n$, we can look for a position t' for the second party in B' near enough t_2 for t' to be inside $C(p_{j_i}, d(p_{j_i}, t_1))$ for $i = k+1, ..., n$ and for t_1 to be inside $C(p_{j_i}, d(p_{j_i}, t'))$ for $i = 1, ..., k$, being this the area in B of maximum gaining when the second party is in t'.

It is then fulfilled that (t_1, t') is also an equilibrium position by proposition 1. #

3.2.2. Examples

The presented model has the advantage that, as we saw in remark 1 of proposition 2, there may be positions of the voters where there are infinite situations of equilibrium for the parties. This enhances the possibilities of variation for the political positions of the rivals. Nevertheless, we will see in this section that there are also examples of a lack of equilibrium positions or unique equilibrium for one of the parties, a common characteristic in political competition games without restrictions.

Example 1:

Examining the situation with the following initial positions of the political parties X_1, X_2 and the two voters p_1, p_2: $X_1 = (0,0)$, $X_2 = (4,0)$, $p_1 = (0,3)$, $p_2 = (0,-3)$, $r = 1$, $r' = \frac{29}{10}$, it occurs that $d(p_1, B) = d(p_2, B) = 2 < d(p_1, B') = d(p_2, B') = \frac{21}{10}$, and that $B_1^1 \cap B_2^1 = \phi$, (see Figure 6), then the condition in proposition 2 is not fulfilled.

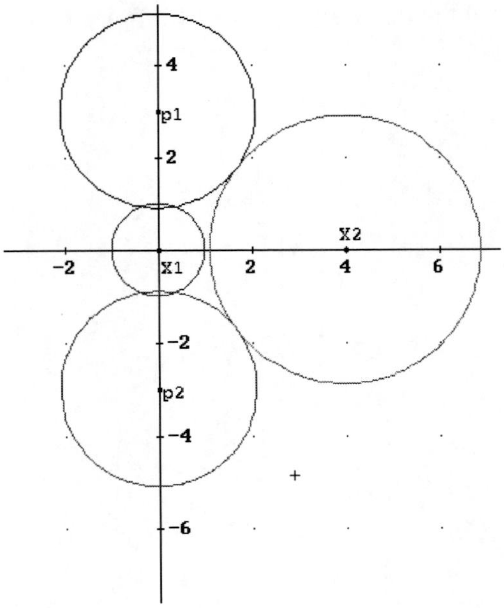

Figure 6. Example of a situation of lack of equilibrium.

We can also see that whenever $C(p_1, R)$ intersects with $C(p_2, S)$ for certain $R \geq d(p_1, B')$, $S \geq d(p_2, B')$ (so they can intersect with B'), there is some point of the intersection in B.

Then, for any position t_2 of the second party on B', it occurs that $C(p_1, d(p_1, t_2)) \cap C(p_2, d(p_2, t_2))$ has some point of B, since $R = d(p_1, t_2) \geq d(p_1, B')$, $S = d(p_2, t_2) \geq d(p_2, B')$, and $C(p_1, d(p_1, t_2)) \cap C(p_2, d(p_2, t_2)) \neq \phi$ (t_2 is in the intersection). Then, locating the second party in some of these points of B, the two points of the set will be achieved in that intersection, and then any position (t_1, t_2) with $\prod^1(t_1, t_2) < 2$ is not in equilibrium. But if $\prod^1(t_1, t_2) = 2$ the position (t_1, t_2) is not in equilibrium either, since then $\prod^2(t_1, t_2) = 0$, and since $C(p_1, d(p_1, t_1)) \cap B' \neq \phi$ or $C(p_2, d(p_2, t_1)) \cap B' \neq \phi$, then the second party being in $C(p_1, d(p_1, t_1)) \cap B'$ or in $C(p_2, d(p_2, t_1)) \cap B'$, whichever is not empty, would win either p_1 or p_2 and it would improve its gain.

Therefore, there is no equilibrium in this example.

Example 2:

Let's see an example that does not fulfil the condition established in proposition 2, and nevertheless there are (infinite) equilibrium positions, demonstrating that the condition is not necessary:

If $X_1 = (0, 0)$, $X_2 = (3, 0)$, $p_1 = (3, 9)$, $p_2 = (3, -9)$, $r = 1$, $r' = \frac{1}{2}$, it occurs that:

$d(p_1, B) = d(p_2, B) = 3\sqrt{10} - 1 < d(p_1, B') = d(p_2, B') = \frac{17}{2}$, and that $B_1^1 \cap B_2^1 = \phi$,

(see Figure 7), then the condition of proposition 2 does not occur.

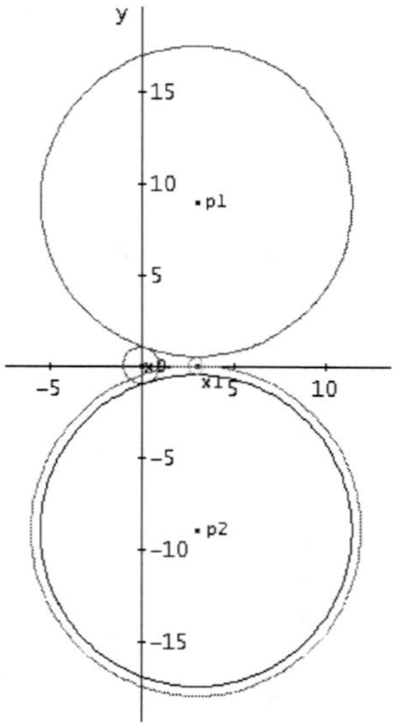

Figure 7. Example of infinite equilibrium positions not complying with a sufficient condition.

If the second party is on X_2, and the first one in, for example, B_1^1 (see Figure 8), then the first party certainly has p_1, and it cannot have both by moving since $C(p_1,9) \cap C(p_2,9) = \{X_2\}$, and $X_2 \notin B$, then in these positions each party has a voter (p_2 is the pay off for the second party, with the first one being on B_1^1: see Figure 7), and they cannot improve their situation by moving, then they are equilibrium positions.

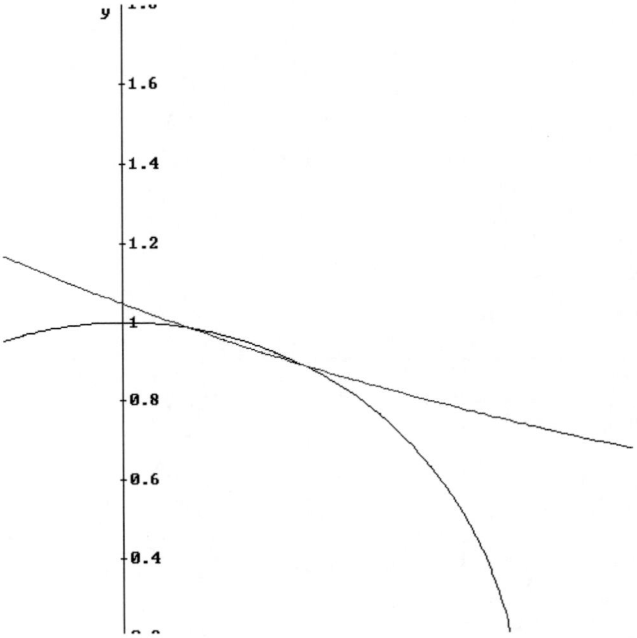

Figure 8. Detail of the area B_1^1 where the first party should be situated.

Remark: In general, for any $n \geq 2$ there are examples of positions of n points where there is no equilibrium, or there is equilibrium without the sufficient condition occurring: it is sufficient to take $n-2$ points whose maximum distance to B is less than the minimum to B', such that $B_i^1 = B$ for all those points, and the other two points in the positions of the two previous examples (with the flexibility neighbourhoods for the two parties given in these examples).

Example 3:

Let's see an example where there is a unique equilibrium position for the first party, and where there is no equilibrium position in which both parties are at the boundary of their neighbourhoods. This contrasts with the result of Abellanas et al. (2006), which shows that for a fixed position of one of the parties there is always a position for the other that maximizes its gain at the boundary limit of its neighbourhood:

If we take $X_1 = (0,0)$, $X_2 = (\sqrt{7},0)$, $p_1 = (0,3)$, $p_2 = (0,-3)$, $r = r' = 1$, it occurs that $d(p_1, B) = d(p_2, B) = 2 < d(p_1, B') = d(p_2, B') = 3$, and $B_1^1 \cap B_2^1 = \{X_1\} \neq \phi$ (see Figure 9). In this example, the only equilibrium positions are (X_1, t') with $t' \in B'$, since they satisfy the condition in proposition 2, and occurring that in any position (t_1, t_2) with $t_1 \neq X_1$ and with $\prod^1(t_1, t_2) = 2$, it will occur that, for example, $d(p_1, t_1) > 3$, so after locating the second party in $B(p_1, d(p_1, t_1)) \cap B'$, it will win p_1 and it will improve its gain; therefore, (t_1, t_2) is not in equilibrium.

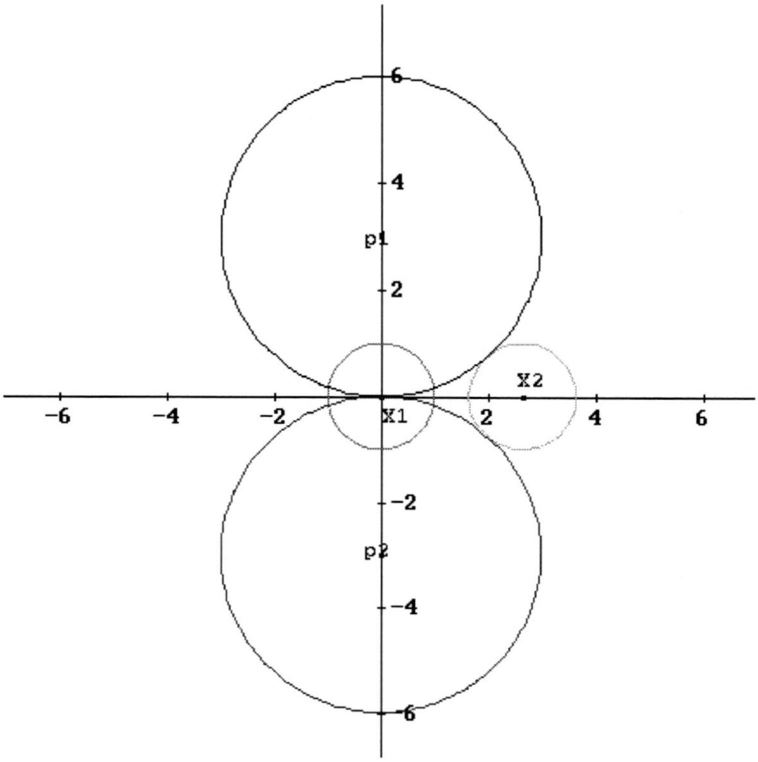

Figure 9. Example of unique position (not in the boundary of the neighbourhood) for one of the two parties.

3.3. Conclusions

There are a number of localization studies in the framework of the public economy that study various problems using discrete or continuous approaches (Eiselt et al., 1993; Ghosh and Harche, 1993; Hakimi, 1986; Hakimi, 1990; Mehrez and Stulman, 1982; Church, 1984). In the equilibrium analysis of most competitive multidimensional games, it is found that such positions do not exist except for singular cases, so there exist no positions for the players that guarantee that the competitor cannot increase his gain by moving. A discrete two-dimensional political competition model has been proposed and addressed with geometric strategies that find the equilibrium positions, if they exist. The most relevant news in the presented model is that it reflects the political reality of many countries, permitting the positions of the parties, in certain topics of interest, to vary to a certain extent in order to obtain a larger number of followers. This situation is common in politics in which the positions of the parties are not inflexible and they adapt to the preferences of the voters. They look for equilibrium positions within these flexible neighbourhoods. As established in subsection 3.2, there are cases in which the existence of equilibrium positions are not only unique but are also regions in the plane; that is, there are infinite positions, a condition that provides more possibilities to the game, and it contrasts with the results of unique positions in equilibrium where both parties should adopt the same position.

The study of the existence and locations of the positions of equilibrium has been developed here with a wider scope by applying techniques from computational geometry because of the discrete nature of the game as presented.

APPENDIX A: DEVELOPMENT OF THE ALGORITHM

Let C be the boundary circumference of the circle centred in the point p where p can move. We will suppose that the point q is outside of C. Let p' and p'' be the tangency points between C and the tangent lines to C traced from q (p' the right tangency point and p'' the left one seen from q).

We will suppose that points v_i are exterior to C. For each point v_i let us consider x'_i and x''_i as the intersection points between C and the circumference centred in v_i that goes through q. x'_i is the right intersection point and x''_i is the left one seen from v_i (Figure 2).

Procedure:

- Step 1: Find p' and p'' and consider a counter c' with initial value c' = 0.
- Step 2: Let L be an empty list and let m be a counter with initial value m =0. For each point v_i find the intersection between C and the circumference centred in v_i that goes through q.
    - 2.1. If there is no intersection because the circumference C is contained in the circle centred in v_i that goes through q, increase m by one unit.
    - 2.2. If there is no intersection because the corresponding circles are disjoint, preserve the value of m.
    - 2.3. If the intersection has two points outside of the visible part of C from q, then increase m by one unit.
    - 2.4. Otherwise:
        - 2.4.1. If both points belong to the visible part of C from q, then include them in the list L.
        - 2.4.2. If x'_i belongs to the visible part of C from q and x''_i doesn't, then include x'_i in L.
        - 2.4.3. If x''_i belongs to the visible part of C from q and x'_i doesn't, then include x''_i in L and increase c' by one unit.
- Step 3. Arrange the points of L according to the angle with respect to p (clockwise).
- Step 4. Let c =c'+m and x = p'. Trace the list L doing the following for each element:
    - 4.1. If it is a x'_i element, let c=c+1, and if c>m then let m=c and x= x'_i .
    - 4.2. If it is a x''_i element, let c=c-1.

Remark: If x'_i and x''_i coincide because the corresponding circumferences are tangent, we consider x'_i previous to x''_i in the list L.

Once we end the execution of the algorithm, the counter m indicates the maximum number of points v_i that point p can obtain closer to it than to q, if we locate at a point of the arc of C whose initial extreme is the point stored in the variable x and the last extreme is the following to that point in the list L.

The worst-case time complexity of the algorithm is $O(nlogn)$. That is because we can execute step 1 in constant time. Step 2 requires a constant number of operations for each point v_i; therefore, it is executed in linear time. Step 3 requires $O(nlogn)$ operations in the worst case because it is considered an arrangement of a list of n elements, and step 4 requires a quantity of operations that is proportional to the number of elements of the list L that is linear in the worst case.

REFERENCES

Abellanas, M., Lillo, I., López, M., Rodrigo, J. (2006) Electoral strategies in a dynamical democratic system: geometric models. *European Journal of Operational Research 175*, 870–878.

Ahn, H., Cheng, S., Cheong, O., Golin, M. and Oostrum, R. (2001). *Competitive Facility Location along a Highway.* 7th Annual International Computing and Combinatory Conference, vol. 2108 of *LNCS*.

Ansolabehere, S., Snyder, J. (2000) Valence politics and equilibrium in spatial election models. *Public Choice 103*, 327–336.

Aurenhammer, R., Klein, R. Voronoi diagrams. (2000) In: Sack, J.-R. and Urrutia, J. (eds.), *Handbook of Computational Geometry*. Elsevier, Amsterdam.

Banks, J., Sundaram, R.K. (1993) Moral hazard and adverse selection in a model of repeated elections. In: Barnett, W.A., Hinich, M.J., and Schoefield, N.J. (eds.), *Political Economy: Institutions, Competition and Representation*. Cambridge: Cambridge University Press.

Bentley J.L., Ottmann T.A. (1979) Algorithms for reporting and counting geometric intersections. *IEEE Trans Comput. C-28*, 643-647.

de Berg, M., van Kreveld, M., Overmars, M., Schwarzkopf, O. (1997) *Computational Geometry, Algorithms and Applications*. Springer. New York.

CIS: Sociological Investigations Center of Spain: Opinion surveys. Study nº 2615. Available online: http://www.cis.es.

Consolidated General Government Budget (1997–2006). Available online: *http://www.igae.meh.es/Internet/Cln_Principal/ClnPresupuesto/PresupuestosGeneralesEst/PresupuestosGeneralesEstado/.*

Church, R. (1984) The planar maximal covering location problem. *Journal of Regional Science 24*(2), 185–201.

Downs, A. (1957) *An Economic Theory of Democracy*. Harper and Row.

Edelsbrunner H., Guibas L., Pach J., Pollack R., Seidel R., Sharir M. (1992) Arrangements of curves in the plane-topology, combinatorics, and algorithms. *Theoretical Computer Science 92*, 319-336.

Eisets, H.A., Laporte, G., Thisse, J.H. (1993) Competitive location models: a framework and bibliography. *Transportation Science 27*, 44–54.

Enelow, J., Hinich, M. (1982) Nonspatial candidate characteristics and electoral competition. *Journal of Politics 44*, 115–130.

Ghosh, A., Harche, F. (1993) Location-allocation models in the private sector: progress, problems and prospects. *Location Science 1*, 81–106.

Hakimi, S.L. (1986) P-median theorems for competitive location. *Annals of Operations Research 5*, 79–88.

Hakimi, S.L. (1990) Locations with spatial interaction: competitive locations and games. In: Mirchandani PB, Francis RL, editors. *Discrete Location Theory*. New York: Wiley.

Hinich, M.H., Munger, M.G. (1995) *Ideology and the Theory of Political Choice*. Ann Arbor: University of Michigan Press.

Kramer, G.H. (1973) On a class of equilibrium conditions for majority rule. *Econometrica 42*, 285–297.

Laver, M., Shepsle, K.A. (1996). *Making and Breaking Governments*. Cambridge: Cambridge University Press.

Lillo, I. López, M. Rodrigo, J. (2007) A Geometric study of the Nash equilibrium in a weighted case. *Applied Mathematical Sciences 55*, Vol 1 2715-2725. Bulgaria.

Lodregan, J., Romer, T. *Polarization, incumbency, and the personal vote.* In Barnett, W.A., Hinich, M.J., and Schoefield, N.J. (1993) *Political Economy: Institutions, Competition and Representation.* Cambridge: Cambridge University Press.

McKelvin, R.D. (1976) Intransitivities in multidimensional voting models and implications for agenda control. *Journal of Economic Theory 12*, 472–482.

McKelvey, R.D. (1986) Covering, dominance, and the institution-free properties of social choice. *American Journal of Political Science 30*, 283–314.

Mehrez, A., Stulman, A. (1982) The maximal covering location problem with facility placement on the entire plane. *Journal of Regional Science*, 22(3), 361–365.

Nash, J. (1951) *Non-cooperative games*. *Annals of Mathematics 54*, 286–295.

Okabe, A., Boots, B., Sugihara, K., Chiu, S. (2000) *Spatial Tessellations Concepts and Applications of Voronoi Diagrams.* John Wiley & Sons. Chichester.

Plott, C.R. (1967) A notion of equilibrium and its possibility under majority rule. *American Economic Review 57*, 787–806.

Preparata F. & Salmos M. (1985). *Computational Geometry. An introduction*. Springer-Verlag. New York.

Roemer, J. (2001) *Political Competition.* Harvard University Press.

Serra, D., Revelle, C. (1994) Market capture by two competitors: the preemptive location problem. *Journal of Regional Science 34*(4), 549–561.

Sharir M., Agarwal P. (1995) *Davenport-Schinzel Sequences and Their Geometric Applications*. Cambridge University Press.

Smid, M. (1997) Closest-point problems in computational geometry, In Sack, J.-R., and Urrutia, J. (eds.) *Handbook of Computational Geometry.* Elsevier, Amsterdam.

In: Computational Mathematics
Editor: Peter G. Chareton, pp.85-115

ISBN:978-1-60876-271-2

Chapter 4

COHERENCE – HOMOTOPIES OF HIGHER ORDER

Nikita Shekutkovski*
Institute of Mathematics, Faculty of Mathematics & Natural Sciences, Sts. Cyril and Methodius University, Skopje, Republic of Macedonia

INTRODUCTION

The notion of an inverse system is well known and widely used in mathematics.The category of inverse systems pro-HTop is the fundamental tool in establishing of proper homotopy and shape theory. Using homotopies of higher orders, we will present notions of coherent map, coherent homotopy and coherent inverse system, corresponding categories and their relations.

This manuscript is organized as follows

- Introduction
- Coherent systems and coherent maps
- Level coherent category
- Coherent shift and coherent category
- Relations of coherent categories
- Appendix: Strict ordering vs ordering for directed sets

In Section 1. is presented the construction of the category pro-HTop with objects arbitrary inverse systems (usually, by HTop is denoted the category of topological spaces and homotopy classes), and of the coherent category CPHTop with objects strictly commutative inverse systems and coherent maps of order ∞.

The inverse systems could be strictly commutative or commutative up to homotopy. We introduce the notion of coherent inverse system, the system which is commutative up to homotopy, and homotopies are connected with homotopies of higher and higher order (coherence of order ∞).

* e-mail: nikita@iunona.pmf.ukim.edu.mk

The notion of a coherent map of order ∞ between such system is introduced and the notion of level homotopy of two coherent maps

This allows in Section 2. to deal with the level coherent category $\mathrm{Coh}^{\mathrm{A}}$ over a directed set A.

Explicitly is constructed a level category pro-$\mathrm{H}^2\mathrm{Top}^{\mathrm{N}}$ of arbitrary inverse sequences and strong fundamental sequences (maps between sequences of order of coherence 2)

It is shown that the subcategory pro-$\mathrm{H}^2\mathrm{Top}^{\mathrm{N}}$ having as objects coherent inverse sequences is isomorphic to $\mathrm{Coh}^{\mathrm{N}}$ – level coherent category over natural numbers.

As an application, the following theorem is proven. If in a strictly commutative inverse sequence the spaces are replaced by homotopy equivalent spaces, one obtains an inverse sequence, which is commutative only up to homotopy. The two inverse sequences are obviously isomorphic in the category pro-HTop.It is proven that two inverse sequences are isomorphic in pro-$\mathrm{H}^2\mathrm{Top}^{\mathrm{N}}$, and consequently in $\mathrm{Coh}^{\mathrm{N}}$. Moreover, they are isomorphic in the category Coh, constructed in Section 3.

In Section 3. by introducing the notion of coherent shift it is defined the notion of homotopy between two coherent maps of coherent inverse systems and is formed the category Coh of coherent inverse systems and coherent maps defined using homotopies of all orders (i.e. the order of coherence is ∞).

In Section 4, different types of coherent categories are considered: Coh - category of coherent inverse systems and coherent maps, CPHTop - category of strictly commutative inverse systems and coherent maps. It is showt that CPHTop is a full subcategory of the category Coh [13].

In the Appendix it is shown that in the construction of coherent category CPHTop of commutative inverse systems we can use the strict ordering $<$ for directed sets as well as more usual ordering $\leq$. The resulting coherent categories are isomorphic.

Keywords: Inverse sequence, Coherent maps, Level maps, Coherent category, Shape, Compact metric space, Strong fundamental sequence

AMS Subject Classification: 55P55, 54C56

1. Coherent Systems and Coherent Maps

Let $(A,<)$ be an arbitrary directed set i.e. an ordered set with the property: for any $a_0,a_1 \in A$ there exists $a_2 \in A$ such that $a_0,a_1 < a_2$. $(A,<)$ will be cofinite i.e. each $a \in A$ has only a finite number of predecessors.

An *inverse system* $\underline{X} = (X_a, p_{a_0a_1}, A)$ consists of topological spaces X_a and of bonding maps $p_{a_0a_1}: X_{a_1} \to X_{a_0}$ for any pair $a_0 < a_1$, and $p_{a_0a_1}p_{a_1a_2} \cong p_{a_0a_2}$ for $a_0 < a_1 < a_2$. If the equality $p_{a_0a_1}p_{a_1a_2} = p_{a_0a_2}$ holds we call the inverse system *commutative* .

The map of inverse systems $(f_b): \underline{X} \to \underline{Y} = (Y_b, q_{b_0b_1}, B)$ consists of C1) a strictly increasing function $f: B \to A$ and for any $b \in B$ of a map $f_b: X_{f(b)} \to Y_b$, and for any pair $b_0 < b_1$ of a map $f_{b_0 b_1}: I \times X_{f(b_1)} \to Y_{b_0}$ such that $f_{b_o b_1}(0,x) = f_{b_o} p_{f(b_0 b_1)}$, $f_{b_o b_1}(1,x) = q_{b_0 b_1} f_{b_1}(x)$.

Category of inverse system pro-HTop is used in shape theory by replacing the topological space by an inverse system: the system of its neighbourhoods, by a resolution or Cech system of the space.

The category pro-HTop is not a satisfactory tool for the construction of strong shape theory. One needs a stronger category of inverse systems. The notion of a map of systems is replaced by a notion of a coherent map i.e. not only the maps f_{b_0}, f_{b_1}, for $b_0 < b_1$ are connected by a homotopy denoted by $f_{b_0 b_1}: I \times X_{f(b_1)} \to Y_{b_0}$ (in fact, connecting $q_{b_0 b_1} f_{b_1}$ and $f_{b_o} p_{f(b_0 b_1)}$). For example for $n = 2$ and three indices $b_0 < b_1 < b_2$ the maps f_{b_0}, f_{b_1} and f_{b_2}, and the homotopies $f_{b_0 b_1}$, $f_{b_1 b_2}$ and $f_{b_0 b_2}$ are connected by a homotopy of second order denoted by $f_{b_0 b_1 b_2}: \Delta^2 \times X_{f(b_2)} \to Y_{b_0}$.

The precise definition for all natural numbers is given by use of $n-$simplex: $\Delta^1 = I$ and $\Delta^n = \{(t_1,\dots,t_n) \mid 1 \ge t_1 \ge \dots \ge t_n \ge 0\}$.

A special coherent map of commutative inverse systems $(f_{\underline{b}}): \underline{X} \to \underline{Y} = (Y_b, q_{\underline{b}}, B)$ consists: of C1) and of C2) for $n \ge 2$ and any strictly increasing sequence, $\underline{b} = (b_0, b_1, \dots, b_n)$, $b_0 < b_1 < \dots < b_n$, of a map $f_{\underline{b}}: \Delta^n \times X_{f(b_n)} \to Y_{b_0}$ such that

$$f_{\underline{b}}(t,x) = \begin{cases} q_{b_0 b_1} f_{b_1 \dots b_n}(t_2,\dots,t_n,x)), & t_1 = 1 \\ f_{b_0 \dots \hat{b}_i \dots b_n}(t_1,\dots,\hat{t}_i,\dots,t_n,x), & t_i = t_{i+1} \\ f_{b_0 \dots b_{n-1}}(t_1,\dots,t_{n-1}, p_{f(b_{n-1} b_n)}(x)), & t_n = 0 \end{cases}.$$

The special coherent map $(f_{\underline{b}}): \underline{X} \to \underline{Y}$ is defined only for commutative inverse systems. Coresponding category CPHTop is succesfully used for construction of strong shape and the theory uses only the commutative inverse systems - the system of neighbourhoods and resolutions [7]. It is shown that category CPHTop is equivalent to the category obtained by localization of pro-Top at level homotopy equivalences.

There are inverse systems that are not commutative. The main example is Cech system of a topological space. In order to use Cech sytem in strong shape theory, one needs to extend the coherent category CPHTop to include this type of systems..

The first ideas about the existence of coherent category of this type and of order infinity are presented in [2].

In order to define a coherent map (in which appear homotopies of all orders) between two such systems it is natural such systems to have also bonding maps which are connected by homotopies of all orders.

Definition: A triple $\mathcal{X} = (X_a, p_{\underline{a}}, A)$ is called *a coherent inverse system over* a directed set A if for $n \ge 2$, and $a_0 < a_1 < \dots < a_n$, $\underline{a} = (a_0, a_1, \dots, a_n)$, the map $p_{\underline{a}}: I^{n-1} \times X_{a_n} \to X_{a_0}$ satisfy the following two conditions $p_{\underline{a}}(\tau_1,\dots,\tau_{j-1},0,\tau_{j+1},\dots,\tau_{n-1},x) = p_{a_0 \dots \hat{a}_j \dots a_n}(\tau_1,\dots,\hat{\tau}_j,\dots,\tau_{n-1},x)$ $p_{\underline{a}}(\tau_1,\dots,\tau_{j-1},1,\tau_{j+1},\dots,\tau_{n-1},x) = p_{a_0 \dots a_j}(\tau_1,\dots,\tau_{j-1}, p_{a_j \dots a_n}(\tau_{j+1},\dots,\tau_{n-1},x))$.

For example in the special case $n = 2$, the map $p_{a_0 a_1 a_2}: I \times X_{a_2} \to X_{a_0}$ is just a homotopy connecting $p_{a_0 a_2}$ and $p_{a_0 a_1} p_{a_1 a_2}$.

The coherent category of the coherent inverse systems is denoted by Coh. Her we start the construction of this category. We will continue the construction in Section 3.

A coherent map $(f_{\underline{b}}^{j}) : \mathcal{X} \to \mathcal{Y} = (Y_b, q_{\underline{b}}, B)$ consists of 1) and of 2) for $n \geq 2$ and any strictly increasing sequence $\underline{b} = (b_0, b_1, \ldots, b_n)$, $b_0 < b_1 < \ldots < b_n$, there is a set $f_{\underline{b}}$ of 2^{n-1} maps $\{ f_{\underline{b}}^{j_0 j_1 \ldots j_k} \mid 0 = j_0 < j_1 < \ldots < j_k = n \} = f_{\underline{b}}$.

The map $f_{\underline{b}}^{j_0 j_1 \ldots j_k} : I^{n-k} \times \Delta^k \times X_{f(b_n)} \to Y_{b_0}$ must satisfy the following boundary conditions:

$$f_{\underline{b}}^{j_0 j_1 \ldots j_k}(\tau, t, x) = \begin{cases} q_{b_0 \ldots b_{j_1}}(\tau_{01}, f_{b_{j_1} \ldots b_n}^{j_1 - j_1 \ldots j_k - j_1}(\tau_{1k}, t_2, \ldots, t_k, x)), & t_1 = 1 \\ f_{\underline{b}}^{j_0 \ldots \hat{j}_i \ldots j_k}(\tau_{0,i}, 1, \tau_{i,k}, t_1, \ldots, \hat{t}_i, \ldots, t_k, x), & t_i = t_{i+1} \\ f_{b_0 \ldots b_{j_{k-1}}}^{j_0 \ldots j_{k-1}}(\tau_{0,k-1}, t_1, \ldots, t_{n-1}, p_{f(b_{j_{k-1}} \ldots b_n)}(\tau_{k-1,k}, x)), & t_k = 0 \end{cases} \tag{CM1}$$

and for $j_m < j < j_{m+1}$ and $\tau_{j-m} = 0$,

$$f_{\underline{b}}^{j_0 j_1 \ldots j_k}(\tau, t, x) = f_{b_0 \ldots \hat{b}_j \ldots b_n}^{j_0 \ldots j_m j_{m+1} - 1 \ldots j_k - 1}(\tau_1, \ldots, \hat{\tau}_{j-m}, \ldots, \tau_{n-k}, t, x). \tag{CM2}$$

In the above formulas it is identified: $f_{b_0}^{0} \equiv f_{b_0}$ and $f_{b_0 b_1}^{01} \equiv f_{b_0 b_1}$. Also, in the formula for $\tau \in I^{n-k}$ and $0 \leq i \leq i' \leq k$, $\tau_{i,i'}$ is a short notation for $\tau_{i,i'} = \begin{cases} (\tau_{j_i - i + 1}, \ldots, \tau_{j_{i'} - i'}), & j_i - i + 1 \leq j_{i'} - i' \\ * \; , & j_i - i + 1 > j_{i'} - i' \end{cases}$

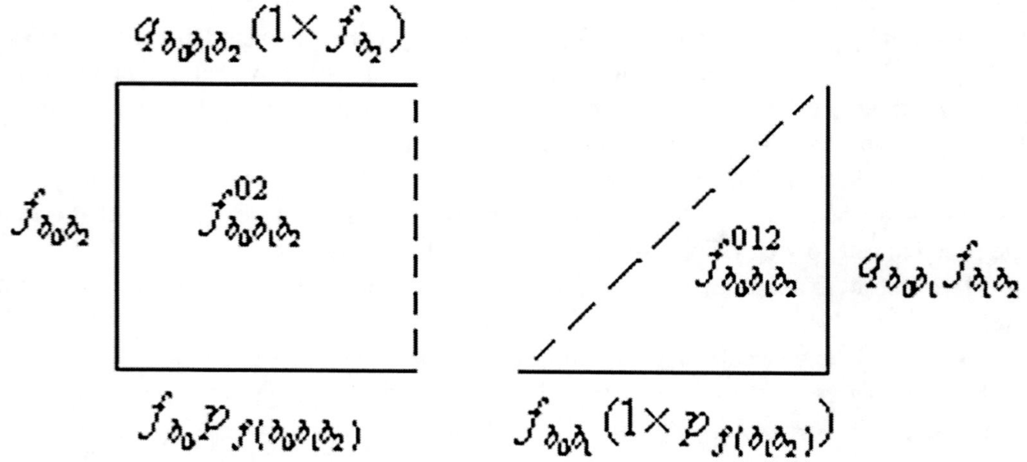

Figure 1. For $n = 2$ and three indices $b_0 < b_1 < b_2$, the maps $f_{b_0 b_1 b_2}^{02} : I \times \Delta^1 \times X_{f(b_2)} \to Y_{b_0}$ and $f_{b_0 b_1 b_2}^{012} : \Delta^2 \times X_{f(b_2)} \to Y_{b_0}$ coincide along the dashed line – i.e for $t_1 = t_2$, $f_{b_0 b_1 b_2}^{012}(t_1, t_1, x) = f_{b_0 b_1 b_2}^{02}(1, t_1, x)$

Remark:

a) Here instead of the usual ordering ≤ it is used the strict ordering <. The advantage is that there is no repeating of indices in the sequences and we can successfully avoid the degeneracy conditions for coherent maps [7]. This is important in dealing with complicated formulas which appear in the coherent theory since the degeneracy condition must always be checked in the same way as the boundary condition is checked. The two approaches are equivalent, as it is shown in Appendix.
b) We use the non-standard simplex instead of a standard one, since the composition formula looks like a multidimensional case of the well known formula for composition of paths. We can always pass from non-standard coordinates $(t_1,\ldots,t_k)$ to standard simplex coordinates $(t_0,t_1,\ldots,t_k)$ by replacing t_i by $t_i+\ldots+t_k$. This will additionally complicate the formulas which follow.

The construction of the three categories Coh,CPHTop and pro-HTop proceeds as follows:

Composition: In pro-HTop, the composition of maps of systems $(f_b):\underline{X}\to\underline{Y}$ and $(g_c):\underline{Y}\to\underline{Z}=(Z_c,r_{c_0c_1},C)$ is a map of systems $(g_cf_{g(c)}):\underline{X}\to\underline{Z}$ given by the strictly increasing function $fg:C\to A$.

To define composition of coherent maps first we define a partition of Δ^n to subpolyhedra K_m, $0\le m\le k$, $K_m=\{(t_1,\ldots,t_k,)\mid\ t_m\ge\frac{1}{2}\ge t_{m+1}\}$.

In CPHTop, the composition of special coherent maps $(f_{\underline{b}}):\underline{X}\to\underline{Y}$ and $(g_{\underline{c}}):\underline{Y}\to\underline{Z}=(Z_c,r_{c_0c_1},C)$ is given by the strictly increasing function $fg:C\to A$ and by maps $(gf)_{\underline{c}}:\Delta^n\times X_{fg(c_n)}\to Z_{c_0}$ defined by $(gf)_{\underline{c}}:(t_1,\ldots,t_n,x)=g_{c_0\ldots c_m}(2t_1-1,\ldots,2t_m-1,f_{g(c_m\ldots c_n)}(2t_{m+1},\ldots,2t_n,x)$ for $t\in K_m$.

In Coh, the composition of coherent maps $f:\mathcal{X}\to\mathcal{Y}$ and $g:\mathcal{Y}\to\mathcal{Z}=(Z_c,r_{\underline{c}},C)$ is given is given by the strictly increasing function $fg:C\to A$ and by maps $(gf)_{\underline{c}}^{\underline{j}}:I^{n-k}\times\Delta^k\times X_{fg(c_n)}\to Z_{c_0}$ defined by $(gf)_{\underline{c}}^{\underline{j}}(\tau,t,x)=g_{c_0\ldots c_{j_m}}^{j_0\ldots j_m}(\tau_{0,m},2t_1-1,\ldots,2t_m-1,f_{g(c_{j_m}\ldots c_n)}^{j_m-j_m\ldots j_k-j_m}(\tau_{m,k},2t_{m+1},\ldots,2t_k,x)$ for $t\in K_m$.

Level homotopy: Two maps of systems $(f_b),(f'_b)$ are *level homotopic* if they are defined by the same strictly increasing function $f:B\to A$ and there exists a of systems (F_b) defined by the same strictly increasing function such that $F_b:I\times X_{f(b)}\to Y_b$ is a homotopy which connects the maps f_b, f'_b i.e. holds $F_b(0,x)=f_b(x),\ F_b(1,x)=f'_b(x)$

Two special coherent maps $(f_{\underline{b}}),(f'_{\underline{b}})$ are *level homotopic* if they are defined by the same strictly increasing function $f:B\to A$ and there exists a special coherent map $(F_{\underline{b}})$ defined by the same strictly increasing function such that $F_{\underline{b}}:\Delta^n\times I\times X_{f(b_n)}\to Y_{b_0}$ is a homotopy which connects the maps $f_{\underline{b}}$, $f'_{\underline{b}}$ i.e. $F_{\underline{b}}(t,0,x)=f_{\underline{b}}(t,x),\ F_{\underline{b}}(t,1,x)=f'_{\underline{b}}(t,x)$

For definition of homotopy of coherent maps $(f_{\underline{b}}^{\underline{j}}):\mathcal{X}\to\mathcal{Y}$ it is used the coherent inverse system $I\times\mathcal{X}=(I\times X_a,1\times p_{\underline{a}},A)$.

Two coherent maps $(f_{\underline{b}}^{\underline{j}}),(f'^{\underline{j}}_{\underline{b}})$ are *level homotopic* if they are defined by the same strictly increasing function $f:B\to A$ and there exists a coherent map $(F_{\underline{b}}^{\underline{j}}):I\times\mathcal{X}\to\mathcal{Y}$ defined by the same strictly increasing function such that $F_{\underline{b}}^{\underline{j}}:I^{n-k}\times\Delta^k\times I\times X_{f(b_n)}\to Y_{b_0}$ is a homotopy which connects the maps $,f'^{\underline{j}}_{\underline{b}}$ i.e. holds $F_{\underline{b}}(\tau,t,0,x)=f_{\underline{b}}(\tau,t,x),\ F_{\underline{b}}(\tau,t,1,x)=f'_{\underline{b}}(\tau,t,x)$.

Theorem 1.1

The relation of level homotopy of coherent maps is an equivalence relation.

Proof: The reflexivity and symmetry of the relation are obvious. Let $f,f':\mathcal{X}\to\mathcal{Y}$ be level homotopic by a homotopy $F:I\times\mathcal{X}\to\mathcal{Y}$ and $f',f'':\mathcal{X}\to\mathcal{Y}$ be level homotopic by a homotopy $F':I\times\mathcal{X}\to\mathcal{Y}$. Then f and f'' are level homotopic by a homotopy $(F''^{\underline{j}}_{\underline{b}})$ defined by

$$F''^{\underline{j}}_{\underline{b}}(\tau,t,s,x)=\begin{cases}F_{\underline{b}}^{\underline{j}}(\tau,t,2s,x),0\le s\le\frac{1}{2}\\ F'^{\underline{j}}_{\underline{b}}(\tau,t,2s-1,x),\frac{1}{2}\le s\le 1\end{cases}.$$

Theorem 1.2

If $f:\mathcal{X}\to\mathcal{Y}$, $g:\mathcal{Y}\to\mathcal{Z}$, $h:\mathcal{Z}\to\mathcal{W}=(W_d,w_{\underline{d}},D)$, are coherent maps, then the coherent maps $h(gf)$ and $(hg)f$ are level homotopic.

Proof: Suppose $f=(f_{\underline{b}}^{\underline{j}})$, $g=(g_{\underline{c}}^{\underline{j}})$ and $h=(h_{\underline{d}}^{\underline{j}})$. In order to obtain an explicit formula for the coherent map $h(gf)$ we define a decomposition of Δ^k into subpolyhedra $K_{l,m}$ for any pair of integers l,m such that $0\le l\le m\le k$, $K_{l,m}=\{(t_1,\dots,t_k)\mid t_l\ge\frac{1}{2}\ge t_{l+1},\ t_m\ge\frac{1}{4}\ge t_{m+1}\}$.

By applying the composition formula twice, for $(t_1,\dots,t_k)\in K_{l,m}$ we have

$$(h(gf))_{\underline{d}}^{\underline{j}}(\tau,t,x)=h_{d_0\dots d_{j_l}}^{j_0\dots j_l}(\tau_{0,l},2t_1-1,\dots,2t_l-1,\ g_{h(d_{j_l}\dots d_{j_m})}^{j_l-j_l\dots j_m-j_l}(\tau_{l,m},4t_{l+1}-1,\dots,4t_m-1,f_{gh(d_{j_m}\dots d_n)}^{j_m-j_m\dots j_k-j_m}(\tau_{m,k},4t_{m+1},\dots,4t_k,x))).$$

In the same way, to obtain an explicit formula for the coherent map $(hg)f$ we define a decomposition of Δ^k into subpolyhedra $Q_{l,m}$ for any pair of integers l,m such that $0\le l\le m\le k$, $Q_{l,m}=\{(t_1,\dots,t_k)\mid t_l\ge\frac{3}{4}\ge t_{l+1},\ t_m\ge\frac{1}{2}\ge t_{m+1}\}$

Then, for $(t_1,\dots,t_k)\in Q_{l,m}$ we have $((hg)f)_{\underline{d}}^{\underline{j}}(\tau,t,x)=h_{d_0\dots d_{j_l}}^{j_0\dots j_l}(\tau_{0,l},4t_1-3,\dots,4t_l-3,$

$$g_{h(d_{j_l}\dots d_{j_m})}^{j_l-j_l\dots j_m-j_l}(\tau_{l,m},4t_{l+1}-2,\dots,4t_m-2,f_{gh(d_{j_m}\dots d_n)}^{j_m-j_m\dots j_k-j_m}(\tau_{m,k},2t_{m+1},\dots,2t_k,x)))$$

We define a partition of $\Delta^k \times I$ into subpolyhedra $M_{l,m}$ for any pair of integers l, m such that $0 \le l \le m \le k$, $M_{l,m} = \{(t_1,\dots,t_k,s) \mid t_l \ge \frac{2+s}{4} \ge t_{l+1},\ t_m \ge \frac{1+s}{2} \ge t_{m+1}\}$

We define a coherent homotopy $H : I \times \mathcal{X} \to \mathcal{W}$ which connects $h(gf)$ and $(hg)f$. This map will be given by the function $fgh : D \to A$ and by the maps $H_{\underline{d}}^{\underline{j}} : I^{n-k} \times \Delta^k \times I \times X_{fgh(d_n)} \to W_{d_0}$ defined for $(t,s) \in M_{l,m}$ by

$$H_{\underline{d}}^{\underline{j}}(\tau,t,s,x) = h_{d_0 \dots d_{j_l}}^{j_0 \dots j_l}(\tau_{0,l}, \frac{4t_1-2-s}{2-s},\dots,\frac{4t_l-2-s}{2-s},$$
$$g_{h(d_{j_l}\dots d_{j_m})}^{j_l-j_l \dots j_m-j_l}(\tau_{l,m}, 4t_{l+1}-1-s,\dots,4t_m-1-s, f_{gh(d_{j_m}\dots d_n)}^{j_m-j_m \dots j_k-j_m}(\tau_{m,k}, \frac{4t_{m+1}}{1+s},\dots,\frac{4t_k}{1+s}, x)))$$

We mention that $K_{l,m} \times 0 = \{(t_1,\dots,t_k,0) \mid (t_1,\dots,t_k,0) \in M_{l,m}\}$ and then, for $(t_1,\dots,t_k) \in K_{l,m}$, it is easily checked that $H_{\underline{d}}^{\underline{j}}(\tau,t,0,x) = (h(gf))_{\underline{d}}^{\underline{j}}(\tau,t,x)$.

Also, $Q_{l,m} \times 1 = \{(t_1,\dots,t_k,1) \mid (t_1,\dots,t_k,1) \in M_{l,m}\}$ and for $(t_1,\dots,t_k) \in Q_{l,m}$, $H_{\underline{d}}^{\underline{j}}(\tau,t,1,x) = ((hg)f)_{\underline{d}}^{\underline{j}}(\tau,t,x)$.

To the end of the proof we will check the well definition and the coherence conditions of the map $H_{\underline{d}}^{\underline{j}}$. To check the well definition suppose $(t_1,\dots,t_k,s) \in M_{l,m} \cap M_{l-1,m}$, in which case $H_{\underline{d}}^{\underline{j}}$ is defined in two ways. Then must $t_l = \frac{2+s}{4}$. If we compute the formula for $t = (t_1,\dots,t_k,s) \in M_{l,m}$ and $t_l = \frac{2+s}{4}$ then we have

$$H_{\underline{d}}^{\underline{j}}(\tau,t,s,x) = h_{d_0 \dots d_{j_{l-1}}}^{j_0 \dots j_{l-1}}(\tau_{0,l-1}, \frac{4t_1-2-s}{2-s},\dots,\frac{4t_{l-1}-2-s}{2-s}, r_{h(d_{j_{l-1}}\dots d_{j_l})}(\tau_{l-1,l},$$
$$g_{h(d_{j_l}\dots d_{j_m})}^{j_l-j_l \dots j_m-j_l}(\tau_{l,m}, 4t_{l+1}-1-s,\dots,4t_m-1-s, f_{gh(d_{j_m}\dots d_n)}^{j_m-j_m \dots j_k-j_m}(\tau_{m,k}, \frac{4t_{m+1}}{1+s},\dots,\frac{4t_k}{1+s}, x)))$$

The same expression is obtained if we compute the formula for $t = (t_1,\dots,t_k,s) \in M_{l-1,m}$ and $t_l = \frac{2+s}{4}$. Similarly, we can check the well definition for $(t_1,\dots,t_k,s) \in M_{l,m} \cap M_{l,m-1}$, and the other cases can be deduced to one of these two cases.

To check the coherence conditions of the map $H_{\underline{d}}^{\underline{j}}$, suppose $t = (t_1,\dots,t_k,s) \in M_{l-1,m}$ and $t_k = 0$. Then $f_{gh(d_{j_m}\dots d_n)}^{j_m-j_m \dots j_k-j_m}(\tau_{m,k}, \frac{4t_{m+1}}{1+s},\dots,\frac{4t_{k-1}}{1+s}, 0, x) = f_{gh(d_{j_m}\dots d_{j_{k-1}})}^{j_m-j_m \dots j_{k-1}-j_m}(\tau_{m,k-1}, \frac{4t_{m+1}}{1+s},\dots,\frac{4t_{k-1}}{1+s}, p_{fgh(d_{j_{k-1}}\dots d_n)}(\tau_{k-1,k}, x))$ and it follows that for $t_k = 0$, $H_{\underline{d}}^{\underline{j}}(\tau,t,s,x) = H_{gh(d_{j_m}\dots d_{j_{k-1}})}^{j_0 \dots j_{k-1}}(\tau_{0,k-1}, t_1,\dots,t_{k-1}, p_{fgh(d_{j_{k-1}}\dots d_n)}(\tau_{k-1,k}, x))$.

Similarly is treated case when $t_1 = 0$.

If $t_i = t_{i+1}$, and $i < l$ then

$$H_{\underline{d}}^{\underline{j}}(\tau,t,s,x) = h_{d_0 \dots d_{j_l}}^{j_0 \dots j_l}(\tau_{0,i}, 1, \tau_{i,l}, \frac{4t_1-2-s}{2-s},\dots,\frac{4t_{i-1}-2-s}{2-s}, \frac{4t_{i+1}-2-s}{2-s},\dots,\frac{4t_l-2-s}{2-s},$$
$$g_{h(d_{j_l}\dots d_{j_m})}^{j_l-j_l \dots j_m-j_l}(\tau_{l,m}, 4t_{l+1}-1-s,\dots,4t_m-1-s, f_{gh(d_{j_m}\dots d_n)}^{j_m-j_m \dots j_k-j_m}(\tau_{m,k}, \frac{4t_{m+1}}{1+s},\dots,\frac{4t_k}{1+s}, x)))$$
$$= H_{d_0 \dots d_{j_l}}^{j_0 \dots \hat{j}_i \dots j_l}(\tau_{0,i}, 1, \tau_{i,k}, t_1,\dots,\hat{t}_i,\dots,t_k, s, x)$$

The cases $l<i<m$ and $m<i<k$ are treated similarly. If $j_i<j<j_{i+1}$, $\tau_{j-i}=0$, and $m\leq i$ then $f^{j_m-j_m\dots j_k-j_m}_{gh(d_{j_m}\dots d_n)}(\tau_{m,k},\frac{4t_{m+1}}{1+s},\dots,\frac{4t_k}{1+s},x)=f^{j_m-j_m\dots j_i-j_m j_{i+1}-1-j_m\dots j_k-1-j_m}_{gh(d_{j_m}\dots\hat{d}_j\dots d_n)}(\tau_{j_m-m+1},\dots,\hat{\tau}_{j-i},\dots,\tau_{n-k},\frac{4t_{m+1}}{1+s},\dots,\frac{4t_k}{1+s},x)$

It follows that for $\tau_{j-i}=0$, $H^{j_0\dots j_k}_{d_0\dots d_n}(\tau,t,s,,x)=H^{j_0\dots j_i j_{i+1}-1\dots j_k-1}_{d_{j_m}\dots\hat{d}_j\dots d_n}(\tau_1,\dots,\hat{\tau}_{j-i},\dots,\tau_{n-k},t,x)$.

The cases $i<l$ and $l\leq i\leq k$ are treated similarly.

Theorem 1.3

If coherent maps $f,f':\mathcal{X}\to\mathcal{Y}$ are level homotopic, and coherent maps $g,g':\mathcal{Y}\to\mathcal{Z}$ are level homotopic then the coherent maps $gf,g'f':\mathcal{X}\to\mathcal{Y}$ are level homotopic.

Proof: Let $f,f':\mathcal{X}\to\mathcal{Y}$ be level homotopic by a homotopy $F:I\times\mathcal{X}\to\mathcal{Y}$ given by strictly increasing function $f:B\to A$. Then the coherent maps $gf,gf':\mathcal{X}\to\mathcal{Z}$ are level homotopic by the homotopy $gF:I\times\mathcal{X}\to\mathcal{Z}$ given by strictly increasing function $fg:C\to A$.

Let $g,g':\mathcal{Y}\to\mathcal{Z}$ be level homotopic by a homotopy $G:I\times\mathcal{Y}\to\mathcal{Z}$ given by the strictly increasing function $g:C\to B$. Then the coherent maps $gf',g'f':\mathcal{X}\to\mathcal{Z}$ are level homotopic by the homotopy $G(1\times f'):I\times\mathcal{X}\to\mathcal{Z}$ given by strictly increasing function $fg:C\to A$.

From Theorem 1.1 follows that the coherent maps $gf,g'f':\mathcal{X}\to\mathcal{Z}$ are level homotopic.

The coherent identity map $1_{\mathcal{Y}}:\mathcal{Y}\to\mathcal{Y}$ id defined by identity function $1_B:B\to B$ and the maps $1^j_{\underline{b}}:I^{n-k}\times\Delta^k\times I\times Y_{b_n}\to Y_{b_0}$. For $n=0$, $1_{b_0}=1_{Y_{b_0}}$ and for $n\geq 0$

$1^j_{\underline{b}}(\tau,t,x)=q_{\underline{b}}(\tau_{01},1,\tau_{12},\dots,1,\tau_{k-1,k},x)$.

Theorem 1.4

The coherent maps f and $f1_{\mathcal{X}}$ are homotopic ; f and $1_{\mathcal{Y}}f$ are homotopic.

Proof: We will prove that f and $1_{\mathcal{Y}}f$ are homotopic, and the other statement is treated in the similar way.

First we define a partition of $\Delta^k\times I$ into subpolyhedra L_l, $l=0,1,\dots,k$, by $L_l=\{(t_1,\dots,t_k,s)\mid t_l\geq\frac{s}{2}+\frac{1}{2}\geq t_{l+1}\}$.

We define a coherent homotopy $F : I \times \mathcal{X} \to \mathcal{Y}$. This map will be given by the function $f : B \to A$ and by the maps $F_{\underline{b}}^{j} : I^{n-k} \times \Delta^{k} \times I \times X_{f(b_n)} \to Y_{b_0}$ defined for $(t,s) \in L_l$ by

$$F_{\underline{b}}^{j}(\tau,t,s,x) = q_{b_0 \dots b_{jl}}(\tau_{0,1}, 1, \tau_{1,2}, \dots, 1, \tau_{l-1,l}, f_{b_{jl} \dots b_{jk}}^{j_l - j_l \dots j_k - j_l}(\tau_{l,k}, \frac{2t_{j_{l+1}}}{1+s}, \dots, \frac{2t_{j_k}}{1+s}, x)).$$

We mention that $K_l \times 0 = \{(t_1,\dots,t_k,0) \mid (t_1,\dots,t_k,0) \in L_l\}$ and then, for $t = (t_1,\dots,t_k) \in K_l$, we have $F_{\underline{b}}^{j}(\tau,t,0,x) = (1_{\mathcal{Y}} f)_{\underline{b}}^{j}(\tau,t,x)$.

Also, $\{(t_1,\dots,t_k,1) \mid (t_1,\dots,t_k) \in \Delta^k\} = L_0$ and $F_{\underline{b}}^{j}(\tau,t,1,x) = f_{\underline{b}}^{j}(\tau,t,x)$.

Homotopy: In categories pro-HTop and CPHTop the general definition of homotopy is introduced in the following way: First for a special coherent map $(f_{\underline{b}}) : \underline{X} \to \underline{Y}$ (or a map of systems $(f_b) : \underline{X} \to \underline{Y}$) and a strictly increasing function F such that $f < F$ we define a notion of shift as a special coherent map given by the function F and by maps $f_{\underline{b}}(1 \times p_{f(b_n)F(b_n)}) : \Delta^k \times X_{F(b_n)} \to Y_{b_0}$ (i.e. the map of systems $f_{\underline{b}} p_{f(b_n)F(b_n)} : X_{F(b_n)} \to Y_{b_0}$).

Two special coherent maps (maps of systems) are *homotopic* if they have shifts which are level homotopic. Formally, we put $p_{f(b)F(b)} = 1_{X_{f(b)}}$ if $f = F$. In this way two level homotopic maps are homotopic i.e. level homotopy is a special case of homotopuy.

CPHTop is the category with objects commutative inverse systems and morphisms homotopy classes of special coherent maps.

pro-HTop is category with objects inverse systems and morphisms homotopy classes of maps of systems.

The general definition of homotopy in category Coh is more involved. It will be presented later, in section 3. In the next section we present the level coherent category category since in the definition we need only level homotopy.

2. Level Coherent Category

If $\mathcal{X} = (X_a, p_{\underline{a}}, A)$ and $\mathcal{Y} = (Y_a, q_{\underline{a}}, A)$ are coherent inverse systems indexed by A, then the coherent map $(f_a) : \mathcal{X} \to \mathcal{Y}$ given by the identity function $1_A : A \to A$ we call level coherent map. The previous theorems prove that all coherent inverse systems indexed by one fixed directed set A and and level homotopy classes of coherent maps form a category wich we usualy denote by Coh^A.

Level coherent maps are a special type of coherent maps. The level coherent map over A, $(f_{\underline{a}}^{j}) : (X_a, p_{\underline{a}}) \to (Y_a, q_{\underline{a}})$ consists of:

1) for any $a \in A$ of a map $f_a : X_a \to Y_a$, and for any pair $a_0 < a_1$ of a map $f_{a_0 a_1} : I \times X_{a_1} \to Y_{a_0}$ such that $f_{a_o a_1}(0,x) = f_{a_o} p_{a_0 a_1}(x)$, $f_{a_o a_1}(1,x) = q_{a_0 a_1} f_{a_1}(x)$

2) for $n \geq 2$ and any strictly increasing sequence $\underline{a} = (a_0, a_1, \ldots, a_n)$, $a_0 < a_1 < \ldots < a_n$. of a set $f_{\underline{a}}$ of 2^{n-1} maps, $\{ f_{\underline{a}}^{j_0 j_1 \ldots j_k} \mid 0 = j_0 < j_1 < \ldots < j_k = n \}$ = $f_{\underline{a}}$.

The map $f_{\underline{a}}^{j_0 j_1 \ldots j_k} : I^{n-k} \times \Delta^k \times X_{a_n} \to Y_{a_0}$ must satisfy the following boundary conditions:

$$f_{\underline{a}}^{j_0 j_1 \ldots j_k}(\tau, t, x) = \begin{cases} q_{a_0 \ldots a_{j_1}}(\tau_{01}, f_{a_{j_1} \ldots a_n}^{j_1 - j_1 \ldots j_k - j_1}(\tau_{1k}, t_2, \ldots, t_k, x)), t_1 = 1 \\ f_{\underline{b}}^{j_0 \ldots \hat{j}_i \ldots j_k}(\tau_{0,i}, 1, \tau_{i,k}, t_1, \ldots, \hat{t}_i, \ldots, t_k, x), t_i = t_{i+1} \\ f_{a_0 \ldots a_{j_{k-1}}}^{j_0 \ldots j_{k-1}}(\tau_{0,k-1}, t_1, \ldots, t_{n-1}, p_{a_{j_{k-1}} \ldots a_n)}(\tau_{k-1,k}, x)), t_n = 0 \end{cases}$$ and for $j_m < j < j_{m+1}$ and

$\tau_{j-m} = 0$, $f_{\underline{a}}^{j_0 j_1 \ldots j_k}(\tau, t, x) = f_{a_0 \ldots \hat{a}_j \ldots a_n}^{j_0 \ldots j_m j_{m+1} - 1 \ldots j_k - 1}(\tau_1, \ldots, \hat{\tau}_{j-m}, \ldots, \tau_{n-k}, t, x)$.

We will call this map a level coherent map over A, and we denote it by $(f_{\underline{a}}^{j}) : (X_a, p_{\underline{a}}) \to (Y_a, q_{\underline{a}})$.

Remark: The definition of coherent map under 1) may be considered as a special case of the general definition under 2). Namely, in 2) if $n = 1$ then the set $f_{a_0 a_1}$ contains only one map $f_{a_0 a}^{01} : \Delta^1 \times X_{a_1} \to Y_{a_0}$ which we denote by $f_{a_0 a_1} : I \times X_{a_1} \to Y_{a_0}$. Also, in 2) for $n = 0$ we have only a map $f_{a_0}^{0} : X_{a_0} \to Y_{a_0}$ which we denote by $f_{a_0} : X_{a_0} \to Y_{a_0}$.

It is important to mention the following: if the level coherent map $(h_{\underline{a}}^{j}) : (X_a, p_{\underline{a}}) \to (Z_a, q_{\underline{a}})$ is the composition of level coherent maps $(g_{\underline{a}}^{j}) : (Y_a, p_{\underline{a}}) \to (Z_a, r_{\underline{a}})$ and $(f_{\underline{a}}^{j}) : (X_a, p_{\underline{a}}) \to (Y_a, q_{\underline{a}})$ then the maps of order 0, $h_{a_0} : X_{a_0} \to Z_{a_0}$ are given by $h_{a_0} = g_{a_0} f_{a_0}$ and the maps of order 1, $h_{a_0 a_1} : I \times X_{a_1} \to Z_{a_0}$ are given by

$$h_{a_0 a_1}(t, x) = \begin{cases} g_{a_0} f_{a_0 a_1}(2t, x), & 0 \leq t \leq \frac{1}{2} \\ g_{a_0 a_1}(2t - 1, f_{a_1}(x)), & \frac{1}{2} \leq t \leq 1 \end{cases}$$

The coherent inverse systems over A and the homotopy classes of the level coherent maps form a category Coh^{A}. We call Coh^{A} - the level coherent category.

The level coherent category over the set of natural numbers N is denoted by Coh^{N} .The objects of Coh^{N} are coherent inverse sequences.

There is a simpler description of the level coherent category Coh^{N} : We will present the level category with objects inverse sequences and with morphisms homotopy classes of strong fundamental sequences - the level category of strong fundamental sequences and inverse systems. The following theorem shows that Coh^{N} is isomorphic to a subcategory of this category having for objects coherent inverse systems.

Theorem 2.1

([11], theorem 2. p.303) The level coherent category Coh^{N} is isomorphic to the level category of strong fundamental sequences and coherent inverse systems.

Let $\underline{X}=(X_n,p_{n,m})$ and $\underline{Y}=(Y_n,q_{n,m})$ be inverse sequences commutative up to homotopy. A strong fundamental sequence from $\underline{X}$ to $\underline{Y}$ consists of a map $f_n\colon X_n\to Y_n$ and a homotopy $f_{n,n+1}\colon I\times X_{n+1}\to Y_n$ such that $f_{n,n+1}(0,x)=f_n p_{n,n+1}(x)$, $f_{n,n+1}(1,x)=q_{n,n+1}f_{n+1}(x)$.

Two strong fundamental sequences $(f_n,f_{n,n+1})$ and $(f_n',f_{n,n+1}')\colon\underline{X}\to\underline{Y}$ are coherently homotopic if there exists a strong fundamental sequence $(F_n,F_{n,n+1})\colon I\times\underline{X}\to\underline{Y}$ (where $I\times\underline{X}=(I\times X_n,1\times p_{n,n+1})$) such that $F_n\colon I\times X_n\to Y_n$ satisfy $F_n(0,x)=f_n(x), F_n(1,x)=f_n'(x)$ and maps $F_{n,n+1}\colon I\times I\times X_{n+1}\to Y_n$ satisfy $F_{n,n+1}(t,0,x)=f_{n,n+1}(t,x)$, $F_{n,n+1}(t,1,x)=f_{n,n+1}'(t,x)$.

This is an equivalence relation and we denote $(f_n,f_{n,n+1})\sim(f_n',f_{n,n+1}')$.

The composition of level strong fundamental sequences $(f_n,f_{n,n+1})\colon\underline{X}\to\underline{Y}$ and $(g_n,g_{n,n+1})\colon\underline{Y}\to\underline{Z}$ is the strong fundamental sequence $(h_n,h_{n,n+1})\colon\underline{X}\to\underline{Z}$ defined by

$$h_n=g_nf_n \text{ and } h_{n,n+1}(t,x)=\begin{cases} g_nf_{n,n+1}(2t,x), & 0\le t\le\frac{1}{2} \\ g_{n,n+1}(2t-1,f_{n+1}(x)), & \frac{1}{2}\le t\le 1\end{cases}.$$

We denote the composition $(h_n,h_{n,n+1})\colon\underline{X}\to\underline{Z}$ by $((gf)_n,(gf)_{n,n+1})\colon\underline{X}\to\underline{Z}$.

From Theorems 2.2, 2.3 and 2.4 it follows that inverse systems commutative up to homotopy and homotopy classes of strong fundamental sequences form a category. We call this category *the level category of strong fundamental sequences and inverse systems.*

Theorem 2.2

Let $(f_n,f_{n,n+1}),(f_n',f_{n,n+1}')\colon\underline{X}\to\underline{Y}$ and $(g_n,g_{n,n+1}),(g_n',g_{n,n+1}')\colon\underline{Y}\to\underline{Z}$ be strong fundamental sequences such that $(f_n,f_{n,n+1})\sim(f_n',f_{n,n+1}')$ and $(g_n,g_{n,n+1})\sim(g_n',g_{n,n+1}')$. Then $((gf)_n,(gf)_{n,n+1})\sim((g'f')_n,(g'f')_{n,n+1})$.

Proof: Let $(f_n,f_{n,n+1})\sim(f_n',f_{n,n+1}')$ by a homotopy $(F_n,F_{n,n+1})$ and $(g_n,g_{n,n+1})\sim(g_n',g_{n,n+1}')$ by a homotopy $(G_n,G_{n,n+1})$.

Then $((gf)_n,(gf)_{n,n+1})\sim((gf')_n,(gf')_{n,n+1})$ by a homotopy $(H_n,H_{n,n+1})$ defined by

$$H_n(s,x)=g_nF_n(s,x) \text{ and } H_{n,n+1}(t,s,x)=\begin{cases} g_nF_{n,n+1}(2t,s,x), & 0\le t\le\frac{1}{2} \\ g_{n,n+1}(2t-1,F_{n+1}(s,x)), & \frac{1}{2}\le t\le 1\end{cases}.$$

Also, $((gf')_n,(gf')_{n,n+1}) \sim ((g'f')_n,(g'f')_{n,n+1})$ by a homotopy $(K_n, K_{n,n+1})$ defined by $K_n(s,x) = G_n(s, f_n'(x))$ and $K_{n,n+1}(t,s,x) = \begin{cases} G_n(s, f'_{n,n+1}(2t,s,x)), & 0 \le t \le \frac{1}{2} \\ G_{n,n+1}(2t-1, f'_{n+1}(s,x)), & \frac{1}{2} \le t \le 1 \end{cases}$.

Since ~ is an equivalence relation, it follows $((gf)_n,(gf)_{n,n+1}) \sim ((g'f')_n,(g'f')_{n,n+1})$.

Theorem 2.3

If $(f_n, f_{n,n+1}): \underline{X} \to \underline{Y}$, $(g_n, g_{n,n+1}): \underline{Y} \to \underline{Z}$, $(h_n, h_{n,n+1}): \underline{Z} \to \underline{W}$ are strong fundamental sequences then $[(h_n, h_{n,n+1})(g_n, g_{n,n+1})](f_n, f_{n,n+1}) \sim (h_n, h_{n,n+1})[(g_n, g_{n,n+1})(f_n, f_{n,n+1})]$.

Proof: These two maps are connected by a homotopy $(H_n, H_{n,n+1})$ defined by $H_n(s,x) = h_n g_n f_n(x)$ and

$$H_{n,n+1}(t,s,x) = \begin{cases} h_n g_n f_{n,n+1}(\frac{4t}{s+1}, x), & 0 \le t \le \frac{s+1}{4} \\ h_n g_{n,n+1}(4t-1-s, f_{n+1}(x)), & \frac{s+1}{4} \le t \le \frac{s+2}{4} \\ h_{n,n+1}(\frac{4t-2-s}{2-s}, g_{n+1} f_{n+1}(x)), & \frac{s+2}{4} \le t \le 1 \end{cases}.$$

The identity map $\underline{X} \to \underline{X}$ is defined by the strong fundamental sequence $(1_{X_n}, p_{n,n+1}): \underline{X} \to \underline{X}$.

Theorem 2.4

Let $(f_n, f_{n,n+1}): \underline{X} \to \underline{Y}$ be a strong fundamental sequence. Then $(1_{Y_n}, q_{n,n+1})(f_n, f_{n,n+1}) \sim (f_n, f_{n,n+1})$ and $(f_n, f_{n,n+1})(1_{X_n}, p_{n,n+1}) \sim (f_n, f_{n,n+1})$.

Proof: The homotopy $(H_n, H_{n,n+1})$ connecting strong fundamental sequences $(1_{Y_n}, q_{n,n+1})(f_n, f_{n,n+1})$ and $(f_n, f_{n,n+1})$ is defined by $H_n(s,x) = f_n(x)$ and

$$H_{n,n+1}(t,s,x) = \begin{cases} f_{n,n+1}(\frac{2t}{s+1}, x), & 0 \le t \le \frac{s+1}{2} \\ q_{n,n+1} f_{n+1}(x), & \frac{s+1}{2} \le t \le 1 \end{cases}.$$

The homotopy $(K_n, K_{n,n+1})$ connecting strong fundamental sequences $(f_n, f_{n,n+1})(1_{X_n}, p_{n,n+1})$ and $(f_n, f_{n,n+1})$ is defined by $K_n(s,x) = f_n(x)$ and

$$K_{n,n+1}(t,s,x) = \begin{cases} f_n p_{n,n+1}(x), & 0 \le t \le \frac{1-s}{2} \\ f_{n,n+1}(\frac{2t-1+s}{1+s}, x), & \frac{1-s}{2} \le t \le 1 \end{cases}.$$

The application of strong fundamental sequences.

The strong fundamental sequences (and commutative inverse systems) are used for defining strong shape theory for metric compacts [3], [5].

Here we present another application: In a commutative inverse sequence by a natural replacement of spaces with homotopy equivalent spaces an inverse sequence is obtained, which is not commutative. It is proved that the two inverse sequences are isomorphic in the category of strong fundamental sequences .

Let $\underline{X} = (X_n, p_{n,m})$ be a commutative inverse sequence such that maps $p_{n,m}: X_m \to X_n$, $n < m$ satisfy $p_{n,n+1}p_{n+1,n+2} = p_{n,n+2}$

Let $f_n: X_n \to Y_n$ be a homotopy equivalence with a homotopy inverse $g_n: Y_n \to X_n$ and let $g_n f_n \sim 1_{X_n}$ by a homotopy H_n.

$$\begin{array}{ccccc} X_1 & \leftarrow & X_2 & \leftarrow \ldots \\ f_1 \downarrow \uparrow g_1 & & f_2 \downarrow \uparrow g_2 & \\ Y_1 & & Y_2 & \end{array}.$$

We can define an inverse sequence $\underline{Y} = (Y_n, q_{n,m})$, in a natural way: $q_{n,m} = f_n p_{n,m} g_m$ for $n < m$. Then $q_{k,n}q_{n,m} = f_k p_{k,n} g_n f_n p_{n,m} g_m \sim f_k p_{k,m} g_m = q_{k,m}$.

The inverse sequence $\underline{Y} = (Y_n, q_{n,m})$ although is not strictly commutative, it is commutative up to homotopy, since $q_{k,m} \sim q_{k,n}q_{n,m}$.

It is easily verified that $f_n p_{n,n+1} \sim q_{n,n+1} f_{n+1}$ i.e. the maps $f_n: X_n \to Y_n$ define a morphism $(f_n): \underline{X} \to \underline{Y}$ in pro-HTop. Also, $g_n q_{n,n+1} \sim p_{n,n+1} g_{n+1}$ and it follows that the maps $g_n: Y_n \to X_n$ define a morphism $(g_n): \underline{Y} \to \underline{X}$ in pro-HTop.

Moreover, since $g_n f_n \sim 1_X$ and $f_n g_n \sim 1_Y$ we have $(g_n)(f_n) = 1_{\underline{X}}$ and $(f_n)(g_n) = 1_{\underline{Y}}$ i.e. the inverse systems $\underline{X}$ and $\underline{Y}$ are isomorphic in pro-HTop.

In the following theorem it is proved that the inverse systems $\underline{X}$ and $\underline{Y}$ are isomorphic in a stronger way i.e. they are isomorphic in the category where the morphisms are strong homotopy classes of strong fundamental sequences. The proof is not trivial and uses the Vogt theorem on homotopy equivalence [14].

In [1], the authors considered the problem presented here and gave a positive answer too in their oherent category. The relation of their coherent theory and the category Coh is still unknown.

Theorem 2.5

([11] th 1. p 297) Let $\underline{X} = (X_n, p_{n,m})$ be a commutative inverse system, let $f_n: X_n \to Y_n$ be a homotopy equivalence with a homotopy inverse $g_n: Y_n \to X_n$. Let $\underline{Y} = (Y_n, q_{n,m})$ be the inverse system defined by putting $q_{n,m} = f_n p_{n,m} g_m$, for $n < m$. Then $\underline{X}$ and $\underline{Y}$ are isomorphic in the level category of strong fundamental sequences (and inverse sequences commutative up to homotopy).

Moreover, in the situation above, we can build a coherent inverse sequence (system) $\mathcal{Y}$ over the inverse sequence $\underline{Y}=(Y_n, q_{n,m})$, and show that $\underline{X}$ and $\mathcal{Y}$ are isomorphic in the coherent level category $\mathrm{Coh}^{\mathrm{N}}$.

We will consider a more general situation: Let $\underline{X}=(X_a, p_{a_0a_1})$ be a commutative inverse system over a directed set A. Let $f_a: X_a \to Y_a$ be a homotopy equivalence with a homotopy inverse $g_a: Y_a \to X_a$ and $g_a f_a \sim 1_{X_a}$ by a homotopy H_a.

We can define a map $q_{a_0a_1}: Y_{a_1} \to Y_{a_0}$ for any pair of indices $a_0 < a_1$ by $q_{a_0a_1} = f_{a_0} p_{a_0a_1} g_{a_1}$. For every strictly increasing sequence $\underline{a}=(a_0, a_1, \ldots, a_n)$, $a_0 < a_1 < \ldots < a_n$, $n \geq 2$, we can define a map $q_{\underline{a}}: I^{n-1} \times Y_{a_n} \to Y_{a_0}$ by $q_{\underline{a}}(\tau_1, \ldots, \tau_{n-1}, x) = f_{a_0} p_{a_0a_1} H_{a_1}(\tau_1, p_{a_1a_2} H_{a_2}(\tau_2, \ldots, p_{a_{n-2}a_{n-1}} H_{a_{n-1}}(\tau_{n-1}, p_{a_{n-1}a_n} g_{a_n}(x))))$ where $\tau = (\tau_1, \ldots, \tau_{n-1}) \in I^{n-1}, x \in Y_{a_n}$.

Then, $\mathcal{Y} = (Y_a, q_{\underline{a}}, A)$ is a coherent inverse system.

Theorem 2.6

Let $\underline{X}=(X_n, p_{n,m})$ be a commutative inverse system, let $f_n: X_n \to Y_n$ be a homotopy equivalence with a homotopy inverse $g_n: Y_n \to X_n$. Let $\underline{Y}=(Y_n, q_{n,m})$ be the inverse system defined by $q_{n,m} = f_n p_{n,m} g_m$,for $n < m$. Then there is a coherent inverse system $(Y_m, q_{\underline{m}})$ such that for $\underline{m}=(n,m)$, $q_{\underline{m}} = q_{n,m}$, and $\underline{X}=(X_n, p_{n,m})$ and $(Y_m, q_{\underline{m}})$ are isomorphic

1) in the category $\mathrm{Coh}^{\mathrm{N}}$ 2) in the category Coh .

Proof:

1) is a consequence of Theorems 1 and 2
2) From 1) there are level coherent maps $f: \underline{X} \to (Y_m, q_{\underline{m}})$ and $g: (Y_m, q_{\underline{m}}) \to \underline{X}$ such that $gf \sim 1_{\underline{X}}$ by a level homotopy and $fg \sim 1_{(Y_m, q_{\underline{m}})}$ by a level homotopy. Level coherent maps f and g are coherent maps, also. Since two coherent maps are coherently homotopic by a level coherent homotopy (a special type of coherent homotopy), it follows that $\underline{X}$ and $(Y_m, q_{\underline{m}})$ are isomorphic in the category Coh.

In the previous theorem, we can consider a more general situation [12]: Instead of a commutative inverse sequence $\underline{X}=(X_n, p_{n,m})$, we can consider a coherent inverse sequence $(X_m, p_{\underline{m}})$.

Theorem 2.7

Let $(X_m, p_{\underline{m}})$ be a coherent inverse sequence, let $f_m : X_m \to Y_m$ be a homotopy equivalence with a homotopy inverse $g_m : Y_m \to X_m$. Let $\underline{Y} = (Y_n, q_{n,m})$ be the inverse system defined by $q_{n,m} = f_n p_{n,m} g_m$,for $n < m$. Then there is a coherent inverse system $(Y_m, q_{\underline{m}})$ such that for $\underline{m} = (n,m)$, $q_{\underline{m}} = q_{n,m}$, and $(X_m, p_{\underline{m}})$ and $(Y_m, q_{\underline{m}})$ are isomorphic

1) in the level coherent category $\mathrm{Coh}^{\mathrm{N}}$ 2) in the lcoherent category Coh .

Proof: First we construct the coherent inverse system $(Y_m, q_{\underline{m}})$ such that for $\underline{m} = (n,m)$, $q_{\underline{m}} = q_{n,m}$.

We will consider a more general situation: Let $(X_a, p_{\underline{a}}, A)$ be a commutative inverse system over a directed set A. Let $f_a : X_a \to Y_a$ be a homotopy equivalence with a homotopy inverse $g_a : Y_a \to X_a$ and $g_a f_a \sim 1_{X_a}$ by a homotopy H_a.

We can define a map $q_{a_0 a_1} : Y_{a_1} \to Y_{a_0}$ for any pair of indices $a_0 < a_1$ by $q_{a_0 a_1} = f_{a_0} p_{a_0 a_1} g_{a_1}$. For $a_0 < a_1 < a_2$, we define a map $q_{a_0 a_1 a_2} : I \times Y_{a_2} \to Y_{a_0}$ by

$$q_{a_0 a_1 a_2}(t_1, x) = \begin{cases} f_{a_0} p_{a_0 a_1 a_2}(2t_1, g_{a_2}(x)), & 0 \le t_1 \le \frac{1}{2} \\ f_{a_0} p_{a_0 a_1} H_{a_1}(2t_1 - 1, p_{a_1 a_2} g_{a_2}(x)), & \frac{1}{2} \le t_1 \le 1 \end{cases}$$

For $n \ge 2$, we define a map $q_{\underline{a}} : I^{n-1} \times Y_{a_n} \to Y_{a_0}$ in the following way. We define a partitioning of the cube I^{n-1} into 2^{n-1} smaller cubes. These cubes are defined for any sequence $0 < i_1 < \ldots < i_k < n$ and $0 \le k \le n-1$ by $\{(t_1, \ldots, t_{n-1}) \mid 0 \le t_{i_1} \le \frac{1}{2}, \ldots, 0 \le t_{i_k} \le \frac{1}{2} \ ; \ \frac{1}{2} \le t_i \le 1, i \notin \{i_1, \ldots, i_k\}\}$.

On this part of the cube I^{n-1} the map $q_{\underline{a}} : I^{n-1} \times Y_{a_n} \to Y_{a_0}$, for the sequence $\underline{a} = (a_0, a_1, \ldots, a_n)$, $a_0 < a_1 < \ldots < a_n$, is defined by

$$q_{\underline{a}}(t_1, \ldots, t_{n-1}, x) = f_{a_0} p_{a_0 a_1 \ldots a_{i_1}}(2t_1, \ldots, 2t_{i_1 - 1}, H_{a_{i_1}}(2t_{i_1} - 1, p_{a_{i_1} \ldots a_{i_2}}(2t_{i_1+1}, \ldots$$
$$\ldots, 2t_{i_2 - 1}, H_{a_{i_2}}(2t_{i_2} - 1, \ldots, p_{a_{i_k} \ldots a_n}(2t_{i_k+1}, \ldots, 2t_{n-1}, g_{a_n}(x)))))$$

We have to verify that these maps are well defined. For $t_{i_\ell} = \frac{1}{2}$, $1 \le \ell \le n-1$, that follows from the equalities

$$p_{a_{i_{\ell-1}} \ldots a_{i_\ell}}(2t_{i_{\ell-1}+1}, \ldots, 2t_{i_\ell - 1}, H_{a_{i_\ell}}(0, p_{a_{i_\ell} \ldots a_{i_{\ell+1}}}(2t_{i_\ell+1}, \ldots, 2t_{i_{\ell+1}-1}, y)) =$$
$$= p_{a_{i_{\ell-1}} \ldots a_{i_\ell}}(2t_{i_{\ell-1}+1}, \ldots, 2t_{i_\ell - 1}, p_{a_{i_\ell} \ldots a_{i_{\ell+1}}}(2t_{i_\ell+1}, \ldots, 2t_{i_{\ell+1}-1}, y))$$
$$= p_{a_{i_{\ell-1}} \ldots a_{i_{\ell+1}}}(2t_{i_{\ell-1}+1}, \ldots, 2t_{i_\ell - 1}, 1, 2t_{i_\ell+1}, \ldots, 2t_{i_{\ell+1}-1}, y)$$

Similarly, for $t_j = \frac{1}{2}$, $i_\ell < j < i_{\ell+1}$, that follows from the following equalities $p_{a_{i_\ell}\ldots a_{i_{\ell+1}}}(2t_{i_\ell+1},\ldots,2t_{j-1},1,2t_{j+1},\ldots,2t_{i_{\ell+1}-1},y) = = p_{a_{i_\ell}\ldots a_j}(2t_{i_\ell+1},\ldots,2t_{j-1},p_{a_j\ldots a_{i_{\ell+1}}}(2t_{j+1},\ldots,2t_{i_{\ell+1}-1},y)$
$= p_{a_{i_\ell}\ldots a_j}(2t_{i_\ell+1},\ldots,2t_{j-1},H_{a_{i_j}}(0,p_{a_j\ldots a_{i_{\ell+1}}}(2t_{j+1},\ldots,2t_{i_{\ell+1}-1},y)$

For $t_{i_\ell} = 1$, $1 \le \ell \le n-1$, from $p_{a_{i_{\ell-1}}\ldots a_{i_\ell}}(2t_{i_{\ell-1}+1},\ldots,2t_{i_\ell-1},H_{a_{i_\ell}}(1,p_{a_{i_\ell}\ldots a_{i_{\ell+1}}}(2t_{i_\ell+1},\ldots,2t_{i_{\ell+1}-1},y) =$ $= p_{a_{i_{\ell-1}}\ldots a_{i_\ell}}(2t_{i_{\ell-1}+1},\ldots,2t_{i_\ell-1},g_{i_\ell}f_{i_\ell}p_{a_{i_\ell}\ldots a_{i_{\ell+1}}}(2t_{i_\ell+1},\ldots,2t_{i_{\ell+1}-1},y)$ it follows that $q_{\underline{a}}(t_1,\ldots,t_{i_\ell-1},1,t_{i_\ell+1},\ldots,t_{n-1},x) = q_{a_0\ldots a_{i_\ell}}(t_1,\ldots,t_{i_\ell-1},q_{a_{i_\ell}\ldots a_n}(t_{i_\ell+1},\ldots,t_{n-1},x))$.

For $t_j = 0$, $i_\ell < j < i_{\ell+1}$, from $p_{a_{i_\ell}\ldots a_{i_{\ell+1}}}(2t_{i_\ell+1},\ldots,2t_{j-1},0,2t_{j+1},\ldots,2t_{i_{\ell+1}-1},y) =$ $p_{a_{i_\ell}\ldots \hat{a}_j\ldots a_{i_{\ell+1}}}(2t_{i_\ell+1},\ldots,2t_{j-1},2t_{j+1},\ldots,2t_{i_{\ell+1}-1},y)$

It follows that $q_{\underline{a}}(t_1,\ldots,t_{j-1},0,t_{j+1},\ldots,t_{n-1},x) = q_{a_0\ldots\hat{a}_j\ldots a_n}(t_1,\ldots,t_{j-1},t_{j+1},\ldots,t_{n-1},x)$.

With this, it is verified that the triple $(Y_a, q_{\underline{a}}, A)$ is a coherent inverse system over a directed set A.

Now, exactly as in [11], Theorem 1, we prove that $\underline{Y} = (Y_n, q_{n,m})$ and $(X_n, p_{n,m})$ are isomorphic in the level category of strong fundamental sequences. Then, from [11], Theorem 1 and Theorem 2 it follows that $(Y_m, q_{\underline{m}})$ and $(X_m, p_{\underline{m}})$ are isomorphic in the level coherent category $\mathrm{Coh}^{\mathrm{N}}$, and in a coherent category Coh.

We finish the section with the following theorem

Theorem 2.8

Let $(X_m, p_{\underline{m}})$ and $(Y_m, q_{\underline{m}})$ be arbitrary coherent inverse sequences and $f = (f_{\underline{m}}^{j}) : (X_n, p_{\underline{n}}) \to (Y_m, q_{\underline{m}})$ a level coherent map. If the maps $f_m : X_m \to Y_m$,and $f'_m : X_m \to Y_m$ are homotopic by a homotopy $F_m : X_m \to Y_m$, then there exists a level coherent map $f' = (f'^{j}_{\underline{m}}) : (X_m, p_{\underline{m}}) \to (Y_m, q_{\underline{m}})$and a level coherent homotopy $F = (F_{\underline{m}}^{j}) : (X_m, p_{\underline{m}}) \to (Y_m, q_{\underline{m}})$ which connects coherent maps f and f'.

Proof: By [11], Theorem 2, there is a category isomorphism between level coherent category $\mathrm{Coh}^{\mathrm{N}}$ and the level category of coherent inverse systems as objects and strong fundamental sequences as morphisms. This functor leaves the objects fixed and associates to the coherent homotopy class of coherent map $(f_{\underline{m}}^{j})$ the strong homotopy class of the strong fundamental sequence $(f_m, f_{m,m+1})$.

To prove the theorem it is enough to construct the strong fundamental sequence $(f'_m, f'_{m,m+1})$, and a strong level homotopy $(F_m, F_{m,m+1})$ connecting $(f_m, f_{m,m+1})$ and $(f'_m, f'_{m,m+1})$.

We define a map $F_{m,m+1} : I \times I \times X_m \to Y_m$, first on $((I \times \{0\}) \cup (\{0,1\} \times I)) \times X_m$ by $F_{m,m+1}(t,0,x) = f_{m,m+1}(t,x)$ $F_{m,m+1}(1,s,x) = q_{m,m+1}F_m(s,x)$ $F_{m,m+1}(0,s,x) = F_m(s, p_{m,m+1}(x))$.

The map is well defined on points $(0,0,x)$, $x \in X_n$, since $f_{m,m+1}(0,x) = f_m p_{m,m+1}(x) = F_m(0, p_{m,m+1}(x))$, and on points $(1,0,x)$, $x \in X_m$, since $f_{m,m+1}(1,x) = q_{m,m+1} f_m(x) = q_{m,m+1} F_m(x)$.

Since there is a retraction $I \times I \to (I \times \{0\}) \cup (\{0,1\} \times I)$ we can extend the map to a map $F_{m,m+1}: I \times I \times X_m \to Y_m$.

We define a map $f'_{m,m+1}: I \times X_m \to Y_m$ by putting $f'_{m,m+1}(t,x) = F_{m,m+1}(t,1,x)$.

Then from the definition, the map $F_{m,m+1}: I \times I \times X_m \to Y_m$ connects the maps $f_{m,m+1}$ and $f'_{m,m+1}$ and is a strong fundamental sequence. From $f'_{m,m+1}(0,x) = F_{m,m+1}(0,1,x) = F_m(1, p_{m,m+1}(x)) = f'_m p_{m,m+1}(x)$ and $f'_{m,m+1}(1,x) = F_{m,m+1}(1,1,x) = q_{m,m+1} F_{m+1}(1,x) = q_{m,m+1} f'_{m,m+1}(x)$ one concludes that $(f'_m, f'_{m,m+1})$ is a strong fundamental sequence. It follows the pair $(F_m, F_{m,m+1})$ is a strong homotopy connecting the strong fundamental sequences $(f_m, f_{m,m+1})$ and $(f'_m, f'_{m,m+1})$.

3. Coherent Shift and Coherent Category

Let $\mathcal{X} = (X_a, p_{\underline{a}}, A)$ be a coherent inverse system and $f: B \to A$ a strictly increasing function. We can consider the coherent inverse system $\mathcal{X}_f = (X_{f(b)}, p_{f(b_0 \ldots b_n)}, B)$.

To any coherent map $(f_{\underline{b}}^{j}): \mathcal{X} \to \mathcal{Y} = (Y_b, q_{\underline{b}}, B)$ we can associate a *level* (coherent) map $(f_{\underline{b}}^{j}): \mathcal{X}_f \to \mathcal{Y}$ and this correspondence is bijective.

In this section we will use the following two propositions several times.

Proposition 3.1

Coherent maps $(f_{\underline{b}}^{j}), (f'^{j}_{\underline{b}}): \mathcal{X} \to \mathcal{Y}$ are level homotopic by the coherent map $(F_{\underline{b}}^{j}): I \times \mathcal{X} \to \mathcal{Y}$ if and only if level maps $(f_{\underline{b}}^{j}), (f'^{j}_{\underline{b}}): \mathcal{X}_f \to \mathcal{Y}$ are level homotopic by the level map $(F_{\underline{b}}^{j}): I \times \mathcal{X}_f \to \mathcal{Y}$.

Proposition 3.2.

Let $(g_{\underline{c}}^{j}): \mathcal{Y} \to \mathcal{Z} = (Z_c, r_{\underline{c}}, C)$ and $(f_{\underline{b}}^{j}): \mathcal{X} \to \mathcal{Y}$ be coherent maps. Then to their composition $((gf)_{\underline{c}}^{j}): \mathcal{X} \to \mathcal{Z}$ corresponds the level coherent map $((gf)_{\underline{c}}^{j}): \mathcal{X}_{fg} \to \mathcal{Z}$,

$\mathcal{X}_{fg} = (X_{fg(c)}, p_{fg(\underline{c})}, C)$ - the composition of level maps $(g_{\underline{c}}^{j}) : \mathcal{Y}_g \to \mathcal{Z}$ and $(f_{\underline{c}}^{j}) : \mathcal{X}_{fg} \to \mathcal{Y}_g$.

For two strictly increasing functions $f, F : B \to A$ we define the notion of a *shifting map.* The shifting map is a level coherent map $p[f.F] : \mathcal{X}_F \to \mathcal{X}_f$ given by maps $p[f,F]_{\underline{b}}^{j} : I^{n-k} \times \Delta^k \times X_{F(b_n)} \to X_{f(b_0)}$ defined by $p[f,F]_{\underline{b}}^{j}(\tau,t,x) = p_{f(b_0 \ldots b_{j_m})F(b_{j_m} \ldots b_n)}((\tau_{0,1},1)(2t_1 - 1),\ldots$ $\ldots,(\tau_{m-1,m},1)(2t_m - 1),(1,\tau_{m,m+1})(1-2t_{m+1}),\ldots,(1,\tau_{k-1,k})(1-2t_k),x)$ for $t \in K_m$. For $n = 0$, the map $p[f,F]_{b_0} : X_{F(b_0)} \to X_{f(b_0)}$ is given by $p[f,F]_{b_0}(x) = p_{f(b_0)F(b_0)}(x)$

To prove the well definition of $p[f,F]$ one has to check the definition for the points $(t_1,\ldots,t_k) \in K_m \cap K_{m-1}$ and this is possible if and only if $t_m = \frac{1}{2}$. Then for $t = (t_1,\ldots,t_k) \in K_{m-1}$ and $t_m = \frac{1}{2}$ we have $p[f,F]_{\underline{b}}^{j}(\tau,t,x) = p_{f(b_0 \ldots b_{j_m})F(b_{j_m} \ldots b_n)}((\tau_{0,1},1)(2t_1 - 1),\ldots,(\tau_{m-2,m-1},1)(2t_{m-1} - 1),0,\ldots$ $\ldots,0,(1,\tau_{m,m+1})(1-2t_{m+1}),\ldots,(1,\tau_{k-1,k})(1-2t_k),x)$ and the same expression is obtained if we compute the formula for $(t_1,\ldots,t_k) \in K_m$ and $t_m = \frac{1}{2}$.

That $p[f,F]$ is a coherent map it is shown in [11] .

Definition: If $f : \mathcal{X} \to \mathcal{Y}$ is a coherent map and $F : B \to A$ is a strictly increasing function such that $f < F$, the composition of the level maps $f : \mathcal{X}_F \to \mathcal{Y}$ and $p[f,F] : \mathcal{X}_F \to \mathcal{X}_f$ - the level map $f \circ p[f,F] : \mathcal{X}_F \to \mathcal{Y}$ we call a *coherent shift* of .

The coresponding coherent map $f \circ p[f,F] : \mathcal{X} \to \mathcal{Y}$ we also call *coherent shift* of f .

Theorem 3.1

If $f, F, \mathfrak{F} : B \to A$ are strictly increasing functions such that $f < F < \mathfrak{F}$. Then level maps $p[f,F] \circ p[F,\mathfrak{F}]$ and $p[f,\mathfrak{F}]$ are level homotopic.

Proof: We define a partition of $\Delta^k \times I$ into subpolyhedra $S_{l,r}$ for any pair of integers l,r such that $0 \le l \le r \le k$, $S_{l,r} = \{(t_1,\ldots,t_k,s) \mid t_l \ge \frac{2+s}{4} \ge t_{l+1},\ t_r \ge \frac{2-s}{4} \ge t_{r+1}\}$.

We define a partition of $\Delta^k \times I$ into subpolyhedra $S_{l,m,r}$ for any triple of integers l,m,r such that $0 \le l \le m \le r \le k$, $S_{l,m,r} = S_{l,r} \cap (K_m \times I)$

$$= \{(t_1,\ldots,t_k,s) \mid t_l \ge \frac{2+s}{4} \ge t_{l+1},\ t_m \ge \frac{1}{2} \ge t_{m+1},\ t_r \ge \frac{2-s}{4} \ge t_{r+1}\}$$

Now we define a level coherent map $P:\mathcal{X}_{\mathcal{F}} \to \mathcal{X}_f$. This map will be given by maps $P_{\underline{b}}^{j}: I^{n-k} \times \Delta^k \times X_{F(b_n)} \to X_{f(b_0)}$ defined for $(t,s) \in S_{l,m,r}$ by

$$P_{\underline{b}}^{j}(\tau,t,s,x) = p_{f(b_0...b_{jl})F(b_{jl}...b_{jr})F(b_{jr}...b_n)}((\tau_{0,1},1)\frac{4t_1-2-s}{2-s},...,(\tau_{l-1,l},1)\frac{4t_l-2-s}{2-s},$$
$$,(1,\tau_{l,l+1})(2+s-4t_{l+1}),...,(1,\tau_{m-1,m})(2+s-4t_m),s,(\tau_{m,m+1},1)(4t_{m+1}-2+s),...$$
$$...,(\tau_{r-1,r},1)(4t_r-2+s),(1,\tau_{r,r+1})\frac{2-4t_{r+1}-s}{2-s},...,(1,\tau_{k-1,k})\frac{2-4t_k-s}{2-s},x)$$

Now we check that $P_{\underline{b}}^{j}$ connects maps $(p[f,F]\circ p[F,\mathcal{F}])_{\underline{b}}^{j}$ and $p[f,\mathcal{F}]_{\underline{b}}^{j}$. If $s=0$, then there exists m such that $(t,0)\in S_{m,m,m}$. Also $(t,0)\in S_{m,m,m}$ if and only if $t\in K_m$ and

$$P_{\underline{b}}^{j}(\tau,t,0,x) = p_{f(b_0...b_{jm})F(b_{jm})F(b_{jm}...b_n)}((\tau_{0,1},1)(2t_1-1),...$$
$$...,(\tau_{m-1,m},1)(2t_m-1),0,(1,\tau_{m,m+1})(1-2t_{m+1}),...,(1,\tau_{k-1,k})(1-2t_k),x) = p[f,\mathcal{F}]_{\underline{b}}^{j}(\tau,t,x).$$

If $s=1$, and $(t,1)\in S_{l,r}\cap(K_m\times\{1\}) = \{(t_1,...,t_k,1) \mid t_l \geq \frac{3}{4} \geq t_{l+1},\ t_m \geq \frac{1}{2} \geq t_{m+1},\ t_r \geq \frac{1}{4} \geq t_{r+1}\}$ we have $P_{\underline{b}}^{j}(\tau,t,1,x) = P[f,F]_{b_0...b_{jm}}^{j_0...j_m}(\tau_{0,m},2t_1-1,...,2t_m-1,P[F,\mathcal{F}]_{b_{jm}...b_n}^{j_m-j_m...j_k-j_m}(\tau_{m,k},2t_{m+1},...,2t_k,x))$ $=(p[f,F]\circ p[F,\mathcal{F}])_{\underline{b}}^{j}(\tau,t,x)$.

Corollary: If $f,f':\mathcal{X}\to\mathcal{Y}$ are coherent maps such that the shifts $f\circ p[f,F]$ and $f'\circ p[f',F]$ are level homotopic, and $\mathcal{F}:B\to A$ is a strictly increasing function such that $F<\mathcal{F}$. Then the shifts $f\circ p[f,\mathcal{F}]$ and $f'\circ p[f',\mathcal{F}]$ are level homotopic.

Proof: By Theorem 1.1, Theorem 1.3 and Theorem 3.1

$$f\circ p[f,\mathcal{F}] \cong f\circ(p[f,F]\circ p[F,\mathcal{F}]) \cong (f\circ p[f,F])\circ p[F,\mathcal{F}]$$
$$\cong (f'\circ p[f',F])\circ p[F,\mathcal{F}] \cong f'\circ(p[f',F]\circ p[F,\mathcal{F}]) \cong f'\circ p[f',\mathcal{F}]$$

Definition: Two coherent maps $f,f':\mathcal{X}\to\mathcal{Y}$ are homotopic if they have coherent shifts which are level homotopic.

Formally, we put $p[f,f]=1_{\mathcal{X}_f}$. In this way if two maps are level homotopic they are homotopic.

Theorem 3.2

If B is cofinite, then the relation of homotopy of coherent maps $f:\mathcal{X}\to\mathcal{Y}=(Y_b,q_{\underline{b}},B)$ is an equivalence relation. (The homotopy class of a f is denoted by $[f]$).

Proof: Reflexivity and symmetry are obvious. We prove transitivity. Let $f,f',f'':\mathcal{X}\to\mathcal{Y}$ be coherent maps. Let $F:I\times\mathcal{X}\to\mathcal{Y}$ be a homotopy connecting $f\circ p[f,F]$ and $f'\circ p[f',F]$, and $F':I\times\mathcal{X}\to\mathcal{Y}$ be a homotopy connecting $f'\circ p[f',F]$ and $f''\circ p[f'',F]$.

We define by induction on $L(b) = \max\{n \mid b_0 < \ldots < b_n = b,\ b_0 \in B, \ldots, b_n \in B\}$ a strictly increasing function $\mathcal{F} : B \to A$ such that $\mathcal{F} > F$ and $\mathcal{F} > F'$.

If $b \in B$ is such that $L(b) = 0$, then there exists an index $\mathcal{F}(b)$ in A such that $\mathcal{F}(b) > F(b), F'(b)$.

Let $\mathcal{F}(b)$ be defined for all $b \in B$ with $L(b) = 0, 1, \ldots, n-1$. Now let $b \in B$ with $L(b) = n$, and let $\{b_1, \ldots, b_r\}$ be all the predecessors of b. Then there exists an index $\mathcal{F}(b)$ in A such that $\mathcal{F}(b) > F(b), F'(b)$ and $\mathcal{F}(b) > \mathcal{F}(b_1), \ldots, \mathcal{F}(b_r)$.

By previous corollary $f \circ p[f, \mathcal{F}]$ and $f' \circ p[f', \mathcal{F}]$ are level homotopic, and $f' \circ p[f', \mathcal{F}]$ and $f'' \circ p[f'', \mathcal{F}]$ are level homotopic. It follows that f and f'' are homotopic.

Remark: In this theorem the first time was used that the directed set B has to be cofinite.

The next theorem is proved in [13]

Theorem 3.3

Let $f : \mathcal{X} \to \mathcal{Y}$ be a coherent map, $g, G : C \to B$ be strictly increasing functions such that $g < G$. Then the composition of the level maps $f : \mathcal{X}_{fg} \to \mathcal{Y}_g$ and $p[fg, fG] : \mathcal{X}_{fG} \to \mathcal{X}_{fg}$ is level homotopic with the composition of the level maps $q[g, G] : \mathcal{Y}_G \to \mathcal{Y}_g$ and $f : \mathcal{X}_{fG} \to \mathcal{Y}_G$ i.e. $f \circ p[fg, fG] : \mathcal{X}_{fG} \to \mathcal{Y}_g$ and $q[g, G] \circ f : \mathcal{X}_{fG} \to \mathcal{Y}_g$ are level homotopic.

Theorem 3.4

If $f, f' : \mathcal{X} \to \mathcal{Y}$ are homotopic and $g, g' : \mathcal{Y} \to \mathcal{Z}$ are homotopic then the maps $gf, g'f' : \mathcal{X} \to \mathcal{Z}$ are homotopic.

Proof: Part 1:

(1.1) If $f, f' : \mathcal{X} \to \mathcal{Y}$ are homotopic, i.e. the coherent maps $f \circ p[f, F] : \mathcal{X} \to \mathcal{Y}$ and $f' \circ p[f', F] : \mathcal{X} \to \mathcal{Y}$ are level homotopic with a homotopy $F : I \times \mathcal{X} \to \mathcal{Y}$, then the coherent map $gF : I \times \mathcal{X} \to \mathcal{Z}$ is a level homotopy between coherent maps $g(f \circ p[f, F])$ and $g(f' \circ p[f', F])$.

Coherent map $g(f \circ p[f, F])$ corresponds to the level map obtained as composition of the level maps: $(g_{\underline{c}}^{j}) : \mathcal{Y}_g \to \mathcal{Z}$, $(f_{\underline{c}}^{j}) : \mathcal{X}_{fg} \to \mathcal{Y}_g$ and $(p[fg, Fg]_{\underline{c}}^{j}) : \mathcal{X}_{Fg} \to \mathcal{X}_{fg}$ i.e. to the level map $g \circ (f \circ p[fg, Fg]) : \mathcal{X}_{Fg} \to \mathcal{Z}$

The coherent map $(gf)\circ p[fg,Fg]$ (a shift of gf) also corresponds to composition of these level maps i.e. to the level map $(g\circ f)\circ p[fg,Fg]:\mathcal{X}_{Fg}\to\mathcal{Z}$.

(1.2) From Theorem 1.2 it follows that level maps $g\circ(f\circ p[f,F])$ and $(g\circ f)\circ p[fg,Fg]$ are level homotopic.

(1.3) In the same way, the level maps $g\circ(f'\circ p[f',F])$ and $(g\circ f')\circ p[f'g,Fg]$ are level homotopic.

From (1.1), (1.2) and (1.3) it follows the level maps $(g\circ f)\circ p[fg,Fg]$ and $(g\circ f')\circ p[f'g,Fg]$ are level homotopic.

Then the coherent maps $(gf)\circ p[fg,Fg]$ and $(gf')\circ p[f'g,Fg]$ are level homotopic i.e. the coherent maps gf and gf' are homotopic.

To prove that coherent maps gf and $g'f'$ are homotopic, it remains to prove that gf' and $g'f'$ are homotopic.

Part 2:

(2.1) Let $g,g':\mathcal{Y}\to\mathcal{Z}$ be homotopic, i.e. the coherent maps $g\circ q[g,G]:\mathcal{Y}\to\mathcal{Z}$ and $g'\circ q[g',G]:\mathcal{Y}\to\mathcal{Z}$ are level homotopic with a homotopy $G:I\times\mathcal{Y}\to\mathcal{Z}$. Then the coherent map $G(1\times f'):I\times\mathcal{X}\to\mathcal{Z}$ (given by strictly increasing function $f'G:C\to A$) is a homotopy between coherent maps $(g\circ q[g,G])f'$ and $(g'\circ q[g',G])f'$.

Coherent map $(g\circ q[g,G])f'$ coresponds to the composition $(g\circ q[g,G])\circ f':\mathcal{X}_{f'G}\to\mathcal{Z}$ of the following level maps: $g:\mathcal{Y}_g\to\mathcal{Z}$, $q[g,G]:\mathcal{Y}_G\to\mathcal{Y}_g$ and $f':\mathcal{X}_{f'G}\to\mathcal{Y}_G$.

The coherent map $g(q[g,G]\circ f')$ also coresponds to the composition $g\circ(q[g,G]\circ f')$

(2.2) From Theorem 1.2, level maps $(g\circ q[g,G])\circ f'$ and $g\circ(q[g,G]\circ f')$ are level homotopic.

(2.3) From Theorem 3.3 $g\circ(q[g,G]\circ f')\cong g\circ(f'\circ p[f'g,f'G])$

(2.4) As in Part 1 of this proof $g\circ(f'\circ p[f'g,f'G])\cong(g\circ f')\circ p[f'g,f'G]$

(2.5) From (2.2), (2.3), (2.4) it follows that $(g\circ q[g,G])\circ f'$ and $(g\circ f')\circ p[f'g,f'G])$ are level homotopic i.e the coherent maps $(g\circ q[g,G])f'$ and $(gf')\circ p[f'g,f'G]$ are level homotopic.

(2.6) In the same way it is proved that $(g'\circ q[g,G])f'$ and $(g'f')\circ p[f'g',f'G]$ are level homotopic.

From (2.1), (2.5) and (2.6) it follows that coherent maps $(gf')\circ p[f'g,f'G]$ and $(g'f')\circ p[f'g',f'G]$ are level homotopic i.e. the

All coherent inverse systems with cofinite directed sets and homotopy classes of coherent maps form the coherent category Coh.

Example: (Special cases of coherent maps) If one of the coherent systems is commutative we can simplify the definition of coherent map and interpret the coherent map as cubical coherent map [10]. A cubical coherent map $(f_{\underline{b}}):\underline{X}\to\mathcal{Y}=(Y_b,q_{\underline{b}},B)$, consists of a strictly increasing function $f:B\to A$, and of maps $f_{\underline{b}}:I^n\times X_{f(b_n)}\to Y_{b_0}$ satisfying the following boundary conditions $f_{\underline{b}}(s_1,\dots,e_j,\dots,s_n,x)=\begin{cases} q_{b_0\dots b_j}(s_1,\dots,s_{j-1},f_{b_j\dots b_n}(s_{j+1},\dots,s_n,x)), & e_j=1\\ f_{b_0\dots\hat{b}_j\dots b_n}(s_1,\dots,\hat{s}_j,\dots,s_n,x), & e_j=0,1\le j\le n-1\\ f_{b_0\dots b_{n-1}}(s_1,\dots,s_{n-1},p_{f(b_{n-1}b_n)}(x)), & e_n=0\end{cases}$

A cubical coherent map $(f_{\underline{b}}):\mathcal{X}\to\underline{Y}$, consists of a strictly increasing function $f:B\to A$, and of maps $f_{\underline{b}}:I^n\times X_{f(b_n)}\to Y_{b_0}$ satisfying the following boundary conditions $f_{\underline{b}}(s_1,\dots,e_j,\dots,s_n,x)=\begin{cases} f_{b_0\dots b_j}(s_1,\dots,s_{j-1},p_{f(b_j\dots b_n)}(s_{j+1},\dots,s_n,x)), & e_j=1\\ f_{b_0\dots\hat{b}_j\dots b_n}(s_1,\dots,\hat{s}_j,\dots,s_n,x), & e_j=0,1\le j\le n-1\\ q_{b_0b_1}(s_1,\dots,s_{n-1},p_{f(b_{n-1}b_n)}(x)), & e_n=0\end{cases}$.

4. Relations of Coherent Categories

Any commutative inverse system $\underline{X}=(X_a,p_{a_0a_1},A)$ can be interpreted as coherent inverse system. $\mathcal{X}=(X_a,p_{a_0\dots a_n},A)$ by putting $p_{a_0\dots a_n}(\tau_1,\dots,\tau_{n-1},x)=p_{a_0a_n}(x)$.

We define the functor $\mathcal{F}$:CPHTop→Coh associating to an inverse system $\underline{X}$, its interpretation as a coherent inverse system. $\mathcal{X}$.

To a coherent map $(f_{\underline{b}}):\underline{X}\to\underline{Y}$ given by maps $f_{\underline{b}}:\Delta^n\times X_{f(b_n)}\to Y_{b_0}$, we associate the coherent map $(f_{\underline{b}}^{j}):\mathcal{X}\to\mathcal{Y}$ given by maps $(f_{\underline{b}}^{j}):I^{n-k}\times\Delta^k\times X_{f(b_n)}\to Y_{b_0}$ defined by $f_{\underline{b}}^{j}(\tau,t_1,\dots,t_k,x)=f_{\underline{b}}(t_1',\dots,t_n',x)$ where $t_{j_i}'=t_i, i=1,2,\dots,k$ $t_j'=0, j_i<j<j_{i+1}$.

Now we define the functor $\mathcal{F}$:CPHTop→Coh on morphisms associating to the homotopy class of $(f_{\underline{b}}):\underline{X}\to\underline{Y}$ the homotopy class of $(f_{\underline{b}}^{j}):\mathcal{X}\to\mathcal{Y}$ i.e. $\mathcal{F}[(f_{\underline{b}})]=[(f_{\underline{b}}^{j})]$

To show that this associating is well defined let $(f_{\underline{b}}),(f'_{\underline{b}}):\underline{X}\to\underline{Y}$ be homotopic by a homotopy $(F_{\underline{b}}):I\times\underline{X}\to\underline{Y}$. Let $(f_{\underline{b}}^{j}),(f'^{j}_{\underline{b}}):\mathcal{X}\to\mathcal{Y}$ and $(F_{\underline{b}}^{j}):I\times\mathcal{X}\to\mathcal{Y}$ be corresponding associated maps.

Let $(H_{\underline{b}}^{j}):I\times\underline{X}\to\underline{Y}$ be a coherent map given by the function $F:B\to A$ and by maps $H_{\underline{b}}^{j}:I^{n-k}\times\Delta^{k}\times I\times X_{F(b_n)}\to Y_{b_0}$ defined for $(t_1,\ldots,t_k)=t\in K_m$ by $H_{\underline{b}}^{j}(\tau,t,s,x)=F_{b_0\ldots b_{j_m}}^{j_0\ldots j_m}(\tau_{0,m},2t_1-1,\ldots,2t_m-1,s,p_{F(b_{j_m}b_n)}(x))$.

To prove that $(H_{\underline{b}}^{j})$ connects the maps $f=(f_{\underline{b}}^{j})$ and $f'=(f'^{j}_{\underline{b}})$, first we mention that for $(\tau,t)\in I^{n-k}\times\Delta^{k}$ we have $p[f,F]_{b_0\ldots b_n}^{j_0\ldots j_k}(\tau,t,x)=p_{f(b_0)F(b_n)}(x)$.

For $(t_1,\ldots,t_k)=t\in K_m$,

$$\begin{aligned}H_{\underline{b}}^{j}(\tau,t,0,x)&=F_{b_0\ldots b_{j_m}}^{j_0\ldots j_m}(\tau_{0,m},2t_1-1,\ldots,2t_m-1,0,p_{F(b_{j_m}b_n)}(x))\\&=f_{b_0\ldots b_{j_m}}^{j_0\ldots j_m}(\tau_{0,m},2t_1-1,\ldots,2t_m-1,p_{f(b_{j_m})F(b_{j_m})}p_{F(b_{j_m}b_n)}(x))\text{ and}\\&=f_{b_0\ldots b_{j_m}}^{j_0\ldots j_m}(\tau_{0,m},2t_1-1,\ldots,2t_m-1,p_{f(b_{j_m})F(b_n)}(x))\\&=(f\circ p[f,F])_{\underline{b}}^{j}(\tau,t,x)\end{aligned}$$

similarly $H_{\underline{b}}^{j}(\tau,t,1,x)=(f'\circ p[f',F])_{\underline{b}}^{j}(\tau,t,x)$.

That $\mathcal{F}$: CPHTop→Coh is a functor it is proved in [9], Section 2, and in [11] it is proven the following theorem.

Theorem 4.1

The functor $\mathcal{F}$: CPHTop→Coh is an embedding of CPHTop in Coh.

Theorem 4.2

CPHTop is a full subcategory of Coh

Proof: We mention that for the proof only the level homotopy of coherent maps is necessary i.e. Theorems 1.1 – 1.4.

To prove the theorem, suppose that the commutative inverse systems $\underline{X}$ and $\underline{Y}$, are interpreted as coherent systems $\mathcal{X}$ and $\mathcal{Y}$ respectively, and $(f_{\underline{b}}^{j}):\mathcal{X}\to\mathcal{Y}$ is an arbitrary coherent map in the category Coh. We have to find a special coherent map $(f'_{\underline{b}}):\underline{X}\to\underline{Y}$ such that it is an image of some coherent map $(f'^{j}_{\underline{b}}):\mathcal{X}\to\mathcal{Y}$ and if and $\mathcal{F}(f'_{\underline{b}})=(f'^{j}_{\underline{b}})$ and $(f_{\underline{b}}^{j})$ are homotopic in the category Coh..

At the same time we will define the desired special coherent map $(f'_{\underline{b}}):\underline{X}\to\underline{Y}$ and the level homotopy $(F_{\underline{b}}^{j}):I\times\mathcal{X}\to\mathcal{Y}$, both defined with same function $f:B\to A$, connecting the coherent maps $(f_{\underline{b}}^{\underline{j}})$ and $\mathcal{F}(f'_{\underline{b}})=(f_{\underline{b}}'^{\underline{j}})$ i.e. $F_{\underline{b}}^{\underline{j}}(\tau,t,0,x)=f_{\underline{b}}^{\underline{j}}(\tau,t,x)$, $F_{\underline{b}}^{\underline{j}}(\tau,t,1,x)=f_{\underline{b}}'^{\underline{j}}(\tau,t,x)$

The construction of maps $F_{\underline{b}}^{\underline{j}}$ and $f'_{\underline{b}}$ is made by induction on n, the length of $\underline{b}=(b_0,b_1,\ldots,b_n)$. For $n=0$ and $b_0\in B$, we put $f'_{b_0}=f_{b_0}$ and we define $F_{b_0}:I\times X_{f(b_0)}\to Y_{b_0}$ by $F_{b_0}(s,x)=f'_{b_0}(x)$.

For $n=1$ and $b_0<b_1$, we put $f'_{b_0b_1}=f_{b_0b_1}$ and we define $F_{b_0b_1}:\Delta^1\times I\times X_{f(b_1)}\to Y_{b_0}$ by $F_{b_0b_1}(t,s,x)=f'_{b_0b_1}(t,x)$.

To define the maps $F_{\underline{b}}^{\underline{j}}$ and $f'_{\underline{b}}$ for $n\geq 2$ we put $\partial^0 I^n=\bigcup_{j=1}^{n}\{(\tau_1,\ldots,\tau_{j-1},0,\tau_{j+1},\ldots,\tau_{n-1})\,|\,0\leq\tau_i\leq 1\}$.

For $1\leq k\leq n-1$ by Q_k^n is denoted the following subset of $I^{n-k}\times\Delta^k\times I$, $Q_k^n=(\partial^0 I^{n-k}\times\Delta^k\times I)\cup(I^{n-k}\times\partial\Delta^k\times I)\cup(I^{n-k}\times\Delta^k\times\{0,1\})$

By R_k^n we denote the retraction $R_k^n:I^{n-k}\times\Delta^k\times I\to Q_k^n$ and by r_k we denote the retraction $r_k:\Delta^k\times I\to(\partial\Delta^k\times I)\cup(\Delta^k\times 0)$.

Now, suppose that $F_{\underline{b}}^{\underline{j}}$ and $f'_{\underline{b}}$ are defined for all for all sequences $\underline{b}$ with length $<n$. To the end of this proof $\underline{b}=(b_0,b_1,\ldots,b_n)$ will be a fixed sequence with length n.

The maps $F_{\underline{b}}^{\underline{j}}:I^{n-k}\times\Delta^k\times I\times X_{f(b_n)}\to Y_{b_0}$ are defined in the following way:

a) on $\partial^0 I^{n-k}\times\Delta^k\times I\times X_{f(b_n)}$ by $F_{\underline{b}}^{\underline{j}}(\tau_1,\ldots,\tau_{j-1},0,\tau_{j+1},\ldots,\tau_{n-k},t,s,x)=F_{b_0\ldots\hat{b}_j\ldots b_n}^{j_0\ldots j_i j_{i+1}-1\ldots j_k-1}(\tau_1,\ldots,\hat{\tau}_j,\ldots,\tau_{n-k},t,s,x)$

b) on $I^{n-k}\times\partial\Delta^k\times I\times X_{f(b_n)}$ will be defined only partially by

$F_{\underline{b}}^{\underline{j}}(\tau,1,t_2,\ldots,t_k,s,x)=q_{b_0b_{j_1}}F_{b_{j_1}\ldots b_n}^{j_1-j_1\ldots j_k-j_1}(\tau,t_2,\ldots,t_k,s,x)$,

$F_{\underline{b}}^{\underline{j}}(\tau,t_1,\ldots,t_{k-1},0,s,x)=F_{b_0\ldots b_{j_{k-1}}}^{j_0\ldots j_{k-1}}(\tau,t_1,\ldots,t_{k-1},0,s,p_{f(b_{j_{k-1}}b_n)}(x))$

c) on $I^{n-k}\times\Delta^k\times\{0,1\}\times X_{f(b_n)}$ by $F_{\underline{b}}^{\underline{j}}(\tau,t,0,x)=f_{\underline{b}}^{\underline{j}}(\tau,t,x)$ and for all $\underline{j}$, except for $\underline{j}=(0,1,\ldots,n)$, by $F_{\underline{b}}^{\underline{j}}(\tau,t,1,x)=f'_{b_{j_0}b_{j_1}\ldots b_{j_k}}(t,x)$.

Further on, the definition of maps $F_{\underline{b}}^{\underline{j}}$ will be completed by induction on k, the length of $\underline{j}=(j_0,\ldots,j_k)$. For $k=1$, i.e. $\underline{j}=(0,n)$, the definition under c) is complete and a map $F_{\underline{b}}^{0,n}:Q_1^n\times X_{f(b_n)}\to Y_{b_0}$ is completely defined. Using the retraction

$R_1^n : I^{n-1} \times \Delta^1 \times I \to Q_1^n$, this map can be extended to a map $H_{\underline{b}}^{0,n} : I^{n-1} \times \Delta^1 \times I \times X_{f(b_n)} \to Y_{b_0}$.

Now, let maps $H_{\underline{b}}^{\underline{j}}$ be defined for all $\underline{j}$ with length $< k$. If $\underline{j}$ has length k and $k < n$, then one can complete the definition in b) i.e. on $I^{n-k} \times \partial\Delta^k \times I \times X_{f(b_n)}$. For $t \in \partial\Delta^k$ with $t_i = t_{i+1}$, $1 \le i \le k-1$ we put $F_{\underline{b}}^{\underline{j}}(\tau, t, s, x) = F_{b_0 \dots b_n}^{j_0 \dots \hat{j}_i \dots j_k}(\tau_{0,i}, 1, \tau_{i,k}, t_1, \dots, \hat{t}_i, \dots, t_k, s, x)$. (*)

By this formula and by a), b) and c) the map $F_{\underline{b}}^{\underline{j}} : Q_k^n \times X_{f(b_n)} \to Y_{b_0}$ is defined and using the retraction $R_k^n : I^{n-k} \times \Delta^k \times I \times I \to Q_k^n$ this map can be extended to a map $F_{\underline{b}}^{\underline{j}} : I^{n-k} \times \Delta^k \times I \times X_{f(b_n)} \to Y_{b_0}$.

In this way, a map $F_{\underline{b}}^{\underline{j}}$ is defined for all $\underline{j}$ with length $< n$, i.e. except for $\underline{j} = (0,1,\dots,n)$.

Finally for $\underline{j} = (0,1,\dots,n)$, by b) and c) and (*) a map $F_{\underline{b}}^{012\dots n} : (\partial\Delta^n \times I) \cup (\Delta^n \times 0) \times X_{f(b_n)} \to Y_{b_0}$ is defined. By use of the retraction $r_n : \Delta^n \times I \to (\partial\Delta^n \times I) \cup (\Delta^n \times 0)$ this map can be extended to a map $F_{\underline{b}}^{012\dots n} : (\Delta^n \times I) \times X_{f(b_n)} \to Y_{b_0}$.

Then the map $f'_{b_0 b_1 \dots b_n} : \Delta^n \times X_{f(b_n)} \to Y_{b_0}$ is defined by $f'_{b_0 b_1 \dots b_n}(t, x) = F_{b_0 b_1 \dots b_n}^{012\dots n}(t, 1, x)$.

The maps $(F_{\underline{b}}^{\underline{j}})$ and $(f'_{\underline{b}})$ are completely defined and from c) $F_{\underline{b}}^{\underline{j}}(\tau, t, 1, x) = f'_{b_{j_0} b_{j_1} \dots b_{j_k}}(t, x) = f'^{\underline{j}}_{\underline{b}}(\tau, t, x)$

It follows that the coherent maps $\mathcal{F}(f'_{\underline{b}}) = (f'^{\underline{j}}_{\underline{b}})$ and $(f_{\underline{b}}^{\underline{j}})$ are level homotopic.

The detailed verification that maps are well defined, that $(f'_{\underline{b}})$ is a special coherent map and that $(F_{\underline{b}}^{\underline{j}})$ is a coherent map is given in [9].

APPENDIX: STRICT ORDERING VS ORDERING FOR DIRECTED SETS

More often, for directed set instead of strct ordering $(A, <)$ it is used ordering $(A, \le)$. Then $(A, \le)$ is directed set if for any $a_0, a_1 \in A$ there exists $a_2 \in A$ such that $a_0, a_1 \le a_2$. $(A, \le)$ will be cofinite i.e. each $a \in A$ has only a finite number of predecessors.

For the appendix we will consider commutative inverse systems over directed sets $(A, \le)$, $\underline{X} = (X_a, p_{a_0 a_1}, A)$ i.e. for $a_0 \le a_1 \le a_2$ the equality $p_{a_0 a_1} p_{a_1 a_2} = p_{a_0 a_2}$ holds, and $p_{aa} = 1_{X_a}$.

In the category pro-Top, a map of inverse systems $(f_b): \underline{X} \to \underline{Y} = (Y_b, q_{b_0 b_1}, B)$ consists of an increasing function $f: B \to A$ and for any $b \in B$ of a map $f_b: X_{f(b)} \to Y_b$ such that and for any pair $b_0 \le b_1$ holds $q_{b_0 b_1} f_{b_1}(x) = f_{b_o} p_{f(b_0 b_1)}$

A morphism in pro-Top is the equivalence class of a map of inverse systems $(f_b): \underline{X} \to \underline{Y}$ defined by: (f_b) and $(f_b'): \underline{X} \to \underline{Y}$ are in the same class if there exists an increasing function $F: B \to A$, such that $F \ge f$, $F \ge f'$ and $f_b p_{f(b)F(b)} = f_b' p_{f'(b)F(b)}$.

If we work with the ordering $\le$, as in [7] and [4], for construction of coherent category of commutative inverse systems we have to consider increasing sequences of indices and increasing fiunctions (instead of strictly increasing sequences of indices and strictly increasing functions). Then beside the coherence conditions for maps $f_{\underline{b}}: \Delta^n \times X_{f(b_n)} \to Y_{b_0}$ we have to add a "degeneracy condition" for repeating of indices i.e if $\underline{b} = (b_o, \ldots, b_{n-1})$ and $\underline{b}^j = (b_o, \ldots, b_j, b_j, \ldots, b_{n-1})$ then $f_{\underline{b}^j}(t_1, \ldots, t_n, x) = f_{\underline{b}}(t_1, \ldots, \hat{t}_j, \ldots, t_n, x)$

The obtained category we denote with CPHTop $_\le$. To the end we will show that CPHTop $_\le$ and CPHTop are naturally isomorphic [8]

Theorem 1

Suppose $\underline{X} = (X_a, p_{a_0 a_1}, A)$ is a commutative inverse system consisting of topological spaces X_a and of bonding maps $p_{a_0 a_1}: X_{a_1} \to X_{a_0}$ for any pair $a_0 < a_1$.

(i) there is an inverse system $\underline{X'} = (X_a', p_{a_0' a_1'}, A')$ such that $\underline{X}$ is a subsystem of $\underline{X'}$ and A' is a directed set without a maximal element.

(ii) The restriction morphism of inverse systems $(r_a): \underline{X'} \to \underline{X}$ presented by the function $r: A \to A'$, $r(a) = a$ and $r_a: X_a \to X_a$, $a \in A$ is an isomorphism in pro-Top

(iii) The special coherent restriction map $(r_{\underline{a}}): \underline{X'} \to \underline{X}$ presented by the function $r: A \to A'$, $r(a) = a$ and by the maps $r_{\underline{a}}: \Delta^n \times X_{a_n} \to X_{a_0}$ is an isomorphism in CPHTop $_\le$

Proof:

(i) If there is no maximal element in A we put $\underline{X'} = \underline{X}$. Now, suppose that A has a maximal element (unique or not). We choose one maximal element $a_{\max}$. We put $A' = A \cup Z^+$, where $Z^+ = \{1,2,3,\ldots\}$, Z^+ is ordered in a usual way and for

$a \in A$, $z \in Z^+$ we put $a \leq z$. For all $z \in Z^+$ we put $X_z = X_{a_{\max}}$. For $a \in A$, $z, z' \in Z^+$, $z \leq z'$ we put $p_{zz'} = 1_{a_{\max}}$, $p_{az} = p_{aa_{\max}}$

It is almost clear, that in this way an inverse system $\underline{X}' = (X'_a, p_{a'_0 a'_1}, A')$ is defined, $\underline{X}$ is a subsystem of $\underline{X}'$ and A' is a directed set without a maximal element.

(ii) We define map of inverse systems $(r'_{a'}) : \underline{X} \to \underline{X}'$ presented by a function $r' : A' \to A$, $r'(a) = a$ for $a \in A$, $r'(z) = a_{\max}$ for $z \in Z^+$ and by maps $r'_a = 1_{X_a}$, for $a \in A$, and $r'_z = 1_{X_{a_{\max}}}$ for $z \in Z^+$. Then $(r'_{a'})(r_a) = 1_{\underline{X}}$ and $(r_a)(r'_{a'}) = 1_{\underline{X}'}$ i.e. (r_a) is an isomorphism in pro-Top.

(iii) We define a special coherent map $r' : \underline{X} \to \underline{X}'$ given by the function $r' : A' \to A$ and for each increasing sequence $(a'_0, \ldots, a'_n) = \underline{a}'$, $n \geq 0$ by a map

$$r'_{\underline{a}'} : \Delta^n \times X_{a'_n} \to X_{a'_0} \text{ defined by } r'_{a'}(t,x) = \begin{cases} p_{a'_0 a'_n}(x), & a'_0, a'_n \in A \\ p_{a'_0 a_{\max}}(x), & a'_0 \in A, a'_n \in Z^+ \\ 1_{a_{\max}}(x), & a'_0, a'_n \in Z^+ \end{cases}$$

Now, $rr' = 1_{\underline{X}}$ and $r'r$ is specially coherently homotopic to $1_{\underline{X}'}$, so the class of spetial coherent homotopy of r is an isomorphism in CPHTop $_{\leq}$.

Remark . To every morphism $\underline{f} : \underline{X} \to \underline{Y}$ in pro-Top we associate a morphism $\underline{f}' : \underline{X}' \to \underline{Y}'$ between inverse systems without maximal element in the following way: Let (f_b) be a representative of $\underline{f}$. We define the function $f' : B' \to A'$ by $f'(b) = f(b)$ for $b \in B$ and $f'(z) = f(b_{\max})$ for $z \in Z^+$, where $b_{\max}$ is the chosen maximal element of B. We put $f'_b = f_b$ for $b \in B$ and $f_z = f_{b_{\max}}$ for $z \in Z^+$. Now f' is the class of the map (f'_b) between inverse systems.

It is almost clear that in this way we obtain a functor pro-Top→pro-Top which is naturally equivalent to the identity functor 1: pro-Top→pro-Top, i.e. the following diagram is commutative in pro-Top

$$\begin{array}{ccc} \underline{X}' & \xrightarrow{\underline{f}'} & \underline{Y}' \\ \downarrow & & \downarrow \\ \underline{X} & \xrightarrow[\underline{f}]{} & \underline{Y} \end{array}$$

Also, every special coherent map $f:\underline{X}\to\underline{Y}$ we can associate a special coherent map $f':\underline{X'}\to\underline{Y'}$ ($\underline{X'},\underline{Y'}$ are inverse systems without maximal element which correspond to $\underline{X},\underline{Y}$, respectively). Let f be given by function ϕ and by maps $f_{\underline{b}}$.Then map f' is given by the function ϕ' and for every $(b'_0,\dots,b'_n)=\underline{b'}\in B'^n$ by a map $f'_{\underline{b'}}:\Delta^n\times X_{\phi'(b'_n)}\to Y$

defined by $$f'_{\underline{b'}}(t,x)=\begin{cases} f_{\underline{b'}}(t,x), & b'_0,b'_n\in B\\ f_{b'_0\dots b_{\max}}(t_0,\dots,t_{j'},x), & b'_{j-1},b'_j\in Z^+\\ f_{b_{\max}}(x), & b'_0,b'_n\in Z^+\end{cases}.$$

It is almost clear that in this way we obtain a functor CPHTop $_{\le}\to$ CPHTop $_{\le}$ which is naturally equivalent to the identity functor 1: CPHTop $_{\le}\to$ CPHTop $_{\le}$, i.e. the following diagram commutes up to special coherent homotopy

$$\begin{array}{ccc} \underline{X'} & \xrightarrow{\quad f'\quad} & \underline{Y'}\\ \downarrow & & \downarrow\\ \underline{X} & \xrightarrow[\quad f\quad]{} & \underline{Y}\end{array}$$

This remark and Theorem 1 shows that it is sufficient to consider only inverse systems over directed set without maximal elements.

Theorem 2

Let $\underline{X}=(X_a,p_{a_0a_1},A)$ and $\underline{Y}=(Y_b,q_{b_0b_1},B)$ be inverse systems of topological spaces and maps. Let A be without maximal elements and let $f:B\to A$ be an increasing function.

(i) There is a strictly increasing function $F:B\to A$ such that $F(b)>f(b),\ b\in B$ and if $b\ge b', b\le b$ then $F(b)=F(b')$;

(ii) If $(f_b):\underline{X}\to\underline{Y}$ is a map of systems given by function $f:B\to A$ then there is a map of systems $(F_b):\underline{X}\to\underline{Y}$ given by function $F:B\to A$ such that (F_b) and (f_b) represent the same morphism in pro-Top;

(iii) If $(f_{\underline{b}}):\underline{X}\to\underline{Y}$ is special coherent map (in CPHTop $_{\le}$) then there is a special coherent map $(F_{\underline{b}}):\underline{X}\to\underline{Y}$ such that $(F_{\underline{b}})$ is specially coherently homotopic (in CPHTop $_{\le}$) to $(f_{\underline{b}})$.

Proof.

(i) We define equivalence relation $\sim$ by $b \sim b'$ if $b \geq b'$ and $b' \geq b$. As always B is cofinite, so every equivalence class contains only a finite number of elements. For $b \in B$ we put $\eta(b) = \max\{n \mid b_0 < \ldots < b_{n-1} < b\}$.

Then $\eta(b) = \eta(b')$ if $b \sim b'$.

If b is an element of B with $\eta(b) = 0$ we can choose $F(b)$ such that $F(b) > f(b)$ and if $b \sim b', F(b) = F(b')$. Now suppose that $F(b)$ is defined for all b with $\eta(b) \leq n - 1$.

Let b be an element of B with $\eta(b) = n$. There are only finitely many elements of B which are predecessors of b, say $b_1, \ldots, b_k$. We choose $\eta(b)$ such that $\eta(b) > f(b), F(b_1), \ldots, F(b_k)$ and if $b \sim b'$ then $F(b) = F(b')$.

(ii) We put $F_b = f_b p_{f(b)F(b)}$.

Then (F_b) and (f_b) represent the same morphism in pro-Top because the following diagram commutes $X_{F(b)}$

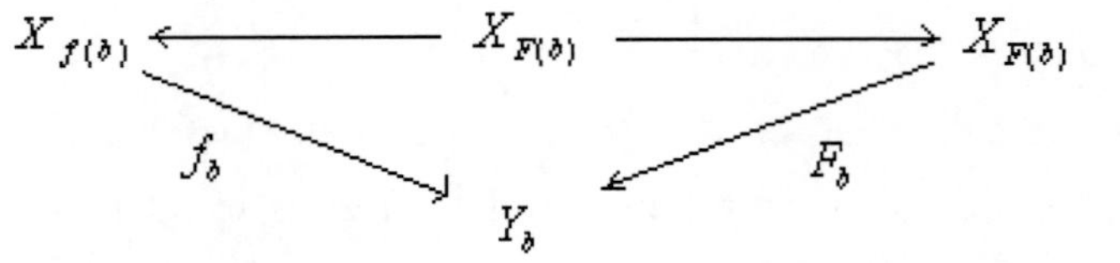

(iii) We define a map $F_{\underline{b}} : \Delta^n \times X_{F(b_n)} \to Y_{b_0}$ by $F_{\underline{b}}(t, x) = f_{\underline{b}}(t, p_{f(b_n)F(b_n)}(x))$

Then $(F_{\underline{b}})$ is a special coherent map which is specially coherently homotopic to $(f_{\underline{b}})$.

We denote by $A^n_{\leq} = \{\underline{a} \mid \underline{a} = (a_0, a_1, \ldots, a_n), a_0 \leq a_1 \leq \ldots \leq a_n\}$, and by $A^n_{<} = \{\underline{a} \mid \underline{a} = (a_0, a_1, \ldots, a_n), a_0 < a_1 < \ldots < a_n\}$.

Here $a < a'$ means that $a \leq a'$ holds, and $a' \leq a$ does not.

Now put $A^{\infty}_{\leq} = \bigcup_{n=1}^{\infty} A^n_{\leq}$, $A^{\infty}_{<} = \bigcup_{n=1}^{\infty} A^n_{<}$

Note that $A^{\infty}_{<} = A^{\infty}_{\leq}$.

Theorem 3

If A' is cofinal in A then the restriction morphism $(r_{\underline{a}}):\underline{X}\to\underline{X}'$, $\underline{X}'=(X_a,p_{a_0a_1},A')$ is isomorphism in CPHTop (CPHTop $_\leq$)

Proof. First we consider the case of CPHTop. The restriction morphism is defined by the function $r:A'\to A$, $r(a)=a$ for all $a\in A'$ and by maps $r_{\underline{a}}:\Delta^n\times X_{a_n}\to X_{a_0}$, $\underline{a}\in A_<^n$.

A' is cofinal in A, so there is an increasing function $h':A\to A'$ such that $h'(a)\geq a$ for every $a\in A$ .Then there is a strictly increasing function $h(a)>h'(a)$ as in theorem 2 (i).

Now, for every $\underline{a}\in A_<^\infty$ we define a map $h_{\underline{a}}:\Delta^n\times X_{h(a_n)}\to X_{a_0}$ with $h_{\underline{a}}(t,x)=p_{a_0h(a_n)}(x)$.

It is almost clear that $(h_{\underline{a}}):\underline{X}'\to\underline{X}$ is a special coherent map. Moreover for composition $(h_{\underline{a}})(r_{\underline{a}})$ we have $(hr)_{\underline{a}}(t,x)=h_{a_0..a_m}(2t_1-1,...2t_m-1,r_{h(a_m..a_n)}(2t_{m+1},...2t_n,x))=p_{a_0h(a_m)}p_{h(a_ma_n)}(x)=p_{a_0h(a_n)}(x)$ for $t\in K_m$. This shows that $(h_{\underline{a}})(r_{\underline{a}})$ is specially coherently homotopic to $1:\underline{X}\to\underline{X}$ .Also, for composition $(r_{\underline{a}})(h_{\underline{a}})$ and $\underline{a}\in A_<'^n$ we have $(rh)_{\underline{a}}(t,x)=r_{a_0..a_m}(2t_1-1,...2t_m-1,h_{a_m..a_n}(2t_{m+1},...2t_n,x))=p_{a_0a_m}p_{a_ma_n}(x)=p_{a_0a_n}(x)$ where $t\in K_m$. This shows that the composition $(r_{\underline{a}})(h_{\underline{a}})$ is equal to $1:\underline{X}'\to\underline{X}'$.

The proof for CPHTop $_\leq$ is the same, we have to replace the function h, sets $A_<^n$ and $A_<'^n$ with h', $A_\leq^n$ and $A_\leq'^n$ respectively.

Let $(A,\leq)$ be a directed set. If $A'\subseteq A$ is a subset which contains exactly one element of each equivalence class of the following relation: a is in relation with a' if $a\leq a'$ and $a'\leq a$, then $(A',\leq)$ is a directed set and $\leq$ in A' is antisymmetric, so it is an ordering. Moreover A' is cofinal in A .

That is why, to define functor CPHTop $\to$ CPHTop$_\leq$ it is enough to consider inverse systems $\underline{X}=(X_a,p_{a_0a_1},A)$ over directed ordered sets A which have no maximal elements, and special coherent maps defined by strictly increasing function.

The functor CPHTop $\to$ CPHTop$_\leq$ maps the inverse system $\underline{X}=(X_a,p_{a_0a_1},A)$ (over directed set A which has no maximal element) in the same system. If $(f_{\underline{b}})$, $\underline{b}\in B_\leq^\infty$ is a special coherent map (in CPHTop$_\leq$), it is defined only with maps $(f_{\underline{b}'}),\underline{b}'\in B_<^\infty$, because for every $\underline{b}\in B_\leq^\infty\setminus B_<^\infty$, $\underline{b}=(b_0,...,b_n)$ as in [LM cohprohom GlMat] (I.6.4) there exists a unique $\underline{b}'\in B_<^\infty$ and a unique sequence $1\leq i_1\leq...\leq i_s\leq n$ such that $\underline{b}=(...((\underline{b}')^{i_1})^{i_2}...)^{i_s}$ and we obtain for $f_{\underline{b}}:\Delta^n\times X_{f(b_n)}\to Y_{b_0}$ $f_{\underline{b}}(t_1,...,t_n,x)=f_{\underline{b}'}(t_1,...,\hat{t}_{i_1},...,\hat{t}_{i_2},.......,\hat{t}_{i_s},...,t_n,x)$

The above reasons show that to every special coherent map $(f_{\underline{b}'})$, $\underline{b}'\in B_<^\infty$ (from CPHTop) we can associate a unique special coherent map $(f_{\underline{b}})$, $\underline{b}\in B_\leq^\infty$ (from CPHTop$_\leq$)

and this associating is bijective .Moreover, as homotopy, composition and the identity map are defined in the same way in $\mathrm{CPHTop}_{\leq}$ and CPHTop, we derive the conclusion that the defined functor CPHTop $\rightarrow$ $\mathrm{CPHTop}_{\leq}$, is an isomorphism.

REFERENCES

[1] J.M. Cordier and J.Porter, *Maps between homotopy coherent diagrams*, Topology and its Applications, 28 (1988), 255-275

[2] Yu.T.Lisica, *Duality theorems and dual shape and coshape categories*, Dokl. Akad. Nauk SSSR 263 (1982) 532-536

[3] Yu. Lisitsa, *On the exactness of the spectral homotopy group sequence in shape theory*, Doklady Akad. Nauk SSSR 236 (1977)

[4] Yu. Lisica, S. Mardešic, *Coherent prohomotopy and strong shape theory*, Glasnik Mat. Ser.III 19(39) (1984), 53-78

[5] Yu. Lisica, S. Mardešic, *Coherent homotopy and strong shape for compact metric spaces,* Glasnik Mat. Vol. 20(40) (1985), 159-167

[6] S. Mardešic, *Coherent homotopy and localization*, Topology and its applications 94 (1999), 253-274

[7] S. Mardešic, *Strong Shape and Homology*, Springer Verlag, 1-489, Berlin, 2000,

[8] N. Shekutkovski, *Category of commutative inverse systems is a subcategory of the category of coherent inverse systems*, Macedonian Academy of Sciences and Arts, Contributions, Sec. Math. Tech. Sci X (1989) 5-19

[9] N. Shekutkovski, *Equivalence of coherent theories*, Topology and its Applications 75 (1997) 113-123

[10] N. Shekutkovski, *Maps between inverse and coherent systems*, Macedonian Academy of Sciences and Arts, Contributions, Sec. Math. Tech. Sci XVIII, 1-2 (1997) 27-37

[11] N. Shekutkovski, *From inverse to coherent systems*, Topology and its applications 113 (2001), 293-307 '

[12] N. Shekutkovski, *A contribution to the theorem on level homotopy equivalences*, Mat. Bilten, 24(L) (2000) 111-120

[13] N. Shekutkovski, *Shift and coherent shift in inverse systems*, Topology and its Applications 140 (2004) 111-130

[14] R.M. Vogt, *A note on homotopy equivalences*, Proc. Amer. Math.Soc. 32 (1972), 627-629

In: Computational Mathematics
Editor: Peter G. Chareton, pp. 117-171

ISBN: 978-1-60876-271-2

Chapter 5

STABLE MFS-BASED SOLUTION TO SINGULAR AND NON-SINGULAR INVERSE PROBLEMS FOR TWO-DIMENSIONAL HELMHOLTZ-TYPE EQUATIONS

Liviu Marin*
Institute of Solid Mechanics, Romanian Academy, Bucharest, Romania

Abstract

We investigate the application of a meshless method for the stable and accurate solution of inverse problems associated with two-dimensional Helmholtz-type equations in domains with smooth and piecewise smooth boundaries, as well as in the presence of boundary singularities. More precisely, the governing equation and available boundary conditions are discretized by the method of fundamental solutions (MFS). It is well known that inverse problems are ill-posed and, consequently, their associated MFS system of linear algebraic equations is ill-conditioned. Therefore, the stability of the solutions to the aforementioned inverse problems is a key issue and this is usually taken into account by employing a regularization method. These difficulties are overcome in the case of the first type of inverse problems investigated herein, i.e. inverse problems for Helmholtz-type equations in two-dimensional domains without boundary singularities, by considering the Tikhonov regularization method (TRM).

The existence of boundary singularities affects adversely the accuracy and convergence of standard numerical methods. Solutions to such problems and/or their corresponding derivatives may have unbounded values in the vicinity of the singularity. In this situation, these difficulties are overcome by combining the TRM with the subtraction from the original MFS solution of the corresponding singular solutions/eigenfunctions, without an appreciable increase in the computational effort and at the same time keeping the same MFS discretization. Examples for both the Helmholtz and the modified Helmholtz equations, as well as the types of inverse problems considered, are presented and carefully investigated.

*E-mail addresses: marin.liviu@gmail.com; liviu@imsar.bu.edu.ro

1 Introduction

Numerous physical applications in engineering and science related to wave propagation and vibration phenomena are naturally modelled by Helmholtz-type equations. These equations are often used to describe the vibration of a structure [1], the acoustic cavity problem [2], the radiation wave [3], the scattering of a wave [4] and the heat conduction in fins [5]. The knowledge of the geometry of the domain of interest, the complete boundary and initial conditions, the wave number and the external sources acting in the solution domain gives rise to *direct problems* for Helmholtz-type equations which have been extensively studied in the literature, see for example [6, 7, 8]. Unfortunately, many engineering problems do not belong to this category. When one or more of these conditions for solving the direct problem are partially or entirely unknown then an *inverse problem* may be formulated to determine the unknowns from specified or measured system responses. It should be noted that most of the inverse problems are ill-posed and hence they are more difficult to solve than direct problems. It is well known that inverse problems are in general unstable [9], in the sense that small measurement errors in the input data may amplify significantly the errors in the solution. A classical example of an inverse problem for Helmholtz-type equations is the *Cauchy problem*. In this case, boundary conditions are incomplete, in the sense that a part of the boundary of the solution domain is over-specified by prescribing on it both the primary field and its normal derivative, whilst the remaining boundary is under-specified and boundary conditions on the latter boundary are unknown and have to be determined. The uniqueness of the Cauchy problem is guaranteed without the necessity of removing the eigenvalues for the Laplacian operator, as it happens in the case of direct problems for the Helmholtz equation [10]. However, the Cauchy problem suffers from the non-existence and instability of the solution.

Over the last decades, many theoretical and numerical studies have been devoted to the Cauchy problem associated with Helmholtz-type equations. Bai [11] developed a boundary element method (BEM)-based acoustic holography technique using the singular value decomposition (SVD) for the reconstruction of sound fields generated by irregularly shaped sources. The vibrational velocity, sound pressure and acoustic power on the vibrating boundary comprising an enclosed space were reconstructed by Kim and Ih [12], who used the SVD to obtain the inverse solution in the least-squares sense and express the acoustic modal expansion between the measurement and source field. A method employing the spherical wave expansion theory and a least-squares minimisation to reconstruct the acoustic pressure field from a vibrating object was developed by Wang and Wu [13], while Wu and Yu [14] extended this technique to the reconstruction of acoustic pressure fields inside the cavity of a vibrating object. DeLillo *et al.* [15, 16] detected the source of acoustical noise inside the cabin of a midsize aircraft from measurements of the acoustical pressure field inside the cabin by solving a linear Fredholm integral equation of the first kind and also extended this study to three-dimensional problems. Marin *et al.* [17, 18] solved the Cauchy problem associated with Helmholtz-type equations by employing the BEM in conjunction with an alternating iterative algorithm consisting of obtaining successive solutions to well-posed mixed boundary value problems and the conjugate gradient method, respectively. These iterative methods were compared with the Tikhonov regularization method (TRM) and SVD by Marin *et al.* [19], whilst the same authors proposed yet another iterative algorithm based on the Landweber-Fridman method in combination with the BEM [20]. Jin and Zheng [21] solved some inverse boundary value problems for the Helmholtz equation using the boundary knot method and a SVD regularization and they also extended this method to some inverse problems associated with the inhomogeneous Helmholtz equation [22]. The numerical solution for the Cauchy problem for two- and three-dimensional Helmholtz-type equations by employing the method of fundamental solutions (MFS), in conjunction with the TRM and SVD, was investigated by Marin and Lesnic [23] and Marin [24], and Jin and Zheng [25], respectively. Some spectral regularization methods and a modified TRM to stabilize the Cauchy problem for the Helmholtz equation at fixed frequency were proposed by Xiong and Fu [26], whilst Jin and Marin [27] used the plane wave method and the SVD to solve stably the same problem. Recently, Wei *et al.* [28] and Qin *et al.* [29, 30] reduced the Cauchy problem associated with Helmholtz-type equations to a moment problem and also provided an error estimate and convergence analysis for the latter.

In many problems governed by partial differential equations, boundary singularities arise when there are sharp re-entrant corners in the boundary, the boundary conditions change abruptly, or there are discontinuities in the material properties. It is well known that these situations give rise to singularities of various types and, as a consequence, the solutions to such problems and/or their corresponding derivatives may have unbounded values in the vicinity of the singularity. Singularities are known to affect adversely the accuracy and convergence of standard numerical methods, such as the BEM, finite element (FEM), finite-difference (FDM), spectral and meshless/meshfree methods. When the computed function is bounded, but has a branch point at the corner, the difficulty is not serious. Grid refinement and higher-order discretizations are common strategies aimed at improving the convergence rate and accuracy of the above-mentioned standard methods [31]. If, however, the form of the singularity is taken into account and is properly incorporated into the numerical scheme then a more effective method may be constructed.

There are important studies regarding the numerical treatment of singularities occurring in Helmholtz-type equations. Time-harmonic waves in a membrane which contains one or more fixed edge stringers or cracks were investigated by Chen *et al.* [1], who employed the dual BEM in order to obtain an efficient solution of the Helmholtz equation in the presence of geometric singularities. Huang *et al.* [3] investigated the electromagnetic field due to a line source radiating in the presence of a two-dimensional composite wedge made of a number of conducting and dielectric materials by employing the Fourier transform path integral method. A hybrid asymptotic/FEM for computing the acoustic field radiated or scattered by acoustically large objects was developed by Barbone *et al.* [4]. Chen and Chen [32] used the dual integral formulation for the Helmholtz equation to determine the acoustic modes of a two-dimensional cavity with a degenerate boundary. Schiff [33] computed the transverse electric (TE) and transverse magnetic (TM) mode eigenvalues for ridged and other waveguides by using super-elements for the FEM, a refined local mesh and basis functions at the corner tip. The method of the auxiliary mapping, in conjunction with the p-version of the FEM, was used by Cai *et al.* [34] and Lucas and Oh [35] in order to remove the pollution effect caused by singularities in the Helmholtz equation. Both Laplace- and Helmholtz-type boundary value problems with singularities were considered by Wu and Han [36], who solved these problems using the FEM and by introducing a sequence of approximations to the boundary conditions at an artificial boundary and then reducing the original problems to boundary value problems away from the singularities. Xu and Chen [37] employed the FDM and higher-order discretized boundary conditions at the edges of perfectly conducting wedges for TE waves to retrieve accurately the field behaviour near a sharp edge. Mantič *et al.* [38] investigated in a comprehensive way the singularity exponents and principal terms associated to multi-material corners in the case of anisotropic potential problems. The treatment of singularities in both isotropic and anisotropic two-dimensional Helmholtz-type equations was considered by Marin *et al.* [39], who also modified the standard BEM to account for the presence of singularities. For an excellent survey on the treatment of singularities in elliptic boundary value problems, we refer the reader to Li and Lu [40] and the references therein.

The MFS was originally introduced by Kupradze and Aleksidze [41] and numerically formulated for the first time by Mathon and Johnston [42]. The main idea of this meshless method consists of approximating the solution of the problem by a linear combination of fundamental solutions with respect to some singularities/source points located outside the domain. Consequently, the original problem is reduced to determining the unknown coefficients of the fundamental solutions and the coordinates of the singularities/source points by requiring the approximation to satisfy the boundary conditions and hence solving a nonlinear problem. This is usually achieved by considering either the static or the dynamic approach. In the first approach, the singularities/source points are pre-assigned and kept fixed throughout the solution process, which is reduced to solving a linear problem. In the dynamic approach, the singularities/source points and the unknown coefficients are determined simultaneously during the solution process, which now results in solving a system of nonlinear equations by using minimization methods. An excellent survey of the MFS over the past three decades was presented by Fairweather and Karageorghis [43].

A wide variety of direct boundary value problems were successfully approached by employing the MFS. Karageorghis and Fairweather [44] solved numerically the biharmonic equation using the MFS and later their method was modified to account for the presence of boundary singularities in both the Laplace and the biharmonic equations by Karageorghis [45] and Poullikkas *et al.* [46]. Moreover, Poullikkas *et al.* [47] investigated the numerical solution of the inhomogeneous harmonic and biharmonic equations by reducing these problems to the homogeneous corresponding cases and subtracting a particular solution of the governing equation. The MFS was formulated for three-dimensional Signorini boundary value problems and it was tested on a three-dimensional electropainting problem related to the coating of vehicle roofs in Poullikkas *et al.* [48]. Karageorghis and Fairweather [49] investigated the use of the MFS for the approximate solution of three-dimensional isotropic materials with axisymmetrical geometry and both axisymmetrical and arbitrary boundary conditions. The application of the MFS to two-dimensional problems of steady-state heat conduction and elastostatics in isotropic and anisotropic bimaterials was addressed by Berger and Karageorghis [50, 51], whilst Poullikkas *et al.* [52] employed the MFS for solving three-dimensional elastostatics problems. Recently, the MFS has been adapted and successfully applied to solving inverse problems associated with the heat equation [53, 54], linear elasticity [55, 56], steady-state heat conduction in functionally graded materials [57], Helmholtz-type equations [23, 24, 25], biharmonic equation [58, 59], source reconstruction in steady-state heat conduction problems [60] etc.

The objective of this chapter is to present and analyse the application of the MFS for obtaining accurate and stable solutions of inverse problems for Helmholtz-type equations in two-dimensional domains with smooth and piecewise smooth boundaries, as well as in the presence of boundary singularities. Since inverse problems are ill-posed, their associated MFS system of linear algebraic equations is ill-conditioned. Therefore, the stability of the solutions to the aforementioned inverse problems is a key issue and this is usually accounted for by employing a regularization method. These difficulties are overcome in the case of the first type of inverse problems investigated herein, i.e. inverse problems associated with Helmholtz-type equations in two-dimensional domains with smooth and piecewise smooth boundaries, by considering the TRM. The existence of boundary singularities affects adversely the accuracy and convergence of standard numerical methods. Solutions to such problems and/or their corresponding derivatives may have unbounded values in the vicinity of the singularity. This situation is overcome by combining the TRM with the subtraction from the original MFS solution of the corresponding singular solutions/eigenfunctions, i.e. using the so-called singularity subtraction technique (SST), without an appreciable increase in the computational effort and at the same time keeping the same MFS discretization. Examples for both the Helmholtz and the modified Helmholtz equations, as well as the types of inverse problems considered, are presented and carefully investigated. The advantages of the proposed method over other methods, such as mesh refinement in the neighbourhood of the singularity, the use of singular BEMs and/or FEMs etc., are the high accuracy which can be obtained even when employing a small number of collocation points and sources, and the simplicity of the computational scheme. A possible drawback of this method is the difficulty in extending it to deal with singularities in three-dimensional problems since such an extension is not straightforward.

2 Mathematical Formulation of the Inverse Problems

For a fixed non-zero real number $k \in \mathbb{R}$ the homogeneous Helmholtz-type equation in the two-dimensional open bounded domain $\Omega \subset \mathbb{R}^2$, with a smooth or a piecewise smooth boundary $\Gamma \equiv \partial\Omega$, reads as

$$\mathcal{L}\mathrm{u}(\mathbf{x}) \equiv \Delta\mathrm{u}(\mathbf{x}) \pm k^2\mathrm{u}(\mathbf{x}) = 0, \quad \mathbf{x} = (\mathrm{x}_1, \mathrm{x}_2) \in \Omega, \tag{1}$$

where $\Delta \equiv \partial^2/\partial\mathrm{x}_1^2 + \partial^2/\partial\mathrm{x}_2^2$ is the two-dimensional Laplace operator. Let $\mathbf{n}(\mathbf{x})$ be the outward unit normal vector at $\mathbf{x} \in \Gamma$ and denote by $\mathrm{q}(\mathbf{x}) \equiv \nabla\mathrm{u}(\mathbf{x}) \cdot \mathbf{n}(\mathbf{x})$ the normal flux associated with the potential $\mathrm{u}(\mathbf{x})$.

In the case of the Helmholtz equation, i.e. $\mathcal{L} = \Delta + k^2$, u is the acoustical field, q is the normal velocity of the sound and $k > 0$ is the frequency of the acoustical field. When $\mathcal{L} = \Delta - k^2$, the partial differential equation (1) reduces to the modified Helmholtz equation which models the heat conduction in a fin, see Kraus *et al.* [5], where u is the dimensionless local fin temperature, $k^2 = h/(k_\mathrm{f}\delta_\mathrm{f})$, h is the surface heat transfer coefficient [W/(m^2 K)], k_f is the thermal conductivity of the fin [W/(m K)] and δ_f is the half-fin thickness [m].

In the direct problem formulation, the knowledge of the potential, u, and/or the normal flux, q, on the whole boundary Γ gives the corresponding Dirichlet, Neumann, or mixed boundary conditions, which enables us to determine the potential in the entire domain Ω. If it is possible to measure both the potential, u, on $\Gamma_\mathrm{D} \subset \Gamma$ and the normal flux, q, on $\Gamma_\mathrm{N} \subset \Gamma$, such that $\Gamma_D, \Gamma_N \neq \emptyset$, $\Gamma_D \cap \Gamma_N \neq \emptyset$ and $\Gamma_D \cup \Gamma_N \subset \Gamma$, then this leads to the mathematical formulation of an inverse problem consisting of equation (1) and the boundary conditions

$$u(\mathbf{x}) = \widetilde{u}(\mathbf{x}), \quad \mathbf{x} \in \Gamma_D, \qquad q(\mathbf{x}) = \widetilde{q}(\mathbf{x}), \quad \mathbf{x} \in \Gamma_N, \tag{2}$$

where $\widetilde{u}$ and $\widetilde{q}$ are prescribed functions on Γ_D and Γ_N, respectively. Moreover, we also assume that the input Dirichlet, $\widetilde{u}|_{\Gamma_D}$, and Neumann, $\widetilde{q}|_{\Gamma_N}$, data have been perturbed as

$$\widetilde{u}^{\varepsilon}|_{\Gamma_D} = \widetilde{u}|_{\Gamma_D} + \delta u, \qquad \widetilde{q}^{\varepsilon}|_{\Gamma_N} = \widetilde{q}|_{\Gamma_N} + \delta q, \tag{3}$$

where $\delta\widetilde{u}$ and $\delta\widetilde{q}$ are Gaussian random variables with mean zero and standard deviations $\sigma_u = \max_{\Gamma_D} |u| \times (p_u/100)$ and $\sigma_q = \max_{\Gamma_N} |q| \times (p_q/100)$, respectively, generated by the NAG subroutine G05DDF, and p_u and p_q are the percentages of additive noise included into the input data $u|_{\Gamma_D}$ and $q|_{\Gamma_N}$ in order to simulate the inherent measurement errors.

In the above formulation of the boundary conditions (2), it can be seen that the boundary $\Gamma_D \cap \Gamma_N$ is over-specified, whilst the boundary $\Gamma \setminus (\Gamma_D \cap \Gamma_N)$ is under-specified since both the potential $u|_{\Gamma \setminus \Gamma_D}$ and the normal flux $q|_{\Gamma \setminus \Gamma_N}$ are unknown and have to be determined on $\Gamma \setminus \Gamma_D$ and $\Gamma \setminus \Gamma_N$, respectively. It should be noted that, if $\Gamma_D = \Gamma_N$ then the inverse problem (1) and (2) becomes the classical Cauchy problem for Helmholtz-type equations. The aforementioned inverse problem is much more difficult to solve both analytically and numerically than the direct problem, since the solution does not satisfy the general conditions of well-posedness. Whilst the Dirichlet, Neumann and mixed direct problems associated with equation (1) do not always have a unique solution due to the eigensolutions [10], the solution of the inverse problem given by equations (1) and (2) is unique based on the analytical continuation property. However, it is well known that if this solution exists then it is unstable with respect to small perturbations in the data on $\Gamma_D \cap \Gamma_N$, see Hadamard [9]. Thus the problem under investigation is ill-posed and we cannot use a direct approach, such as the Gauss elimination method, in order to solve the system of linear equations which arises from the discretization of the partial differential equations (1) and the boundary conditions (2). Therefore, regularization methods are required in order to solve accurately the inverse boundary value problems associated with Helmholtz-type equations.

In the following, two distinct inverse problems for Helmholtz-type equations will be analysed depending upon the singular or non-singular character of this problem. We consider the inverse problem (1) and (2) in the two-dimensional domain $\Omega \subset \mathbb{R}^2$ whose boundary Γ either does or does not contain a singularity at the point $O\ (0,0)$, that may be caused by a change in the boundary conditions at the origin and/or a re-entrant corner at the origin. More precisely, the singularity point, O, is located at the intersection of the Dirichlet, Γ_D, and Neumann, Γ_N, boundary parts, i.e. $\{O\} \subset \overline{\Gamma}_D \cap \overline{\Gamma}_N$.

3 Singular Solutions for the Two-Dimensional Helmholtz-Type Operator

In this section, some well-known results on the solution of homogeneous two-dimensional Helmholtz-type equations using the separation of variables in polar coordinates are revised. For more details, we refer the reader to Marin *et al.* [20]. Let the polar coordinate system (r, θ) be defined in the usual way with respect to the Cartesian coordinates

$(x_1, x_2) = (r\cos\theta, r\cos\theta)$. For $r > 0$, equation (1) written in polar coordinates takes the following form:

$$\frac{\partial^2 u(r,\theta)}{\partial r^2} + \frac{1}{r}\frac{\partial u(r,\theta)}{\partial r} + \frac{1}{r^2}\frac{\partial^2 u(r,\theta)}{\partial \theta^2} + \widetilde{k}^2 u(r,\theta) = 0. \tag{4}$$

Here $\widetilde{k} \in \mathbb{C}$ is a complex number, with the mention that $\widetilde{k} = k$ and $\widetilde{k} = ik$ for the Helmholtz ($\mathcal{L} = \Delta + k^2$) and modified Helmholtz ($\mathcal{L} = \Delta - k^2$) equations, respectively, where $i = \sqrt{-1}$. If we assume that the solution of equation (1) in the domain Ω can be written using the separation of variables

$$u(r,\theta) = f(r)\, g(\theta) \tag{5}$$

then the Helmholtz-type equation (1) can be recast as

$$\left(f''(r) + r^{-1} f'(r) + \widetilde{k}^2 f(r)\right) g(\theta) + r^{-2} f(r) g''(\theta) = 0. \tag{6}$$

If $f(r)\, g(\theta) \neq 0$ then the following ratio must be a constant, denoted here by λ^2:

$$\frac{f''(r) + r^{-1} f'(r) + \widetilde{k}^2 f(r)}{r^{-2} f(r)} = -\frac{g''(\theta)}{g(\theta)} = \lambda^2, \tag{7}$$

and this results in the following two linear homogeneous ordinary differential equations with respect to the variables r and θ, respectively

$$f''(r) + r^{-1} f'(r) + \left(\widetilde{k}^2 - \lambda^2 r^{-2}\right) f(r) = 0, \tag{8}$$

$$g''(\theta) + \lambda^2 g(\theta) = 0. \tag{9}$$

For a given value of λ, the general solutions of equations (8) and (9) are given by

$$f_\lambda(\widetilde{k}r) = \gamma_1 J_\lambda(\widetilde{k}r) + \gamma_2 N_\lambda(\widetilde{k}r), \tag{10}$$

and

$$g_\lambda(\theta) = a\cos(\lambda\theta) + b\sin(\lambda\theta), \tag{11}$$

respectively. Here γ_1, γ_2, a and b are constants, whilst J_λ and N_λ are the Bessel functions of the first kind and the second kind, respectively. Hence the general solution of equation (1) in the form (5) can be written as

$$u(r,\theta) = [\gamma_1 J_\lambda(\widetilde{k}r) + \gamma_2 N_\lambda(\widetilde{k}r)][a\cos(\lambda\theta) + b\sin(\lambda\theta)]. \tag{12}$$

Assume that Ω is a two-dimensional isotropic wedge domain of interior angle $2\,\omega$, where $0 \leq \omega < \pi$, with the tip at the origin, O, of the local polar coordinates system and determined by two straight edges of angles θ_1 and θ_2, such that $\theta_2 - \theta_1 = 2(\pi - \omega)$, see Figs. 1(a)–(c). Therefore, $\Omega = \left\{\mathbf{x} \in \mathbb{R}^2 \,|\, 0 < r < R(\theta), \theta_1 < \theta < \theta_2\right\}$, where $R(\theta)$ is either a bounded continuous function or infinity. Consider now that the boundary value problem given by equation (1) in Ω and homogeneous Neumann and/or Dirichlet boundary conditions prescribed on the wedge edges. On assuming $\operatorname{Re}\lambda \geq 0$ and taking into account the finite character of the solution, u, in a wedge tip neighbourhood, we obtain $\gamma_2 = 0$ in equation (12). Hence the basis function of singular solutions to the aforementioned boundary value problem obtained from expression (12) can be written in the general form as

$$u^{(S)}(r,\theta) = J_{\lambda}(\widetilde{k}r)\left[a\cos(\lambda\theta) + b\sin(\lambda\theta)\right], \tag{13}$$

where a and b are the unknown singular coefficients, whilst λ is referred to as the singularity exponent or eigenvalue. The singularity exponent/eigenvalue, as well as the corresponding singular coefficients, are determined by the geometry and boundary conditions on the boundaries adjacent to the singularity point.

The normal flux, i.e. the normal derivative of the primary field, through a straight radial line defined by an angle θ and associated with the normal vector $\mathbf{n}(\theta) = (-\sin\theta, \cos\theta)$ is given by

$$q^{(S)}(r,\theta) = \frac{1}{r}\frac{\partial}{\partial\theta}u^{(S)}(r,\theta). \tag{14}$$

For the sake of convenience, the singular solution, $u^{(S)}$, and normal flux, $q^{(S)}$, given by equations (13) and (14), respectively, can be recast as:

$$u^{(S)}(r,\theta) = J_{\lambda}(\widetilde{k}r)\{a\cos[\lambda(\theta-\theta_1)] + b\sin[\lambda(\theta-\theta_1)]\}, \tag{15}$$

$$q^{(S)}(r,\theta) = \frac{\lambda}{r}J_{\lambda}(\widetilde{k}r)\{-a\sin[\lambda(\theta-\theta_1)] + b\cos[\lambda(\theta-\theta_1)]\}. \tag{16}$$

In the present chapter, four configurations of homogeneous Neumann (N) and Dirichlet (D) boundary conditions at the wedge edges applied to expressions (15) and (16) are considered. The conditions which allow a nontrivial solution of the resulting system of equations under the assumption $\mathrm{Re}\,\lambda \geq 0$ are listed below, see Marin *et al.* [20]:

Case I: N-N wedge

$$q^{(S)}(r,\theta_1) = q^{(S)}(r,\theta_2) = 0 \Rightarrow b = 0 \;\&\; \sin[\lambda(\theta_2-\theta_1)] = 0 \Rightarrow \lambda = n\frac{\pi}{\theta_2-\theta_1},\quad n \geq 0 \tag{17}$$

Case II: D-D wedge

$$u^{(S)}(r,\theta_1) = u^{(S)}(r,\theta_2) = 0 \Rightarrow a = 0 \;\&\; \sin[\lambda(\theta_2-\theta_1)] = 0 \Rightarrow \lambda = n\frac{\pi}{\theta_2-\theta_1},\quad n \geq 1 \tag{18}$$

Case III: N-D wedge $q^{(S)}(r,\theta_1) = u^{(S)}(r,\theta_2) = 0 \Rightarrow b = 0 \;\&\; \cos[\lambda(\theta_2-\theta_1)] = 0$
$\Rightarrow \lambda = \left(n - \frac{1}{2}\right)\frac{\pi}{\theta_2-\theta_1},\ n \geq 1$

Case IV: D-N wedge $u^{(S)}(r,\theta_1) = q^{(S)}(r,\theta_2) = 0 \Rightarrow a = 0 \;\&\; \cos[\lambda(\theta_2-\theta_1)] = 0$
$\Rightarrow \lambda = \left(n - \frac{1}{2}\right)\frac{\pi}{\theta_2-\theta_1},\ n \geq 1$

From formulae (17) – (3) it can be noticed that the singularity exponents, λ, are real and simple and they coincide in cases I and II, and III and IV, respectively. Using the above results, the general asymptotic expansions for the singular solution of Helmholtz-type equations for a single wedge and corresponding to homogeneous Neumann and Dirichlet boundary conditions on the wedge edges are obtained in the following form:

Case I: N-N wedge

$$u^{(S)}(r,\theta) = \sum_{n=0}^{\infty} a_n u_n^{(NN)}(r,\theta) = \sum_{n=0}^{\infty} a_n J_{\lambda_n}(\widetilde{k}r)\cos[\lambda_n(\theta-\theta_1)], \quad \lambda_n = n\frac{\pi}{\theta_2-\theta_1}, \; n \geq 0 \quad (19)$$

Case II: D-D wedge

$$u^{(S)}(r,\theta) = \sum_{n=1}^{\infty} a_n u_n^{(DD)}(r,\theta) = \sum_{n=1}^{\infty} a_n J_{\lambda_n}(\widetilde{k}r)\sin[\lambda_n(\theta-\theta_1)], \quad \lambda_n = n\frac{\pi}{\theta_2-\theta_1}, \; n \geq 1 \quad (20)$$

Case III: N-D wedge

$$u^{(S)}(r,\theta) = \sum_{n=1}^{\infty} a_n u_n^{(ND)}(r,\theta) = \sum_{n=1}^{\infty} a_n J_{\lambda_n}(\widetilde{k}r)\cos[\lambda_n(\theta-\theta_1)], \lambda_n = \left(n-\frac{1}{2}\right)\frac{\pi}{\theta_2-\theta_1}, \; n \geq 1 \quad (21)$$

Case IV: D-N wedge

$$u^{(S)}(r,\theta) = \sum_{n=1}^{\infty} a_n u_n^{(DN)}(r,\theta) = \sum_{n=1}^{\infty} a_n J_{\lambda_n}(\widetilde{k}r)\sin[\lambda_n(\theta-\theta_1)], \lambda_n = \left(n-\frac{1}{2}\right)\frac{\pi}{\theta_2-\theta_1}, \; n \geq 1 \quad (22)$$

The following particular two-dimensional geometries containing a boundary singularity are investigated herein, see Figs. $1(a)-(c)$:

(*i*) Two-dimensional domain containing an edge crack, i.e. $\theta_1 = -\theta_2 = \pi$.

(*ii*) L-shaped domain, i.e. $\theta_1 = 0$ and $\theta_2 = 3\pi/2$.

(*iii*) Two-dimensional domain containing a V-notch with the re-entrant angle $2\,\omega$, $\omega \in (0,\pi/2)$, i.e. $\theta_1 = -\theta_2 = \pi - \omega$.

It should be mentioned that a domain containing an edge crack can also be considered as the limiting case for a domain containing a V-notch with the re-entrant angle $2\,\omega$, $\omega \in (0,\pi/2)$, in the sense that $\omega = 0$ for the aforementioned geometry. For the sake of completeness and taking into account the symmetry of the domains containing an edge crack or a V-notch, see Figs. 1(a) and (c), we mention that the singularity exponents corresponding to the particular geometries analysed in this chapter are given by:

Cases I & II: N-N and D-D wedges

$$\lambda_n = \begin{cases} n\dfrac{\pi}{\pi-\omega} & \text{for a domain containing a V-notch with } \omega \in [0,\pi/2) \\ \dfrac{2n}{3} & \text{for an L-shaped domain} \end{cases} \quad (23)$$

Cases III & IV: N-D and D-N wedges

$$\lambda_n = \begin{cases} \left(n-\dfrac{1}{2}\right)\dfrac{\pi}{\pi-\omega} & \text{for a domain containing a V-notch with } \omega \in [0,\pi/2) \\ \dfrac{2n-1}{3} & \text{for an L-shaped domain} \end{cases} \quad (24)$$

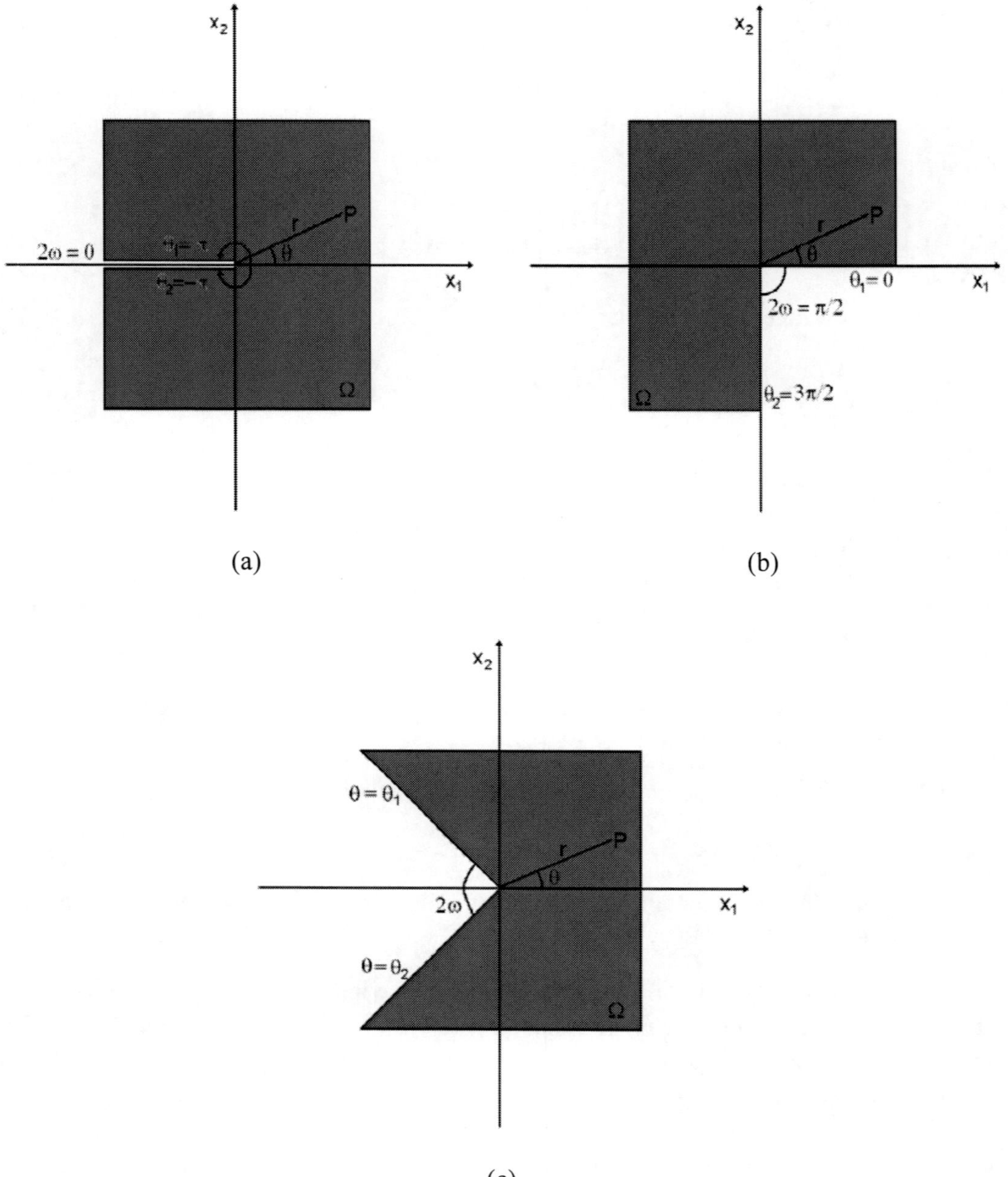

Figure 1: Schematic diagram of the two-dimensional isotropic wedge domain Ω of interior angle 2ω, where (a) $\omega = 0$ (i.e. domain containing an edge crack), (b) $\omega = \pi/4$ (i.e. L-shaped domain), and (c) $\omega \in (0; \pi/2)$ (i.e. domain containing a V-notch).

4 Singular Inverse Problem: Singularity Subtraction Technique

Consider the singular inverse problem (1) and (2) subject to perturbed boundary data in the two-dimensional bounded domain Ω, whose piecewise smooth boundary Γ contains a singularity at the origin $\mathrm{O}(0,0)$, as described in Section 2. For the simplicity of the following explanations, we assume that the singularity point is located at the intersection of the Dirichlet, Γ_{D}, and Neumann, Γ_{N}, boundary parts, i.e. $\{\mathrm{O}\} \subset \overline{\Gamma}_{\mathrm{D}} \cap \overline{\Gamma}_{\mathrm{N}}$, where $\Gamma_{\mathrm{D}}, \Gamma_{\mathrm{N}} \neq \emptyset$, $\Gamma_{\mathrm{D}} \cap \Gamma_{\mathrm{N}} \neq \emptyset$, $\Gamma_{\mathrm{D}} \cup \Gamma_{\mathrm{N}} \subset \Gamma$, and $\overline{\Lambda}$ denotes the closure of the set Λ. However, it should be stressed out that the method presented herein can easily be extended to other local configurations or boundary conditions. Hence the perturbed singular inverse problem considered can be recast as

$$\Delta \mathrm{u}(\mathbf{x}) \pm k^2 \mathrm{u}(\mathbf{x}) = 0, \quad \mathbf{x} \in \Omega \tag{.25}$$

$$\mathrm{u}(\mathbf{x}) = \widetilde{\mathrm{u}}^{\varepsilon}(\mathbf{x}), \quad \mathbf{x} \in \Gamma_{\mathrm{D}} \tag{.26}$$

$$\mathrm{q}(\mathbf{x}) = \widetilde{\mathrm{q}}^{\varepsilon}(\mathbf{x}), \quad \mathbf{x} \in \Gamma_{\mathrm{N}} \tag{.27}$$

In order to avoid numerical difficulties arising from the presence of the singularity in the solution at O, it is convenient to modify the original problem before it is solved by the MFS. Due to the linearity of the Helmholtz and modified Helmholtz operators, as well as the boundary conditions, the superposition principle is valid and the potential solution, u, and normal flux, q, can be written as

$$\mathrm{u}(\mathbf{x}) = \left(\mathrm{u}(\mathbf{x}) - \mathrm{u}^{(\mathrm{S})}(\mathbf{x})\right) + \mathrm{u}^{(\mathrm{S})}(\mathbf{x}) = \mathrm{u}^{(\mathrm{R})}(\mathbf{x}) + \mathrm{u}^{(\mathrm{S})}(\mathbf{x}), \quad \mathbf{x} \in \overline{\Omega} = \Omega \cup \Gamma, \tag{28}$$

$$\mathrm{q}(\mathbf{x}) = \left(\mathrm{q}(\mathbf{x}) - \mathrm{q}^{(\mathrm{S})}(\mathbf{x})\right) + \mathrm{q}^{(\mathrm{S})}(\mathbf{x}) = \mathrm{q}^{(\mathrm{R})}(\mathbf{x}) + \mathrm{q}^{(\mathrm{S})}(\mathbf{x}), \quad \mathbf{x} \in \Gamma, \tag{29}$$

where $\mathrm{u}^{(\mathrm{S})}(\mathbf{x})$ is a particular singular solution of the original problem (.25)–(.27) which satisfies the corresponding homogeneous boundary conditions on the parts of the boundary containing the singularity point O and $\mathrm{q}^{(\mathrm{S})}(\mathbf{x}) \equiv \nabla \mathrm{u}^{(\mathrm{S})}(\mathbf{x}) \cdot \mathbf{n}(\mathbf{x})$ is its normal derivative. If appropriate functions are chosen for the singular solution and its normal derivative then the numerical analysis can be carried out for the regular solution $\mathrm{u}^{(\mathrm{R})}(\mathbf{x})$ and its normal derivative $\mathrm{q}^{(\mathrm{R})}(\mathbf{x}) \equiv \nabla \mathrm{u}^{(\mathrm{R})}(\mathbf{x}) \cdot \mathbf{n}(\mathbf{x})$ only. In terms of the regular solution $\mathrm{u}^{(\mathrm{R})}(\mathbf{x})$, the original inverse problem (.25)–(.27) becomes

$$\Delta \mathrm{u}^{(\mathrm{R})}(\mathbf{x}) \pm k^2 \mathrm{u}^{(\mathrm{R})}(\mathbf{x}) = 0, \quad \mathbf{x} \in \Omega \tag{.30}$$

$$\mathrm{u}^{(\mathrm{R})}(\mathbf{x}) = \widetilde{\mathrm{u}}^{\varepsilon}(\mathbf{x}) - \mathrm{u}^{(\mathrm{S})}(\mathbf{x}), \quad \mathbf{x} \in \Gamma_{\mathrm{D}} \tag{.31}$$

$$\mathrm{q}^{(\mathrm{R})}(\mathbf{x}) = \widetilde{\mathrm{q}}^{\varepsilon}(\mathbf{x}) - \mathrm{q}^{(\mathrm{S})}(\mathbf{x}), \quad \mathbf{x} \in \Gamma_{\mathrm{N}} \tag{.32}$$

The modified boundary conditions (.31) and (.32) introduce additional unknowns into the problem, which are the constants of the particular solution used to represent the singular solution. It should be noted that these constants are similar to the stress intensity factors corresponding to an analogous problem for the Lamé system and, in what follows, they will be referred to as "flux intensity factors". Since the flux intensity factors are unknown at this stage of the problem, they become primary unknowns.

In order to obtain a unique solution to the regular problem (.30)–(.32), it is necessary to specify additional constraints which must be as many as the number of the unknown flux intensity factors, i.e. one for each singular solution/eigenfunction included in the analysis. These extra conditions must be applied in such a way that the cancelation of the singularity in the regular potential solution is ensured. This is achieved by constraining the regular solution and/or its normal derivative directly in a neighbourhood of the singularity point O

$$u^{(R)}(\mathbf{x}) = 0,\ \mathbf{x} \in \Gamma_N \cap B(O;\tau) \quad \text{and/or} \quad q^{(R)}(\mathbf{x}) = 0,\ \mathbf{x} \in \Gamma_D \cap B(O;\tau), \tag{33}$$

where $B(O;\tau) = \{\mathbf{x} \in \mathbb{R}^2 \,|\, \|\mathbf{x} - \mathbf{x}^0\| < \tau\}$, $\tau > 0$ is sufficiently small and $\|\cdot\|$ represents the Euclidean norm.

For example, for the regular inverse problem (.30)–(.32) the singular solution and its normal derivative are expressed, in terms of the polar coordinates (r,θ), as

$$u^{(S)}(\mathbf{x}) \equiv u^{(S)}(r,\theta) = \sum_{n=1}^{n_S} a_n u_n^{(DN)}(r,\theta), \quad q^{(S)}(\mathbf{x}) \equiv q^{(S)}(r,\theta) = \sum_{n=1}^{n_S} a_n q_n^{(DN)}(r,\theta), \tag{34}$$

where $u_n^{(DN)}(r,\theta)$ is given by equation (22), $q_n^{(DN)}(r,\theta)$ is obtained by taking the normal derivative of $u_n^{(DN)}(r,\theta)$ and a_n, $n = 1,\ldots,n_S$, are the unknown flux intensity factors.

5 Standard and Modified Methods of Fundamental Solutions

The fundamental solutions $\mathcal{F}_H$ and $\mathcal{F}_{MH}$ of the Helmholtz ($\mathcal{L} = \Delta + k^2$) and modified Helmholtz ($\mathcal{L} = \Delta - k^2$) equations, respectively, in the two-dimensional case are given by, see e.g. Fairweather and Karageorghis [43],

$$\mathcal{F}_H(\mathbf{x},\mathbf{y}) = \frac{i}{4} H_0^{(1)}(k\|\mathbf{x}-\mathbf{y}\|), \quad \mathbf{x} \in \overline{\Omega}, \quad \mathbf{y} \in \mathbb{R}^2 \setminus \overline{\Omega}, \tag{35}$$

and

$$\mathcal{F}_{MH}(\mathbf{x},\mathbf{y}) = \frac{1}{2\pi} K_0(k\|\mathbf{x}-\mathbf{y}\|), \quad \mathbf{x} \in \overline{\Omega}, \quad \mathbf{y} \in \mathbb{R}^2 \setminus \overline{\Omega}, \tag{36}$$

respectively. Here $\mathbf{x} = (x_1,x_2)$ is either a boundary or a domain point, $\mathbf{y} = (y_1,y_2)$ is a source point, $H_0^{(1)}$ is the Hankel function of the first kind of order zero and K_0 is the modified Bessel function of the second kind of order zero.

According to the MFS approach, the regular solution, $u^{(R)}$, in the solution domain is approximated by a linear combination of fundamental solutions with respect to M source points $\mathbf{y}^j$ in the form

$$u^{(R)}(\mathbf{x}) \approx \sum_{j=1}^{M} c_j \mathcal{F}(\mathbf{x},\mathbf{y}^j), \quad \mathbf{x} \in \overline{\Omega}, \tag{37}$$

where $\mathcal{F} = \mathcal{F}_H$ in the case of the Helmholtz equation, $\mathcal{F} = \mathcal{F}_{MH}$ in the case of the modified Helmholtz equation, $c_j \in \mathbb{R}$, $j = 1,\ldots,M$, are the unknown coefficients. Then the regular normal flux on the boundary Γ can be approximated by

$$q^{(R)}(\mathbf{x}) \approx \sum_{j=1}^{M} c_j \mathcal{G}(\mathbf{x},\mathbf{y}^j), \quad \mathbf{x} \in \Gamma, \tag{38}$$

where $\mathcal{G}(\mathbf{x},\mathbf{y}) \equiv \nabla_{\mathbf{x}}\mathcal{F}(\mathbf{x},\mathbf{y})\cdot\mathbf{n}(\mathbf{x})$, while $\mathcal{G} = \mathcal{G}_{\mathrm{H}}$ in the case of the Helmholtz equation and $\mathcal{G} = \mathcal{G}_{\mathrm{MH}}$ in the case of the modified Helmholtz equation are given by

$$\mathcal{G}_{\mathrm{H}}(\mathbf{x},\mathbf{y}) = -\frac{[(\mathbf{x}-\mathbf{y})\cdot\mathbf{n}(\mathbf{x})]\,k\,i}{4\,\|\mathbf{x}-\mathbf{y}\|}\,\mathrm{H}_1^{(1)}(k\|\mathbf{x}-\mathbf{y}\|), \quad \mathbf{x}\in\Gamma, \quad \mathbf{y}\in\mathbb{R}^2\setminus\overline{\Omega}, \tag{39}$$

and

$$\mathcal{G}_{\mathrm{MH}}(\mathbf{x},\mathbf{y}) = -\frac{[(\mathbf{x}-\mathbf{y})\cdot\mathbf{n}(\mathbf{x})]\,k}{2\pi\,\|\mathbf{x}-\mathbf{y}\|}\,\mathrm{K}_1(k\|\mathbf{x}-\mathbf{y}\|), \quad \mathbf{x}\in\Gamma, \quad \mathbf{y}\in\mathbb{R}^2\setminus\overline{\Omega}, \tag{40}$$

respectively. Here $\mathrm{H}_1^{(1)}$ is the Hankel function of the first kind of order one and K_1 is the modified Bessel function of the second kind of order one.

Assume that the singularity point O is located between the collocation points $\mathbf{x}^{n_D^0} \in \Gamma_D$ and $\mathbf{x}^{n_N^0} \in \Gamma_N$, see also Fig. 12(c), and n_S singular solutions/eigenfunctions $u_n^{(DN)}(r,\theta)$, as well as flux intensities, a_n, are taken into account, such that the additional constraints for the regular solution and/or its normal derivative given by equation (33) read as

$$u^{(R)}(\mathbf{x}^{n_N^0+1-m}) = 0, \quad i = 2m-1 \in \{1,\ldots,n_S\}, \tag{41}$$

and

$$q^{(R)}(\mathbf{x}^{n_D^0-1+m}) = 0, \quad i = 2m \in \{1,\ldots,n_S\}. \tag{42}$$

If n_D collocation points $\mathbf{x}^i$, $i = 1,\ldots,n_D$, and n_N collocation points $\mathbf{x}^{n_D+i}$, $i = 1,\ldots,n_N$, are chosen on the boundaries Γ_D and Γ_N, respectively, such that $N = n_D + n_N$, and the location of the source points $\mathbf{y}^j$, $j = 1,\ldots,M$, is set then the boundary value problem (.30) – (.32), together with the additional conditions (33), recasts as a system of $(N+n_S)$ linear algebraic equations with $(M+n_S)$ unknowns which can be generically written as

$$\mathbf{A}\widetilde{\mathbf{c}} = \mathbf{F}, \tag{43}$$

where $\widetilde{\mathbf{c}} = (c_1,\ldots,c_M,a_1,\ldots,a_{n_S}) \in \mathbb{R}^{M+n_S}$ and the components of the MFS matrix $\mathbf{A} \in \mathbb{R}^{(N+n_S)\times(M+n_S)}$ and right-hand side vector $\mathbf{F} \in \mathbb{R}^{N+n_S}$ are given by

$$A_{ij} = \begin{cases} \mathcal{F}(\mathbf{x}^i,\mathbf{y}^j), & i=1,\ldots,n_D, & j=1,\ldots,M \\ u_{j-M}^{(DN)}(r^i,\theta^i), & i=1,\ldots,n_D, & j=M+1,\ldots,M+n_S \\ \mathcal{G}(\mathbf{x}^i,\mathbf{y}^j), & i=n_D+1,\ldots,n_D+n_N, & j=1,\ldots,M \\ q_{j-M}^{(DN)}(r^i,\theta^i), & i=n_D+1,\ldots,n_D+n_N, & j=M+1,\ldots,M+n_S \\ \mathcal{F}(\mathbf{x}^{n_N^0+1-m-N},\mathbf{y}^j), & i-N=2m-1 \in \{1,\ldots,n_S\}, & j=1,\ldots,M \\ \mathcal{G}(\mathbf{x}^{n_D^0-1+m-N},\mathbf{y}^j), & i-N=2m \in \{1,\ldots,n_S\}, & j=1,\ldots,M \\ 0, & i-N \in \{1,\ldots,n_S\}, & j=M+1,\ldots,M+n_S \end{cases} \tag{44}$$

$$F_i = \begin{cases} \widetilde{u}^{\varepsilon}(\mathbf{x}^i), & i = 1,\ldots,n_D \\ \widetilde{q}^{\varepsilon}(\mathbf{x}^i), & i = n_D+1,\ldots,n_D+n_N \\ 0, & i = n_D+n_N+1,\ldots,n_D+n_N+n_S \end{cases} \tag{45}$$

Here (r^i, θ^i) are the local polar coordinates of the collocation point $\mathbf{x}^i$, $i = 1, \ldots, N$.

It is important to mention that by taking $n_S = 0$ in the *modified MFS* system (43), the latter is reduced to the *standard MFS* system associated with non-singular problems. It should be noted that in order to uniquely determine the solution $\widetilde{\mathbf{c}}$ of the system of linear algebraic equations (43), i.e. the coefficients c_j, $j = 1, \ldots, M$, in approximations (37) and (38) and, eventually, the flux intensity factors a_n, $n = 1, \ldots, n_S$, in the asymptotic expansions (34), the total number of collocation points corresponding to the Dirichlet and Neumann boundary conditions, N, and the number of source points, M, must satisfy the inequality $M \leq N$.

In order to implement the standard or modified MFS, the location of the source points has to be determined and this is usually achieved by considering either the static or the dynamic approach. In the static approach, the source points are pre-assigned and kept fixed throughout the solution process, whilst in the dynamic approach, the source points and the unknown coefficients are determined simultaneously during the solution process, see Fairweather and Karageorghis [43]. For nonlinear systems, the uniqueness of the solution is not always guaranteed and it is computationally much more expensive. In addition, the discretized MFS system is severely ill-posed in the case of inverse problems and thus the dynamic approach transforms the problem into a more difficult nonlinear ill-posed problem. The dynamic approach results in a system of nonlinear equations, which may be solved using minimization methods. Alternatively, Tankelevich *et al.* [61] consider the source points located on a so-called pseudo-boundary, which has the same shape as the boundary of the domain, and the problem is solved for a sequence of such pseudo-boundaries, whilst the optimal pseudo-boundary is taken to be the one for which boundary conditions are satisfied most accurately. From a computational point of view, the dynamic approach might not be appropriate for inverse problems with noisy data. Moreover, Mitic and Rashed [62] have shown that the distribution and number of the source points are not, in general, important under certain conditions, in the sense that the number of sources should reflect the degrees of freedom inherent in the boundary conditions of the problem. Hence the dynamic approach for determining the optimal location of the source points might be unnecessary. Therefore, we decided to employ the static approach in our computations with the source points located on a pseudo-boundary chosen in the same manner as Tankelevich *et al.* [61].

6 Regularization

The MFS can be regarded as a Fredholm integral equation of the first kind with an analytical kernel function which is severely ill-posed according to the theory of integral equations, see e.g. Golberg and Chen [63]. Consequently, as an approximation to the integral operator, the discretization matrix $\mathbf{A}$ is severely ill-conditioned. The accurate and stable solution of equation (43) is very important for obtaining physically meaningful numerical results. Regularization methods are among the most popular and successful methods for solving stably and accurately ill-conditioned matrix equations, see Tikhonov and Arsenin [64]. In the present computations, we use the TRM to solve the matrix equation arising from the MFS discretization.

6.1 The Tikhonov Regularization Method

The Tikhonov regularized solution to the system of linear algebraic equations (43) is sought as [64]

$$\widetilde{\mathbf{c}}_\lambda : \mathcal{T}_\lambda(\widetilde{\mathbf{c}}) = \min_{\widetilde{\mathbf{c}} \in \mathbb{R}^{M+n_S}} \mathcal{T}_\lambda(\widetilde{\mathbf{c}}), \tag{46}$$

where $\mathcal{T}_\lambda$ represents the zeroth-order Tikhonov functional given by

$$\mathcal{T}_\lambda(\cdot) : \mathbb{R}^{M+n_S} \longrightarrow [0,\infty), \qquad \mathcal{T}_\lambda(\widetilde{\mathbf{c}}) = \|\mathbf{A}\,\widetilde{\mathbf{c}} - \mathbf{F}\|^2 + \lambda^2 \|\widetilde{\mathbf{c}}\|^2, \tag{47}$$

and $\lambda > 0$ is the regularization parameter to be chosen. Formally, the Tikhonov regularized solution $\widetilde{\mathbf{c}}_\lambda$ of the problem (46) is given as the solution of the normal equation

$$\left(\mathbf{A}^{\mathrm{tr}}\,\mathbf{A} + \lambda^2\,\mathbf{I}_{M+n_S}\right)\widetilde{\mathbf{c}} = \mathbf{A}^{\mathrm{tr}}\,\mathbf{F}, \tag{48}$$

where $\mathbf{I}_{M+n_S}$ is the identity matrix. If the right-hand side of equation (43) is corrupted by noise, i.e.

$$\|\mathbf{F} - \mathbf{F}^\delta\| \leq \delta, \tag{49}$$

then the following stability estimate holds, see Engl *et al.* [65],

$$\|\widetilde{\mathbf{c}}_\lambda - \widetilde{\mathbf{c}}_\lambda^\delta\| \leq \frac{\delta}{\lambda}, \tag{50}$$

where

$$\widetilde{\mathbf{c}}_\lambda = \left(\mathbf{A}^{\mathrm{tr}}\,\mathbf{A} + \lambda^2\,\mathbf{I}_{M+n_S}\right)^{-1}\mathbf{A}^{\mathrm{tr}}\,\mathbf{F}. \tag{51}$$

6.2 The L-Curve Method

The performance of regularization methods depends crucially on the suitable choice of the regularization parameter. One extensively studied criterion is the discrepancy principle, see e.g. Morozov [66]. Although this criterion is mathematically rigorous, it requires a reliable estimation of the amount of noise added into the data which may not be available in practical problems. Heuristic approaches are preferable in the case when no *a priori* information about the noise is available. For the TRM, several heuristic approaches have been proposed, including the L-curve criterion, see Hansen [67], and the generalized cross-validation, see Wahba [68]. In this paper, we employ the L-curve criterion to determine the optimal regularization parameter, λ_{opt}.

If we define on a logarithmic scale the curve $\{(\|\mathbf{A}\,\widetilde{\mathbf{c}}_\lambda - \mathbf{F}\|, \|\widetilde{\mathbf{c}}_\lambda\|) \,|\, \lambda > 0\}$ then this typically has an L-shaped form and hence it is referred to as L-curve. According to the L-curve criterion, the optimal regularization parameter corresponds to the corner of the L-curve since a good tradeoff between the residual and solution norms is achieved at this point. Numerically, the L-curve method is robust and stable with respect to both uncorrelated and highly correlated noise. Furthermore, this criterion works effectively with certain classes of practical problems, see Hansen [67] and Chen *et al.* [69]. For a discussion of the theoretical aspects of the L-curve criterion, we refer the reader to Hanke [70] and Vogel [71].

Several algorithms for locating the corner of the L-curve have been reported in the literature, see e.g. Hansen [67], Guerra and Hernandez [72], Kaufman and Neumaier [73] and

Castellanos *et al.* [74]. The first procedure is based on fitting a parametric cubic spline to the discrete points and then taking the point corresponding to the maximum curvature of the L-curve to be its corner [67]. The second algorithm employs a conic to fit the set of discrete points [72], whilst the third one is based on using a linear-linear scale and inverting the axis [73]. All these procedures need to check the monotonicity condition for the sequences of the residual and solution norms, and discard those points where the monotonicity condition is not fulfilled. The last algorithm, namely the triangle method, is based on geometric considerations [74]. In the present chapter, we mainly employ the first algorithm. However, the curvature of the parametric spline is very sensitive to the distribution of the collocation points and occasionally the located corner is not suitable, see Hansen [67]. Therefore, visual inspection is used as an auxiliary procedure.

7 Numerical Results

In this section, the performance of the standard and modified MFS approaches described in Section 5 and associated with non-singular and singular inverse problems, respectively, for Helmholtz-type equations is presented. To do so, we solve numerically the inverse boundary value problem $(.25)-(.27)$ for Helmholtz-type equations in two-dimensional domains with and without boundary singularities.

7.1 Accuracy Errors

In what follows, we denote by $u^{(num)}$ and $q^{(num)}$ the numerical values for the solution and normal flux, respectively, obtained using the least-squares method (LSM), i.e. by a direct inversion method, and by subtracting the first $n_S \geq 0$ singular solutions/eigenfunctions, with the convention that when $n_S = 0$ then the numerical solution and normal flux are obtained using the standard MFS, i.e. without removing the singularity. It should be mentioned that formally the LSM solution, $\widetilde{\mathbf{c}}_{LSM}$, of the MFS system of linear algebraic equations (43) is given as the solution of the normal equation (48) with $\lambda = 0$, i.e.

$$\left(\mathbf{A}^{tr}\mathbf{A}\right)\widetilde{\mathbf{c}} = \mathbf{A}^{tr}\mathbf{F}, \tag{52}$$

in the sense that

$$\widetilde{\mathbf{c}}_{LSM} = \left(\mathbf{A}^{tr}\mathbf{A}\right)^{-1}\mathbf{A}^{tr}\mathbf{F}. \tag{53}$$

In order to measure the accuracy of the numerical approximation for the solution, $u^{(num)}$, and normal flux, $q^{(num)}$, with respect to their corresponding analytical values, $u^{(an)}$, and , $q^{(an)}$, respectively, we define the *relative root mean-square* (*RMS*) *errors* by

$$e_u(\Gamma_j) = \sqrt{\sum_{j=1}^{N_j}\left(u^{(num)}(\mathbf{x}^j) - u^{(an)}(\mathbf{x}^j)\right)^2 \Big/ \sum_{j=1}^{N_j}\left(u^{(an)}(\mathbf{x}^j)\right)^2} \tag{54}$$

$$e_q(\Gamma_j) = \sqrt{\sum_{j=1}^{N_j}\left(q^{(num)}(\mathbf{x}^j) - q^{(an)}(\mathbf{x}^j)\right)^2 \Big/ \sum_{j=1}^{N_j}\left(q^{(an)}(\mathbf{x}^j)\right)^2} \tag{55}$$

where N_j is the number of collocation points on the boundary $\Gamma_j \subset \Gamma$.

Furthermore, we also define the *normalized errors*

$$\mathrm{err}\,(u(\mathbf{x})) = \frac{|u^{(num)}(\mathbf{x}) - u^{(an)}(\mathbf{x})|}{\max_{\mathbf{y}\in\widetilde{\Gamma}} |u^{(an)}(\mathbf{y})|}, \quad \mathrm{err}\,(q(\mathbf{x})) = \frac{|q^{(num)}(\mathbf{x}) - q^{(an)}(\mathbf{x})|}{\max_{\mathbf{y}\in\widetilde{\Gamma}} |q^{(an)}(\mathbf{y})|}, \tag{56}$$

for the solution and normal flux, respectively, where $\widetilde{\Gamma}$ denotes the set of boundary collocation points, since on using these errors divisions by zero and very high errors at points where the solution and/or normal flux have relatively small values are avoided.

In addition, we introduce an error that measures the inaccuracies in the numerical results obtained for the flux intensity factors, namely the *absolute error* defined by

$$\mathrm{Err}(a_j) = |a_j^{(num)} - a_j|. \tag{57}$$

Here $a_j^{(num)}$ represents the numerical value for the exact flux intensity factor a_j, provided that the latter is available.

7.2 Non-Singular Inverse Problems

7.2.1 Examples

In the case of the *non-singular inverse problems* for the Helmholtz-type equations investigated i n this chapter, the solution domains under consideration, Ω, accessible boundaries, Γ_D and Γ_N, and corresponding analytical solutions for the potential are given as follows:

Example 1. Modified Helmholtz equation ($\mathcal{L} = \Delta - k^2, k = 2$) in a disk, see Fig. 2(a):

$$\Omega = \left\{\mathbf{x} \in \mathbb{R}^2 \,\middle|\, x_1^2 + x_2^2 < r^2\right\}, \quad r = 1.0 \tag{.58}$$

$$\Gamma_N = \{\mathbf{x} \in \Gamma | 0 \le \theta(\mathbf{x}) \le \pi\} \subset \Gamma_D = \{\mathbf{x} \in \Gamma | 0 \le \theta(\mathbf{x}) \le 3\pi/2\} \tag{.59}$$

$$\begin{aligned} &u^{(an)}(\mathbf{x}) = \exp(\mathbf{x}\cdot\xi), \\ &\xi = (\xi_1, \xi_2): \quad \xi_1^2 + \xi_2^2 = k^2, \quad \xi_1 = 1.0, \quad \xi_2 = -\sqrt{k^2 - \xi_1^2} = -\sqrt{3} \end{aligned} \tag{.60}$$

Example 2. Helmholtz equation ($\mathcal{L} = \Delta + k^2, k = 1$) in an annulus, see Fig. 2(b):

$$\Omega = \left\{\mathbf{x} \in \mathbb{R}^2 \,\middle|\, {r_i}^2 < x_1^2 + x_2^2 < {r_o}^2\right\}, \quad r_i = 1.0, \quad r_o = 2.0 \tag{.61}$$

$$\Gamma_D = \Gamma_N = \{\mathbf{x} \in \Gamma | \rho(\mathbf{x}) = r_i\} \tag{.62}$$

$$\begin{aligned} &u^{(an)}(\mathbf{x}) = \cos(\mathbf{x}\cdot\xi), \quad \mathbf{x} \in \overline{\Omega} \\ &\xi = (\xi_1, \xi_2): \quad \xi_1^2 + \xi_2^2 = k^2, \quad \xi_1 = 0.5, \quad \xi_2 = \sqrt{k^2 - \xi_1^2} = \frac{\sqrt{3}}{2} \end{aligned} \tag{.63}$$

Example 3. Helmholtz equation ($\mathcal{L} = \Delta + k^2, k = 2$) in a rectangle, see Fig. 2(c):

$$\Omega = (-2r, 2r) \times (-r, r), \quad r = 1.0 \tag{.64}$$

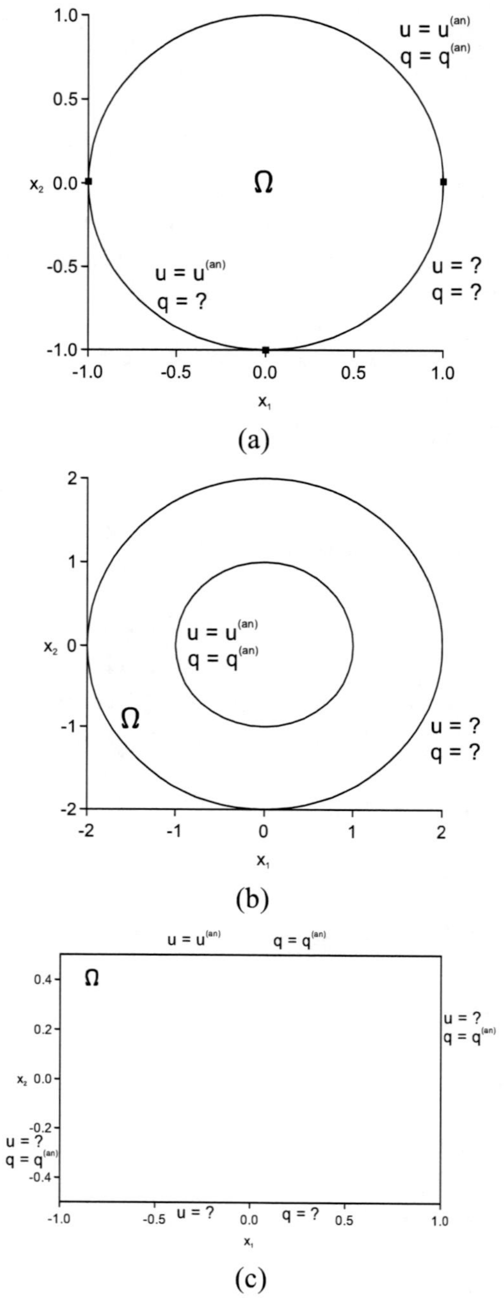

Figure 2: Schematic diagram of the geometry and boundary conditions for the non-singular inverse problems investigated, namely (a) Example 1: Disk; (b) Example 2: Annulus; and (c) Example 3: Rectangle.

$$\Gamma_D = [-2r, 2r] \times \{r\} \subset \Gamma_N = \{\pm r\} \times [-r, r] \cup [-2r, 2r] \times \{r\} \quad (.65)$$

$$\begin{aligned} &u^{(an)}(\mathbf{x}) = \cos(\mathbf{x} \cdot \xi), \quad \mathbf{x} \in \overline{\Omega} \\ &\xi = (\xi_1, \xi_2): \quad \xi_1^2 + \xi_2^2 = k^2, \quad \xi_1 = \sqrt{2}, \quad \xi_2 = \sqrt{k^2 - \xi_1^2} = \sqrt{2} \end{aligned} \quad (.66)$$

It can be seen from Examples $1-3$ that non-singular inverse problems for both the Helmholtz and the modified Helmholtz equations in simply and doubly connected domains with smooth and piecewise smooth boundaries are considered in this chapter. Moreover, three possible situations for such problems are analysed in Examples $1-3$, namely $\mathrm{meas}(\Gamma_N) < \mathrm{meas}(\Gamma_D)$, $\mathrm{meas}(\Gamma_N) = \mathrm{meas}(\Gamma_D)$ and $\mathrm{meas}(\Gamma_N) > \mathrm{meas}(\Gamma_D)$, respectively. These non-singular inverse problems have been solved using a uniform distribution of both the boundary collocation points $\mathbf{x}^i$, $i = 1, \ldots, N$, and the source points $\mathbf{y}^j$, $j = 1, \ldots, M$, without subtracting any singular solutions/eigenvalues, i.e. $n_S = 0$, with the mention that the latter were located on a so-called pseudo-boundary, Γ_S, which has the same shape as the boundary Γ of the solution domain and is situated at the distance $d = 2$ from Γ, see e.g. Tankelevich *et al.* [61]. More precisely, we have taken $N \in \{40, 80, 160\}$, $N = 120$ and $N = 116$ for Examples 1, 2 and 3, respectively.

7.2.2 Effect of the TRM

Before presenting the numerical results, it is interesting to investigate how regularization methods improve the accuracy of the numerical results. To do this, we consider the non-singular inverse problem given by Example 1 only. If the LSM is applied to solving the aforementioned inverse problem subject to noisy boundary data then the numerical solution retrieved by this direct inversion method and given by relation (53) is not only inaccurate, but also unstable. This aspect is strongly related to the ill-posedness of the inverse problem [9] and hence the ill-conditioning of the standard MFS system, see equation (43) with $n_S = 0$. Figs. 3(a) and (b) present the analytical and LSM-based numerical potential on $\Gamma \setminus \Gamma_D$ and normal flux on $\Gamma \setminus \Gamma_N$ when the Dirichlet data $u|_{\Gamma_D}$ was perturbed by $p_u = 1\%$. From these figures, as well as Figs. 3(c) and (d) that show the associated normalized errors $\mathrm{err}(u(\mathbf{x}))$, $\mathbf{x} \in \Gamma \setminus \Gamma_D$, and $\mathrm{err}(q(\mathbf{x}))$, $\mathbf{x} \in \Gamma \setminus \Gamma_N$, respectively, it can be seen that the numerical solutions for both the potential and its normal derivative are highly oscillatory and unbounded, and hence they represent very inaccurate approximations to the corresponding analytical values. Thus standard methods could not yield accurate results for noisy data and, consequently, regularization should be employed to retrieve stable numerical solutions when the input data are contaminated by noise.

7.2.3 Choice of the Optimal Regularization Parameter

Figs. 4(a) and (b) illustrate the relative RMS errors $e_u(\Gamma_C)$ and $e_q(\Gamma_C)$, respectively, given by relations (54) and (55), as functions of the regularization parameter λ, obtained with the TRM, $N = 80$ collocation points, $M = 80$ source points and various levels of noise added into the Dirichlet data $u|_{\Gamma_D}$, for the non-singular inverse problem given by Example 1. Here $\Gamma_C = (\Gamma \setminus \Gamma_D) \cap (\Gamma \setminus \Gamma_N)$. From these figures it can be seen that both errors $e_u(\Gamma_C)$ and $e_q(\Gamma_C)$ decrease as the level of noise p_u added into the input Dirichlet data decreases for all

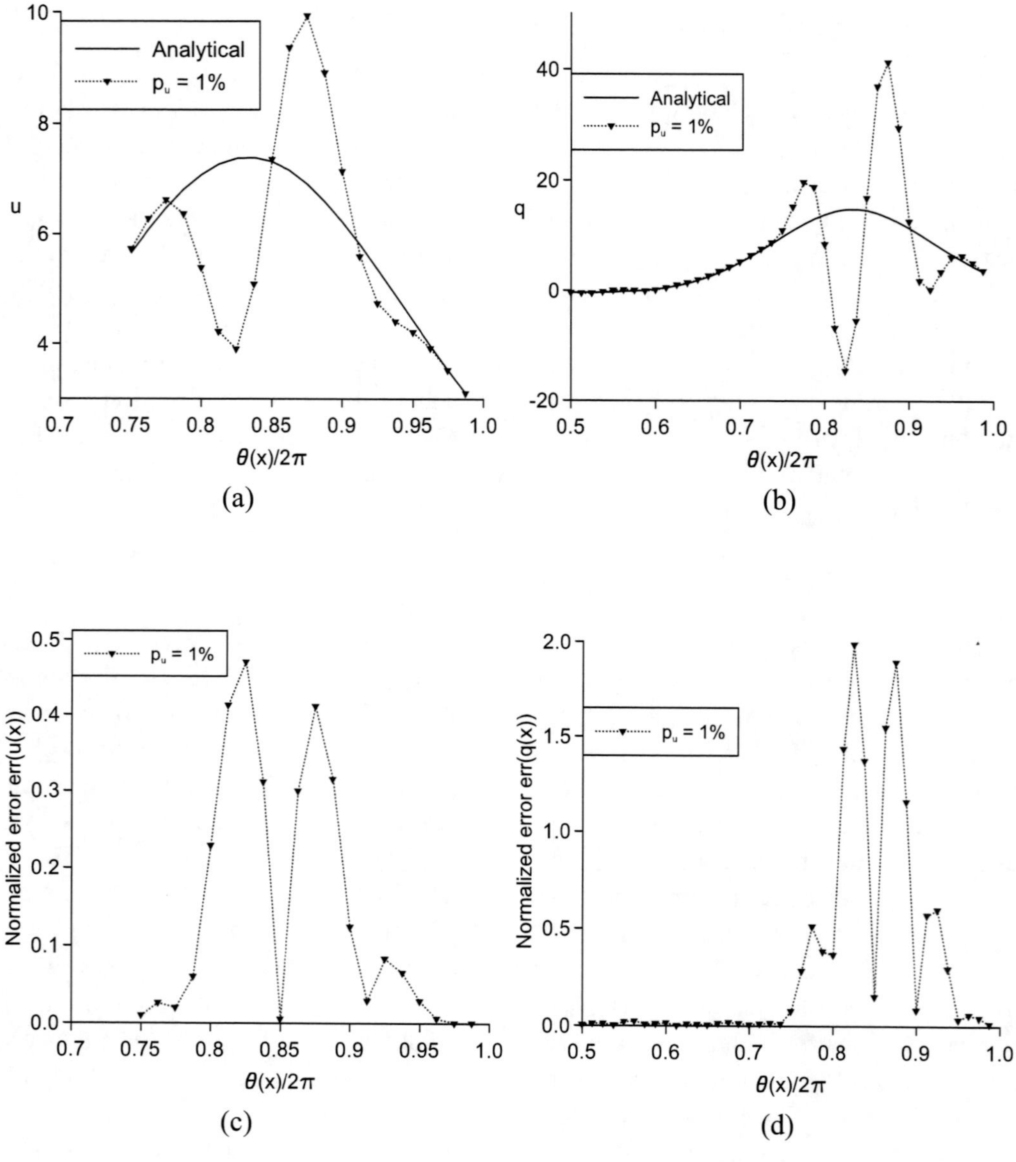

Figure 3: Analytical and numerical solutions for (a) $\mathrm{u}\,|_{\Gamma\setminus\Gamma_\mathrm{D}}$ and (b) $\mathrm{q}|_{\Gamma\setminus\Gamma_\mathrm{N}}$, and the corresponding normalized errors (c) $\mathrm{err}\,(\mathrm{u}(\mathbf{x}))$, $\mathbf{x}\in\Gamma\setminus\Gamma_\mathrm{D}$, and (d) $\mathrm{err}(\mathrm{q}(\mathbf{x}))$, $\mathbf{x}\in\Gamma\setminus\Gamma_\mathrm{N}$, obtained using the LSM and $p_\mathrm{u} = 1\%$ noise added into the Dirichlet data $\mathrm{u}\,|_{\Gamma_\mathrm{D}}$, for Example 1.

regularization parameters λ; as expected, $e_u(\Gamma_C) < e_q(\Gamma_C)$ for all regularization parameters λ and a fixed amount p_u of noise added into the input potential $u|_{\Gamma_D \cap \Gamma_N}$, i.e. the numerical results obtained for the normal flux are more inaccurate than those retrieved for the potential solution on the under-specified boundary Γ_C. Fig. 4(c) presents on a log-log scale the

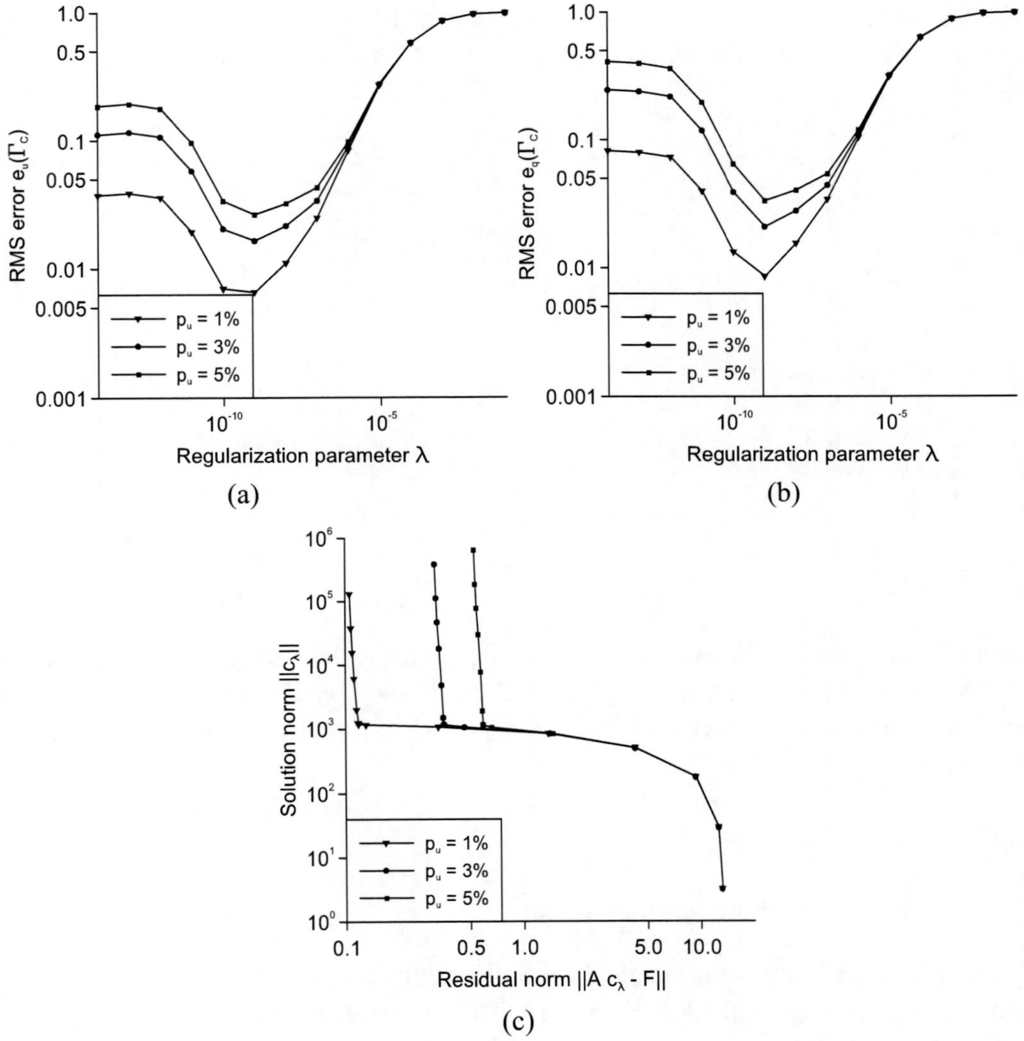

Figure 4: The RMS errors (a) $e_u(\Gamma_C)$ and (b) $e_q(\Gamma_C)$, and (c) the corresponding L-curve, obtained using the TRM and various levels of noise added into the Dirichlet data $u|_{\Gamma_D}$, namely $p_u \in \{1\%, 3\%, 5\%\}$, for Example 1.

L-curves obtained using the TRM and various levels of noise added into the input potential data, in the case of Example 1. From Figs. 4(a)–(c), it can be seen that, for various levels of noise, the "corner" of the L-curve occurs at about the same value of the regularization parameter λ where the minimum in the accuracy errors $e_u(\Gamma_C)$ and $e_q(\Gamma_C)$ is attained, i.e. $\lambda_{opt} = 10^{-9}$ for $p_u \in \{1\%, 3\%, 5\%\}$. Hence the choice of the optimal regularization parameter λ_{opt} according to the L-curve criterion is fully justified. Moreover, the importance of a rigorous and accurate selection of the optimal regularization parameter, λ_{opt}, is also clearly emphasized in Figs. 5(a) and (b), which present the numerical solutions for the potential $u|_{\Gamma\setminus\Gamma_D}$ and normal flux $q|_{\Gamma\setminus\Gamma_D}$, respectively, retrieved using the TRM, $p_u = 1\%$ noise added into the Dirichlet data $u|_{\Gamma_D}$ and various values of the regularization parameter λ, namely $\lambda < \lambda_{\text{opt}}$, $\lambda = \lambda_{\text{opt}}$ and $\lambda > \lambda_{\text{opt}}$, for Example 1, in comparison with their corresponding analytical values. Similar results have been obtained for the other non-singular inverse problems investigated in this chapter and therefore they are not presented here.

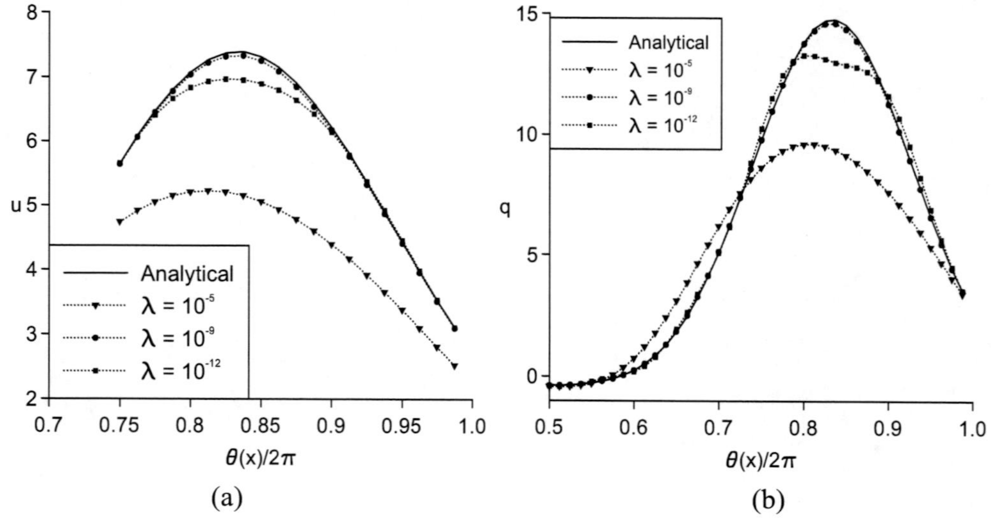

Figure 5: Analytical and numerical solutions for (a) $u|_{\Gamma\setminus\Gamma_D}$ and (b) $q|_{\Gamma\setminus\Gamma_N}$, obtained using the TRM, $p_u = 1\%$ noise added into the Dirichlet data $u|_{\Gamma_D}$ and various values of the regularization parameter λ, namely $\lambda = 10^{-12} < \lambda_{\text{opt}}$, $\lambda = 10^{-9} = \lambda_{\text{opt}}$ and $\lambda = 10^{-5} > \lambda_{\text{opt}}$, for Example 1.

7.2.4 Numerical Stability of the Method

The numerical stability of the standard MFS algorithm described in Section 5, in conjunction with the L-curve method of Hansen [67] for selecting the optimal value for the regularization parameter, λ, is investigated next for the non-singular inverse problems given by Examples 1 – 3, with the corresponding MFS discretizations mentioned in Section 7.2.1, by varying the levels of noise added into the Dirichlet and/or Neumann data as $p_u, p_q \in \{1\%, 3\%, 5\%\}$.

Figs. 6(a) and (b) present the numerical potential solution on the boundary $\Gamma\setminus\Gamma_D$ and normal flux on the boundary $\Gamma\setminus\Gamma_N$, respectively, obtained using the standard MFS algorithm, in conjunction with the TRM, for various levels of Gaussian random noise added into the Dirichlet data $u|_{\Gamma_D}$, in the case of the inverse problem for the modified Helmholtz equation investigated in Example 1, in comparison with their analytical values. From these figures it can be seen that, for all amounts of noise added into $u|_{\Gamma_D}$, both the numerical potential solution on $\Gamma\setminus\Gamma_D$ and the normal flux on $\Gamma\setminus\Gamma_N$ represent excellent approximations for their analytical counterparts, being at the same time exempted from high and unbounded

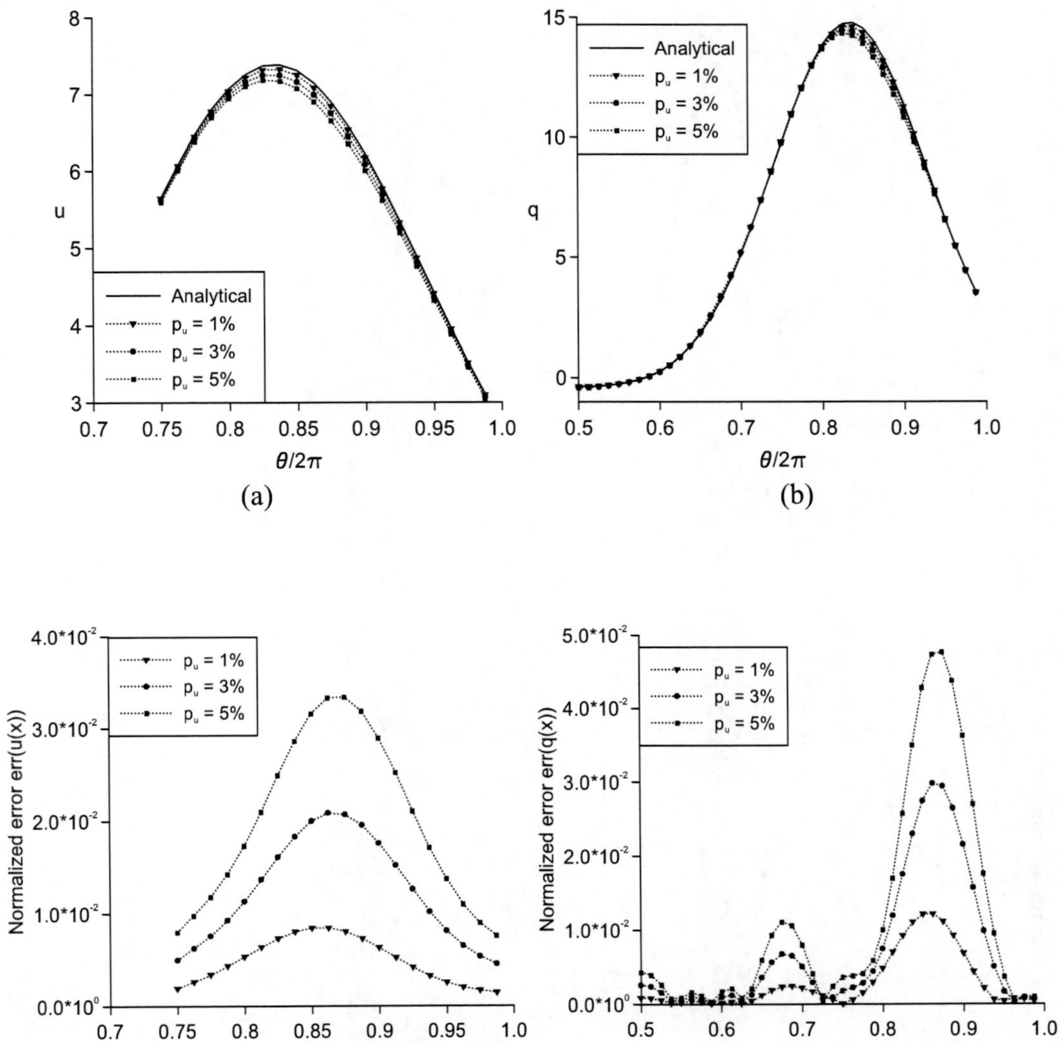

Figure 6: Analytical and numerical solutions for (a) $u|_{\Gamma\setminus\Gamma_D}$ and (b) $q|_{\Gamma\setminus\Gamma_N}$, and the corresponding normalized errors (c) $\mathrm{err}(u(\mathbf{x}))$, $\mathbf{x}\in\Gamma\setminus\Gamma_D$, and (d) $\mathrm{err}(q(\mathbf{x}))$, $\mathbf{x}\in\Gamma\setminus\Gamma_N$, obtained using the TRM and various levels of noise added into the Dirichlet data $u|_{\Gamma_D}$, namely $p_u\in\{1\%,3\%,5\%\}$, for Example 1.

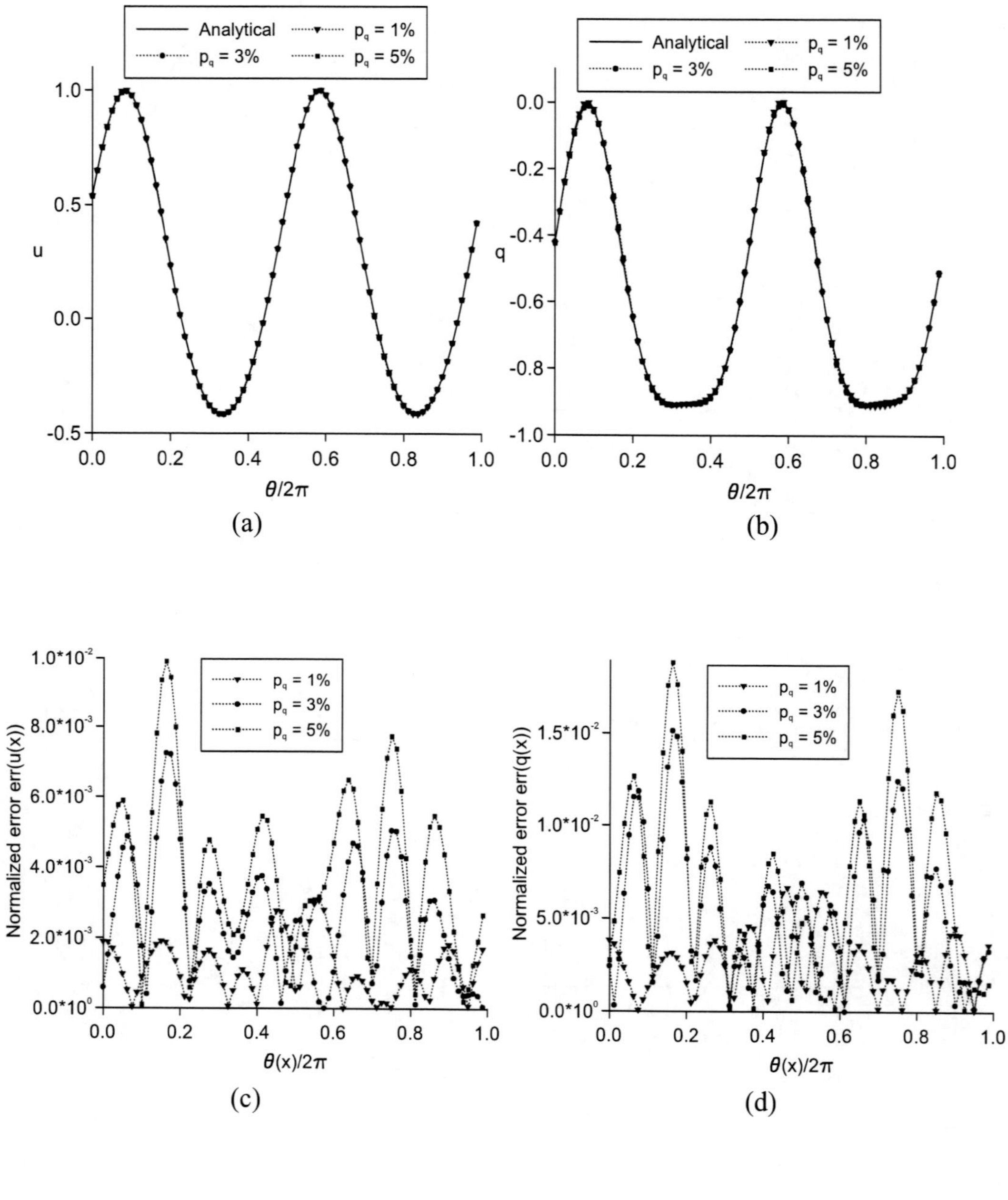

Figure 7: Analytical and numerical solutions for (a) $u|_{\Gamma\setminus\Gamma_D}$ and (b) $q|_{\Gamma\setminus\Gamma_N}$, and the corresponding normalized errors (c) $\mathrm{err}(u(\mathbf{x}))$, $\mathbf{x}\in\Gamma\setminus\Gamma_D$, and (d) $\mathrm{err}(q(\mathbf{x}))$, $\mathbf{x}\in\Gamma\setminus\Gamma_N$, obtained using the TRM and various levels of noise added into the Neumann data $q|_{\Gamma_N}$, namely $p_q\in\{1\%,3\%,5\%\}$, for Example 2.

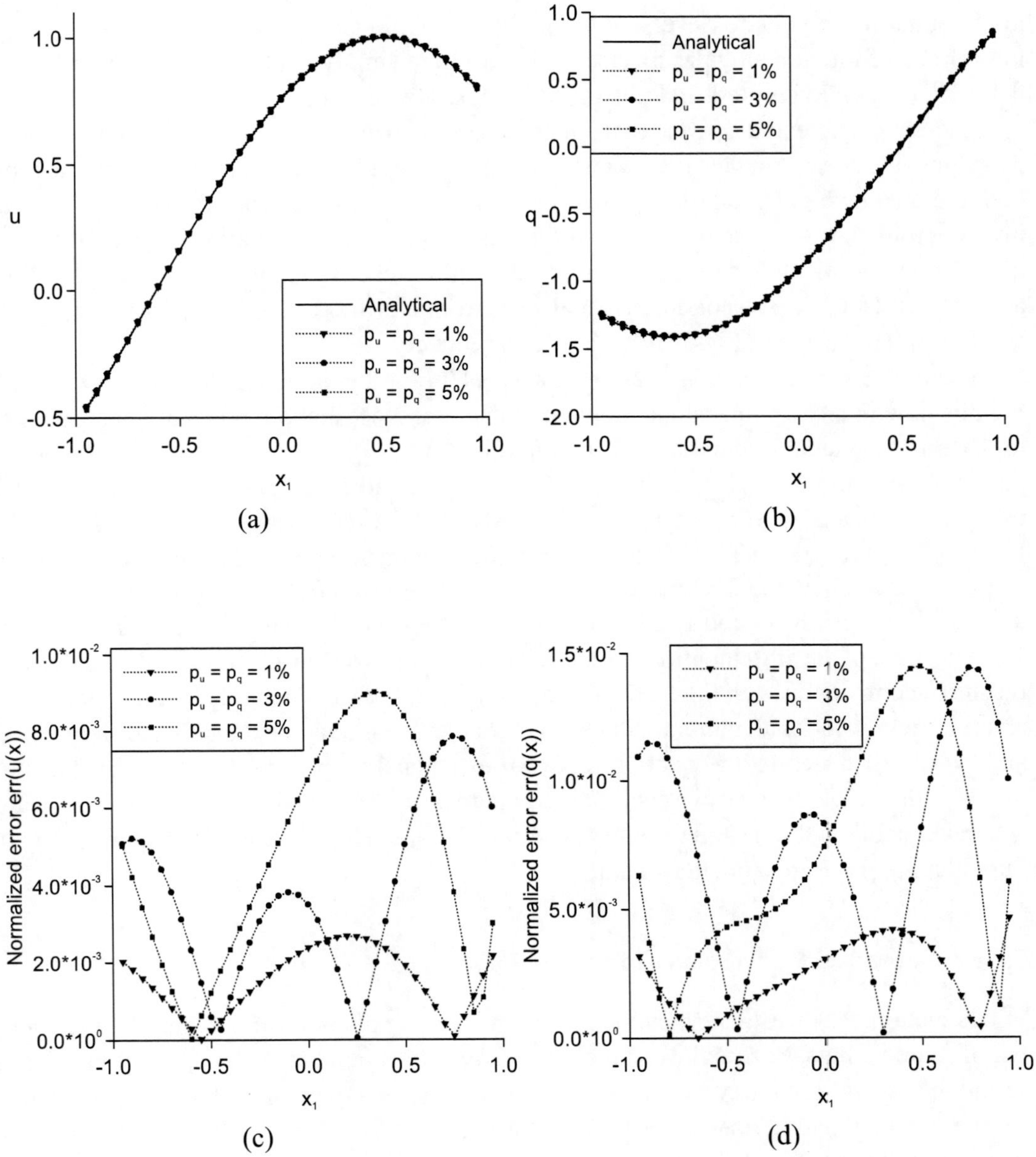

Figure 8: Analytical and numerical solutions for (a) $\mathrm{u}|_{[-2\mathrm{r},2\mathrm{r}]\times\{-\mathrm{r}\}}$ and (b) $\mathrm{q}|_{[-2\mathrm{r},2\mathrm{r}]\times\{-\mathrm{r}\}}$, and the corresponding normalized errors (c) $\mathrm{err}(\mathrm{u}(\mathbf{x}))$, $\mathbf{x} \in [-2\mathrm{r}, 2\mathrm{r}] \times \{-\mathrm{r}\}$, and (d) $\mathrm{err}(\mathrm{q}(\mathbf{x}))$, $\mathbf{x} \in [-2\mathrm{r}, 2\mathrm{r}] \times \{-\mathrm{r}\}$, obtained using the TRM and various levels of noise added into both the Dirichlet data $\mathrm{u}|_{\Gamma_\mathrm{D}}$ and the Neumann data $\mathrm{q}|_{\Gamma_\mathrm{N}}$, namely $p_\mathrm{u} = p_\mathrm{q} \in \{1\%, 3\%, 5\%\}$, for Example 3.

oscillations. Similar conclusions can be drawn from Figs. 6(c) and (d) which show the normalized errors $\mathrm{err}(\mathrm{u}(\mathbf{x}))$, $\mathbf{x} \in \Gamma \setminus \Gamma_{\mathrm{D}}$, and $\mathrm{err}(\mathrm{q}(\mathbf{x}))$, $\mathbf{x} \in \Gamma \setminus \Gamma_{\mathrm{N}}$, respectively, corresponding to the numerical results retrieved for the potential solution $\mathrm{u}|_{\Gamma\setminus\Gamma_{\mathrm{D}}}$ and normal flux $\mathrm{q}|_{\Gamma\setminus\Gamma_{\mathrm{N}}}$, in the case of the non-singular inverse problem in the simply connected two-dimensional domain given by Example 1 and illustrated in Figs. 6(a) and (b).

Accurate, stable and convergent numerical results with respect to the amount of noise added into the Neumann data $\mathrm{q}|_{\Gamma_{\mathrm{N}}}$, namely $p_{\mathrm{q}} \in \{1\%, 3\%, 5\%\}$, have also been obtained for both the unknown potential $\mathrm{u}|_{\Gamma\setminus\Gamma_{\mathrm{D}}}$ and normal flux $\mathrm{q}|_{\Gamma\setminus\Gamma_{\mathrm{N}}}$, in the case of the non-singular inverse problem associated with the two-dimensional Helmholtz equation in a doubly connected domain, as considered in Example 2. These results are illustrated in Figs. 7(a) and (b), whilst the corresponding normalized errors $\mathrm{err}(\mathrm{u}(\mathbf{x}))$, $\mathbf{x} \in \Gamma \setminus \Gamma_{\mathrm{D}}$, and $\mathrm{err}(\mathrm{q}(\mathbf{x}))$, $\mathbf{x} \in \Gamma \setminus \Gamma_{\mathrm{N}}$, are shown in Figs. 7(c) and (d), respectively.

Example 3 is related again to an inverse problem for the modified Helmholtz equation, but this time in a two-dimensional domain with a piecewise smooth boundary, and subject to perturbed boundary potential $\mathrm{u}|_{\Gamma_{\mathrm{D}}}$ and normal flux $\mathrm{q}|_{\Gamma_{\mathrm{N}}}$. The analytical and numerical potentials and normal fluxes on the boundaries $\Gamma \setminus \Gamma_{\mathrm{D}}$ and $\Gamma \setminus \Gamma_{\mathrm{N}}$, obtained for Example 3 using the standard MFS together with the TRM and the L-curve method, are illustrated in Figs. 8(a) and (b), respectively. These results are also quantitatively described in Figs 8(c) and (d), which show the associated normalized error $\mathrm{err}(\mathrm{u}(\mathbf{x}))$, $\mathbf{x} \in \Gamma \setminus \Gamma_{\mathrm{D}}$, and $\mathrm{err}(\mathrm{q}(\mathbf{x}))$, $\mathbf{x} \in \Gamma \setminus \Gamma_{\mathrm{N}}$, respectively, retrieved by the standard MFS+TRM algorithm and they emphasize the effect of the aforementioned regularization method on the accuracy of the numerical results in comparison with the LSM. From Figs. $6-8$ we can conclude that the numerical solutions retrieved for all non-singular inverse problems associated with both the Helmholtz and the modified Helmholtz equations in two-dimensional simply and doubly connected domains with smooth or piecewise smooth boundaries, as described by Examples $1-3$, are accurate, stable and convergent with respect to the amount of noise added into the input Dirichlet $\mathrm{u}|_{\Gamma_{\mathrm{D}}}$ and/or Neumann data $\mathrm{q}|_{\Gamma_{\mathrm{N}}}$.

7.2.5 Numerical Convergence of the Method

In this section, we investigate the influence of the number of collocation points, N, the number of source points, M, and the distance, d, between the boundary Γ of the solution domain Ω and the pseudo-boundary Γ_{S}, on the convergence and accuracy of the numerical solutions for the potential and normal flux. To do so, we consider the non-singular inverse problem given by Example 1 and set $p_{\mathrm{u}} = 5\%$, as well as two of the aforementioned parameters, while at the same time varying the remaining parameter and accounting for the restriction $\mathrm{M} \leq \mathrm{N}$.

We first analyse the convergence of the standard MFS, in conjunction with the TRM, with respect to the number of collocation points, N, and hence we set $\mathrm{M} = \mathrm{N}$ and $d = 2$. In Figs. 9(a) and (b) we present the analytical and numerical values for the potential $\mathrm{u}|_{\Gamma\setminus\Gamma_{\mathrm{D}}}$ and its normal derivative $\mathrm{q}|_{\Gamma\setminus\Gamma_{\mathrm{N}}}$, respectively, obtained using the TRM, $\lambda = \lambda_{\mathrm{opt}}$ given by the L-curve criterion and various numbers of collocation points $\mathrm{N} \in \{40, 80, 160\}$. It can be seen from these figures that the numerical results for both the unknown potential $\mathrm{u}|_{\Gamma\setminus\Gamma_{\mathrm{D}}}$ and normal flux $\mathrm{q}|_{\Gamma\setminus\Gamma_{\mathrm{N}}}$ represent very accurate approximations for their analytical counterparts, for all the MFS discretization. Furthermore, the numerical potential $\mathrm{u}^{(\mathrm{num})}|_{\Gamma\setminus\Gamma_{\mathrm{D}}}$

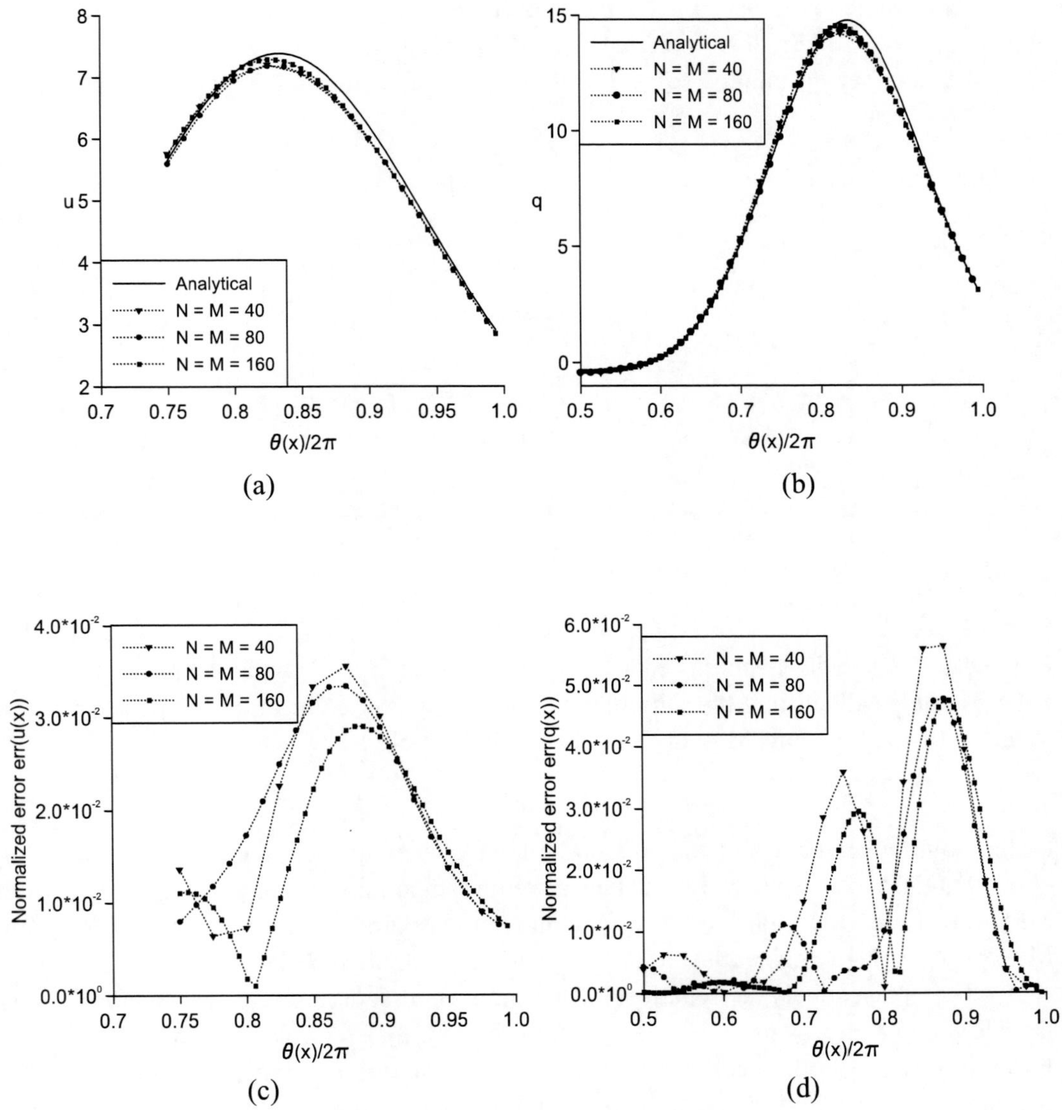

Figure 9: Analytical and numerical solutions for (a) $u|_{\Gamma\setminus\Gamma_D}$ and (b) $q|_{\Gamma\setminus\Gamma_N}$, and the corresponding normalized errors (c) $\mathrm{err}(u(\mathbf{x})), \mathbf{x}\in\Gamma\setminus\Gamma_D$, and (d) $\mathrm{err}(q(\mathbf{x})), \mathbf{x}\in\Gamma\setminus\Gamma_N$, obtained using the TRM, $p_u = 5\%$ noise added into the Dirichlet data $u|_{\Gamma_D}$, $M = N$ source points and various numbers of collocation points $N \in \{40, 80, 160\}$, for Example 1.

and normal flux $q^{(num)}|_{\Gamma \setminus \Gamma_N}$ converge to the analytical potential $u^{(an)}|_{\Gamma \setminus \Gamma_D}$ and normal flux $q^{(an)}|_{\Gamma \setminus \Gamma_N}$, respectively, as the MFS discretization is refined. Also, for all numbers of collocation points employed, the numerical results retrieved for the unknown normal flux on the under-specified boundary Γ_C are more inaccurate than those obtained for the unknown potential on Γ_C, i.e. $\text{err}(u(\mathbf{x})) < \text{err}(q(\mathbf{x}))$, $\mathbf{x} \in \Gamma_C$, see Figs. 9(c) and (d).

We now set the number of collocation points and distance between the boundary

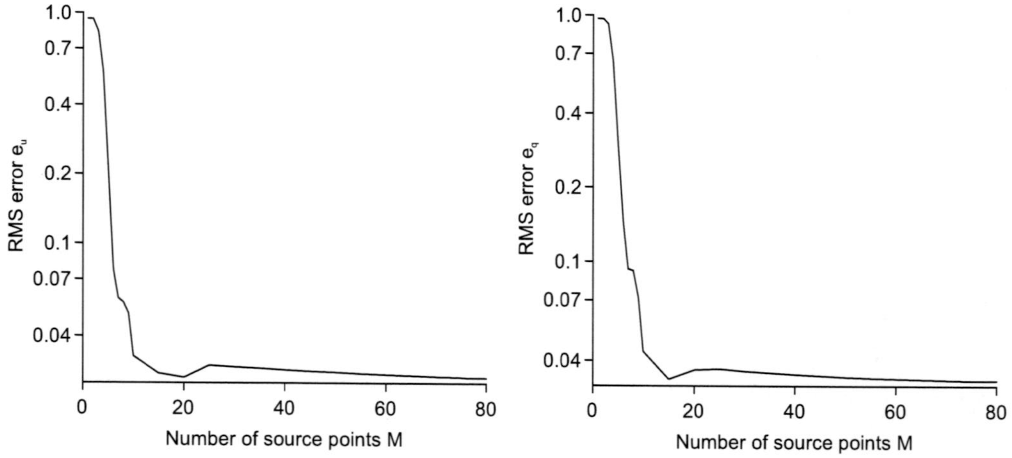

(a) (b)

Figure 10: The RMS errors (a) $e_u(\Gamma \setminus \Gamma_D)$, and (b) $e_q(\Gamma \setminus \Gamma_N)$, obtained using the TRM, $p_u = 5\%$ noise added into the Dirichlet data $u|_{\Gamma_D}$, $N = 80$ collocation points and various numbers of source points $M \in \{1, 2, \ldots, 80\}$, for Example 1.

of the solution domain and the pseudo-boundary on which the sources are located as $N = 80$ and $d = 2$, respectively, and hence we investigate the convergence of the proposed MFS+TRM algorithm with respect to the number of source points by varying the latter as $M \in \{1, 2, \ldots, N\}$. Figs. 10(a) and (b) illustrate the relative RMS errors $e_u(\Gamma \setminus \Gamma_D)$ and $e_q(\Gamma \setminus \Gamma_N)$, respectively, as functions of the number of source points, M, obtained using the TRM and $\lambda = \lambda_{opt}$ given by the L-curve criterion, for Example 1. It can be seen from these figures that both accuracy errors tend to zero as the number of source points, M, increases and, in addition, these errors do not decrease substantially for $M \geq 20$. These results indicate the fact that the standard MFS, together with the TRM, provides accurate and convergent numerical solutions with respect to increasing the number of source points, with the mention that only a small number of source points is required in order for a good accuracy of the numerical potential on $\Gamma \setminus \Gamma_D$ and normal flux on $\Gamma \setminus \Gamma_N$ to be achieved.

Next, we analyse the convergence of the proposed numerical method with respect to the position of the source points by setting $N = M = 80$, while at the same time varying d. Figs. 11(a) and (b) present the RMS errors $e_u(\Gamma \setminus \Gamma_D)$ and $e_q(\Gamma \setminus \Gamma_N)$, respectively, as functions of the distance d, obtained using the TRM and $\lambda = \lambda_{opt}$ chosen according to the L-curve criterion. From these figures it can be seen that both these errors decrease very rapidly until the distance between the boundary Γ of the domain Ω and the pseudo-boundary Γ_S attains the value $d = 2$, after which $e_u(\Gamma \setminus \Gamma_D)$ and $e_q(\Gamma \setminus \Gamma_N)$ reach a plateau region so that any

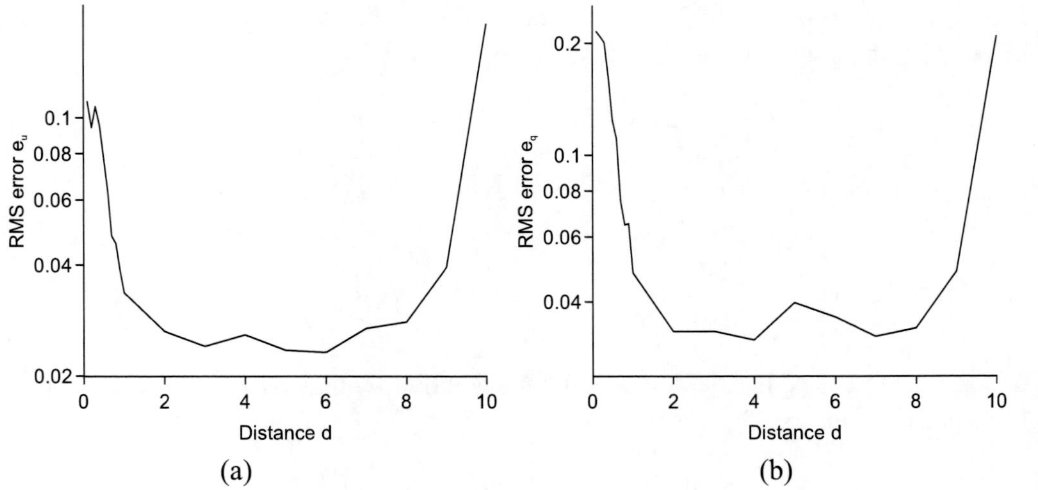

Figure 11: The RMS errors (a) $e_u(\Gamma \setminus \Gamma_D)$, and (b) $e_q(\Gamma \setminus \Gamma_N)$, obtained using the TRM, $p_u = 5\%$ noise added into the Dirichlet data $u|_{\Gamma_D}$, $N = 80$ collocation points, $M = 80$ source points and various values of the distance d between the boundary Γ and the pseudo-boundary Γ_S, for Example 1.

value $2 < d < 8$ does not significantly improve the accuracy of the numerical results for Example 1. However, the numerical results for both the unknown potential $u|_{\Gamma \setminus \Gamma_D}$ and the normal flux $q|_{\Gamma \setminus \Gamma_N}$ deteriorate significantly for $d \geq d_{\text{lim}}$, where the threshold value for d, in the case of Example 1, is given by $d_{\text{lim}} \approx 8$. This result is in accordance with the findings for the numerical solution of inverse problems in functionally graded materials using the MFS, in conjunction with the TRM, see e.g. Marin [57], and contrary to the theoretical results [43, 63]. However, this inconvenience can be overcome by employing the SVD instead of the TRM, as suggested by Jin and Zheng [21, 22, 25], Jin and Marin [60] etc.

Although not presented here, it is reported that similar results have been obtained for the non-singular inverse problems associated with the two-dimensional Helmholtz-type equations given by Examples 2 and 3. Hence we can conclude that the standard MFS, together with the TRM and the L-curve criterion, is convergent with respect to increasing the number of collocation points, the number of source points and the distance between the source points and the boundary of the solution domain.

7.3 Singular Inverse Problems

7.3.1 Examples

For the *singular inverse problems* associated with both the Helmholtz and the modified Helmholtz equations analysed herein, the solution domains under consideration, Ω, accessible boundaries, Γ_D and Γ_N, and corresponding analytical solutions for the potential are given as follows:

Example 4. N-D singularity for the modified Helmholtz equation ($\mathcal{L} = \Delta - k^2$, $k = 1$) in a rectangle containing an edge crack OA, see Fig. 12(a):

$$\Omega = \mathrm{ABCD} = (-1,1) \times (0,1) \tag{.67}$$

$$\Gamma_\mathrm{D} = \mathrm{AB} \cup \mathrm{CD} \cup \mathrm{DO} = \{-1,1\} \times (0,1) \cup (-1,0) \times \{0\} \tag{.68}$$

$$\Gamma_\mathrm{N} = \mathrm{OA} \cup \mathrm{AB} \cup \mathrm{CD} = (0,1) \times \{0\} \cup \{-1,1\} \times (0,1) \tag{.69}$$

$$\mathrm{u}^{(\mathrm{an})}(\mathbf{x}) = \mathrm{u}_1^{(\mathrm{ND})}(\mathbf{x}) - 1.3\,\mathrm{u}_2^{(\mathrm{ND})}(\mathbf{x}) + 2.0\,\mathrm{u}_3^{(\mathrm{ND})}(\mathbf{x}) - 1.7\,\mathrm{u}_4^{(\mathrm{ND})}(\mathbf{x}), \quad \mathbf{x} \in \overline{\Omega} \tag{.70}$$

Example 5. D-D singularity for the modified Helmholtz equation ($\mathcal{L} = \Delta - k^2$, $k = 1$) in an L-shaped domain, see Fig. 12(b):

$$\Omega = \mathrm{OABCDE} = (-1,1) \times (0,1) \cup (-1,0) \times (-1,0] \tag{.71}$$

$$\begin{aligned} \Gamma_\mathrm{D} = \mathrm{OA} \cup \mathrm{AB} \cup \mathrm{CD} \cup \mathrm{DE} \cup \mathrm{EO} = (0,1) \times \{0\} \cup \{1\} \times (0,1) \\ \cup \{-1\} \times (-1,1) \cup (-1,0) \times \{-1\} \cup \{0\} \times (-1,0) \end{aligned} \tag{.72}$$

$$\Gamma_\mathrm{N} = \mathrm{AB} \cup \mathrm{CD} = \{1\} \times (0,1) \cup \{-1\} \times (-1,1) \tag{.73}$$

$$\mathrm{u}^{(\mathrm{an})}(\mathbf{x}) = 2.0\,\mathrm{u}_1^{(\mathrm{DD})}(\mathbf{x}) - 1.5\,\mathrm{u}_2^{(\mathrm{DD})}(\mathbf{x}) + 1.3\,\mathrm{u}_4^{(\mathrm{DD})}(\mathbf{x}), \quad \mathbf{x} \in \overline{\Omega} \tag{.74}$$

Example 6. D-N singularity for the Helmholtz equation ($\mathcal{L} = \Delta + k^2$, $k = 2$) in a rectangle containing a V-notch with the re-entrant angle $2\,\omega = \pi/3$, see Fig. 12(c):

$$\Omega = \mathrm{OABCD} = (-1,1) \times (0,1) \setminus \Delta\mathrm{ODD}' \tag{.75}$$

$$\Gamma_\mathrm{D} = \mathrm{OA} \cup \mathrm{AB} \cup \mathrm{CD} = (0,1) \times \{0\} \cup \{1\} \times (0,1) \cup \{-1\} \times (\sin\omega, 1) \tag{.76}$$

$$\begin{aligned} \Gamma_\mathrm{N} = \mathrm{AB} \cup \mathrm{CD} \cup \mathrm{DO} = \{1\} \times (0,1) \\ \cup \{-1\} \times (\sin\omega, 1) \cup \{(-\rho, \rho\sin\omega)\,|\,0 < \rho < 1\} \end{aligned} \tag{.77}$$

$$\mathrm{u}^{(\mathrm{an})}(\mathbf{x}) = \mathrm{u}_2^{(\mathrm{DN})}(\mathbf{x}) - 1.5\,\mathrm{u}_3^{(\mathrm{DN})}(\mathbf{x}) + 1.2\,\mathrm{u}_4^{(\mathrm{DN})}(\mathbf{x}), \quad \mathbf{x} \in \overline{\Omega} \tag{.78}$$

It should be mentioned that Examples 4 – 6 analysed in this chapter contain a singularity at the point O$(0,0)$. Moreover, this singularity is caused by the nature of the analytical solutions considered, i.e. the analytical solutions are given as linear combinations of the first four singular solutions/eigenfunctions satisfying homogeneous boundary conditions on the edges of the wedge, as well as by a sharp corner in the boundary (Examples 5 and 6) or by an abrupt change in the boundary conditions at O (Examples 4 and 6), see Figs. 12(a)-(c). For all examples considered, it can be seen that the boundary $\Gamma_\mathrm{D} \cap \Gamma_\mathrm{N}$ is over-specified by prescribing on it both the solution, $\mathrm{u}\,|_{\Gamma_\mathrm{D} \cap \Gamma_\mathrm{N}}$, and the normal flux, $\mathrm{q}|_{\Gamma_\mathrm{D} \cap \Gamma_\mathrm{N}}$, whilst the boundary BC is under-specified since neither the solution, $\mathrm{u}\,|_{\mathrm{BC}}$, nor the normal flux, $\mathrm{q}|_{\mathrm{BC}}$, is known and has to be determined.

The singular inverse problems investigated herein have been solved using a uniform

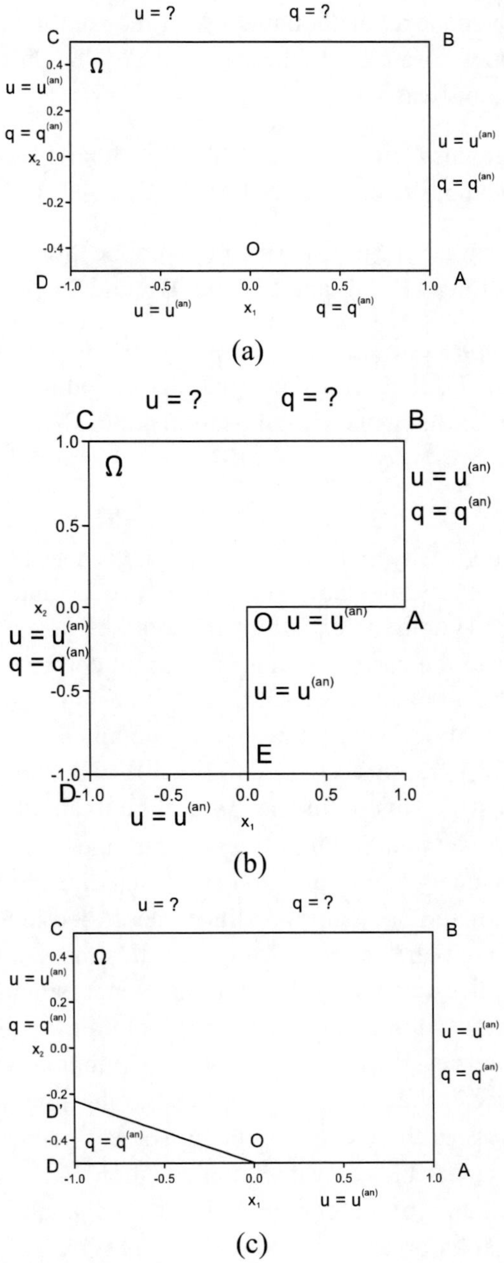

Figure 12: Schematic diagram of the geometry and boundary conditions for the singular inverse problems investigated, namely (a) Example 4: N-D singularity in a domain containing an edge crack OD; (b) Example 5: D-D singularity in an L-shaped domain; and (c) Example 6: D-N singularity in a domain containing a V-notch with the re-entrant angle $2\omega = \pi/3$.

distribution of both the boundary collocation points $\mathbf{x}^i$, $i = 1,\dots,N$, and the source points $\mathbf{y}^j$, $j = 1,\dots,M$, with the mention that the latter were located on a so-called pseudo-boundary, Γ_S, which has the same shape as the boundary Γ of the solution domain and is situated at the distance $d = 2$ from Γ, see e.g. Tankelevich *et al.* [61]. Furthermore, the number of boundary collocation points was set to:

(*i*) $N = 120$ for Examples 4 and 6, such that $N/3 = 40$ and $N/6 = 20$ collocation points are situated on each of the boundaries BC and OA, AB, CD and DO, respectively;

(*ii*) $N = 154$ for Example 5, such that 19 and 39 collocation points are situated on each of the boundaries OA, AB, DE and EO, and BC and CD, respectively.

In addition, for all examples corresponding to the above singular inverse problems associated with two-dimensional Helmholtz-type equations, the number of source points, M, was taken to be equal to that of the boundary collocation points, N, i.e. $M = N$.

7.3.2 Effect of the SST

Example 4 contains a singularity at the boundary point O caused by both the abrupt change in the boundary conditions and the nature of the analytical solution, see equation (.70), in the case of the modified Helmholtz equation. It should be noted that this singularity is of a form which is similar to the case of a sharp re-entrant corner of angle $\omega = 0$. This may be seen by extending the domain $\Omega = (-1,1) \times (0,1)$ using symmetry with respect to the x_1-axis, see also Fig 1(a). In this way, a problem is obtained for a square domain containing a crack, namely $\widetilde{\Omega} = (-1,1) \times (-1,1) \setminus [0,1] \times \{0\}$ with zero flux boundary conditions along the crack $[0,1] \times \{0\}$. This problem may also be treated by considering the domain $\widetilde{\Omega}$ described above, with the mention that the singular eigenvectors (19) corresponding to Neumann-Neumann boundary conditions along the crack must be used. However, the original domain Ω and the mixed boundary conditions described in equations (.68) and (.69) have been considered in our analysis, i.e. $2\omega = \pi$ as illustrated in Fig 12(a).

If the LSM is applied to solving the aforementioned singular inverse problem subjected to noisy data without subtracting any singular solutions/eigenfunctions ($n_S = 0$) then the numerical solution retrieved by this direct solution method is not only inaccurate, but also unstable. This aspect, which is strongly related to the singular character of the problem and the ill-posedness of the inverse problem, see e.g. Hadamard [9], and hence the ill-conditioning of the standard MFS system, i.e. equation (43) with $n_S = 0$, can be clearly noticed from Figs. 13(a) and (b) that present the analytical and LSM-based numerical normal flux and potential solution on the wedges DO and OA, respectively, when both the Dirichlet $u|_{\Gamma_D \cap \Gamma_N} = u|_{AB \cup CD}$ and the Neumann data $q|_{\Gamma_D \cap \Gamma_N} = q|_{AB \cup CD}$ were perturbed by $p_u = p_q = 3\%$ noise, in the case of Example 4. As expected, the numerical values retrieved for the potential solution and normal flux on the under-specified boundary BC are highly oscillatory and they represent very inaccurate approximations for their corresponding analytical values, see Figs. 13(c) and (d).

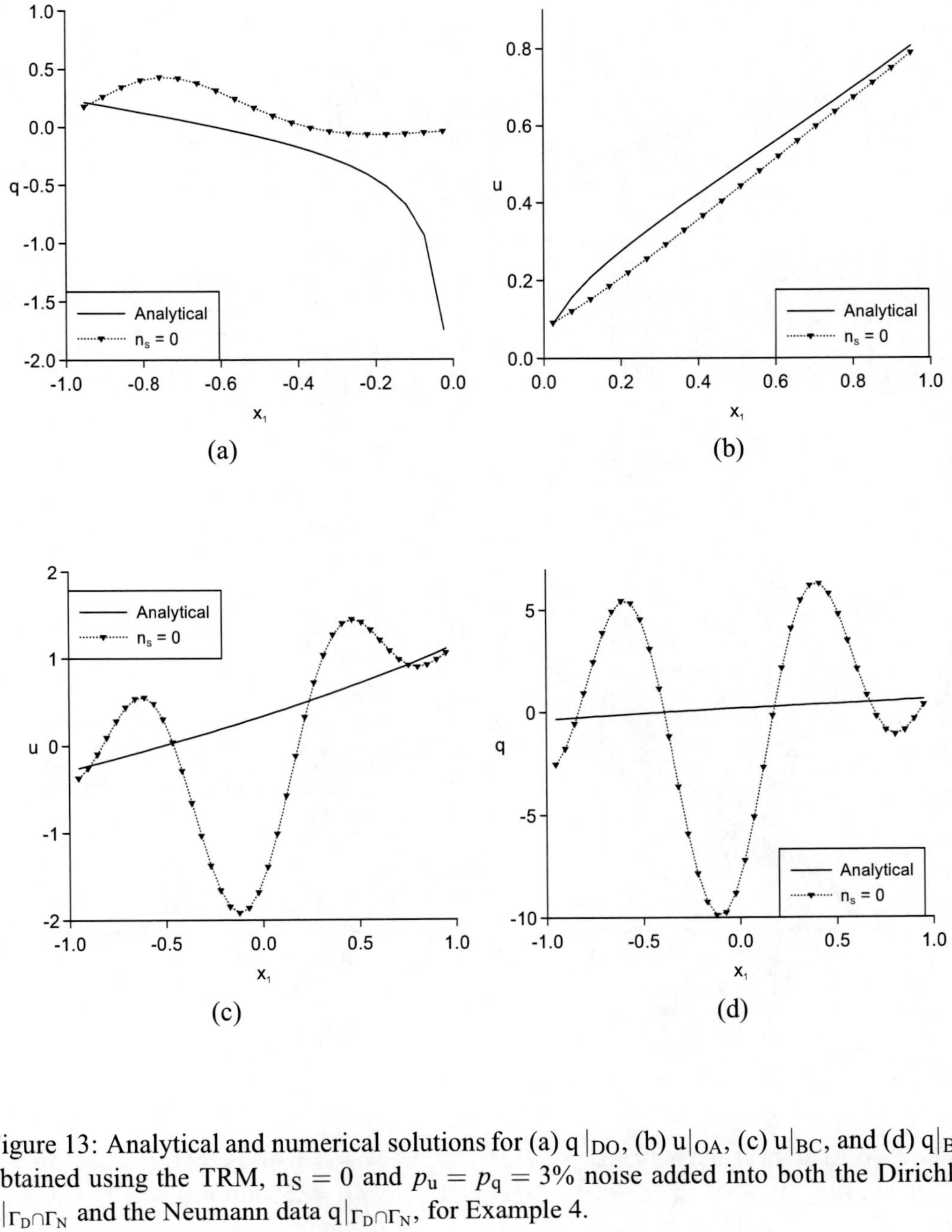

Figure 13: Analytical and numerical solutions for (a) $q\,|_{DO}$, (b) $u|_{OA}$, (c) $u|_{BC}$, and (d) $q|_{BC}$, obtained using the TRM, $n_S = 0$ and $p_u = p_q = 3\%$ noise added into both the Dirichlet $u|_{\Gamma_D \cap \Gamma_N}$ and the Neumann data $q|_{\Gamma_D \cap \Gamma_N}$, for Example 4.

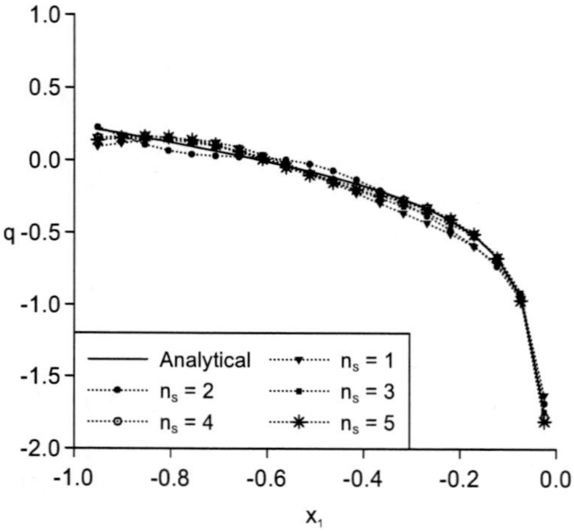

(a)

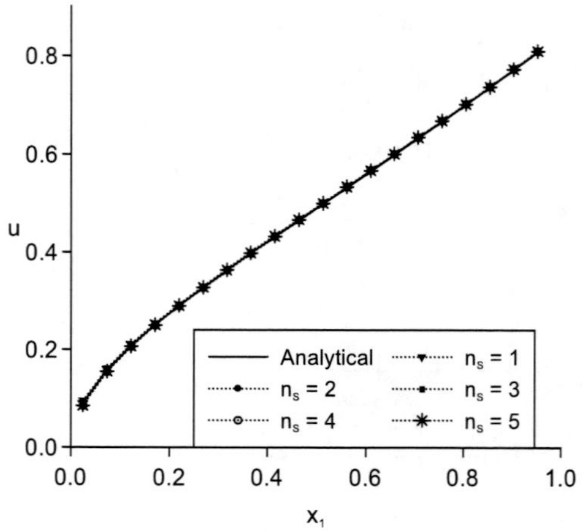

(b)

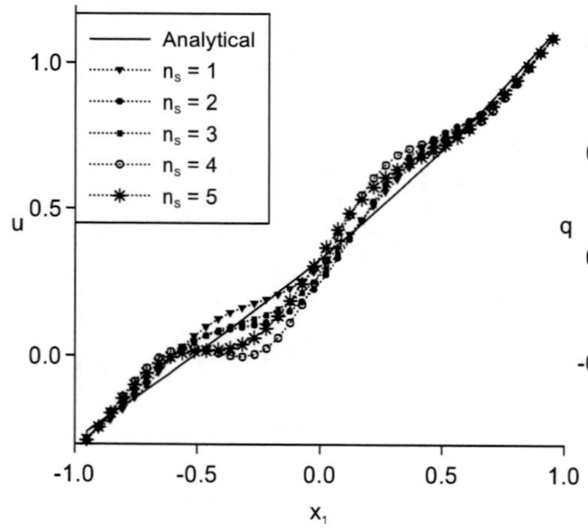

(c)

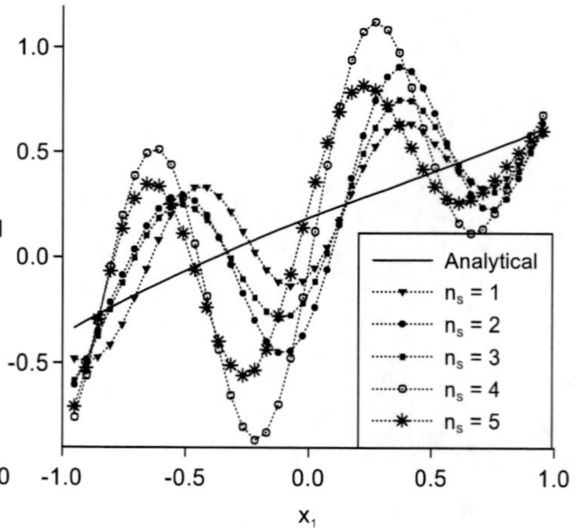

(d)

Figure 14: Analytical and numerical solutions for (a) $q\,|_{DO}$, (b) $u|_{OA}$, (c) $u|_{BC}$, and (d) $q|_{BC}$, obtained using the LSM, $n_S \in \{1,2,3,4,5\}$ and $p_u = p_q = 3\%$ noise added into both the Dirichlet $u|_{\Gamma_D \cap \Gamma_N}$ and the Neumann data $q|_{\Gamma_D \cap \Gamma_N}$, for Example 4.

Figs. 14(a) and (b) illustrate a comparison between the analytical and numerical solutions for $q|_{DO}$ and $u|_{OA}$, respectively, obtained with $p_u = p_q = 3\%$ and by removing various numbers of singular potential solutions/eigenfunctions, namely $n_S \in \{1,2,3,4,5\}$, for the N-D singular problem given by Example 4. It can be seen from these figures that the numerical results for both the flux and the potential solution are considerably improved, even if only the first singular potential solution/eigenfunction corresponding to Dirichlet-Neumann boundary conditions on $(-1,1) \times \{0\}$ is removed, i.e. $n_S = 1$. The same pattern is observed if one continues to remove higher-order singular potential solutions/eigenfunctions in the modified MFS, i.e. $n_S \geq 2$, as can be seen from Figs. 14(a) and (b). Moreover, the removal of $n_S \geq 3$ singular potential solutions/eigenfunctions from the standard MFS ensures the retrieval of reasonably accurate numerical solutions for the flux on DO and the potential solution on OA, respectively, see Figs. 14(a) and (b). However, very inaccurate and highly oscillatory solutions have been obtained for both the unknown potential solution and normal flux on the under-specified boundary BC and these are presented in Figs. 14(c) and (d), respectively.

Although not presented, it is reported that similar results have been obtained for the other singular inverse problems described by Examples 5 and 6. At this stage, we may conclude that in order to retrieve accurate and stable numerical solutions for singular inverse problem associated with Helmholtz-type equations, the use of the SST in the modified MFS approach only is not sufficient, as clearly shown in Figs 14(a)–(d).

7.3.3 Effect of the TRM

If only the TRM is employed to solve the resulting MFS system of linear algebraic equations (43) without removing any singular solutions/eigenfunctions, i.e. $n_S = 0$, then again very inaccurate numerical results have been retrieved for both the potential solution and the normal flux. These results are presented in Figs. 15(a) and (b) which illustrate the analytical and numerical normal flux and potential solution on the wedges DO and OA, respectively, when both the Dirichlet $u|_{\Gamma_D \cap \Gamma_N} = u|_{AB \cup CD}$ and the Neumann data $q|_{\Gamma_D \cap \Gamma_N} = q|_{AB \cup CD}$ were perturbed by $p_u = p_q = 3\%$ noise and $n_S = 0$, in the case of Example 4. However, the numerical results for the unknown potential $u|_{BC}$ and normal flux $q|_{BC}$, retrieved using the TRM and $n_S = 0$, approximate their corresponding analytical values in a more stable and accurate manner than the numerical results for the potential $u|_{BC}$ and normal flux $q|_{BC}$, retrieved using the LSM and $n_S = 0$, see Figs. 15(c) and (d), and 13(c) and (d), respectively. By comparing Figs. 13 and 15, we can conclude that both the SST and the TRM should be employed in order to solve stably and accurately the inverse problem given by Example 4.

Indeed, if these two techniques are used together then the difficulties caused by the ill-posedness of the inverse problem, as well as the boundary singularity at O, can be overcome. Figs. 16(a) and (b) present the analytical and numerical results for the normal flux on DO and potential solution on OA, respectively, retrieved by solving the MFS+SST system of linear algebraic equations (41) using the TRM, in conjunction with the L-curve criterion for choosing the optimal regularization parameter, $p_u = p_q = 3\%$ and $n_S \in \{1,2,3,4,5\}$, in the case of Example 4. From these figures it can be noticed that the effect of the SST in the presence of the TRM is remarkable. When TRM is employed then even the removal of the first two singular solutions/eigenfunctions, i.e. $n_S = 2$, provides a very accurate numerical

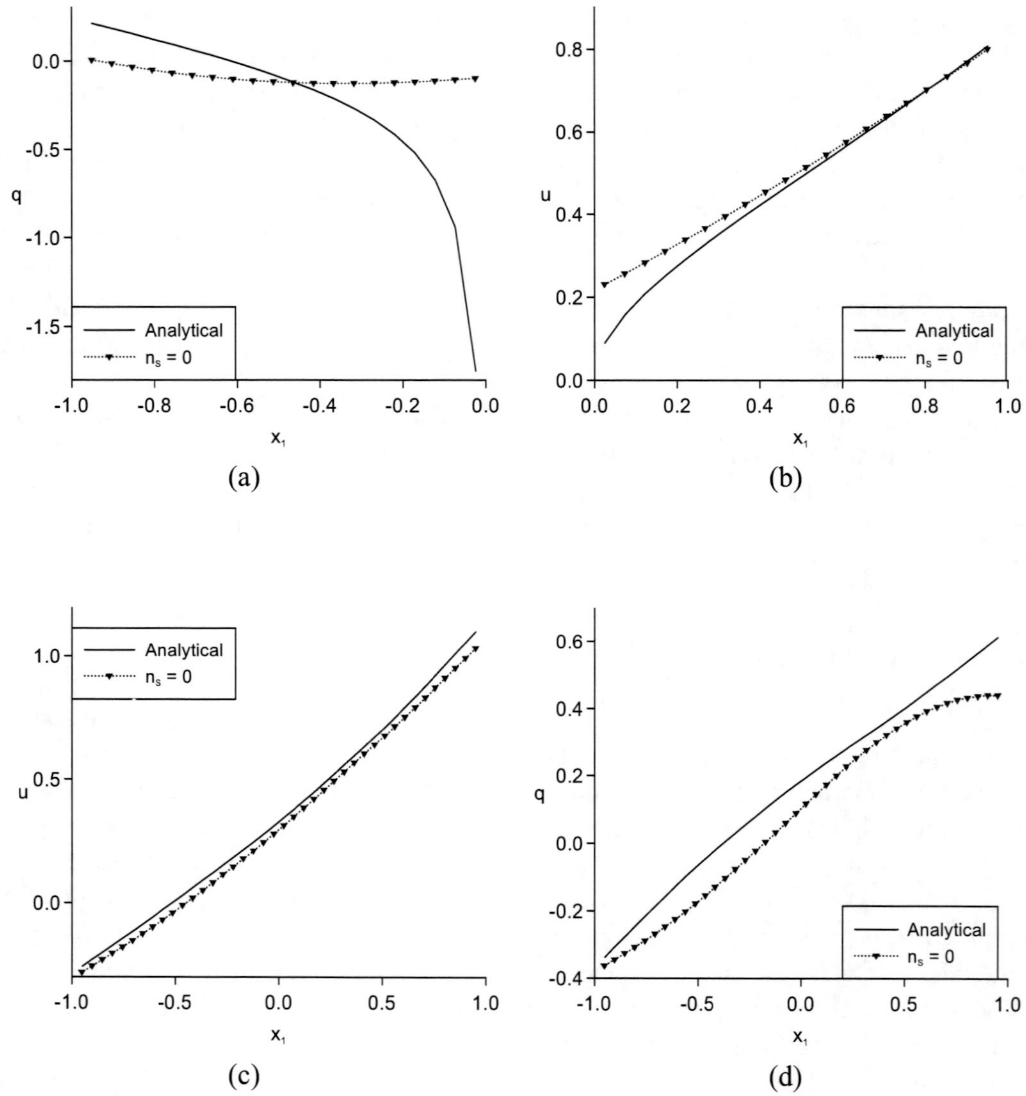

Figure 15: Analytical and numerical solutions for (a) $q\,|_{\mathrm{DO}}$, (b) $u|_{\mathrm{OA}}$, (c) $u|_{\mathrm{BC}}$, and (d) $q|_{\mathrm{BC}}$, obtained using the TRM, $n_S = 0$ and $p_u = p_q = 3\%$ noise added into both the Dirichlet $u|_{\Gamma_D \cap \Gamma_N}$ and the Neumann data $q|_{\Gamma_D \cap \Gamma_N}$, for Example 4.

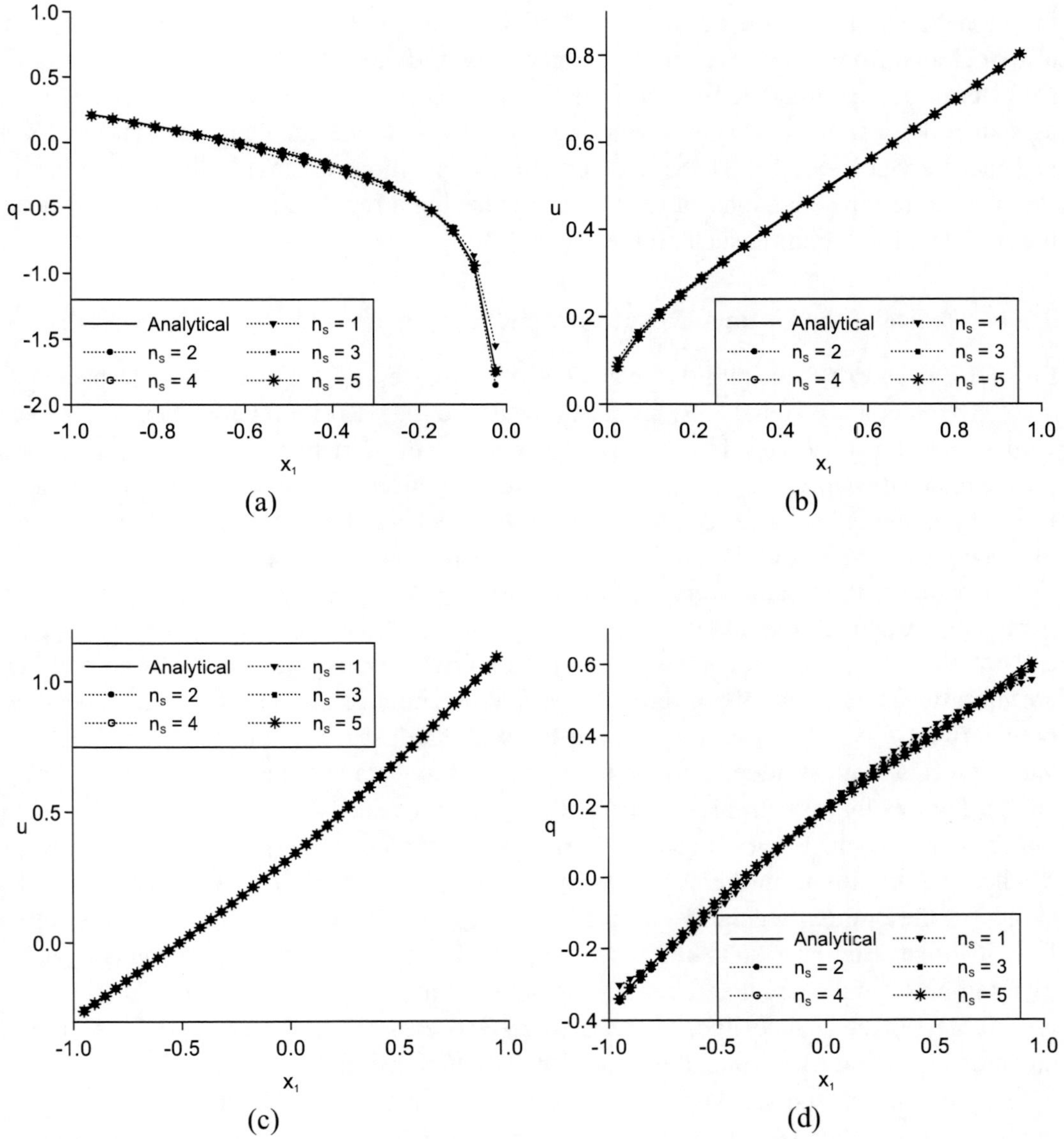

Figure 16: Analytical and numerical solutions for (a) $q\,|_{\mathrm{DO}}$ and (b) $u|_{\mathrm{OA}}$, and the corresponding normalized errors (c) $\mathrm{err}(q(\mathbf{x}))$, $\mathbf{x} \in \mathrm{DO}$, and (d) $\mathrm{err}(u(\mathbf{x}))$, $\mathbf{x} \in \mathrm{OA}$, obtained using the TRM, $n_S \in \{1,2,3,4,5\}$ and $p_u = p_q = 3\%$ noise added into both the Dirichlet $u\,|_{\Gamma_D \cap \Gamma_N}$ and the Neumann data $q|_{\Gamma_D \cap \Gamma_N}$, for Example 4.

approximation for the potential solution $u|_{OA}$, which is also bounded and exempted from oscillations. Similar estimations are also valid for the numerical normal flux $q|_{DO}$, with the mention that, as expected, the numerical results obtained for the normal flux on DO are more inaccurate than those retrieved for the potential solution on OA. The same conclusion can also be drawn from Figs. 16(c) and (d) which present the results shown in Figs. 16(a) and (b) in terms of the normalized errors $\mathrm{err}(q(\mathbf{x}))$, $\mathbf{x} \in \mathrm{DO}$, and $\mathrm{err}(u(\mathbf{x}))$, $\mathbf{x} \in \mathrm{OA}$, respectively, as defined by formula (54). On comparing Figs. 13 – 16, we can conclude that the TRM provides very accurate MFS+SST-based numerical solutions to singular inverse problems for Helmholtz-type equations, at the same time having a regularizing/stabilizing effect on the MFS+SST solutions to such problems.

7.3.4 Choice of the Optimal Regularization Parameter

Figs. 17(a) and (b) illustrate the relative RMS errors $e_u(BC)$ and $e_q(BC)$, respectively, given by relations (54) and (55), as functions of the regularization parameter λ, obtained with $n_S = 5$ and various levels of noise added into both the Dirichlet $u|_{\Gamma_D \cap \Gamma_N} = u|_{AB \cup CD}$ and the Neumann data $q|_{\Gamma_D \cap \Gamma_N} = q|_{AB \cup CD}$, for the singular inverse problem given by Example 4. From these figures it can be seen that both errors $e_u(BC)$ and $e_q(BC)$ decrease as the levels of noise p_u and p_q added into the input Dirichlet and Neumann data, respectively, decrease for all regularization parameters λ; furthermore, $e_u(BC) < e_q(BC)$ for all regularization parameters λ and a fixed amount $p_u = p_q$ of noise added into the input potential $u|_{\Gamma_D \cap \Gamma_N}$ and normal flux $q|_{\Gamma_D \cap \Gamma_N}$, respectively, i.e. the numerical results obtained for the normal flux are more inaccurate than those retrieved for the potential solution on the under-specified boundary BC. Fig. 17(c) shows on a log-log scale the L-curves obtained for $n_S = 5$ and various levels of noise added into the input normal flux data in the case of Example 4. By comparing this figure with Figs. 17(a) and (b), it can be seen, for various levels of noise, that the corner of the L-curve occurs at about the same value of the regularization parameter λ where the minimum in the accuracy errors $e_u(BC)$ and $e_q(BC)$ is attained. Hence the choice of the optimal regularization parameter λ_{opt} according to the L-curve criterion is fully justified. Similar results have been obtained for the singular inverse problems given by Examples 5 and 6 and therefore they are not presented here.

Table 1 presents the values of the relative RMS errors $e_u(BC)$ and $e_q(BC)$ defined by equations (54) and (55), respectively, obtained by both methods, namely the LSM and the TRM, with $n_S = 5$ and various levels of noise added into both the input potential $u|_{\Gamma_D \cap \Gamma_N}$ and normal flux data $q|_{\Gamma_D \cap \Gamma_N}$, namely $p_u = p_q \in \{1\%, 3\%, 5\%\}$, in the case of Example 4, as well as the optimal values for the regularization parameter λ. By considering this table, as well as Figs. 13 – 16, we can conclude that the use of regularization methods, in conjunction with the SST+MFS approach, is fully justified and provides both stable and accurate numerical results not only on the edges sharing the singularity point, but also on the under-specified boundary BC.

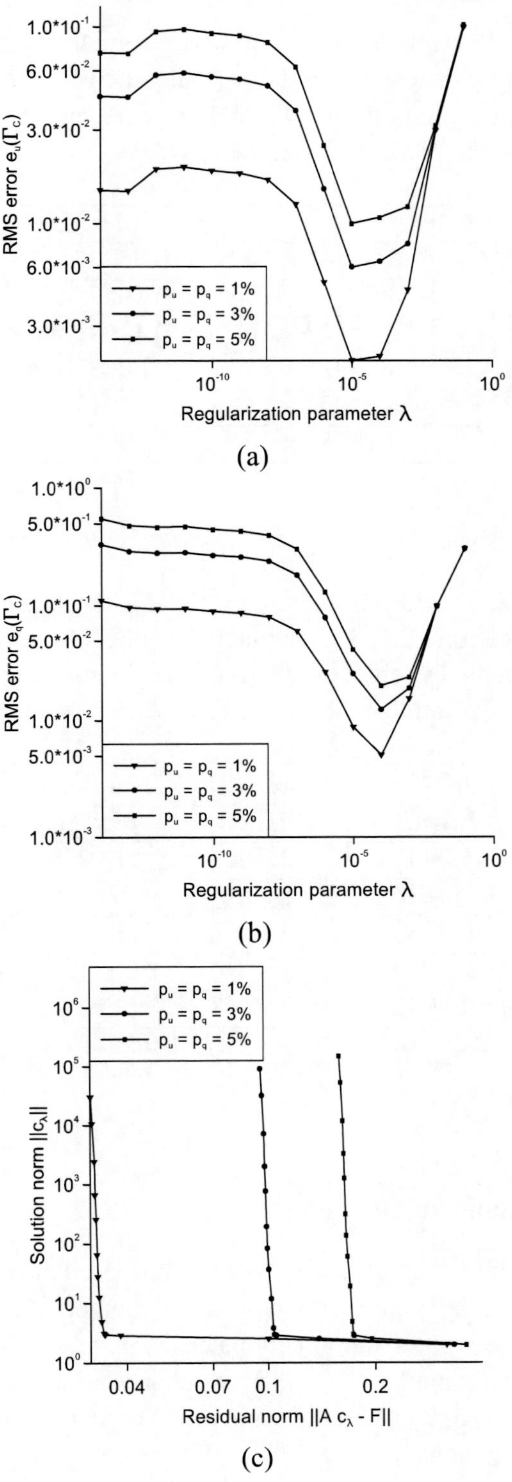

Figure 17: The RMS errors (a) $e_u(BC)$ and (b) $e_q(BC)$, and (c) the corresponding L-curve, obtained using the TRM, $n_S = 5$ and various levels of noise added into both the Dirichlet $u|_{\Gamma_D \cap \Gamma_N}$ and the Neumann data $q|_{\Gamma_D \cap \Gamma_N}$, namely $p_u = p_q \in \{1\%, 3\%, 5\%\}$, for Example 4.

Table 1: The relative RMS errors, $e_u(BC)$ and $e_q(BC)$, and the values for the corresponding optimal regularization parameter, λ_{opt}, obtained using the LSM and TRM, $n_S = 5$ and various levels of noise added into both the Dirichlet $u|_{\Gamma_D \cap \Gamma_N}$ and the Neumann data $q|_{\Gamma_D \cap \Gamma_N}$, namely $p_u = p_q \in \{1\%, 3\%, 5\%\}$, for Example 4.

Method	$p_u = p_q$	$e_u(BC)$	$e_q(BC)$	λ_{opt}
LSM	1%	0.31448×10^{-1}	0.35287×10^{0}	–
	3%	0.94332×10^{-1}	0.10585×10^{1}	–
	5%	0.15722×10^{0}	0.17641×10^{1}	–
TRM	1%	0.20923×10^{-2}	0.50682×10^{-2}	1.0×10^{-4}
	3%	0.64077×10^{-2}	0.12313×10^{-1}	1.0×10^{-4}
	5%	0.10755×10^{-1}	0.19894×10^{-1}	1.0×10^{-4}

Table 2: The numerically retrieved values, $a_j^{(num)}$, for the flux intensity factors and the corresponding absolute errors, $Err(a_j)$, obtained using the TRM , $n_S \doteq 5$ and various levels of noise added into both the Dirichlet $u|_{\Gamma_D \cap \Gamma_N}$ and the Neumann data $q|_{\Gamma_D \cap \Gamma_N}$, namely $p_u = p_q \in \{1\%, 3\%, 5\%\}$, for Example 4.

$p_u = p_q$	$a_1^{(num)}$	$a_2^{(num)}$	$a_3^{(num)}$	$a_4^{(num)}$
	$Err(a_1)$	$Err(a_2)$	$Err(a_3)$	$Err(a_4)$
1%	1.0001	−1.3068	1.9228	−1.3850
	0.79×10^{-4}	0.68×10^{-2}	0.77×10^{-1}	0.31×10^{0}
3%	0.9968	−1.2868	1.9741	−1.2741
	0.31×10^{-2}	0.13×10^{-1}	0.25×10^{-1}	0.42×10^{0}
5%	0.9937	−1.2668	2.0254	−1.1633
	0.62×10^{-2}	0.33×10^{-1}	0.25×10^{-1}	0.53×10^{0}

7.3.5 Numerical Stability of the Method

In order to investigate the numerical stability of the proposed modified MFS algorithm described in Section 5, in conjunction with the L-curve method of Hansen [67] for selecting the optimal value for the regularization parameter, λ, in what follows we consider the inverse problems given by Examples 4 – 6, the corresponding MFS discretizations mentioned in Section 7.3.1 and $n_S = 5$, whilst at the same time varying the levels of noise added into the Dirichlet and/or Neumann data as $p_u, p_q \in \{1\%, 3\%, 5\%\}$.

Figs. 18(a) and (b) present the numerical normal flux on the boundary $DO = (-1,0) \times \{0\}$ and potential solution on the boundary $OA = (0,1) \times \{0\}$, respectively, obtained using the MFS+SST algorithm for various levels of Gaussian random noise added into both the Dirichlet $u|_{\Gamma_D \cap \Gamma_N}$ and the Neumann data $q|_{\Gamma_D \cap \Gamma_N}$, in the case of Example 4, in comparison with their analytical values. From these figures it can be seen that, for all amounts, $p_u = p_q$,

Table 3: The relative RMS errors, $e_u(BC)$ and $e_q(BC)$, and the values for the corresponding optimal regularization parameter, λ_{opt}, obtained using the LSM and TRM, $n_S = 5$ and various levels of noise added into the potential solution $u|_{\Gamma_D}$, namely $p_u \in \{1\%, 3\%, 5\%\}$, for Example 5.

Method	p_u	$e_u(BC)$	$e_q(BC)$	λ_{opt}
LSM	1%	0.55255×10^{-2}	0.93942×10^{-1}	–
	3%	0.16576×10^{-1}	0.28181×10^{0}	–
	5%	0.27626×10^{-1}	0.46969×10^{0}	–
TRM	1%	0.31899×10^{-3}	0.19134×10^{-2}	1.0×10^{-5}
	3%	0.96014×10^{-3}	0.54738×10^{-2}	1.0×10^{-5}
	5%	0.16016×10^{-2}	0.90389×10^{-2}	1.0×10^{-5}

of noise added into $u|_{\Gamma_D \cap \Gamma_N}$ and $q|_{\Gamma_D \cap \Gamma_N}$, both the numerical potential solution on OA and the normal flux on DO represent excellent approximations for their analytical counterparts, being at the same time exempted from high and unbounded oscillations in the vicinity of the singularity. Similar conclusions can be drawn from Figs. 18(c) and (d) which show the normalized errors $\mathrm{err}(q(\mathbf{x}))$, $\mathbf{x} \in DO$, and $\mathrm{err}(u(\mathbf{x}))$, $\mathbf{x} \in OA$, respectively, corresponding to the numerical results retrieved for the normal flux $q|_{DO}$ and potential solution $u|_{OA}$ in the case of the singular inverse problem given by Example 4 and illustrated in Figs. 18(a) and (b). The numerical potential solution and normal flux on the under-specified boundary BC, as well as their corresponding normalized errors, obtained using the regularized MFS+SST, $n_S = 5$ and $p_u = p_q \in \{1\%, 3\%, 5\%\}$, for Example 4, are illustrated in Figs. 19(a)–(d). From these figures we can conclude that the numerical results for the potential solution and normal flux on the under-specified boundary BC are also excellent approximations for their corresponding exact values and, in addition, they are convergent and stable with respect to decreasing the amount of noise added into the input boundary potential $u|_{\Gamma_D \cap \Gamma_N}$ and normal flux $q|_{\Gamma_D \cap \Gamma_N}$. Accurate, stable and convergent numerical results with respect to $p_u = p_q$ have also been obtained for the flux intensity factors a_j, $j = 1, \ldots, 4$, and these, together with their associated absolute errors $\mathrm{Err}(a_j)$, $j = 1, \ldots, 4$, are presented in Table 2.

Example 5 is related again to the modified Helmholtz equation and contains a singularity at the origin O, which is caused by a sharp corner in the boundary, as well as the nature of the analytical potential solution corresponding to this problem, i.e. the analytical potential solution is given as a linear combination of the first four singular potential solutions satisfying homogeneous Dirichlet boundary conditions on the edges of the wedge, whilst the boundary potential measurements $u|_{\Gamma_D}$ are perturbed. The analytical and numerical fluxes on the boundaries $EO = \{0\} \times (-1, 0)$ and $OA = (0, 1) \times \{0\}$ obtained for Example 5 by subtracting $n_S = 5$ singular potential solutions/eigenfunctions are illustrated in Figs. 20(a) and (b), respectively. Although not presented, it is worth mentioning that the numerical flux obtained on $EO \cup OA$ using the standard MFS, i.e. $n_S = 0$, exhibits very high oscillations in the neighbourhood of the singular point and hence it represents an inaccurate approximation for the analytical flux. These results are also quantitatively described in Figs 20(c) and (d), which show the corresponding normalized error $\mathrm{err}(q(\mathbf{x}))$ for $\mathbf{x} \in EO$ and $\mathbf{x} \in OA$, respectively, retrieved by the TRM and the MFS+SST algorithm. The effect of the afore-

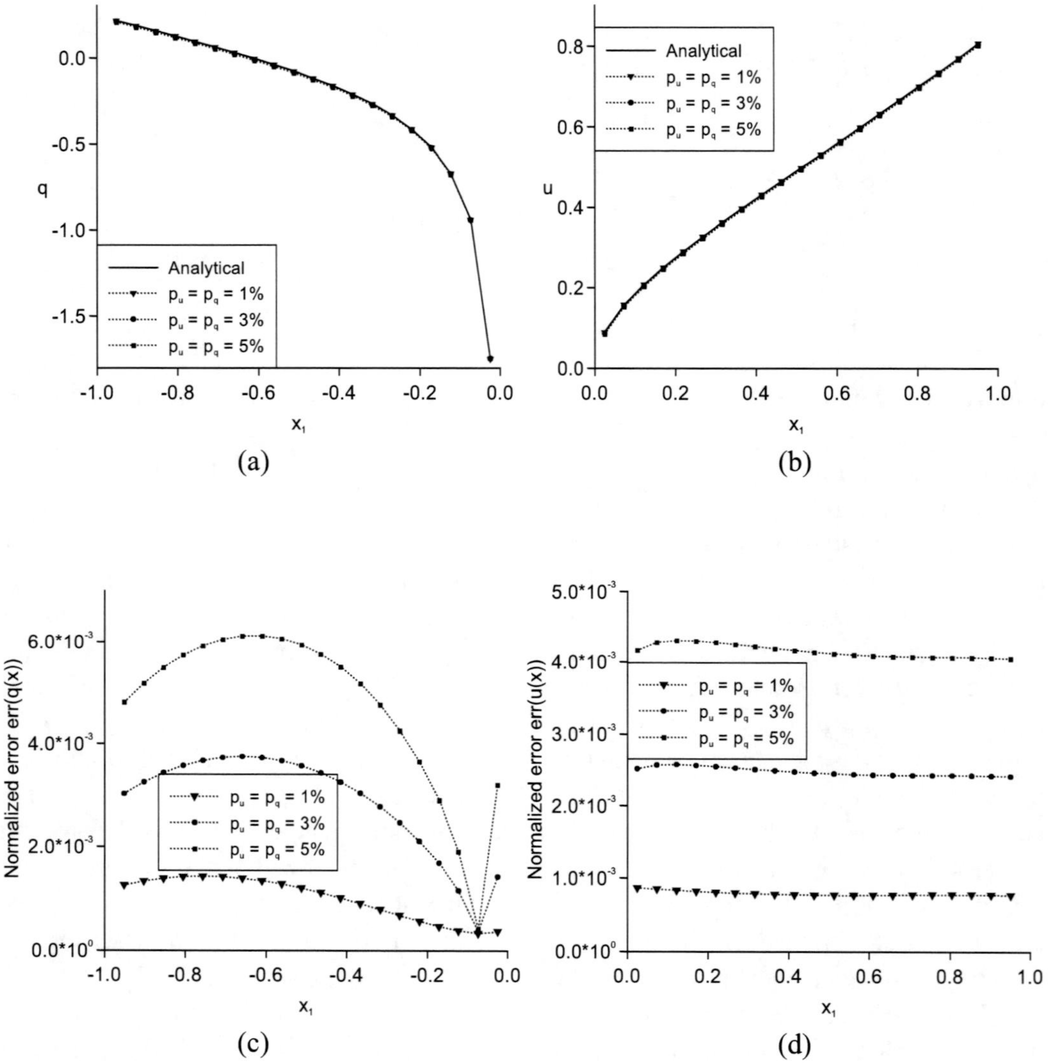

Figure 18: Analytical and numerical solutions for (a) $q|_{DO}$ and (b) $u|_{OA}$, and the corresponding normalized errors (c) $err(q(\mathbf{x}))$, $\mathbf{x} \in DO$, and (d) $err(u(\mathbf{x}))$, $\mathbf{x} \in OA$, obtained using the TRM, $n_S = 5$ and various levels of noise added into both the Dirichlet $u|_{\Gamma_D \cap \Gamma_N}$ and the Neumann data $q|_{\Gamma_D \cap \Gamma_N}$, namely $p_u = p_q \in \{1\%, 3\%, 5\%\}$, for Example 4.

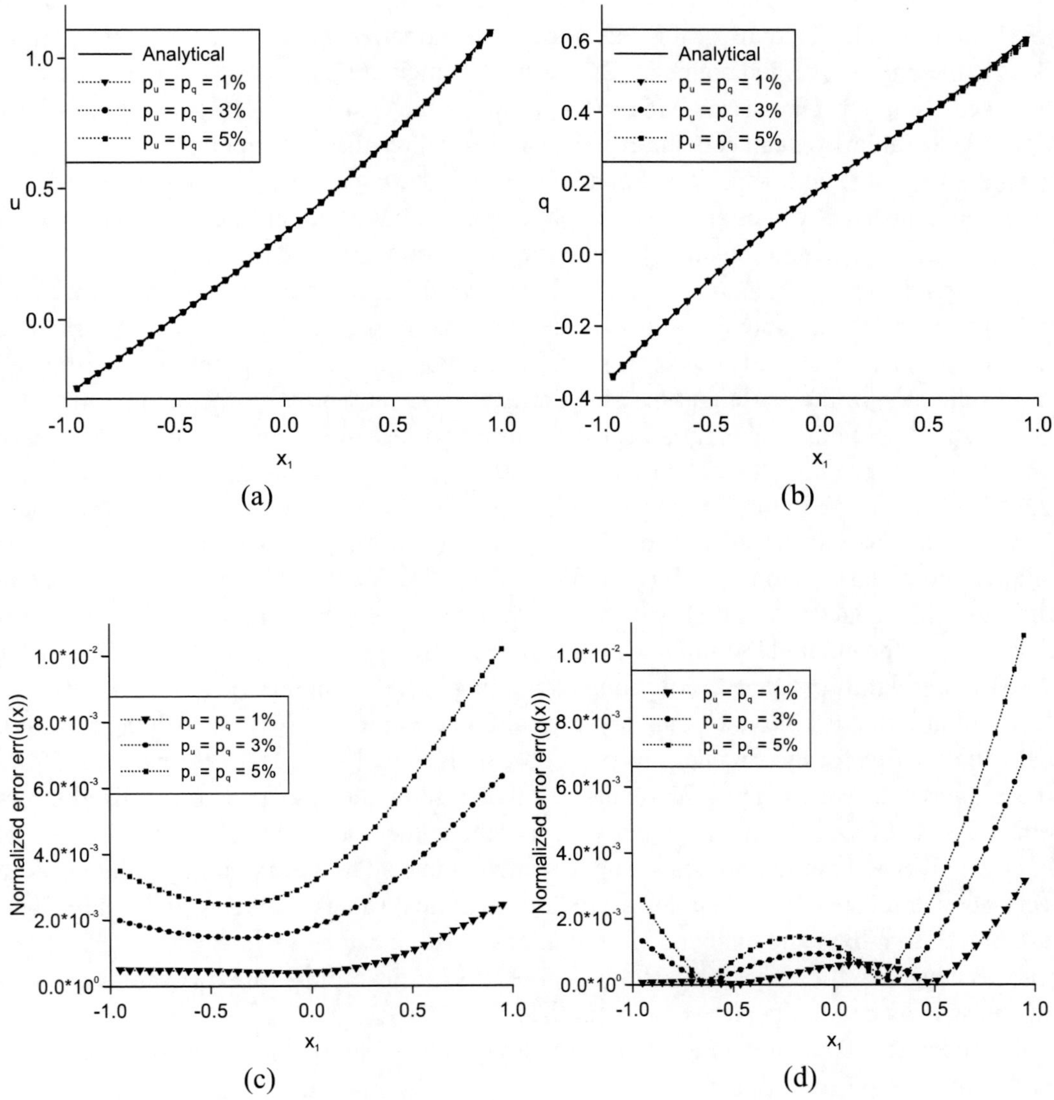

Figure 19: Analytical and numerical solutions for (a) $u|_{BC}$ and (b) $q|_{BC}$, and the corresponding normalized errors (c) $err(u(\mathbf{x}))$, $\mathbf{x} \in BC$, and (d) $err(q(\mathbf{x}))$, $\mathbf{x} \in BC$, obtained using the TRM, $n_S = 5$ and various levels of noise added into both the Dirichlet $u|_{\Gamma_D \cap \Gamma_N}$ and the Neumann data $q|_{\Gamma_D \cap \Gamma_N}$, namely $p_u = p_q \in \{1\%, 3\%, 5\%\}$, for Example 4.

mentioned regularization method on the accuracy of the numerical results in comparison with the LSM is clearly shown in Table 3, which presents the relative RMS errors, $e_u(BC)$ and $e_q(BC)$, and the values for the corresponding optimal regularization parameter, λ_{opt}, obtained on the under-specified boundary BC using the LSM and TRM, $n_S = 5$ and various levels of noise added into $u|_{\Gamma_D}$, for Example 5. The numerical potential solution and normal flux on the under-specified boundary BC, as well as their corresponding normalized errors, obtained using the regularized MFS+SST, are illustrated in Figs. 21(a)-(d). From these figures and Table 3 we can conclude that the numerical results for the potential solution and normal flux on the under-specified boundary BC are also excellent approximations for their corresponding exact values and, in addition, they are convergent and stable with respect to decreasing the amount of noise added into the input boundary potential $u|_{\Gamma_D}$.

Consider now the singular inverse problem for the two-dimensional Helmholtz equation, as given by Example 6, with perturbed boundary normal flux on Γ_N. This singular Cauchy problem is actually the most severe one among the inverse problems investigated in this chapter, in the sense that the singularity at O is caused by all factors that may occur in such a situation, namely a sharp re-entrant corner and abrupt change in boundary conditions on the side $DO \cup OA$ (here $u|_{OA}$ and $q|_{DO}$ are prescribed), see Figs. 1(c) and 22(c), as well as the nature of the analytical potential solution corresponding to this problem, i.e. the analytical potential solution is given as linear a combination of the first four singular potential solutions satisfying homogeneous Dirichlet and Neumann boundary conditions on the edges OA and DO, respectively. Figs. 22(a) and (b) present the numerical solutions for the potential solution $u|_{DO}$ and normal flux $q|_{OA}$, respectively, retrieved by the TRM along with the L-curve criterion, subtracting $n_S = 5$ singular functions and various levels of noise added into the boundary normal flux $q|_{\Gamma_N}$, in comparison with their analytical counterparts, for the singular inverse problem given by Example 6. It can be seen from these figures, as well as Figs. 22(c) and (d), which show the associated normalized errors $\mathrm{err}(u(\mathbf{x}))$, $\mathbf{x} \in DO$, and $\mathrm{err}(q(\mathbf{x}))$, $\mathbf{x} \in OA$, that the numerical results for both the potential solution $u|_{DO}$ and normal flux $q|_{OA}$ on the wedges adjacent to the singularity point O are in very good agreement with their corresponding analytical values and being at the same time exempted from high and unbounded oscillations. Also, it can be seen from Figs. 22(a)–(d) that, as expected, the errors in the numerical results for the normal flux $q|_{OA}$ are larger than those associated with the numerical potential solution $u|_{DO}$.

Accurate, stable and convergent results have also been obtained for the unspecified potential solution $u|_{BC}$ and normal flux $q|_{BC}$ when the modified MFS described in Section 5, in conjunction with the TRM and the L-curve criterion, is employed to numerically solve the singular inverse problem given by Example 6 subjected to perturbed Neumann data, as can be observed form Figs. 23(a) and (b), respectively. From these figures, as well as Figs. 23(c) and (d) which present the normalized errors $\mathrm{err}(u(\mathbf{x}))$ and $\mathrm{err}(q(\mathbf{x}))$ for $\mathbf{x} \in BC$, respectively, we can conclude that the numerical results for the potential solution on the under-specified boundary BC are also excellent approximations for their corresponding exact values. Moreover, the numerical normal fluxes $q|_{BC}$ are more inaccurate than the associated potential solutions $u|_{BC}$, as can be noticed by comparing Figs. 23(a) and (c), or Figs. 23(b) and (d), and this is a direct consequence of the severity of the singular inverse problem given by Example 6. In addition, from these figures it can be concluded that both numerical potential solutions and normal fluxes on the under-specified boundary BC are

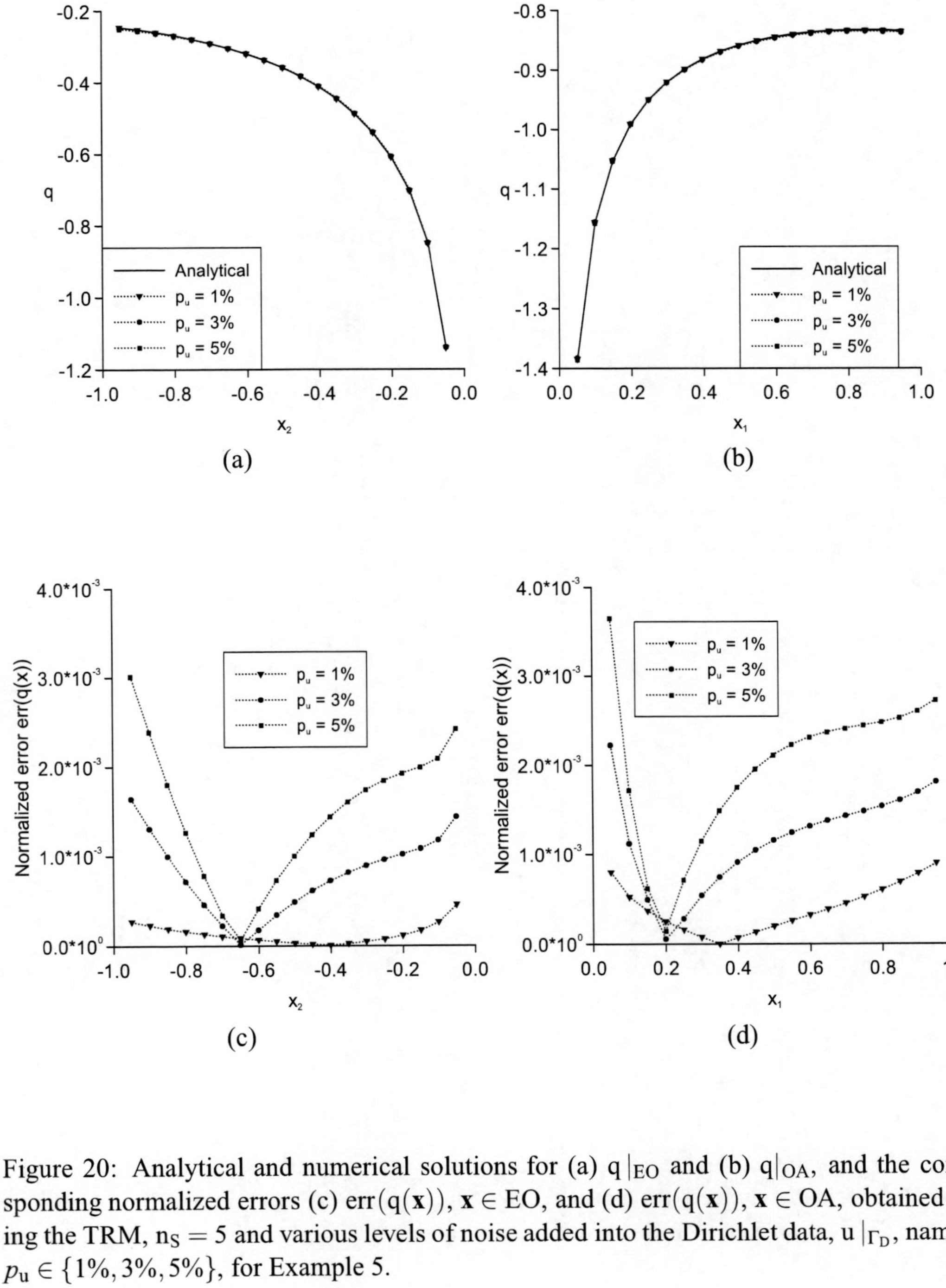

Figure 20: Analytical and numerical solutions for (a) $q|_{EO}$ and (b) $q|_{OA}$, and the corresponding normalized errors (c) $\text{err}(q(\mathbf{x}))$, $\mathbf{x} \in EO$, and (d) $\text{err}(q(\mathbf{x}))$, $\mathbf{x} \in OA$, obtained using the TRM, $n_S = 5$ and various levels of noise added into the Dirichlet data, $u|_{\Gamma_D}$, namely $p_u \in \{1\%, 3\%, 5\%\}$, for Example 5.

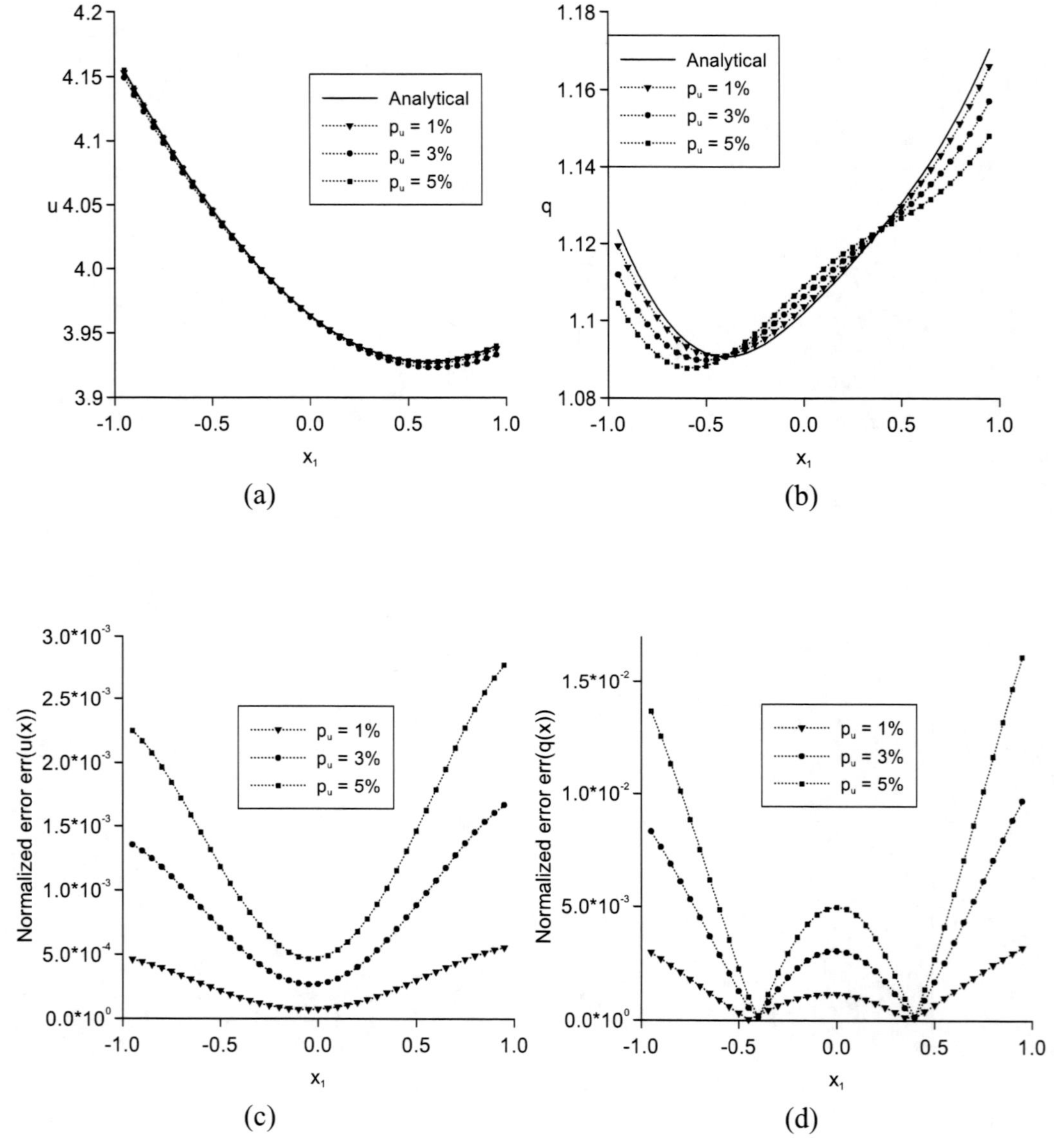

Figure 21: Analytical and numerical solutions for (a) $u|_{BC}$ and (b) $q|_{BC}$, and the corresponding normalized errors (c) $\mathrm{err}(u(\mathbf{x}))$, $\mathbf{x} \in BC$, and (d) $\mathrm{err}(q(\mathbf{x}))$, $\mathbf{x} \in BC$, obtained using the TRM, $n_S = 5$ and various levels of noise added into the Dirichlet data, $u|_{\Gamma_D}$, namely $p_u \in \{1\%, 3\%, 5\%\}$, for Example 5.

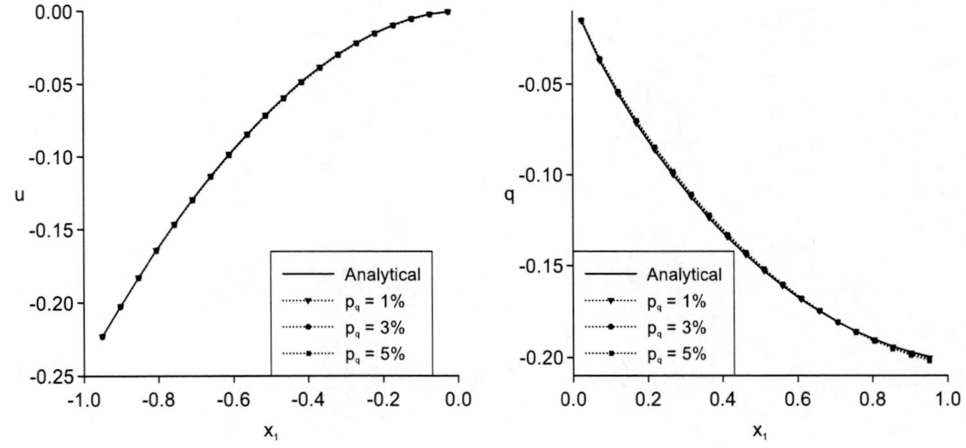

(a)

(b)

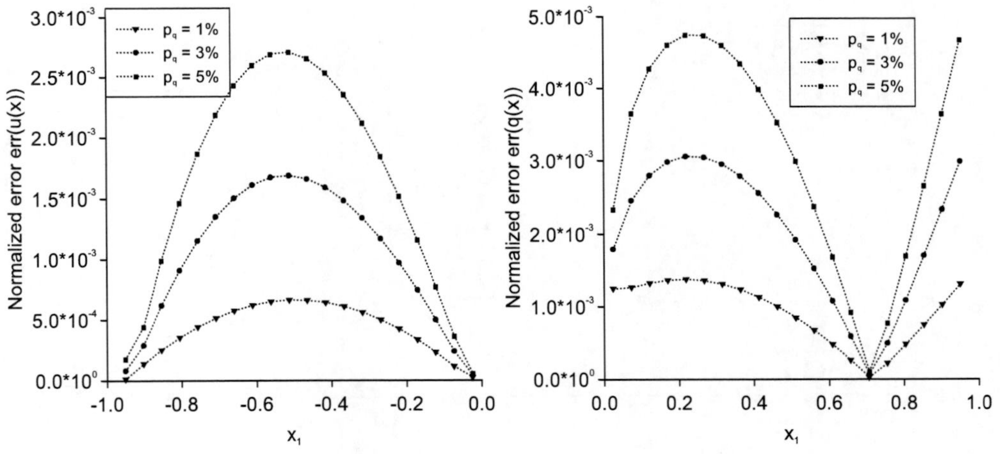

(c)

(d)

Figure 22: Analytical and numerical solutions for (a) $u|_{DO}$ and (b) $q|_{OA}$, and the corresponding normalized errors (c) $err(u(\mathbf{x}))$, $\mathbf{x} \in DO$, and (d) $err(q(\mathbf{x}))$, $\mathbf{x} \in OA$, obtained using the TRM, $n_S = 5$ and various levels of noise added into the Neumann data, $u|_{\Gamma_N}$, namely $p_q \in \{1\%, 3\%, 5\%\}$, for Example 6.

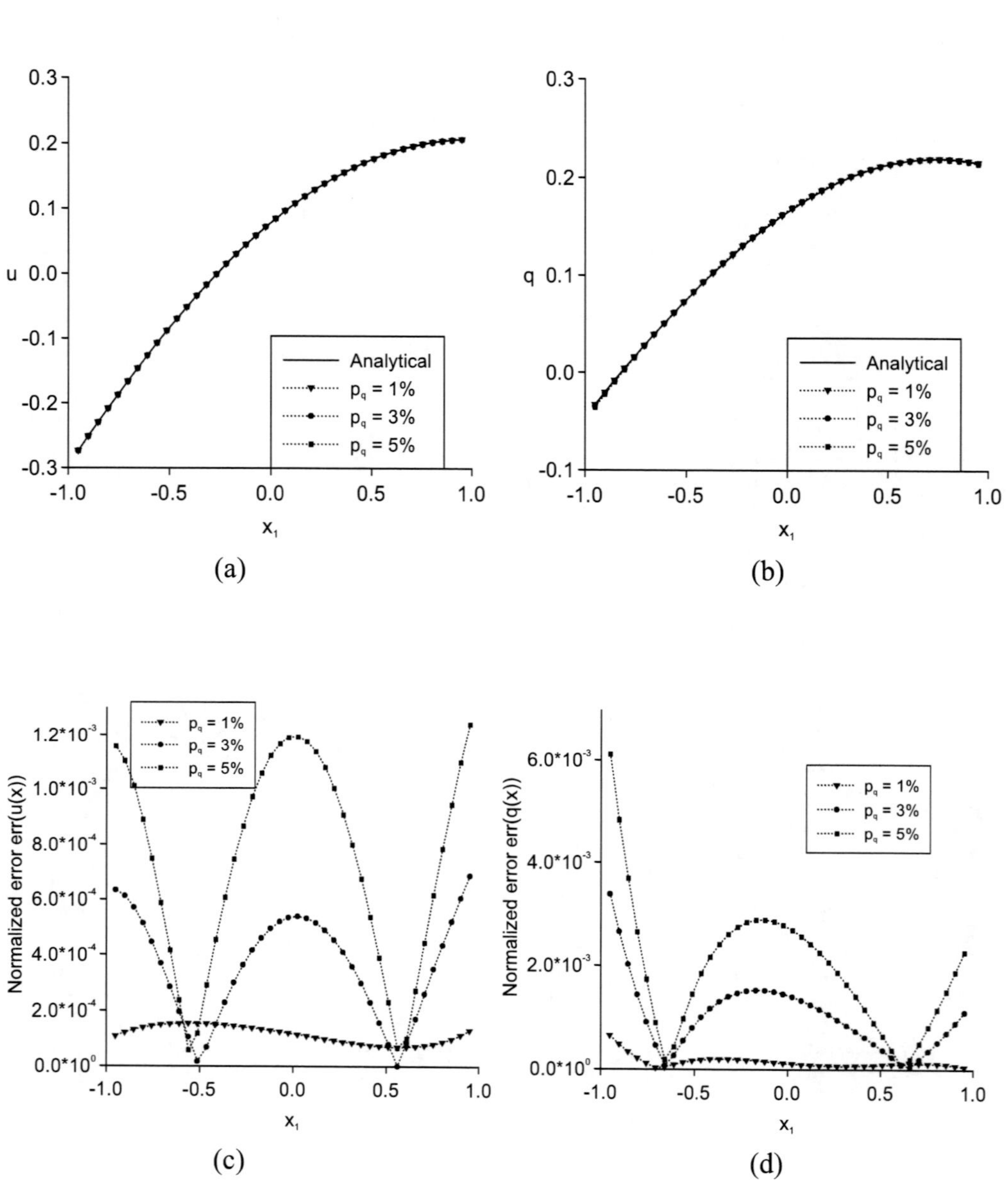

Figure 23: Analytical and numerical solutions for (a) $u|_{BC}$ and (b) $q|_{BC}$, and the corresponding normalized errors (c) $err(u(\mathbf{x}))$, $\mathbf{x} \in BC$, and (d) $err(q(\mathbf{x}))$, $\mathbf{x} \in BC$, obtained using the TRM, $n_S = 5$ and various levels of noise added into the Neumann data, $u|_{\Gamma_N}$, namely $p_q \in \{1\%, 3\%, 5\%\}$, for Example 6.

Table 4: The numerically retrieved values, $a_j^{(num)}$, for the flux intensity factors and the corresponding absolute errors, $Err(a_j)$, obtained using the TRM, $n_S = 5$ and various levels of noise added into the potential solution $u|_{\Gamma_D}$, namely $p_u \in \{1\%, 3\%, 5\%\}$, for Example 6.

p_q	$a_1^{(num)}$	$a_2^{(num)}$	$a_3^{(num)}$	$a_4^{(num)}$
	$Err(a_1)$	$Err(a_2)$	$Err(a_3)$	$Err(a_4)$
1%	-0.00011	0.99301	-1.2408	1.4686
	0.11×10^{-3}	0.69×10^{-2}	0.25×10^{0}	0.28×10^{0}
3%	0.00014	0.98064	-1.8284	1.3366
	0.14×10^{-3}	0.19×10^{-1}	0.32×10^{0}	0.13×10^{0}
5%	0.00039	0.96826	-1.0239	1.6565
	0.39×10^{-0}	0.31×10^{-1}	0.47×10^{0}	0.45×10^{0}

convergent and stable with respect to decreasing the amount of noise added into the input boundary temperature $u|_D$. Similar conclusions can be drawn from Table 4, which tabulates the numerical flux intensity factors, $a_j^{(num)}$, $j = 1, \ldots, 4$, and the corresponding absolute errors, $Err(a_j), j = 1, \ldots, 4$, obtained using the TRM, $n_S = 5$ and various levels of noise added into the normal flux $q|_{\Gamma_N}$, namely $p_q \in \{1\%, 3\%, 5\%\}$, for Example 6.

8 Conclusion

In this chapter, the MFS was applied for solving, in an accurate and stable manner, inverse problems associated with Helmholtz-type equations in two-dimensional domains without and with boundary singularities. For the first type of inverse problems analysed herein, the resulting ill-conditioned standard MFS system of linear algebraic equations was stabilised/regularized by using the TRM, while the choice of the optimal regularization parameter was based on the L-curve criterion. Three benchmark examples involving both the Helmholtz and the modified Helmholtz equations in simply and doubly connected, smooth and piecewise smooth geometries were investigated. The numerical results obtained show that the proposed method is convergent with respect to increasing the number of MFS collocation points, the number of MFS sources and the distance between the boundary of the solution domain and the pseudo-boundary were the source points are located and stable with respect to decreasing the amount of noise added into the input Dirichlet and/or Neumann data. Moreover, the method is efficient, easy to adapt to three-dimensional Cauchy problems associated with Helmholtz-type equations.

It is well-known that the existence of boundary singularities affect adversely the accuracy and convergence of standard numerical methods. Consequently, the MFS solutions to such problems and/or their corresponding derivatives, obtained by a direct inversion of the MFS system (i.e. by the LSM or the equivalent normal equation), may have unbounded values in the vicinity of the singularity. This inconvenience was overcome by subtracting from the original MFS solution the corresponding singular solutions, as given by the asymptotic expansion of the solution near the singularity point, and at the same time employing the TRM, in conjunction with Hansen's L-curve method for choosing the optimal regularization parameter. Hence, in addition to the original MFS unknowns, new unknowns were introduced, namely the so-called flux intensity factors. Consequently, the original MFS system was modified by considering a number of additional equations which equals the number of flux intensity factors introduced and specifically imposes the type of singularity analysed in the vicinity of the singularity point. The proposed MFS+SST was analysed for inverse problems associated with both the Helmholtz and the modified Helmholtz equations in two-dimensional domains containing an edge crack or a V-notch, as well as an L-shaped domain. From the numerical results presented in this chapter, we can conclude that the advantages of the proposed method over other methods, such as mesh refinement in the neighbourhood of the singularity, the use of singular BEMs and/or FEMs etc., are the high accuracy which can be obtained even when employing a small number of collocation points and sources, and the simplicity of the computational scheme. A possible drawback of the present method is the difficulty in extending the method to deal with singularities in three-dimensional problems since such an extension is not straightforward.

References

[1] J.T. Chen, M.T. Liang, I.L. Chen, S.W. Chyuan, K.H. Chen, Dual boundary element analysis of wave scattering from singularities, *Wave Motion* **30** (1999) 367–381.

[2] J.T. Chen, F.C. Wong, Dual formulation of multiple reciprocity method for the acoustic mode of a cavity with a thin partition, *Journal of Sound and Vibration* **217** (1998) 75–95.

[3] C. Huang, Z. Wu, R.D. Nevels, Edge diffraction in the vicinity of the tip of a composite wedge, *IEEE Transactions on Geoscience and Remote Sensing* **31** (1993) 1044–1050.

[4] P.A. Barbone, J.M. Montgomery, O. Michael, I. Harari, Scattering by a hybrid asymptotic/finite element. *Computer Methods in Applied Mechanics and Engineering* **164** (1998) 141–156.

[5] A.D. Kraus, A. Aziz, J. Welty, *Extended Surface Heat Transfer* (Wiley, 2001).

[6] Y. Niwa, S. Kobayashi, M. Kitahara, Determination of eigenvalue by boundary element method. In: *Development in Boundary Element Methods* (P.K. Banerjee, R. Shaw, editors), Applied Science Publisher, New York, 1982, Chapter 7.

[7] A.J. Nowak, C.A Brebbia, Solving Helmholtz equation by boundary elements using multiple reciprocity method. In: *Computer Experiment in Fluid Flow* (G.M. Calomagno, C.A. Brebbia, editors), CMP/Springer Verlag, Berlin, 1989, 265–270.

[8] J.P. Agnantiaris, D. Polyzer, D. Beskos, Three-dimensional structural vibration analysis by the dual reciprocity BEM, *Computational Mechanics* **21** (1998) 372–381.

[9] J. Hadamard, *Lectures on Cauchy Problem in Linear Partial Differential Equations* (Oxford University Press, 1923).

[10] G. Chen, J. Zhou, *Boundary Element Methods* (Academic Press, 1992).

[11] M.R. Bai, Application of BEM-based acoustic holography to radiation analysis of sound sources with arbitrarily shaped geometries. *Journal of the Acoustical Society of America* **92** (1992) 533–549.

[12] B.K. Kim, J.G. Ih, On the reconstruction of the vibro-acoustic field over the surface enclosing an interior space using the boundary element method. *Journal of the Acoustical Society of America* **100** (1996) 3003–3016.

[13] Z. Wang, S.R. Wu, Helmholtz equation-least-squares method for reconstructing the acoustic pressure field. *Journal of the Acoustical Society of America* **102** (1997) 2020–2032.

[14] S.R. Wu, J. Yu, Application of BEM-based acoustic holography to radiation analysis of sound sources with arbitrarily shaped geometries. *Journal of the Acoustical Society of America* **104** (1998) 2054–2060.

[15] T. DeLillo, V. Isakov, N. Valdivia, L. Wang, The detection of the source of acoustical noise in two dimensions. *SIAM Journal on Applied Mathematics* **61** (2001) 2104–2121.

[16] T. DeLillo, V. Isakov, N. Valdivia, L. Wang, The detection of surface vibrations from interior acoustical pressure. *Inverse Problems* **19** (2003) 507–524.

[17] L. Marin, L. Elliott, P.J. Heggs, D.B. Ingham, D. Lesnic, X. Wen, An alternating iterative algorithm for the Cauchy problem associated to the Helmholtz equation. *Computer Methods in Applied Mechanics and Engineering* **192** (2003) 709–722.

[18] L. Marin, L. Elliott, P.J. Heggs, D.B. Ingham, D. Lesnic, X. Wen, Conjugate gradient-boundary element solution to the Cauchy problem for Helmholtz-type equations. *Computational Mechanics* **31** (2003) 367–377.

[19] L. Marin, L. Elliott, P.J. Heggs, D.B. Ingham, D. Lesnic, X. Wen, Comparison of regularization methods for solving the Cauchy problem associated with the Helmholtz equation. *International Journal for Numerical Methods in Engineering* **60** (2004) 1933–1947.

[20] L. Marin, L. Elliott, P.J. Heggs, D.B. Ingham, D. Lesnic, X. Wen, BEM solution for the Cauchy problem associated with Helmholtz-type equations by the Landweber method. *Engineering Analysis with Boundary Elements* **28** (2004) 1025–1034.

[21] B. Jin, Y. Zheng, Boundary knot method for some inverse problems associated with the Helmholtz equation. *International Journal for Numerical Methods in Engineering* **62** (2005) 1636–1651.

[22] B. Jin, Y. Zheng, Boundary knot method for the Cauchy problem associated with the inhomogeneous Helmholtz equation. *Engineering Analysis with Boundary Elements* **29** (2005) 925–935.

[23] L. Marin, D. Lesnic, The method of fundamental solutions for the Cauchy problem associated with two-dimensional Helmholtz-type equations, *Computers & Structures* **83** (2005) 267–278.

[24] L. Marin, A meshless method for the numerical solution of the Cauchy problem associated with three-dimensional Helmholtz-type equations, *Applied Mathematics and Computation* **165** (2005) 355–374.

[25] B. Jin, Y. Zheng, A meshless method for some inverse problems associated with the Helmholtz equation, *Computer Methods in Applied Mechanics and Engineering* **195** (2006) 2270–2280.

[26] X.-T. Xiong, C.-L. Fu, Two approximate methods of a Cauchy problem for the Helmholtz equation. *Computational and Applied Mathematics* **26** (2007) 285–307.

[27] B. Jin, L. Marin, The plane wave method for inverse problems associated with Helmholtz-type equations. *Engineering Analysis with Boundary Elements* **32** (2008) 223–240.

[28] T. Wei, H.H. Qin, R. Shi, Numerical solution of an inverse 2D Cauchy problem connected with the Helmholtz equation. *Inverse Problems* **24** (2008) art. no. 035003.

[29] H.H. Qin, D.W. Wen, Tikhonov type regularization method for the Cauchy problem of the modified Helmholtz equation. *Applied Mathematics and Computation* , in press (2009).

[30] H.H. Qin, T. Wei, R. Shi, Modified Tikhonov regularization method for the Cauchy problem of the Helmholtz equation. *Journal of Computational and Applied Mathematics* **224** (2009) 39–53.

[31] T. Apel, A.-M. Sändig, J.R. Whiteman, Graded mesh refinement and error estimates for finite element solution of elliptic boundary value problems in non-smooth domains, *Mathematical Methods in the Applied Sciences* **19** (1996) 63–85.

[32] J.T. Chen, K.H. Chen, Dual integral formulation for determining the acoustic modes of a two-dimensional cavity with a degenerate boundary, *Engineering Analysis with Boundary Elements* **21** (1998) 105–116.

[33] B. Schiff, Eigenvalues for ridged and other waveguides containing corners of angle $3\pi/2$ or 2π by the finite element method, *IEEE Transactions on Microwave Theory and Techniques* **39** (1991) 1034–1039.

[34] W. Cai, H.C. Lee, H.S. Oh, Coupling of spectral methods and the *p*-version for the finite element method for elliptic boundary value problems containing singularities, *Journal of Computational Physics* **108** (1993) 314–326.

[35] T.R. Lucas, H.S. Oh, The method of auxiliary mapping for the finite element solutions of elliptic problems containing singularities, *Journal of Computational Physics* **108** (1993) 327–342.

[36] X. Wu, H. Han, A finite-element method for Laplace- and Helmholtz-type boundary value problems with singularities, *SIAM Journal on Numerical Analysis* **134** (1997) 1037–1050.

[37] Y.S. Xu, H.M. Chen, Higher-order discretised boundary conditions at edges for TE waves, *IEEE Proceedings – Microwave Antennas and Propagation* **146** (1999) 342–348.

[38] V. Mantič, F. París, J. Berger, Singularities in 2D anisotropic potential problems in multi-material corners. Real variable approach, *International Journal of Solids and Structures* **40** (2003) 5197–5218.

[39] L. Marin, D. Lesnic, V. Mantič, Treatment of singularities in Helmholtz-type equations using the bounadry element method, *Journal of Sound and Vibration* **278** (2004) 39–62.

[40] Z.-C. Li, T.T. Lu, Singularities and treatments of elliptic boundary value problems, *Mathematical and Computer Modelling* **31** (2000) 97–145.

[41] V.D. Kupradze, M.A. Aleksidze, The method of functional equations for the approximate solution of certain boundary value problems, *USSR Computational Mathematics and Mathematical Physics* **4** (1964) 82–126.

[42] R. Mathon, R.L. Johnston, The approximate solution of elliptic boundary value problems by fundamental solutions, *SIAM Journal on Numerical Analysis* **14** (1977) 638–650.

[43] G. Fairweather, A. Karageorghis, The method of fundamental solutions for elliptic boundary value problems, *Advances in Computational Mathematics* **9** (1998) 69–95.

[44] A. Karageorghis, G. Fairweather, The method of fundamental solutions for the numerical solution of the biharmonic equation, *Journal of Computational Physics* **69** 1987) 434–459.

[45] A. Karageorghis, Modified methods of fundamental solutions for harmonic and biharmonic problems with boundary singularities, *Numerical Methods for Partial Differential Equations* **8** (1992) 1–19.

[46] A. Poullikkas, A. Karageorghis, G. Georgiou, Methods of fundamental solutions for harmonic and biharmonic boundary value problems, *Computational Mechanics* **21** (1998) 416–423.

[47] A. Poullikkas, A. Karageorghis, G. Georgiou, The method of fundamental solutions for inhomogeneous elliptic problems, *Computational Mechanics* **22** (1998) 100–107.

[48] A. Poullikkas, A. Karageorghis, G. Georgiou, The numerical solution of three-dimensional Signorini problems with the method of fundamental solutions, *Engineering Analysis with Boundary Elements* **25** (2001) 221–227.

[49] A. Karageorghis, G. Fairweather, The method of fundamental solutions for axisymmetric elasticity problems, *Computational Mechanics* **25** (2000) 524–532.

[50] J.R. Berger, A. Karageorghis, The method of fundamental solutions for heat conduction in layered materials, *International Journal for Numerical Methods in Engineering* **45** (1999) 1681–1694.

[51] J.R. Berger, A. Karageorghis, The method of fundamental solutions for layered elastic materials, *Engineering Analysis with Boundary Elements* **25** (2001) 877–886.

[52] A. Poullikkas, A. Karageorghis, G. Georgiou, The numerical solution for three-dimensional elastostatics problems, *Computers & Structures* **80** (2002) 365–370.

[53] Y.C. Hon, T. Wei, A fundamental solution method for inverse heat conduction problems, *Engineering Analysis with Boundary Elements* **28** (2004) 489–495.

[54] N.S. Mera, The method of fundamental solutions for the backward heat conduction problem, *Inverse Problems in Science and Engineering* **13** (2005) 79–98.

[55] L. Marin, D. Lesnic, The method of fundamental solutions for the Cauchy problem in two-dimensional linear elasticity, *International Journal of Solids and Structures* **41** (2004) 3425–3438.

[56] L. Marin, A meshless method for solving the Cauchy problem in three-dimensional elastostatics, *Computers & Mathematics with Applications* **50** (2005) 73–92.

[57] L. Marin, Numerical solutions of the Cauchy problem for steady-state heat transfer in two-dimensional functionally graded materials, *International Journal of Solids and Structures* **42** (2005) 4338–4351.

[58] L. Marin, D. Lesnic, The method of fundamental solutions for inverse boundary value problems associated with the two-dimensional biharmonic equation, *Mathematical and Computer Modelling* **42** (2005) 261–278.

[59] A. Zeb, D.B. Ingham, D. Lesnic, The method of fundamental solutions for a biharmonic boundary determination. *Computational Mechanics* **42** (2008) 371–379.

[60] B. Jin, L. Marin, The method of fundamental solutions for inverse source problems associated with the steady-state heat conduction, *International Journal for Numerical Methods in Engineering* **69** (2007) 1570–1589.

[61] R. Tankelevich, G. Fairweather, A. Karageorghis, Potential field based geometric modeling using the method of fundamental solutions, *International Journal for Numerical Methods in Engineering* **68** (2006) 1257–1280.

[62] P. Mitic, Y.F. Rashed, Convergence and stability of the method of meshless fundamental solutions using an array of randomly distributed source, *Engineering Analysis with Boundary Elements* **28** (2004) 143–153.

[63] M.A. Golberg, C.S. Chen, The method of fundamental solutions for potential, Helmholtz and diffusion problems, in: M.A. Golberg (Ed.), *Boundary Integral Methods: Numerical and Mathematical Aspects*, WIT Press and Computational Mechanics Publications, Boston, 1999, pp. 105–176.

[64] A.N. Tikhonov, V.Y. Arsenin, *Methods for Solving Ill-Posed Problems* (Nauka, 1986).

[65] H.W. Engl, M. Hanke, A. Neubauer, *Regularization of Inverse Problems* (Kluwer Academic, 2000).

[66] V.A. Morozov, On the solution of functional equations by the method of regularization, *Soviet Mathematical Doklady* **7** (1966) 414–417.

[67] P.C. Hansen, *Rank-Deficient and Discrete Ill-Posed Problems: Numerical Aspects of Linear Inversion*, SIAM, Philadelphia, 1998.

[68] G. Wahba, Practical approximate solutions to linear operator equations when the data are noisy, *SIAM Journal on Numerical Analysis* **14** (1977) 651–667.

[69] L.Y. Chen, J.T. Chen, H.K. Hong, C.H. Chen, Application of Cesàro mean and the L-curve for the deconvolution problem, *Soil Dynamics Earthquake Engineering* **14** (1995) 361–373.

[70] M. Hanke, Limitations of the L-curve method in ill-posed problems, *BIT* **36** (1996) 287–301.

[71] C.R. Vogel, Non-convergence of the L-curve regularization parameter selection method, *Inverse Problems* **12** (1996) 535–547.

[72] V. Guerra, V. Hernandez, Numerical aspects in locating the corner of the L-curve, in: M. Lassonde (Ed.), *Approximation, Optimization and Mathematical Economics* , Springer-Verlag, Heidelberg, 2001, pp. 121–131.

[73] L. Kaufman, A. Neumaier, PET regularization by envelope guided conjugate gradients, *IEEE Transactions on Medical Imaging* **15** (1996) 385–389.

[74] J.L. Castellanos, S. Gomez, V. Guerra, The triangle method for finding the corner of the L-curve, *Applied Numerical Mathematics* **43** (2002) 359–373.

In: Computational Mathematics
Editor: Peter G. Chareton, pp. 173-200
ISBN 978-1-60876-271-2

Chapter 6

VANDERMONDE SYSTEMS: THEORY AND APPLICATIONS

Giuseppe Fedele
Department of Electronics, Computer and System Science,
University of Calabria
Via Pietro Bucci 42C, 87036, Rende, Italy

1 Introduction

The Vandermonde matrix, defined by $\widetilde{V}_n(i,j) = x_j^{i-1}, x_j \in \mathbb{C}$, is ubiquitous in mathematics and engineering. Its use includes, for example, polynomial interpolation, coding theory and signal processing, where the matrix for the discrete Fourier transform is a Vandermonde matrix. There is an extensive literature on numerically solving systems of linear equations when the matrix is given by a Vandermonde matrix. The objective of this chapter is twofold: to gain structural properties and explicit formulas useful to further analytic works in many problems directly connected to the inversion of the Vandermonde matrix and to the least-squares problems; to gain, as a consequence, fast and efficient algorithms. The primal system $\widetilde{V}_n x = b$ represents a moment problem, which arises when determining the weights for a quadrature rule, while the matrix $V_n = \widetilde{V}_n^T$ involved in the dual system $V_n c = f$, plays an important role in polynomial interpolation [36]. The special structure of V_n allows the use of ad-hoc algorithms that require $O(n^2)$ elementary operations for solving a Vandermonde system. The most celebrated of them is the one by Björck and Pereyra [4]; the high accuracy of the solutions it gives has been theoretically justified [37]. A generalization of Vandermonde matrices is obtained by replacing the monomials by a family of orthogonal polynomials which satisfy a three-term recurrence relation. The corresponding matrix is known as Vandermonde-like matrix. The interest in this generalization stems from the fact that these Vandermonde-like matrices generally have much smaller condition numbers than the classical Vandermonde ones [9, 10, 48]. Using a new algorithm for fast evaluation of the Schur function it is possible to obtain new $O(n^2)$ Björck and Pereyra-type algorithms for solving generalized Vandermonde linear systems [41]. In [38] Kailath and Olshevsky use the displacement structure concept to introduce a new class of matrices, designated as Chebyshev-Vandermonde like matrices, generalizing ordinary Chebyshev-Vandermonde matrices. Furthermore, the fact that the displacement structure is inherited by Schur complements leads to a fast $O(n^2)$ implementation of Gaussian elimination with partial pivoting for Chebyshev-Vandermonde like matrices. Many explicit formulas and computational schemes for the entries of V_n^{-1} have also been given [40, 23]. Bounds or estimates of the norm of both V_n and V_n^{-1} are also interesting, for example to investigate the conditioning of the polynomial interpolation problem. Answer to these problems have been given first for special configurations of the nodes and recently for general ones [55, 34]. Another field of interest of the Vandermonde matrices is in least-squares problem.

In many practical applications the Polynomial Least-Squares Problem (PLSP) must be solved [3]. It is formulated as follows: given a set of points

$$\Theta = \{(x_i, f_i),\ i = 1, 2, \ldots, n\},$$

find a polynomial $p(x)$ of degree less than or equal to $m-1$ with coefficients $c_1, c_2, \ldots, c_m$ such that the least-squares criterion

$$\varepsilon(c_1, c_2, \ldots, c_m) = \sum_{i=1}^{n} (p(x_i) - f_i)^2$$

is minimized. In general, m is much smaller than n. The problem can be reformulated as follows

$$\min_c ||Vc - f||_2, \tag{1}$$

where V is a Vandermonde matrix of order $n \times m$, depending on the observation points x_i

$$V(i, j) = x_i^{j-1}, \quad i = 1, 2, \ldots, n,\ j = 1, 2, \ldots, m, \tag{2}$$

and $f \in \mathbb{R}^n$ and $c \in \mathbb{R}^m$ are vectors containing experimental data f_i and coefficients c_i, respectively. Finding the solution of problem (1) is equivalent to compute the best approximate solution [3] of the over-determined linear system $Vc = f$. This is given by

$$\begin{aligned} c &= (V^T V)^{-1} V^T f \\ &= V^{\dagger} f, \end{aligned}$$

where $V^{\dagger}$ is the Moore-Penrose pseudo-inverse of V, that is well-defined if $x_i \neq x_j$. The numerical solution of PLSP is usually ill-conditioned. The most reliable algorithms either use orthogonal transformations, for example the well-known QR method, or compute the Cholesky factorization of the normal matrix $B = V^T V$. The choice between the two methods is still an open problem. Golub and Van Loan [34] reported some guidelines for the above choice, based on the value of the ratio n/m and on that of the residual $\rho = \min_c ||Vc - f||_2$, but they recognized that both methods can give inaccurate solutions when applied to problems with large value of $\kappa_2(V)$. Rectangular Vandermonde matrices on "not structured" nodes have been considered in [14] where fast algorithms for the Cholesky factorization of the normal matrix $B = V^T V$ and for the QR factorization of V have been given.

To avoid the difficulties with ill-conditioned system of equations that occur in the least-squares applications [3], the expansion in orthogonal polynomials is often used. Although orthogonal polynomials over discrete sets were considered as early as the middle of the nineteenth century by Chebyshev, comparatively little attention had been paid to them until now. They appear naturally in combinatorics, genetics, statistics and various areas in applied mathematics (see, for example [45]).

The chapter is organized as follows. In Section 2 interpolation formulas due to Lagrange and Newton are reviewed. Some results obtained studying the convergence properties of the Lagrange polynomial interpolation, through the investigation of the Lebesgue function for equidistant interpolation points, are presented. In Section 3 an identity on the fundamental symmetric functions on a set of n complex elements is shown with some properties on different sets of nodes. Such an identity is applied to the inversion of the Vandermonde matrix and the eigensystem problem. In Section 4 explicit solutions of both polynomial interpolation problem and the least-squares problem with Gauss-Lobatto points, in terms of the explicit Cholesky factorization and discrete orthogonal polynomials, are provided.

2 Polynomial Interpolation

The problem of constructing a continuously defined function from given discrete data is unavoidable whenever one wishes to manipulate the data in a way that requires information not explicitly included in the data. The relatively easiest, and in many applications often most desired, approach to solve the problem is interpolation, where an approximating function is constructed in such a way as to agree perfectly with the usually unknown original function at the given measurements points. In this section we review some interpolation formulas due to Lagrange and Newton and some properties about several set of nodes. We show some results obtained studying the convergence properties of the Lagrange polynomial interpolation. The behavior of the Lebesgue function for equidistant interpolation points is investigated showing that it can be expressed in terms of Jacobi's polynomial; moreover, for large n, we found the value of x where the Lebesgue constant is obtained.

If we decide to approximate a function $f \in C[a,b]$ (the space of all continuous functions $f:[a,b] \to \mathbb{R}$) by a polynomial, formally we are interested in solving the following problem

Problem: Given data (x_i, y_i), $i = 0,1,\ldots,n$, find a polynomial p of minimal degree which satisfies the equations

$$p(x_i) = y_i, \quad i = 0,1,\ldots,n. \tag{3}$$

We note that there are as many conditions as coefficients, and the following theorem shows that they determine p uniquely.

Theorem 1. *Let $n+1$ distinct real numbers $x_0, x_1, \ldots, x_n$ and associated values $y_0, y_1, \ldots, y_n$ be given. Then there exists a unique polynomial p of degree at most n such that $p(x_i) = y_i$, $i = 0,1,\ldots,n$.*

Two well known formulas for the polynomial interpolation are the Lagrange and the Newton ones. Nevertheless, there are many formulas or algorithms for calculating the polynomial $p(x)$ that are equivalent in exact arithmetic. They differ, however, in the accuracy that is obtained in the presence of rounding errors, and in the amount of work that is done when they are applied.

2.1 Lagrange and Newton form

Lagrange recognized the connection between the polynomial that interpolates a function at $n+1$ given points $X = \{x_i,\ i = 0,1,\ldots,n\}$ and the $n+1$ polynomials that vanish on the distinct subset of n points from X. Such polynomials $\ell_i(X,x),\ \ i = 0,1,\ldots,n$ satisfy

$$\ell_i(X,x_j) = \begin{cases} 1 & i = j \\ 0 & i \neq j \end{cases}, \quad i,j = 0,1,\ldots,n \tag{4}$$

and assume the form

$$\ell_i(X,x) = \frac{\prod_{k=0,k\neq i}^{n} (x - x_k)}{\prod_{k=0,k\neq i}^{n} (x_i - x_k)}. \tag{5}$$

We refer to the polynomials $\ell_i(X,x)$ as *fundamental polynomials* of degree exactly n [28]. It immediately follows that the function

$$p_n(x) = \sum_{i=0}^{n} f(x_i)\ell_i(X,x) \tag{6}$$

satisfies the required interpolation conditions (3). We want to derive an alternative notation for fundamental polynomials which will be useful in the sequel for the error analysis of the Lagrange interpolation. Let $\omega(x)$ be

$$\omega(x) = (x-x_0)(x-x_1)\cdots(x-x_n). \tag{7}$$

Then applying the product rule for derivatives and evaluating the result at x_i one gets the denominator of $\ell_i(X,x)$

$$\omega'(x_i) = (x_i - x_0)\cdots(x_i - x_{i-1})(x_i - x_{i+1})\cdots(x_i - x_n). \tag{8}$$

This identity gives a new representation for the fundamental polynomials, namely,

$$\ell_i(X,x) = \frac{\omega(x)}{(x-x_i)\omega'(x_i)}. \tag{9}$$

The Lagrange polynomial interpolation now becomes

$$p_n(x) = \sum_{i=0}^{n} \frac{f(x_i)\omega(x)}{(x-x_i)\omega'(x_i)}. \tag{10}$$

Observe that $\omega(x)/(x-x_i)$ is a polynomial of degree n with the coefficient of x^n equal to 1. Thus the separate contributions from the fundamental polynomials that determine the coefficient of x^n in (10) are easy to identify. If the symbol $f[x_0,x_1,\ldots,x_n]$ represents the coefficient of x^n in $p_n(X,x)$ then

$$f[x_0,x_1,\ldots,x_n] \equiv \sum_{i=0}^{n} \frac{f(x_i)}{\omega'(x_i)}. \tag{11}$$

Such symbols are defined to be a *divided difference* of order n.

Suppose that $p_m(x)$ is the Lagrange polynomial of degree m that interpolates $f(x)$ at the points $x_0, x_1, \ldots, x_m$ and that $m = 0, 1, \ldots, n$. By (11) the coefficient of x^m is $f[x_0, x_1, \ldots, x_m]$. The difference $p_m(x) - p_{m-1}(x)$ can be expressed as

$$p_m(x) - p_{m-1}(x) = c(x - x_0)(x - x_1)\cdots(x - x_{m-1}) \tag{12}$$

where

$$c = f[x_0, x_1, \ldots, x_m]. \tag{13}$$

If we systematically replace the differences in the identity

$$p_n(x) = p_0(x) + [p_1(x) - p_0(x)] + \cdots + [p_n(x) - p_{n-1}(x)] \tag{14}$$

using (12) then we have the Newton polynomial

$$\begin{aligned} p_n(x) &= f(x_0) + f[x_0, x_1](x - x_0) + f[x_0, x_1, x_2](x - x_0)(x - x_1) \\ &+ \cdots + f[x_0, x_1, \ldots, x_n](x - x_0)(x - x_1)\cdots(x - x_n). \end{aligned} \tag{15}$$

The divided differences that appear in (15) admit the convenient calculation:

$$\begin{aligned} f[x_0, x_1] &= \tfrac{f(x_1) - f(x_0)}{x_1 - x_0}, \\ f[x_0, x_1, \ldots, x_m] &= \tfrac{f[x_1, x_2, \ldots, x_m] - f[x_0, x_1, \ldots, x_{m-1}]}{x_m - x_0}. \end{aligned} \tag{16}$$

Remarks:

- The interpolation matrix for the Lagrange form is the identity matrix and the coefficients in the basis expansion are given by the data values. This makes the Lagrange form ideal for situations in which many experiments with the same data nodes but different data values need to be performed. However, evaluation (as well as differentiation or integration) is more expensive. A remarkable advantage of Lagrange interpolation is that the quantities that have to be computed in $O(n^2)$ operations do not depend on the data. This feature permits the interpolation of as many functions as desired in $O(n)$ operations each one the $\ell_i(x)$ are known, whereas Newton interpolation requires the re-computation of the divided difference for each new function. Another advantage is that it does not depend on the order in which the nodes are arranged. In the Newton formula we have such a dependence: for larger values of n, most orderings lead to numerical instability. For stability it is necessary to select the points, for example, in Leja sequence [49].

- A major advantage of the Newton form is its efficiency in the case of adaptive interpolation, i.e., when an existing interpolant needs to be refined by adding more data. Due to the recursive nature, the new interpolant can be determined by updating the existing one. Therefore, for the Newton form we have a balance between stability of computation and ease of evaluation.

From here it is commonly concluded that the Lagrange form of $p_n(x)$ is mainly a theoretical tool for proving theorems. For computations, it is generally recommended that one should instead use Newton's formulas [2].

2.2 Lebesgue constant

Let $p_n^*(f)$ be the best approximation of f in $C[a,b]$ measured in the maximum norm, that is, p_n^* minimizes the error $\max_{a\leq x\leq b}|f(x)-q_n(x)|$ among all n-degree $p_n(x)$. For a prescribed set of $n+1$ interpolation points $X=\{x_k,\ k=0,1,\ldots,n\}$ we rewrite this error as

$$f(x)-p_n(f)(x)=[f(x)-p_n^*(f)(x)]+[p_n^*(f)(x)-p_n(f)(x)]. \tag{17}$$

The first difference on the right does not exceed

$$e_n^* = \max_{a\leq x\leq b}|f(x)-p_n^*(f)(x)|. \tag{18}$$

Thus e_n^* measures the best min-max value among all n-degree polynomials. Furthermore, since every polynomial interpolates itself, i.e. $p_n^*(f)=p_n(p_n^*)(x)$, the second difference on the right admits a Lagrange form

$$p_n^*(f)(x)-p_n(f)(x)=\sum_{k=0}^{n}[p_n^*(x_k)-p_n(f)(x_k)]\,\ell_k(X,x), \tag{19}$$

and hence $|p_n^*(f)(x)-p_n(f)(x)|\leq e_n^*\max_{a\leq x\leq b}\sum_{k=0}^{n}|\ell_k(X,x)|$, we conclude that

$$|f(x)-p_n(f)(x)|\leq e_n^*(f)\left[1+\max_{a\leq x\leq b}\sum_{k=0}^{n}|\ell_k(X,x)|\right]. \tag{20}$$

As before we bound the error with a product of two term: the first, $e_n^*(f)$ depends solely on the smoothness properties of f; the second, involving the so-called Lebesgue constant

$$\Lambda_n(X)=\max_{a\leq x\leq b}\sum_{k=0}^{n}|\ell_k(X,x)| \tag{21}$$

depends solely on the distribution of the interpolation points. The quantity

$$\lambda_n(X,x)=\sum_{k=0}^{n}|\ell_k(X,x)| \tag{22}$$

involved in (21) is called Lebesgue function. From (20) it is clear that the Lebesgue constant is closely connected with convergence and divergence of the Lagrangian interpolation polynomials [52, 5].

The importance of the study of the Lebesgue functions for polynomial interpolation was demonstrated in [42] through the investigation of their local maxima. In the last four decades, such an analysis has received constant attention from researchers for some sets of nodes which are of special importance in the interpolation theory, such as equidistant nodes [5, 54, 50, 44], Chebyshev roots [51, 6] and extrema or others ones. We start with the formulation of elementary properties of $\lambda_n(X,x)$, see e.g. [42].

- For $n\geq 2$ the function $\lambda_n(X,x)$ is a piecewise polynomial satisfying $\lambda_n(X,x)\geq 1$ with equality only at the nodes $x_k,\ k=0,1,\ldots,n$.

- $\lambda_n(X,x)$ has precisely one local maximum on (x_{k-1},x_k), $k=1,2,\ldots,n$.
- Let $Z=\{z_k\}_{k=0}^n$ and $Y=\{y_k\}_{k=0}^n$ be two sets of interpolation nodes related by an affine transformation $y_k=\alpha z_k+\beta$. Then

$$\lambda_n(Z,z)=\lambda_n(Y,\alpha z+\beta).$$

We investigate the behavior of the Lebesgue function in the case of equidistant nodes. For a survey of the existing estimation for different sets of nodes see [5, 6, 7, 51].

2.2.1 Lebesgue constant for equidistant nodes

Although the set of equally spaced points

$$E=\left\{x_k=-1+\frac{2k}{n},\ k=0,1,\ldots,n\right\}$$

represents a bad choice for Lagrange interpolation, considerable literature exists regarding the behavior of Lebesgue function corresponding to such nodes. This interest is due to the fact that this choice frequently occurs in many applications. A first result, due to Turetskii [54], proposes an asymptotic expression for the largest maximum

$$\Lambda_n(E)\sim\frac{2^{n+1}}{en\log(n)},\quad n\to\infty. \tag{23}$$

Schönhage [50] derived an asymptotic expression for Λ_n that is a little bit more precise than (23), namely

$$\Lambda_n(E)\sim\frac{2^{n+1}}{en\left[\log(n)+\gamma\right]},\quad n\to\infty \tag{24}$$

where $\gamma=0.577215665$ is the Eulero-Mascheroni constant. In [44], Mills and Smith improved expression (24) by finding an asymptotic expansion of the following form

$$\log\Lambda_n(E)=(n+1)\log 2-\log n-\log\log n-1+\sum_{k=1}^{m}\frac{A_k}{[\log n]^k},\quad n\to\infty, \tag{25}$$

where

$$A_1=-\gamma,\ A_2=\gamma^2/2-\pi^2/12,\ A_3=-\gamma^3/3+\gamma\pi^2/6-\zeta(3)/3,\ldots$$

and $\zeta(3)=\sum_{r=1}^{\infty}r^{-3}$ is the well known Apéry constant. Note that it appears to be a difficult problem to obtain an explicit general formula for the coefficient A_k.

Here, without loss of generality, we shall consider the following set of nodes

$$\overline{E}=\left\{x_k=\frac{k-1}{n-1}\right\},\quad k=1,2,\ldots,n \tag{26}$$

obtained through the change of variable $x\to 2x-1$ which maps the interval $I=[-1,1]$ into $\overline{I}=[0,1]$.

As a consequence, formula (22) becomes

$$\lambda_n(\overline{E},x) = \sum_{k=1}^{n} |\ell_k(\overline{E},x)|. \tag{27}$$

The first proposition gives an alternative formulation of the Lagrange fundamental polynomials at equidistant nodes in terms of binomial coefficients which is simpler than ones known in literature.

Proposition 1.

$$\ell_k(\overline{E},x) = (-1)^{n+k} \binom{nx-x}{k-1} \binom{nx-x-k}{n-k}, \quad k = 1,2,\ldots,n \tag{28}$$

where $\binom{p}{q}$ *is the binomial coefficient [40].*

As it was pointed out in [5], $\lambda_n(\overline{E},x)$ takes its maximum when $x \in \left[0, \frac{1}{n-1}\right]$, therefore our attention will be focused on the behavior of the Lebesgue function in the first subinterval. Following the same approach as [6] we derive

$$\lambda_n(\overline{E},x) = \ell_1(\overline{E},x) + \sum_{k=2}^{n} (-1)^k \ell_k(\overline{E},x), \quad x \in \left[0, \frac{1}{n-1}\right]. \tag{29}$$

The next result proposes two alternative expressions of the Lebesgue function restricted to such a subinterval [19].

Proposition 2.

$$\lambda_n(\overline{E},x) = (-1)^{n+1} \binom{nx-x-1}{n-1} \left[2 - {}_2F_1(1-n, x-nx; 1+x-nx; -1)\right], \tag{30}$$

$$\lambda_n(\overline{E},x) = (-1)^{n+1} 2 \binom{nx-x-1}{n-1} - P_{n-1}^{(-nx+x,-n)}(3), \tag{31}$$

where ${}_2F_1(a,b;c;w) = \sum_{k=0}^{\infty} \frac{(a)_k (b)_k}{(c)_k} \frac{w^k}{k!}$ *is the hypergeometric function and* $P_n^{(\alpha,\beta)}$ *is the Jacobi polynomial.*

In [19], an interesting formula of the asymptotic behavior of the Lebesgue constant is discussed, by showing the existence of a relationship between Λ_n, combinatoric theory and Jacobi polynomials.

Proposition 3.

$$\Lambda_n(\overline{E}) = (-1)^{(n+1)} \binom{\frac{1}{\log(n-1)+\gamma} - 1}{n-1} - P_{n-1}^{\left(-\frac{1}{\log(n-1)+\gamma}, -n\right)}(3), \quad n \to \infty, \tag{32}$$

where the asymptotic value of x, corresponding to the maximum of the Lebesgue function in the first subinterval, is given by

$$x^* = \frac{1}{(n-1)\left[\log(n-1)+\gamma\right]}, \quad n \to \infty. \tag{33}$$

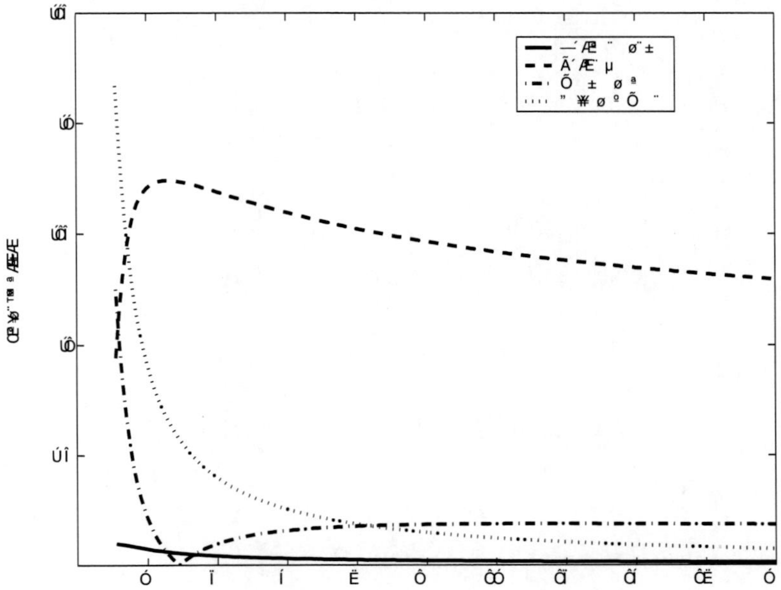

Figure 1: Comparison of Lebesgue constant estimations.

In order to show the accuracy of the expression (32) of the Lebesgue constant with respect to the other formulas proposed in literature and here cited, an extensive numerical comparisons of the relative errors are performed. In Figure 1 the numerical results are shown up to $n = 200$. The proposed estimation gives a better degree of accuracy than the other methods, except that in a small interval $n \in [25, 32]$.

2.3 The Björck and Pereyra algorithm

In 1970 Björck and Pereyra [4] derived an efficient algorithm for solving the interpolation problem for any set of nodes. We briefly recall the idea under this algorithm. The first step is to calculate the Newton form representation of the interpolating polynomial:

$$p(x) = c_0 + \sum_{k=1}^{n} c_k \left(\prod_{i=0}^{k-1} (x - x_i) \right) \tag{34}$$

where the coefficients c_k are the divided difference

$$c_k = f[x_0, x_1, \ldots, x_k] \tag{35}$$

and can be computed by the recurrence relation

$$f[x_i, x_{i+1}, \ldots, x_{i+k}] = \frac{f[x_{i+1}, \ldots, x_{i+k}] - f[x_i, \ldots, x_{i+k-1}]}{x_{i+k} - x_i}. \tag{36}$$

They may be determined as follows:

$$
\begin{array}{l}
for\ k=0,1,\ldots,n-1\\
\quad for\ i=n,n-1,\ldots,k+1\\
\quad\quad c_i=\frac{c_i-c_{i-1}}{x_i-x_{i-k-1}}\\
\quad end\\
end
\end{array}
$$

The next task is to generate the coefficients $a_k,\ k=0,1,\ldots,n$ of

$$p(x)=\sum_{j=0}^{n} a_j x^j \tag{37}$$

from $c_k,\ k=0,1,\ldots,n$. If we define the following set of polynomials

$$
\begin{aligned}
p_n(x) &= c_n,\\
p_k(x) &= c_k+(x-x_k)p_{k+1}(x) \quad k=n-1,n-2,\ldots,0.
\end{aligned} \tag{38}
$$

Noting that $p_0(x)=p(x)$ we can compute the coefficients $a_i^{(k)}$ by (37) and (38):

$$
\begin{array}{l}
a_n^{(n)}=c_n\\
for\ k=n-1,n-2,\ldots,0\\
\quad a_k^{(k)}=c_k-x_k a_{k+1}^{(k+1)}\\
\quad for\ i=k+1,k+2,\ldots,n-1\\
\quad\quad a_i^{(k)}=a_i^{(k+1)}-x_k a_{i+1}^{(k+1)}\\
\quad end\\
\quad a_n^{(k)}=a_n^{(k+1)}\\
end
\end{array}
$$

Note that $a_i=a_i^{(0)},\ \ i=0,1,\ldots,n.$
Given

$$x_0,x_1,\ldots,x_n$$

with distinct entries and

$$f(0),f(1),\ldots,f(n),$$

the following algorithm calculates the solution to the Vandermonde system $V_{n+1}\cdot a=f$.

$$
\begin{array}{l}
for\ k=0:n-1\\
\quad for\ i=n:-1:k+1\\
\quad\quad f(i)=\frac{f(i)-f(i-1)}{x(i)-x(i-k-1)}\\
\quad end\\
end\\
for\ k=n-1:-1:0\\
\quad for\ i=k:n-1\\
\quad\quad f(i)=f(i)-f(i+1)x(k)\\
\quad end\\
end
\end{array}
$$

It can be interesting to give a matrix formulation of the Björck and Pereyra algorithm. Define the lower bidiagonal matrix

$$L_k(\alpha) = \left(\begin{array}{c|cccc} I_k & & 0^T & & \\ \hline & 1 & & \cdots & 0 \\ & -\alpha & 1 & & \\ 0 & \vdots & \ddots & \ddots & \\ & & & \ddots & \ddots \\ & 0 & & -\alpha & 1 \end{array}\right)$$

the diagonal matrix

$$D_k = diag\left(\underbrace{1,\ldots,1}_{k+1}, x_{k-1} - x_0, \ldots, x_n - x_{n-k-1}\right)$$

and the lower and upper triangular matrices respectively

$$L = D_{n-1}^{-1} L_{n-1}(1) \ldots D_0^{-1} L_0(1)$$

$$U = L_0(x_0)^T \ldots L_{n-1}(x_{n-1})^T$$

one has

$$c = L \cdot f,$$

$$a = U \cdot c.$$

Therefore the Björck and Pereyra algorithm for the solution of $V_{n+1} \cdot a = f$ is equivalent to the $L \cdot U$ factorization of V_{n+1} [32, 46].

3 A Property of the Elementary Symmetric Functions

In this section we give an identity on the fundamental symmetric functions on the set X_n of n complex elements [15, 16, 17]. Such an identity could be seen as a generalization of both the results stated in [23] and of the idea under the Traub algorithm [53]. Its use in the inversion of the Vandermonde matrix and the eigensystem problem is discussed.

Let us consider a set of n distinct numbers: $X_n = \{x_1, x_2, \ldots, x_n\}$.

Definition 1. Let $\sigma(n,q)$ be the function recursively defined as follows:

$$\begin{array}{ll} \sigma(n,q) = \sigma(n-1,q) + x_n \sigma(n-1,q-1), & n,q \text{ integer}, \\ \sigma(n,0) = 1, & n = 0,1,\ldots, \\ (q<0) \text{ or } (n<0) \text{ or } (q>n) \rightarrow \sigma(n,q) = 0. & \end{array} \quad (39)$$

The generating function of $\sigma(n,q)$ [20, 23], is:

$$B_n(x)=\prod_{i=1}^{n}(x-x_i)=\sum_{r=0}^{n}(-1)^{r+n}\sigma(n,n-r)x^r. \tag{40}$$

Therefore $\sigma(n,q)$ is the qth order elementary symmetric function associated to the set X_n which is the sum of all products of q distinct elements chosen from X_n:

$$\begin{cases}\sigma(n,q)=\sum\limits_{1\le\pi_1<\cdots<\pi_q\le n} x_{\pi_1}x_{\pi_2}\cdots x_{\pi_q}, & q=1,\ldots,n,\\ \sigma(n,0)=1.\end{cases} \tag{41}$$

By eq. (40) the following equality holds:

$$\sum_{r=0}^{n}(-1)^r x_i^r\sigma(n,n-r)=0, \quad i=1,2,\ldots,n. \tag{42}$$

Eq. (40) is a polynomial identity which means that, when functions σ are known, it is possible to find the elements $x_i,\ i=1,2,\ldots,n$ as roots of the polynomial $B_n(x)$.

Here we give an identity concerning the functions $\sigma(n,q)$ and $\phi(n,q)=(-1)^{n+q}B_n'(x_q)$, where $\phi(n,q)$ is recursively defined as follows:

$$\begin{cases}\phi(n+1,q)=\phi(n,q)(x_{n+1}-x_q), & 1\le q\le n,\\ \phi(n+1,n+1)=\prod\limits_{k=1}^{n}(x_{n+1}-x_k), & n\ge 1,\\ \phi(2,1)=\phi(2,2)=x_2-x_1. \end{cases} \tag{43}$$

Consider the function [20]

$$h(n,i,j,q)=\sum_{k=1}^{n}\sum_{r=0}^{n-k}(-1)^{k+r}x_i^{k+q}x_j^r\sigma(n,n-k-r)+(-1)^{j+1}x_j^{q+1}\phi(n,j)\delta_{i,j}. \tag{44}$$

We have the following result [20]:

Theorem 2. *For $n\ge 2$ and $q\ge -1$ one has $h(n,i,j,q)=0,\ i,j=1,2,\ldots,n.$*

The following theorem gives an explicit expression for a system of a matrix eigenvectors when the eigenvalues are known.

Theorem 3. *Let A be a $n\times n$ matrix with distinct eigenvalues $\Lambda=\{\alpha_1,\alpha_2,\ldots,\alpha_n\}$ where $\alpha_k\in C,\ \ k=1,2,\ldots,n$ and $U=\{u_1,u_2,\ldots,u_n\}$, with $u_k\in C^n,\ \ k=1,2,\ldots,n$, the corresponding right-eigenvectors. A system of right-eigenvectors is*

$$w_i=\sum_{k=1}^{n}\sum_{r=0}^{n-k}(-1)^{r+k+1}\sigma(n,n-k-r)\alpha_i^{k-1}A^r h, \quad i=1,2,\ldots,n \tag{45}$$

where $h\in C^n$; $h\neq 0$; $h^T u_k\neq 0,\ \ k=1,2,\ldots,n.$

Proof. Let

$$h = \sum_{j=1}^{n} \gamma_j u_j. \quad (46)$$

Note that

$$\left(h^T u_k \neq 0,\ \ k = 1,2,\ldots,n\right) \to (\gamma_k \neq 0,\ \ k = 1,2,\ldots,n) \quad (47)$$

and

$$\phi(n,k) \neq 0, \quad k = 1,2,\ldots,n \quad (48)$$

due to the fact that $\alpha_i \neq \alpha_j,\ \ i \neq j;\ i,j = 1,2,\ldots,n.$

Therefore w_j can be expressed as:

$$w_j = \sum_{j=1}^{n} \gamma_j \theta(n,i,j) u_j, \quad j = 1,2,\ldots,n \quad (49)$$

where

$$\theta(n,i,j) = \sum_{k=1}^{n} \sum_{r=0}^{n-k} (-1)^{r+k+1} \alpha_i^{k-1} \alpha_j^r \sigma(n, n-k-r), \quad i,j = 1,2,\ldots,n. \quad (50)$$

By substituting $q = -1$ in theorem 2, we have:

$$\theta(n,i,j) = (-1)^{j+1} \phi(n,j) \delta_{i,j}, \quad i,j = 1,2,\ldots,n. \quad (51)$$

Therefore we have:

$$w_i = (-1)^{i+1} \gamma_i \phi(n,i) u_i, \quad i = 1,2,\ldots,n, \quad (52)$$

and (45) follows. □

The next theorem gives the explicit expression for the generic elements of the inverse of V_n.

Theorem 4. *Let W_n be the inverse of V_n. Then the generic element $w_n(i,j)$ of W_n is:*

$$w_n(i,j) = \frac{\psi(n,i,j)}{\phi(n,j)}, \quad i,j = 1,2,\ldots,n \quad (53)$$

where

$$\psi(n,i,j) = (-1)^{i+j} \sum_{r=0}^{n-i} (-1)^r x_j^r \sigma(n, n-i-r), \quad i,j = 1,2,\ldots,n. \quad (54)$$

Proof. It must be shown that

$$(V_n W_n)_{i,j} = \delta_{i,j}, \quad i,j = 1,2,\ldots,n. \quad (55)$$

But

$$(V_n W_n)_{i,j} = \sum_{k=1}^{n} \sum_{r=0}^{n-k} (-1)^{k+r+j} x_i^{k-1} x_j^r \sigma(n, n-k-r) \frac{1}{\phi(n,j)} \quad (56)$$

therefore, by substituting $q = -1$ in theorem 2 the (55) follows. □

Note that the (53) gives also the following factorization of W_n:

$$W_n = M \cdot V_n^T \cdot D \tag{57}$$

where

$$M(i,j) = (-1)^{i+j+1}\sigma(n, n+1-i-j), \quad i = 1,2,\ldots,n; \; j = 1,2,\ldots,n+1-i, \tag{58}$$

$$D = diag\left\{\frac{(-1)^i}{\phi(n,i)}\right\}_{i=1,2,\ldots,n}. \tag{59}$$

As regards the computational aspects of the proposed inversion formula, it is convenient to relate directly the values of the elements of the inverse relevant to the $(i-1)^{th}$ row to those of the ith row and then to proceed by recursion.

Proposition 4. *The function* $\psi(n,i,j)$ *in (54) can be recursively computed as:*

$$\begin{array}{l} \psi(n,i-1,j) = x_j\psi(n,i,j) - (-1)^{i+j}\sigma(n,n+1-i) \\ \psi(n,n,j) = (-1)^{n+j}, \quad i = n, n-1, \ldots, 2; \; j = 1,2,\ldots,n. \end{array} \tag{60}$$

From (60) the following recursion on w_n holds:

$$\begin{array}{l} w_n(i-1,j) = x_j w_n(i,j) - (-1)^{n+i} w_n(n,j)\sigma(n,n+1-i), \\ \qquad i = n, n-1, \ldots, 2; \; j = 1,2,\ldots,n \\ w_n(n,j) = (-1)^{n+j}\frac{1}{\phi(n,j)}, \quad j = 1,2,\ldots,n. \end{array} \tag{61}$$

Proposition 5. *Eq.* (54) *is equivalent to:*

$$\psi(n,i,j) = (-1)^{i+j}\upsilon_j(n,n-i), \quad i,j = 1,2,\ldots,n \tag{62}$$

where $\upsilon_j(n,n-i)$ *is the* $(n-i)^{th}$ *order elementary symmetric function associated to the set* $X_n - \{x_j\}$.

By using (60) and (62) we propose two algorithms for the inverse of Vandermonde matrix, summarized as follows:

Algorithm PEF

1. Compute the quantities $\sigma(n,s), \quad s = 0,1,\ldots,n$ by (41).
2. Compute the quantities $\phi(n,j), \quad j = 1,2,\ldots,n$ by (43).
3. Compute $\psi(n,i,j), \quad i,j = 1,2,\ldots,n$ by recursion (60).
4. Compute the jth column $\left[\frac{\psi(n,i,j)}{\phi(n,j)}\right]_{1\le i\le n}, \quad j = 1,2,\ldots,n.$

Algorithm EF

1. Compute the quantities $\phi(n,j), \quad j = 1,2,\ldots,n$ by (43).
2. Compute $\psi(n,i,j), \quad i,j = 1,2,\ldots,n$ by using (62).

3. Compute the jth column $\left[\frac{\psi(n,i,j)}{\phi(n,j)}\right]_{1\leq i\leq n}$, $j=1,2,\ldots,n$.

The first algorithm computes all n^2 entries of V in $5.5n^2-3.5n$ flops, the latter in $n^3-0.5n^2+2.5n-1$ flops. It is easy to prove that the algorithm PEF is identical to the Parker inversion algorithm and this is confirmed by all numerical experiments. However a simple modification of PEF algorithm allow us to derive a modified Parker that seems to be better than the Parker's one. From eq. (54) it turns out that for $i=1,2$ the $\psi(n,i,j)$ can be directly expressed as:

$$\psi(n,1,j)=(-1)^{j+1}\frac{\sigma(n,n)}{x_j},\quad j=1,2,\ldots,n$$
$$\psi(n,2,j)=-\psi(n,1,j)\sum_{\substack{q=1\\q\neq j}}^{n}\frac{1}{x_q},\quad j=1,2,\ldots,n.$$

Although the PM algorithm differs from the Parker algorithm only in this apparently "nonessential detail" [33], our numerical experiments show that the small difference between the two algorithms is meaningful, from the numerical point of view [23].

The proposed algorithm comes from an original inversion formula which is obviously equivalent to the existing ones, but is more flexible and allows to obtain special algorithms for particular set of nodes [21, 22]. The present results seem to be a good starting point for further analytic works in many problems directly connected to the inverse of the Vandermonde matrix. Our formula is obviously useful when closed expressions of $\sigma(n,q)$ and $\phi(n,j)$ are known. Here we report expressions of $\sigma(n,q)$ and $\phi(n,j)$ for different sets of nodes that are of practical interest in the applications [21, 23, 22]:

- Equidistant nodes in $[1,n]$: $E_{[1,n]}=\{x_k=k,\quad k=1,2,\ldots,n\}$

$$\phi^{-1}(n,j)=\frac{1}{(n-j)!(j-1)!},\quad j=1,2,\ldots,n \tag{63}$$

$$\sigma(n,q)=\begin{bmatrix}n+1\\n+1-q\end{bmatrix},\quad q=0,1,\ldots,n \tag{64}$$

where $\begin{bmatrix}i\\j\end{bmatrix}$ are the Stirling numbers of the first kind [40].

- Equidistant nodes in $[0,1]$: $E_{[0,1]}=\left\{x_k=\frac{k-1}{n-1},\quad k=1,2,\ldots,n\right\}$

$$\phi^{-1}(n,j)=\frac{(n-1)^{n-1}}{(n-j)!(j-1)!},\quad j=1,2,\ldots,n \tag{65}$$

$$\sigma(n,q)=\frac{1}{(n-1)^q}\begin{bmatrix}n\\n-q\end{bmatrix},\quad q=0,1,\ldots,n \tag{66}$$

- Equidistant nodes in $[-1,1]$: $E_{[-1,1]}=\left\{x_k=\frac{2k-n-1}{n-1},\quad k=1,2,\ldots,n\right\}$

$$\phi^{-1}(n,j)=\frac{\left(\frac{n-1}{2}\right)^{n-1}}{(n-j)!(j-1)!},\quad j=1,2,\ldots,n \tag{67}$$

$$\sigma(n,q)=\left(\frac{2}{n-1}\right)^q\sum_{k=0}^{q}(-1)^k\left(\frac{n+1}{2}\right)^k\binom{n+k-q}{k}\begin{bmatrix}n+1\\n+1+k-q\end{bmatrix},\quad q=0,1,\ldots,n \tag{68}$$

- Chebyshev nodes: $T_1 = \left\{x_k = \cos\left[\frac{2k-1}{2n}\pi\right],\quad k = 1,2,\ldots,n\right\}$

$$\phi^{-1}(n,j) = (-1)^{n+1}\frac{2^{n-1}}{n}\sin\left(\frac{2j-1}{2n}\pi\right),\quad j = 1,2,\ldots,n \tag{69}$$

$$\sigma(n,2q) = \frac{(-1)^q}{2^{2q}}\frac{\binom{2q}{q}\binom{n}{2q}}{\binom{n-1}{q}},\quad q = 0,1,\ldots,\left\lfloor\frac{n}{2}\right\rfloor \tag{70}$$

- Extended Chebyshev nodes: $T_2 = \left\{x_k = -\frac{\cos\left(\frac{2k-1}{2n}\pi\right)}{\cos\left(\frac{\pi}{2n}\right)},\quad k = 1,2,\ldots,n\right\}$

$$\phi^{-1}(n,j) = \frac{2^{n-1}}{n}\cos\left(\frac{\pi}{2n}\right)^{n-1}\sin\left(\frac{2j-1}{2n}\pi\right),\quad j = 1,2,\ldots,n \tag{71}$$

$$\sigma(n,2q) = \frac{(-1)^q}{2^{2q}}\frac{1}{\cos\left(\frac{\pi}{2n}\right)^{2q}}\binom{2q}{q}\frac{\binom{n}{2q}}{\binom{n-1}{q}},\quad q = 0,1,\ldots,\left\lfloor\frac{n}{2}\right\rfloor \tag{72}$$

- Gauss-Lobatto Chebyshev (extrema): $T_3 = \left\{x_k = -\cos\left[\frac{k-1}{n-1}\pi\right],\quad k = 1,2,\ldots,n\right\}$

$$\phi^{-1}(n,j) = \begin{cases} \frac{2^{n-3}}{n-1} & j = 1,n \\ \\ \frac{2^{n-2}}{n-1} & j = 2,\ldots,n-1 \end{cases} \tag{73}$$

$$\sigma(n,2q) = (-1)^q\frac{1}{2^{2q}}\binom{n-q}{q}\frac{n^2-n-2q}{(n-q-1)(n-q)},\quad q = 0,1,\ldots,\left\lfloor\frac{n}{2}\right\rfloor \tag{74}$$

- Fekete nodes: such nodes are roots of the polynomial

$$P_n(x) = (x^2-1)L'_{n-1}(x) \tag{75}$$

where $L_n(x)$ is the nth Legendre polynomial in x.

$$\sigma(n,2q) = (-1)^q\frac{\binom{n-2}{q-1}\binom{n}{2q}}{\binom{2n-3}{2q-1}},\quad q = 0,1,\ldots,\left\lfloor\frac{n}{2}\right\rfloor \tag{76}$$

Unfortunately the function $\phi(n,j)$ is not rational for $j = 2,3,\ldots,n-1$.

4 Polynomial Approximation with Gauss-Lobatto Points

4.0.1 The interpolation problem

Polynomial interpolation on several set of nodes has received much attention over the past decade [43]. Theoretically, any discretization grid can be used to construct the interpolation polynomial. However, the interpolated solution between discretization points are accurate only if the individual building blocks behave well between points. Lagrangian polynomials with a uniform grid suffer for the effect of the Runge phenomenon: small data near the center of the interval are associated with wild oscillations in the interpolant, on the order 2^n times bigger, near the edges of the interval, [12, 35]. The best choice is to use nodes that are clustered near the edges of the interval with an asymptotic density proportional to $(1-x^2)^{-1/2}$ as $n \to \infty$, [2]. The family of Chebyshev points, obtained by projecting equally spaced points on the unit circle down to the unit interval $[-1, 1]$ has such density properties. In [1] it is proved that interpolation on the Chebyshev polynomial extrema minimizes the diameter of the set of the vectors of the coefficients of all possible polynomials interpolating the perturbed data. Although the set of Gauss-Lobatto Chebyshev nodes failed to be a good approximation to the optimal interpolation set, such set is of considerable interest since the norm of corresponding interpolation operator is less than the norm of the operator induced by interpolation at the Chebyshev roots [8]. Vandermonde matrices on Gauss-Lobatto Chebyshev nodes have a (relative) moderate growth, versus n, of the condition number $\kappa_2(V_n)$, [30, 36]. Figure 2 shows the κ_2 comparison between the Vandermonde matrix on the Chebyshev nodes ($V_n(T_1)$), Chebyshev extesa nodes ($V_n(T_2)$) and Gauss-Lobatto Chebyshev nodes ($V_n(T_3)$). The choice of Gauss-Lobatto Chebyshev nodes is due to the high use in numerical applications: in polynomial interpolation [21], approximation theory [26, 27], spectral methods [18, 56], method for estimating the length of a parametric curve using only samples of points [29], method for estimating surface area [47], quadrature formulas, etc. Considering the high use of the Gauss-Lobatto nodes it is seemed natural to investigate their properties in approximation theory.

In [21] it is shown that the inverse of the Vandermonde matrix on Gauss-Lobatto Chebyshev nodes can be factorized as:

$$W_n = \frac{1}{n-1} K \cdot Q \cdot H \cdot F \tag{77}$$

where

$$K = diag\{2^{i-1}\}_{i=1,2,\dots,n}, \tag{78}$$

$$\begin{cases} Q(1, n-2) = 2, & \\ Q(i, n+1-i) = (-1)^i & i = 1, 2, \dots, n, \\ Q(1, n-2j-2) = -Q(1, n-2j), & j = 1, 2, \dots, \lceil \frac{n-4}{2} \rceil, \\ Q(i, n+1-i-2j) = -Q(i, n+3-i-2j) + & \\ \quad -Q(i-1, n+2-i-2j), & i = 2, 3, \dots, n, \\ & j = 1, 2, \dots, j^*, \\ Q(i, 1) = Q(i, 1)/2, & i = 1, 2, \dots, n \end{cases} \tag{79}$$

with

$$j^* = \begin{cases} \lfloor \frac{n-i}{2} \rfloor, & \text{n even}, \\ \lceil \frac{n-1-i}{2} \rceil, & \text{n odd}, \end{cases}$$

$$H(i,j) = \cos\left[\frac{(i-1)(j-1)}{n-1}\pi\right], \quad i = 1,\ldots,n, \; j = 1,\ldots,n, \tag{80}$$

$$\begin{cases} F(1,1) = -\frac{1}{2}, \\ F(i,i) = (-1)^i, \qquad i = 2,\ldots,n-1, \\ F(n,n) = (-1)^n \frac{1}{2}. \end{cases} \tag{81}$$

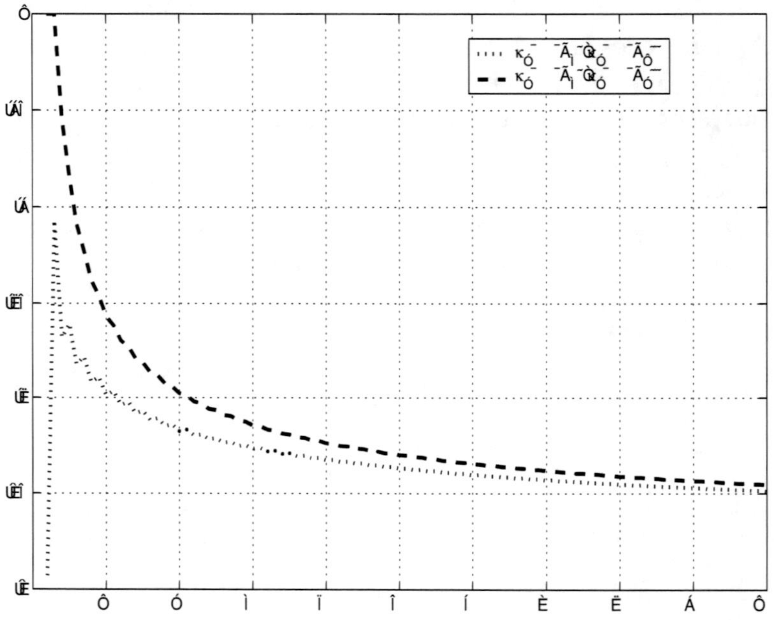

Figure 2: Plot of the ratios $\frac{\kappa_2(V_n(T_3))}{\kappa_2(V_n(T_1))}$ and $\frac{\kappa_2(V_n(T_3))}{\kappa_2(V_n(T_2))}$.

4.0.2 The least-squares problem: explicit Moore-Penrose pseudo-inverse formula

In the case of the least-squares problem on Gauss-Lobatto Chebyshev nodes an explicit solution can be found in [27] where it is shown that the inverse of the normal matrix, namely $B = V^T V$, on this set of nodes can be represented as the sum of two symmetric matrices: a full rank matrix which admits a Cholesky factorization and a 2-rank matrix. Moreover the following explicit formula for the Moore-Penrose pseudo-inverse of V is given

$$V^\dagger = (Q_1 - Q_2)H, \tag{82}$$

where

$$\begin{cases} Q_1(i,i) = \frac{2^{i-1}}{n-1}, & i = 1,2,\ldots,m, \\ Q_1(2i,2j) = (-1)^{i+j}\frac{2^{2i}}{n-1}\frac{2j-1}{2j-2i}\binom{i+j-2}{j-i-1}, & \\ \quad i = 1,2,\ldots,\lfloor\frac{m}{2}\rfloor,\ j = i+1,i+2,\ldots,\lfloor\frac{m}{2}\rfloor, & \\ Q_1(2i-1,2j-1) = (-1)^{i+j}\frac{2^{2i-1}}{n-1}\frac{2j-2}{2j-2i}\binom{i+j-3}{j-i-1}, & \\ \quad i = 1,2,\ldots,\lceil\frac{m}{2}\rceil,\ j = i+1,i+2,\ldots,\lceil\frac{m}{2}\rceil, & \end{cases} \tag{83}$$

$$\begin{cases} Q_2(2i,2j) = \frac{(-1)^{\frac{m+2i-Mod[m,2]}{2}}}{(n-1)(n+m-1-Mod[m,2])}2^{2i}\binom{\frac{m+2i-2-Mod[m,2]}{2}}{2i-1}, \\ \quad i = 1,2,\ldots,\lfloor\frac{m}{2}\rfloor,\ j = 1,2,\ldots,\lfloor\frac{m}{2}\rfloor, \\ Q_2(2i-1,2j-1) = \frac{(-1)^{\frac{m+2i+Mod[m,2]}{2}}}{(n-1)(n+m-2+Mod[m,2])}2^{2i-1}\binom{\frac{m+2i-4+Mod[m,2]}{2}}{2i-2}, \\ \quad i = 1,2,\ldots,\lceil\frac{m}{2}\rceil,\ j = 2,3,\ldots,\lceil\frac{m}{2}\rceil, \\ Q_2(2i-1,1) = \frac{(-1)^{\frac{m+2i+Mod[m,2]}{2}}}{(n-1)(n+m-2+Mod[m,2])}2^{2i-2}\binom{\frac{m+2i-4+Mod[m,2]}{2}}{2i-2}, \\ \quad i = 1,2,\ldots,\lceil\frac{m}{2}\rceil \end{cases} \tag{84}$$

and

$$H(i,j) = (-1)^{i+j}\cos\left[\frac{(i-1)(j-1)}{n-1}\pi\right], \quad i = 1,2,\ldots,m,\ j = 1,2,\ldots,n. \tag{85}$$

4.0.3 The least-squares problem: discrete orthogonal polynomials

Let f and g be real valued functions defined on the set of n nodes T_3 and introduce the inner product:

$$\langle f,g\rangle = \sum_{k=1}^{n} f(x_k)g(x_k). \tag{86}$$

A family $P = \{p_1(x),p_2(x),\ldots,p_n(x)\}$ of polynomials, where $degree(p_k) = k-1$, is orthogonal with respect to $\langle f,g\rangle$ if the following properties hold:

$$\langle p_k,p_q\rangle = 0, \qquad k \neq q;\ k,q = 1,2,\ldots,n,$$

$$\langle p_k,p_k\rangle = \rho_k \neq 0, \quad k = 1,2,\ldots,n.$$

If $\rho_k = 1,\ k = 1,2,\ldots,n$, then polynomials p_k are said to be orthonormal on the set of nodes T_3. Define the measure μ on $[-1,1]$ by

$$\int_{-1}^{1} f(x)d\mu = \frac{\Gamma(\alpha+\beta+2)}{2^{\alpha+\beta+1}\Gamma(\alpha+1)\Gamma(\beta+1)}\int_{-1}^{1} f(x)(1-x)^{\alpha}(1+x)^{\beta}dx + Mf(-1) + Nf(1). \tag{87}$$

The analytical determination of the recursion coefficients for the orthogonal polynomials belonging to the measure (87), is "virtually" intractable [31], although analytic expressions of the orthogonal polynomials themselves in terms of Jacobi polynomials or hypergeometric functions are known [39].

It is also well known that Gauss-Lobatto quadrature rule

$$\frac{1}{\pi}\int_{-1}^{1}\frac{f(x)}{\sqrt{1-x^2}}dx+\frac{1}{2(n-1)}\left[f(x_1)+f(x_n)\right] = \frac{1}{n-1}\sum_{i=1}^{n} f(x_i)-\frac{1}{2^{2n-3}(2n-2)!}f^{(2n-2)}(\xi), \quad (88)$$

where $x_i = \cos\left(\frac{i-1}{n-1}\pi\right)$, $i = 1,2,\ldots,n$, is exact for polynomials of degree at most $2n-3$, [13]. If (f,g) is the inner product defined as follows:

$$(f,g)=\frac{1}{\pi}\int_{-1}^{1}\frac{f(x)g(x)}{\sqrt{1-x^2}}dx+\frac{1}{2(n-1)}\left[f(x_1)g(x_1)+f(x_n)g(x_n)\right], \quad (89)$$

then

$$\langle f,g\rangle = (n-1)(f,g), \quad degree(f)+degree(g)\leq 2n-3. \quad (90)$$

Note that the explicit expression for the orthogonal polynomials with respect to the inner product (f,g), can be derived as the orthogonal polynomials on the interval $[-1,1]$ with respect to the measure μ, when $\alpha=\beta=-\frac{1}{2}$ and $M=N=\frac{1}{2(n-1)}$, [39]:

$$p_k(x)=\frac{(n+k-2)^2}{(n-1)^2}P_{k-1}^{(-1/2,-1/2)}(x)-\frac{(n+k-2)}{(k-1)(n-1)^2}x\frac{d}{dx}\left[P_{k-1}^{(-1/2,-1/2)}(x)\right], \quad (91)$$

where $P_n^{(\alpha,\beta)}(x)$ are the Jacobi polynomials. By using the following property of the Jacobi polynomials:

$$\frac{d}{dx}\left[P_n^{(\alpha,\beta)}(x)\right]=\frac{1}{2}(n+\alpha+\beta+1)P_{n-1}^{(\alpha+1,\beta+1)}(x),$$

eq. (91) becomes:

$$p_k(x)=\frac{(n+k-2)^2}{(n-1)^2}P_{k-1}^{(-1/2,-1/2)}(x)-\frac{(n+k-2)}{2(n-1)^2}xP_{k-2}^{(1/2,1/2)}(x). \quad (92)$$

Since eq. (90) is valid only for $degree(f)+degree(g)\leq 2n-3$, then the inner product $\langle p_n,p_n\rangle$ cannot be inferred. The orthogonal polynomials with respect to the inner product $\langle f,g\rangle$ on the Gauss-Lobatto nodes can also be connected to Chebyshev polynomials of the first and second kind by using the explicit expression of Jacobi polynomials in term of Chebyshev polynomials:

$$\begin{aligned} P_n^{(-1/2,-1/2)}(x) &= \tfrac{1}{2^{2n}}\binom{2n}{n}T_n(x), \\ P_n^{(1/2,1/2)}(x) &= \tfrac{1}{2^{2n}}\binom{2n+1}{n}U_n(x). \end{aligned} \quad (93)$$

In [26] an explicit expression of the discrete orthogonal polynomials on Gauss-Lobatto Chebyshev nodes is given as

$$
\begin{aligned}
q_k(x) = (n+k-3)x^{k-1} + \sum_{q=1}^{\lfloor \frac{k-1}{2} \rfloor} (-1)^q \tfrac{1}{q2^{2q}} \binom{k-q-2}{q-1} \\
\left[(k-1)n + (k-1)(k-3) + 2q\right] x^{k-2q-1}, \quad k = 1,2,\ldots,n.
\end{aligned} \tag{94}
$$

Moreover the explicit formulas for the inner products $\langle q_k, q_k \rangle$, $k = 2,3,\ldots,n$ is given by the following

Theorem 5. *The set of polynomials* (94) *satisfies:*

$$
\langle q_k, q_k \rangle = \frac{(n-1)(n+k-1)(n+k-3)}{2^{2k-3}}, \quad k = 2,3,\ldots,n-1, \tag{95}
$$

$$
\langle q_n, q_n \rangle = \frac{(n-1)^2(2n-3)}{2^{2n-5}}. \tag{96}
$$

The explicit three term relation follows from the explicit formula of the orthogonal polynomials, eq. (94):

$$
q_k(x) = \alpha_k x q_{k-1}(x) + \gamma_k q_{k-2}(x), \quad k = 4,5,\ldots,n \tag{97}
$$

where

$$
\begin{aligned}
\alpha_k &= \tfrac{n+k-3}{n+k-4} \qquad k = 4,5,\ldots,n, \\
\gamma_k &= -\tfrac{n+k-2}{4(n+k-4)} \quad k = 4,5,\ldots,n.
\end{aligned} \tag{98}
$$

4.0.4 Numerical properties

As far as the expansion in terms of orthogonal polynomials is concerned, we model a given function $y = f(t)$ as a linear combination of m discrete orthogonal polynomials on Gauss-Lobatto Chebyshev nodes,

$$
\tilde{f} = \sum_{j=1}^{m} c_j q_j(t) \tag{99}
$$

and study the least-squares problem to determine the coefficients $c_1, c_2, \ldots, c_m$ such that the Euclidean norm of the error $\tilde{f} - f$ is minimized,

$$
||\tilde{f} - f||^2 = \sum_{i=1}^{n} |\tilde{f}(x_i) - f(x_i)|^2. \tag{100}
$$

Since $q_1, q_2, \ldots, q_m$ are linearly independent, then the least-squares approximation problem has a unique solution [3],

$$
f^* = \sum_{j=1}^{m} c_j^* q_j. \tag{101}
$$

The coefficient c_j^* that are called orthogonal coefficients are computed by

$$c_j^* = \frac{\langle f, q_j \rangle}{\langle q_j, q_j \rangle}, \quad j = 1, 2, \ldots, m. \tag{102}$$

We use the explicit formulas for α_k and γ_k to construct the coefficients of the orthogonal polynomials $q_j,\ j = 1, 2, \ldots, m$ and the values of the polynomials at the grid points. The proposed method for the solution of the problem (101) costs $4\,mn$ flops (algorithm **OP**).

To implement the explicit formula for the Moore-Penrose pseudo-inverse, first note that the matrix $V^\dagger$ can be rewritten as

$$V^\dagger = \frac{1}{n-1} S(\hat{Q}_1 - \hat{Q}_2) H \tag{103}$$

where

$$\begin{cases} S(2i,2i) = \frac{1}{n+m-1-Mod[m,2]}, & i = 1, 2, \ldots, \left\lfloor \frac{m}{2} \right\rfloor, \\ S(2i-1,2i-1) = \frac{1}{n+m-2+Mod[m,2]}, & i = 1, 2, \ldots, \left\lceil \frac{m}{2} \right\rceil, \end{cases} \tag{104}$$

$$\hat{Q}_1 = (n-1) S^{-1} Q_1 \tag{105}$$

and

$$\hat{Q}_2 = (n-1) S^{-1} Q_2. \tag{106}$$

Moreover the matrix $\bar{Q_1} = (n-1) Q_1$ can be constructed using the following properties:

$$\begin{cases} \bar{Q_1}(1,1) = 1, & \\ \bar{Q_1}(i,i) = 2\bar{Q_1}(i-1,i-1), & i = 2, 3, \ldots, m, \\ \bar{Q_1}(1,2j-1) = (-1)^{j+1} 2, & j = 2, 3, \ldots, \left\lceil \frac{m}{2} \right\rceil, \\ \bar{Q_1}(i,j) = 2\bar{Q_1}(i-1,j-1) - \bar{Q_1}(i,j-2), & \\ \qquad i = 2, 3, \ldots, m,\ j = i+2, i+3, \ldots, m,\ i+j \text{ even}. & \end{cases} \tag{107}$$

It is important to underline that both the entries of $\hat{Q}_1$ and $\hat{Q}_2$ are integer numbers which can be stored without rounding. To construct $\hat{Q}_2$, let c_1 and c_2 be defined as

$$c_1 = (-1)^m m - Mod[m+1, 2],$$

$$c_2 = (-1)^m m + Mod[m, 2]$$

and consider the vector v defined as follows:

$$\begin{cases} v(1) = (-1)^{\frac{m}{2}+1+\frac{Mod[m,2]}{2}} 2, \\ v(2i) = \left(1 + \frac{c_1}{2i-1}\right) v(2i-1), & i = 1, 2, \ldots, \left\lfloor \frac{m}{2} \right\rfloor, \\ v(2i+1) = \left(1 - \frac{c_2}{2i}\right) v(2i), & i = 1, 2, \ldots, \left\lfloor \frac{m}{2} \right\rfloor,\ 2i+1 \le m, \end{cases}$$

then

$$\hat{Q}_2(i,j) = v(i), \qquad i = 1,2,\ldots,m,\ j = 2,3,\ldots,m,\ i+j \text{ even},$$

$$\hat{Q}_2(2i-1,1) = \tfrac{v(2i-1)}{2}, \quad i = 1,2,\ldots,\left\lceil \tfrac{m}{2} \right\rceil.$$

Note that the product $b = H \cdot f$ can be efficiently constructed by considering the symmetric properties of the matrix H. Define two vectors f_1 and f_2 as:

$$f_1 = \left\{ f(i),\ i = 1,2,\ldots,\left\lfloor \frac{n}{2} \right\rfloor \right\},$$

$$f_2 = \left\{ f(n+1-i),\ i = 1,2,\ldots,\left\lfloor \frac{n}{2} \right\rfloor \right\},$$

then

$$b(i) = \sum_{j=1}^{\lfloor \frac{n}{2} \rfloor} H(i,j)(f_1(j) + (-1)^{i+1} f_2(j)), \quad i = 1,2,\ldots,m,$$

and if n is odd then

$$b(2i-1) = b(2i-1) + (-1)^{i+1} f\left(\frac{n+1}{2}\right), \quad i = 1,2,\ldots,\left\lceil \frac{m}{2} \right\rceil.$$

The proposed method costs $2mn + 2.5m^2 + O(m)$ flops (algorithm **MP**).

Here we report some numerical experiments, that aim at investigating the effectiveness of our formulas. We compare our algorithms OP and MP with that proposed in [11] (**CB**) that costs $10mn$ flops. These algorithms have been implemented in Mathematica [57], that allows arbitrary precision number. For some values of m and n, we have generated one thousand vectors f, with entries uniformly distributed in $[-1,1]$, and have computed the exact solution of the problem (1) using extended precision of 128 digits. For each algorithm we have computed the maximum component wise relative errors

$$E_{OP} = \max_{1 \le i \le m} \frac{|\hat{c}_i^{OP} - c_i|}{|c_i|},$$

$$E_{MP} = \max_{1 \le i \le m} \frac{|\hat{c}_i^{MP} - c_i|}{|c_i|},$$

$$E_{CB} = \max_{1 \le i \le m} \frac{|\hat{c}_i^{CB} - c_i|}{|c_i|}$$

where $\hat{c}_i^{OP}$, $\hat{c}_i^{MP}$ and $\hat{c}_i^{CB}$ are the approximate solutions (computed in machine precision) of *OP*, *MP* and *CB* algorithm respectively.

The mean and the maximum of E_{OP}, E_{MP} and E_{CB} are reported in Table 1. Table 1 reports also the fraction of trials in which *MP* algorithm gives equal or more accurate results that the *OP* algorithm.

m	**OP**		**MP**		**CB**			
	max	mean	max	mean	max	mean	MP vs OP	MP vs CB
3	6.79-13	5.78-15	7.67-13	4.57-15	1.38-11	2.19-13	0.56	0.99
4	3.74-13	5.21-15	2.65-13	4.13-15	5.75-11	7.10-13	0.56	1.00
5	2.31-12	1.05-14	6.81-12	1.29-14	4.05-10	1.26-12	0.60	1.00
6	1.31-11	2.79-14	1.16-11	2.15-14	6.98-10	1.52-12	0.69	1.00
7	2.20-12	2.48-14	7.68-13	1.22-14	1.19-10	1.70-12	0.73	1.00
8	3.58-12	3.99-14	2.40-12	1.50-14	1.34-10	1.47-12	0.80	1.00
9	3.58-11	1.10-13	1.92-11	4.20-14	1.27-09	3.46-12	0.83	0.99
10	1.63-11	6.34-14	3.06-12	1.67-14	4.97-10	2.37-12	0.88	1.00
50	2.71-10	2.77-12	1.36-11	7.67-14	3.05+03	2.94+01	0.99	1.00

Table 1: Maximum and mean value of E_{OP}, E_{MP} and E_{CB} over 1000 runs, $n = 1000$, $f \in [-1, 1]$. Success rate of MP algorithm.

References

[1] Belforte, G., Gay, P., & Monegato, G., (2000). Some new properties of Chebyshev polynomials. *J. Comput. Appl. Math., 117*, 175-181.

[2] Berrut, J.P., & Trefethen, N. (2004). Baricentric Lagrange interpolation. *Siam Rev., 46, 3*, 501-517.

[3] Björck, A. *Numerical Methods for Least Squares Problems.* SIAM, 1996.

[4] Björck, A., & Pereyra, V. (1970). Solution of Vandermonde systems of linear equations. *Math. of Computation, 24*, 893-903.

[5] Brutman, L. (1997). Lebesgue functions for polynomial interpolation - a survey. *Annals of Numerical Mathematics*, 111-127.

[6] Brutman, L., Gopengauz, I., & Toledano, D. (2001). On the integral of the Lebesgue function induced by interpolation at the Chebyshev nodes. *Acta Math. Hungar., 90*, 12-28.

[7] Brutman, L., & Toledano, D. (1997). An extremal problem of Erdős in interpolation theory. *Computers and Mathematics with Appl., 34*, 37-47.

[8] Brutman, L. (1984), A note on polynomial interpolation at the Chebyshev extrema nodes. *Journal of Approx. Theory, 42*, 283-292.

[9] Calvetti, D., & Reichel, L. (1992). A Chebyshev-Vandermonde solver. *Linear Algebra Appl., 172*, 219-229.

[10] Calvetti, D., & Reichel, L. (1993). Fast inversion of Vandermonde-like matrices involving orthogonal polynomials. *BIT, 33, 3*, 473-484.

[11] Conte, S.D., & De Boor, c. *Elementary Numerical Analysis, 2nd ed.* McGraw-Hill, 1972.

[12] Björck, A., & Dahlquist, G. *Numerical Methods.* Prentice-Hall, Englewood Cliffs, NJ, 1974.

[13] Davis, P., & Rabinowitz, P. *Methods of Numerical Integration, 2nd ed.* Academic Press, New York, 1984.

[14] Demeure, C.J. (1989). Fast QR factorization of Vandermonde matrices. *Linear Algebra Appl., 122*, 165-194.

[15] El-Mikkawy, M. (2003). Explicit inverse of a generalized Vandermonde matrix. *Applied Math. and Computation, 146*, 643651.

[16] El-Mikkawy, M. (2003). On a connection between symmetric polynomials, generalized Stirling numbers and the Newton general divided difference interpolation polynomial. *Applied Math. and Computation, 138*, 375-385.

[17] El-Mikkawy, M. (2003). On a connection between the Pascal, Vandermonde and Stirling matrices - II. *Applied Math. and Computation, 146*, 759-769.

[18] Elnagar, G.N., & Razzaghi, M. (1997). A collocation-type method for linear quadratic optimal control problems. *Opt. Contr. Appl. & Methods, 18*, 227-235.

[19] Eisinberg, A., Fedele, G., & Franzé, G. (2004). On the Lebesgue constant for Lagrange interpolation on equidistant nodes. *Analysis Theory and Application, 20, 4*, 323-331.

[20] Eisinberg, A., & Fedele, G. (2005). A property of the elementary symmetric functions. *Calcolo, 42*, 31-36.

[21] Eisinberg, A., & Fedele, G. (2005). Vandermonde systems on Gauss-Lobatto Chebyshev nodes. *Applied Math. and Computation, 170*, 633-647.

[22] Eisinberg, A., Fedele, G., & Imbrogno, C. (2006). Vandermonde systems on equidistant nodes in $[0, 1]$. *Applied Math. and Computation, 172*, 971-984.

[23] Eisinberg, A., & Fedele, G. (2006). On the inversion of the Vandermonde matrix. *Applied Math. and Computation, 174*, 1384-1397.

[24] Eisinberg, A., & Fedele, G. (2007). Accurate floating point summation: a new approach. *Applied Math. and Computation, 189*, 410-424.

[25] Eisinberg, A., & Fedele, G. (2007). Discrete orthogonal polynomials on equidistant nodes. *Int. Math. Forum, 2*, 1007-1020.

[26] Eisinberg, A., & Fedele, G. (2007). Discrete orthogonal polynomials on Gauss-Lobatto Chebyshev nodes. *J. Approx. Theory, 144*, 238-246.

[27] Eisinberg, A., & Fedele, G. (2007). Explicit solution of the polynomial least-squares approximation problem on Chebyshev extrema nodes. *Linear Algebra Appl., 422*, 553-562.

[28] Fejer, L. (1916). Ueber Interpolation. *Vorgelegt in der Sitzung, 15*, 66-91.

[29] Floater, M.S., & Rasmussen, A.F. (2006). Point-based methods for estimating the length of a parametric curve. *Journal of Comp. and Applied Math., 196, 2*, 512-522.

[30] Gautschi, W. (1974). Norms estimates for inverses of Vandermonde matrices. *Numerische Math., 23*, 337-347.

[31] Gautschi, W. *Orthogonal Polynomials.* Oxford University Press, 2004.

[32] Gohberg, I., & Koltracht, I. (1997). Triangular factors of Cauchy and Vandermonde matrices. *Integral Equation and Operator Theory, 29*, 313-319.

[33] Gohberg, I., & Olshevsky, V. (1997).The fast generalized Parker-Traub algorithm for inversion of Vandermonde and related matrices. *Journal of Complexity, 13*, 208-234.

[34] Golub, G.H., & Van Loan, C.F. *Matrix Computations.* Johns Hopkins University Press, 1996.

[35] Henrici, P. *Essentials of Numerical Analysis.* Wiley, New York, 1982.

[36] Higham, N.J. *Accuracy and Stability of Numerical Algorithms.* SIAM, 1996.

[37] Higham, N.J. (1987). Error analysis of the Björck-Pereyra algorithms for solving Vandermonde systems. *Numerische Math., 50*, 613-632.

[38] Kailath, T., & Olshevsky, V. (1995). Dispacement structure approach to Chebyshev-Vandermonde and related matrices. *Integral Equation and Operator Theory, 22*, 65-92.

[39] Koornwinder, T.H. (1984). Orthogonal Polynomials with weight function $(1-x)^{\alpha}(1+x)^{\beta}+M\delta(x+1)+N\delta(x-1)$. *Canad. Math. Bull., 27, 2*, 205-214.

[40] Knuth, D.E. *The Art of Computer Programming. Vol.2, 2nd ed.* Addison-Wesley, Reading Mass., 1981.

[41] Koev, P.S. *Accurate and Efficient Computations with Structured Matrices.* University of California, Berkeley, 2002.

[42] Luttmann, F.W., & Rivlin, T.J. (1965). Some numerical experiments in the theory of polynomial interpolation. *IBM J. Res. Develop., 9*, 187-191.

[43] Meijering, E. (2002). A Chronology of Interpolation: From Ancient Astronomy to Modern Signal and Image Processing. *Proc. of IEEE, 90*, 319-342.

[44] Mills, T.M., & Smith, S.J. (1992). The Lebesgue constant for Lagrange interpolation on equidistant nodes. *Numerische Math., 61*, 111-115.

[45] Nikiforov, A.F., Suslov, S.K., & Uvarov, V.B. *Classical orthogonal polynomials of a discrete variable.* Springer-Verlag, Berlin, 1991.

[46] Oruç, H., & Phillips, G.M. (2000). Explicit factorization of the Vandermonde matrix. *Linear Algebra Appl., 315*, 113-123.

[47] Rasmussen, A.F., & Floater, M.S. *A point-based method for estimating surface area.* SIAM Conf. on Geom. Design and Comp., Phoenix, 2005.

[48] Reichel, L. (1991). Fast QR Decomposition of Vandermonde-Like Matrices and Polynomial Least Squares Approximation. *SIAM J. Matrix Anal. Appl., 3*, 552-564.

[49] Reichel, L. (1990). Newton interpolation at Leja points. *BIT, 30*, 332-346.

[50] Schonhage, A. (1961). Fehlerfortpflanzung bei Interpolation. *Numerische Math., 3*, 62-71.

[51] Shivakumar, P.N., & Wong, R. (1982). Asymptotic expansion of the Lebesgue constants associated with polynomial interpolation. *Math. of Computation, 39*, 195-200.

[52] Szegö, G. *Orthogonal polynomials.* American Mathematical Society, 1985.

[53] Traub, J.F. (1966). Associated polynomials and uniform methods for the solution of linear problems. *SIAM Rev., 8*, 277-301.

[54] Turetskii, A.H. *Theory of Interpolation in Problem Form.* Minsk, 1968.

[55] Tyrtyshnikov, E. (1994). How bad are Hankel matrices. *Numerische Math., 67*, 261-269.

[56] Valdettaro, L., Rieutord, M., Braconnier, T., & Frayssé, V. (2007). Convergence and round-off errors in a two-dimensional eigenvalue problem using spectral methods and Arnoldi-Chebyshev algorithm. *Journal of Comp. and Applied Math., 205, 1*, 382-393.

[57] Wolfram, S. *Mathematica: a System for Doing Mathematics by Computers, 2nd ed.*, Addison-Wesley, 1991.

In: Computational Mathematics
Editor: Peter G. Chareton, pp. 201-211
ISBN: 978-1-60876-271-2

Chapter 7

A COMPARATIVE STUDY OF DIFFERENT SEMILOCAL CONVERGENCE RESULTS APPLIED TO KEPLER'S EQUATION

M. A. Diloné[a] and J. M. Gutiérrez[b*]
[a] INAFOCAM, Instituto Nacional de Actualización,
Formación y Capacitación del Magisterio, Dominican Republic
[b] Dpt. Mathematics and Computation, University of La Rioja, Spain

Abstract

In this chapter we consider different semilocal convergence results applied to Newton's method to numerically solve nonlinear equations. Let $\{x_n\}_{n=0}^{\infty}$ be the Newton sequence to approximate the root x^* of a nonlinear equation $f(x) = 0$. A semilocal convergence result imposes conditions on the starting point x_0 in order to guarantee the convergence of the Newton sequence to the root x^*. Here we consider the well known Kepler's equation

$$E = e\sin E + M,$$

as a function test in order to compare some well known semilocal convergence results. So we consider Kantorovich's theorem ([7]), Smale's α-theory ([8]) and some of their variants ([4], [5], [9]). These kind of results are not only convergence theorems. In fact, they can be seen as existence and uniqueness results. So we generalize the work of Zhen Shi-Ming ([11]) to obtain the values of the eccentricity e for which Kepler's equation has a solution just by applying one of the aforementioned semilocal convergence results. In addition, we compare the regions of existence and uniqueness of solution given by these theorems.

1 Introduction

The aim of this chapter is to use Kepler's equation as a function test in order to compare different convergence results for Newton's method, defined iteratively by the algorithm

$$x_{n+1} = x_n - \frac{f(x_n)}{f'(x_n)}, \quad n \geq 0. \tag{1}$$

*E-mail address: jmguti@unirioja.e

Under the adequate convergence conditions, this method approximates the solution x^* of a nonlinear equation $f(x) = 0$. The method (1) and its extension to the solution of systems of nonlinear equation, form the basis of the most frequently used techniques for solving nonlinear algebraic equations.

In this chapter we consider the function $f(E)$ given by the famous Kepler's equation

$$f(E) = E - e\sin E - M, \tag{2}$$

where the eccentricity of the orbit e and the mean anomaly M are known. Throughout this chapter we assume $0 < e < 1$ and $0 \leq M \leq \pi$. Then the Newton's sequence related with this equation is

$$E_{n+1} = E_n - \frac{f(E_n)}{f'(E_n)}, \quad n \geq 0, \tag{3}$$

with an adequate starting point E_0.

We are aware that our results could be considerably improved by following other techniques and by taking into account the special properties of Kepler's equation (see [1] or [2] for a comprehensive information). Consequently, this chapter should be regarded as a comparative study of different semilocal convergence theories for Newton's method. In fact, we consider the specific studies given by Kantorovich ([6]) and Smale ([8]) and some of their variants([4], [5], [9]). These results guarantee the convergence of Newton's method to a solution of (2). In addition they can be seen as results of existence and uniqueness of solutions for Kepler's equation.

By using the aforementioned theories for the semilocal convergence of Newton's method, we obtain different constants κ such that if the eccentricity $e \leq \kappa$ then Kepler's equation (2) has a solution. For the different theories considered in this chapter we compare these constants and we analyze the regions of existence and uniqueness of solution for Kepler's equation (2). In this way, we extend the work of [11], where Smale's α-theory is used to establish a result on the existence of solution for Kepler's equation.

We introduce the notation

$$\delta = \frac{e}{1-e}. \tag{4}$$

Throughout this chapter, constant δ plays an important role.

2 Kantorovich's Theory Applied to Kepler's Equation

Historically speaking, in 1829, Cauchy presented the first semilocal convergence result for the Newton's method. In [10] we can find a version of the seminal Cauchy's theorem. This result, initially established for real functions of a real variable, was generalized by Kantorovich ([6], [7]) in his well known theorem enunciated for operators defined in Banach's spaces. Kantorovich realized that techniques in Functional Analysis can be used to solve a problem in Numerical Analysis. The next results shows the well known Kantorovich's theorem, in its scalar version, and the particular case of its application to Kepler's equation.

Theorem 1 (Kantorovich' theorem, [7]). *Let $f : I \subset \mathbb{R} \to \mathbb{R}$ be a differentiable function defined on an open interval I. Let us assume that $x_0 \in I$ is such that $f(x_0) \neq 0$, $f'(x_0) \neq 0$ and*

1. $|f(x_0)/f'(x_0)| \leq \beta$,
2. $|(f'(x)-f'(y))/f'(x_0)| \leq \gamma|x-y|$, $\forall x,y \in I$,
3. $h = \beta\gamma \leq 1/2$,
4. $(x_0 - t^*, x_0 + t^*) \subseteq I$, *where*

$$t^* = \frac{1-\sqrt{1-2h}}{\gamma}, \quad t^{**} = \frac{1+\sqrt{1-2h}}{\gamma},$$

are the roots of the majorizing polynomial

$$\Phi_K(t) = \frac{1}{2}\gamma t^2 - t + \beta.$$

Then Newton's method (1) *starting at* x_0 *is convergent to a solution* x^* *of the equation* $f(x) = 0$. *In addition,* $x^* \in [x_0 - t^*, x_0 + t^*]$, *and it is the only solution into the interval* $(x_0 - t^{**}, x_0 + t^{**})$.

Corollary 1. *If* $e \leq 0.4142$ *then Kepler's equation has a solution. This solution is located into the interval* $[M - t^*, M + t^*]$, *and it is unique into the interval* $(M - t^{**}, M + t^{**})$, *where*

$$t^* = \frac{1-\sqrt{1-2\delta^2}}{\delta},$$
$$t^{**} = \frac{1+\sqrt{1-2\delta^2}}{\delta},$$

are the roots of the majorizing polynomial

$$\phi_K(t) = \frac{1}{2}\delta t^2 - t + \delta, \tag{5}$$

with δ *defined in* (4).

In addition, Newton's method, defined by (3) *and starting at* $E_0 = M$, *converges to this solution.*

Proof. We take $E_0 = M$ in Kantorovich's theorem and we calculate the parameters β and γ appearing in it. In this case we have $\gamma = \beta = \delta$, with δ defined in (4). Consequently, the condition $h \leq 1/2$ is fulfilled when $\delta \leq \sqrt{2}/2$ and therefore $e \leq \sqrt{2} - 1 \approx 0.4142$.

The rest of the proof follows immediately from Kantorovich's theorem. □

There exists several variants of Kantorovich's theorem that also provides conditions for the convergence of Newton's method. In the foregoing, we introduce one of this variants and we obtain the corresponding result for Kepler's equation.

In [4], [5] a new semilocal convergence theorem for Newton's method was introduced. A Lipschitz condition for the second derivative is assumed instead of the Lipschitz condition for the first derivative that appears in Kantorovich's theorem. Both results have been compared in [4] and [5], where we can find examples where one of these results works better than the other and viceversa. Now we show the particular results obtained for Kepler's equation (2).

Theorem 2 ([4], [5]). *Let $f : I \subset \mathbb{R} \to \mathbb{R}$ be a twice differentiable function defined on an open interval I. Let us assume that $x_0 \in I$ is such that $f(x_0) \neq 0$, $f'(x_0) \neq 0$ and*

1. $|f(x_0)/f'(x_0)| \leq \beta$,
2. $|f''(x_0)/f'(x_0)| \leq \gamma$,
3. $|(f''(x) - f''(x_0))/f'(x_0)| \leq k|x - x_0|$, $\forall x \in I$,
4. *The polynomial* $\Phi_G(t) = \beta - t + \frac{\gamma}{2}t^2 + \frac{k}{6}t^3$ *has positive roots,* t^* *and* t^{**}*, with* $t^* \leq t^{**}$,
5. $(x_0 - t^*, x_0 + t^*) \subseteq I$.

Then Newton's method (1) starting at x_0 is convergent to a solution x^* of the equation $f(x) = 0$. In addition, $x^* \in [x_0 - t^*, x_0 + t^*]$, and it is the only solution into the interval $(x_0 - t^{**}, x_0 + t^{**})$.

Corollary 2. *If $e \leq 0.3759$ then Kepler's equation has a solution. This solution is located into the interval $[M - t^*, M + t^*]$, and it is unique into the interval $(M - t^{**}, M + t^{**})$, where t^* and t^{**} are the positive roots of the majorizing polynomial*

$$\phi_G(t) = \delta - t + \frac{1}{2}\delta t^2 + \frac{1}{6}\delta t^3, \tag{6}$$

with δ defined in (4). In addition, Newton's method, defined by (3) and starting at $E_0 = M$ converges to this solution.

Proof. We take $E_0 = M$ in Theorem 2 and we calculate the parameters β, γ and k appearing in it. It is not difficult to see that $\beta = \gamma = k = \delta$, with δ defined in (4). Consequently, the majorizing polynomial for Kepler's equation is defined by (6). Notice that polynomial (6) has an unique minimum $\hat{t} > 0$:

$$\hat{t} = -1 + \sqrt{1 + \frac{2}{\delta}}.$$

Then the equation $\phi_G(t) = 0$ has positive solutions if an only if $\phi_G(\hat{t}) \leq 0$, that is

$$-8 - 3\delta + 18\delta^2 + 15\delta^3 \leq 0.$$

This inequality happens when $\delta \leq 0.6023$ or, equivalently, when $e \leq 0.3759$. Consequently, when $e \leq 0.3759$, Kepler's equation (2) has a solution and the rest of the proof is a direct application of Theorem 2. □

3 Smale's α-Theory Applied to Kepler's Equation

In this section we apply another theory to Kepler's equation (2). The known as α-theory was introduced by H. M. Kim and S. Smale. In fact, Kim applied the α-theory to polynomial equations in one variable in his PhD thesis entitled *"Computational Complexity of the Euler Type Algorithms for the Roots of Polynomials"*, published in February, 1986 and in an article published in 1988 with title *"On approximate zeroes and root finding algorithms"* (see [3] for more details). In addition, Smale [8] in his book entitled *"New Directions in Pure, Applied and Computational Mathematics"*, published in 1986, introduced the α-theory for system of equations. One of the main differences between Kantorovich's and Smale's theories is that the first one assumes conditions on an open ball around the initial point x_0 whereas the second theory is based only in data at the initial point x_0. The other main difference is that Kantorovich's assumptions involve only the first derivative of an operator but Smale's assumptions involve all the derivatives of such operator.

As in the case of Kantorovich's theory, the α-theory can be regarded as a way of obtaining results about the existence and uniqueness of solution of a nonlinear equation and not only as a semilocal convergence result for Newton's method. This fact was took into account by Zhen Shi-Ming [11] to obtain an interesting result for the existence of solution for Kepler's equation.

Theorem 3 (Smale's α-theory, [8]). *Let $f: I \subset \mathbb{R} \to \mathbb{R}$ be a infinitely differentiable function defined on an open interval I. Let us assume that $x_0 \in I$ is such that $f(x_0) \neq 0$, $f'(x_0) \neq 0$ and*

1. $|f(x_0)/f'(x_0)| \leq \beta$,
2. $\sup_{k\geq 2}(|f^{(k)}(x_0)/(k!f'(x_0))|)^{1/(k-1)} \leq \gamma$,
3. $\alpha = \beta\gamma \leq 3 - 2\sqrt{2}$,
4. $(x_0 - t^*, x_0 + t^*) \subseteq I$, *where t^* is the smallest positive root of the majorizing function*

$$\Phi_S(t) = \beta - t + \frac{\gamma t^2}{1 - \gamma t}. \tag{7}$$

Then Newton's method (1) *starting at x_0 is convergent to a solution x^* of the equation $f(x) = 0$. In addition, $x^* \in [x_0 - t^*, x_0 + t^*]$, and it is the only solution into the interval $(x_0 - t^{**}, x_0 + t^{**})$, where t^{**} is the other positive root of the majorizing function* (7).

Corollary 3 (Zhen Shi-Ming, [11]). *If $e \leq 0.3594$ then Kepler's equation has a solution. This solution is located into the interval $[M - t^*, M + t^*]$, and it is unique into the interval $(M - t^{**}, M + t^{**})$, where t^* and t^{**} are the roots of the majorizing function*

$$\phi_S(t) = \delta - t + \frac{\gamma t^2}{1 - \gamma t}, \tag{8}$$

with δ defined in (4) and

$$\sup_{k\geq 2}(\delta/k!)^{(1/(k-1))} = \gamma. \tag{9}$$

In addition, Newton's method, defined by (3) *and starting at $E_0 = M$, converges to this solution.*

Proof. We take $E_0 = M$ in Smale's theorem 3 and we calculate the parameters β and γ. It is easy to show that $\beta = \delta$ but the calculus of γ given by (9) is not as straightforward. Let us denote $a_k = (\delta/k!)^{1/(k-1)}$, $k \geq 2$. Then, the sequence $\{a_k\}$ satisfies $\lim_{k\to\infty} a_k = 0$ and $a_2 \leq a_3 \leq \cdots \leq a_\ell \geq a_{\ell+1} \geq \cdots$ for some ℓ to determine. Consequently, $\gamma = a_\ell$. But the value of ℓ depends on the value of δ.

To avoid this inconvenience, and following [11], we have that in any case $\delta\gamma \le \alpha_0$, with $\alpha_0 = 3 - 2\sqrt{2}$. Then for each $k \ge 2$,

$$\delta\gamma = \left(\frac{\delta^k}{k!}\right)^{1/(k-1)} \le \alpha_0,$$

and consequently,

$$\delta \le (\alpha_0^{k-1} k!)^{1/k}.$$

Now, let us denote $\delta_k = (\alpha_0^{k-1} k!)^{1/k}$, for $k \ge 2$. Notice that $\delta_2 \ge \delta_3$, but $\delta_3 \le \delta_4$ and, in general, $\delta_k \le \delta_{k+1}$ for $k \ge 3$.

Then, if $\delta \le \delta_3 = [6\alpha_0^2]^{1/3} = 0.5610688$, that is

$$e = \frac{\delta}{1+\delta} \le \frac{\delta_3}{1+\delta_3} \approx 0.3594,$$

the hypotheses of Smale's theorem 3 are fulfilled an the rest of the proof is a consequence of such theorem. □

Wang and Zhao, [9], pointed that the second condition in Smale's theorem 3 is too restrictive. They have shown that it is sufficient to demand that

$$\frac{1}{k!}|f^{(k)}(x_0)/f'(x_0)| \le \gamma_k, \quad k \ge 2, \tag{10}$$

and the equation $\Phi_W(t) = 0$ has, at least, a positive solution, where

$$\Phi_W(t) = \beta - t + \sum_{k\ge 2} \gamma_k t^k. \tag{11}$$

Theorem 4 (Wang-Zhao's α-theory, [9]). *Let $f : I \subset \mathbb{R} \to \mathbb{R}$ be a infinitely differentiable function defined on an open interval I. Let us assume that $x_0 \in I$ is such that $f(x_0) \ne 0$, $f'(x_0) \ne 0$ and*

1. $|f(x_0)/f'(x_0)| \le \beta$,
2. *condition* (10) *holds,*
3. *the equation $\Phi_W(t) = 0$ with $\Phi_W(t)$ defined in* (11), *has two positive solutions $t^* \le t^{**}$,*
4. $(x_0 - t^*, x_0 + t^*) \subseteq I$.

Then Newton's method (1) *starting at x_0 is convergent to a solution x^* of the equation $f(x) = 0$. In addition, $x^* \in [x_0 - t^*, x_0 + t^*]$, and it is the only solution into the interval $(x_0 - t^{**} x_0 + t^{**})$.*

Notice that $\gamma_k \le \gamma^{k-1}$ for $k \ge 2$ and then $\Phi_W(t) \le \Phi_S(t)$ for $t \ge 0$. Then $\Phi_S(t) = 0$ implies $\Phi_W(t) = 0$, but not in the opposite way. So, the Whang-Zhao theory can be applied in situations where the Smale theory is not applicable as. This situation happens,

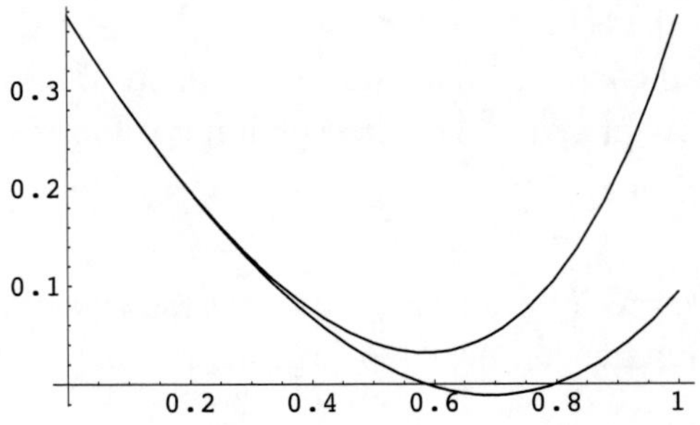

Figure 1: Graphics of the majorizing functions $\phi_S(t)$ and $\phi_W(t)$ obtained when $f(x) = 8\exp(x) - 11$.

for instance, in the example given by the function $f(x) = 8\exp(x) - 11$ and $t_0 = 0$. In this case,

$$\Phi_S(t) = \frac{3}{8} - t + \frac{t^2}{2-t}, \quad \Phi_W(t) = -\frac{5}{8} - 2t + \exp(t).$$

Notice that $\Phi_S(t) = 0$ has no solution whereas $\Phi_W(t) = 0$ has positive solutions (see Figure 1).

The next result is a consequence of the application of Wang-Zhao's α-theory to Kepler's equation (2).

Corollary 4. *If $e \leq 0.3594$ then Kepler's equation has a solution. This solution is located into the interval $[M - t^*, M + t^*]$, and it is unique into the interval $(M - t^{**}, M + t^{**})$, where t^* and t^{**} are the roots of the majorizing function*

$$\phi_W(t) = \delta\exp(t) - (1+\delta)t, \tag{12}$$

with δ defined in (4).

In addition, Newton's method, defined by (3) starting at $E_0 = M$, converges to this solution.

Proof. As in the previous cases, just by taking $E_0 = M$ in Wang-Zhao's theorem 4 we obtain $\beta = \delta$ and $\gamma_k = \delta/k!$. Then the majorizing function is $\phi_W(t)$, given in (12).

Notice that $\phi_W(t)$ has a unique positive minimum $\hat{t} = \ln(1 + 1/\delta)$. Then the equation $\phi_W(t) = 0$ has positive solutions if and only if $\phi_W(\hat{t}) < 0$. This inequality happens when

$$1 - \ln\left(1 + \frac{1}{\delta}\right) < 0 \Rightarrow \delta \leq \frac{1}{\exp(1) - 1}.$$

Consequently, if $e \leq 0.3678$ the conditions of Wang-Zhao theorem 4 are satisfied and the proof follows immediately. □

4 Conclusion

Taking into account all the results obtained in the previous sections, we conclude that for Kepler's equation (2), the best results are deduced from Kantorovich's theory. In fact, regarding the values e of the eccentricity, Kantorovich's theorem gives us the higher value amongst the considered theorems. With more detail, in Table 1 we show the different theories analyzed in this chapter, and the corresponding values for the parameter δ, defined in (4), and e.

Table 1: Values of δ and $e = \delta/(1+\delta)$ that guarantees the existence of solution for Kepler's equation by applying different semilocal convergence techniques for Newton's method

Theory	δ	e
Kantorovich (Corollary 1)	$\sqrt{2}/2$	0.4142
Gutiérrez-Huang (Corollary 2)	0.6023	0.3759
Smale (Corollary 3)	$\sqrt[3]{102-72\sqrt{2}}$	0.3594
Whang-Zhao (Corollary 4)	$1/(\exp(1)-1)$	0.3678

We can see that, for the case of Kepler's equation (2) the more restrictive situation appears under Smale's α-theory and the more advantageous under Kantorovich's theory. To be more precise, if we compare the four majorizing functions that appear in Table 2 we have:

$$\phi_K(0) = \phi_G(0) = \phi_S(0) = \phi_W(0) = \delta$$

and

$$\phi_K(t) < \phi_G(t) < \phi_S(t) < \phi_W(t), \quad t > 0. \tag{13}$$

Table 2: Majorizing functions for the different semilocal convergence theories considered in this chapter, depending on the parameter δ defined in (4).

Theory	Majorizing function
Kantorovich (Corollary 1)	$\phi_K(t) = \delta - t + \frac{1}{2}\delta t^2$
Gutiérrez-Huang (Corollary 2)	$\phi_G(t) = \delta - t + \frac{1}{2}\delta t^2 + \frac{1}{6}\delta t^3$
Smale (Corollary 3)	$\phi_S(t) = \delta - t + \frac{\gamma t^2}{1-\gamma t}, \quad \gamma = \sup_{k\geq 2}(\delta/k!)^{1/(k-1)}$
Whang-Zhao (Corollary 4)	$\phi_W(t) = \delta \exp t - (1+\delta)t$

Let us denote t_K^*, t_G^*, t_S^* and t_W^* the smallest positive root of the majorizing functions existence radius obtained by Kantorovich's, Gutiérrez's Smale's and Wang-Zhao's theories respectively. As we have seen previously, these values provide the radius of the interval around $E_0 = M$ where the root is located. From this point of view, the smaller t^* is, the better the root will be located.

In the same way, t_K^{**}, t_G^{**}, t_S^{**} and t_W^{**} will denote the second positive root of the corresponding majorizing functions. These values give the radius of the interval around $E_0 = M$

where the root is unique. From this point of view, we are interested now in the higher value of t^{**}.

Taking into account (13) we see that in both cases the best results are achieved from $\phi_K(t)$ and next from $\phi_G(t)$, $\phi_W(t)$ and $\phi_S(t)$ in this same order.

In Figures 2 and 3 we show graphically the behavior of these majorizing functions for two specific values of the eccentricity e, corresponding to two real cases: the planet Mercury ($e = 0.205$) and the asteroid Hungaria ($e = 0.073$).

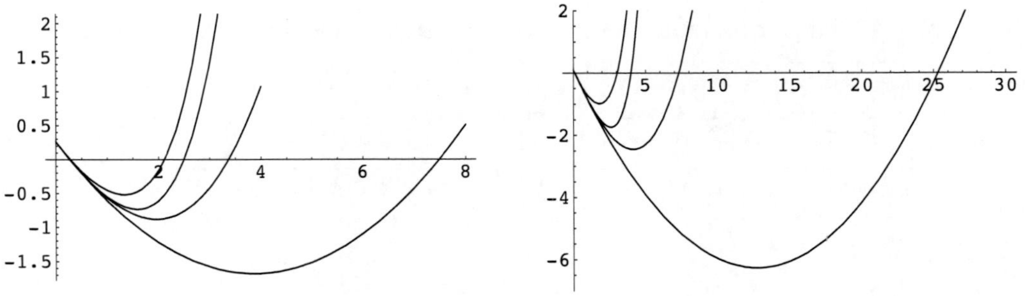

Figure 2: Graphics of the majorizing functions $\phi_S(t)$, $\phi_W(t)$, $\phi_G(t)$ and $\phi_K(t)$ for the cases of the planet Mercury (left) and the asteroid Hungaria (right).

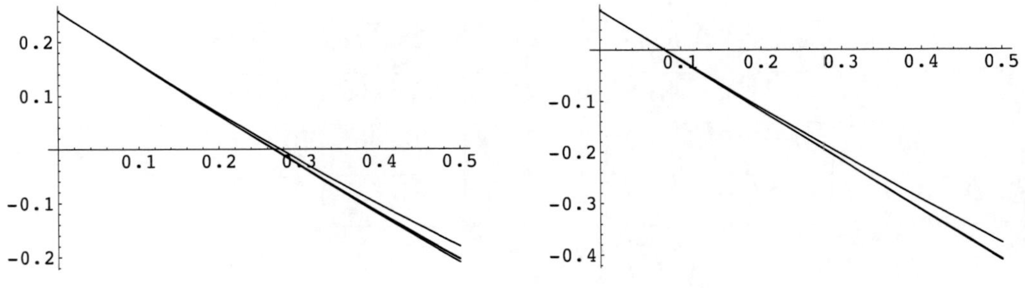

Figure 3: Detail of the graphics of the majorizing functions $\phi_S(t)$, $\phi_W(t)$, $\phi_G(t)$ and $\phi_K(t)$ for the cases of the planet Mercury (left) and the asteroid Hungaria (right), showing the first roots t^*.

In Tables 3 and 4 we analyze the values of the roots t^* and t^{**} for the planet Mercury ($e = 0.205$) and the asteroid Hungaria ($e = 0.073$). As we can see, in these cases we cannot appreciate a significant difference in the values of t^* but in the case of t^{**}, the differences are more relevant.

Finally, in Tables 5 and 6 we analyze numerically, for different values of the eccentricity e, the behavior of the existence radius t^* and the uniqueness radius t^{**}. As we can see, the differences between the existence radius t^* increase when e increases. However the differences between the uniqueness radius t^{**} increase when e decreases.

Table 3: Existence radius for the four majorizing functions given in Table2.

Cases	t_K^*	t_G^*	t_W^*	t_S^*
Planet Mercury ($e = 0.205$)	0.268	0.269	0.263	0.269
Asteroid Hungaria ($e = 0.073$)	0.079	0.079	0.079	0.079

Table 4: Uniqueness radius for the four majorizing functions given in Table2.

Cases	t_K^{**}	t_G^{**}	t_W^{**}	t_S^{**}
Planet Mercury ($e = 0.205$)	7.459	3.363	7.330	2.496
Asteroid Hungaria ($e = 0.073$)	25.094	7.271	79.176	3.993

Table 5: Evolution of the existence radius for different values of the eccentricity e.

e	t_K^*	t_G^*	t_W^*	t_S^*
0.05	0.052705	0.052706	0.052706	0.053054
0.1	0.111806	0.111832	0.111833	0.113400
0.15	0.179307	0.179483	0.179491	0.183295
0.2	0.258343	0.259118	0.259171	0.266465
0.25	0.354249	0.357119	0.357403	0.369312
0.3	0.477412	0.487869	0.489402	0.505545
0.35	0.653407	0.702381	0.716639	0.724562

Table 6: Evolution of the uniqueness radius for different values of the eccentricity e.

e	t_K^{**}	t_G^{**}	t_W^{**}	t_S^{**}
0.05	37.9473	9.25181	4.49976	3.33209
0.1	17.8882	5.93227	3.57715	2.80848
0.15	11.154	4.40664	2.99359	2.45821
0.2	7.74166	3.45089	2.54264	2.14796
0.25	5.64575	2.75074	2.15329	1.87743
0.3	4.18925	2.17266	1.78134	1.62167
0.35	3.06088	1.60847	1.34972	1.31745

References

[1] P. Colwell: *Solving Kepler's equation over three ceturies.* Willmann-Bell, Inc. Virginia. 1993.

[2] J. M. A. Danby and T. M. Burkardt: *The solution of Kepler equation I.* Celestial Mechanics, **Vol. 31**, (1983) pp. 95–107.

[3] J. P. Dedieu and M. H. Kim: *Newton's method for analytic system of equations with constant rank derivatives.* Journal of Complexity, **Vol. 18**, (2002) pp. 187–209.

[4] J. M. Gutiérrez: *A new semilocal convergence theorem for Newton's method.* J. Comput. Appl. Math., **Vol. 79**, (1997) pp. 131–145.

[5] Z. Huang: *A note on the Kantorovich theorem for Newton iteration.* J. Comput. Appl. Math., **Vol. 47**, (1993) 211–217.

[6] L. V. Kantorovich, *On Newton's method for functional equations,* Dokl. Akad. Nauk SSSR, **Vol. 59** (1948) 1237–1240.

[7] L. V. Kantorovich and G. P. Akilov: *Functional analysis.* Pergamon Press. Oxford. 1982.

[8] S. Smale: *Newton's method estimates from data at one point.* The Merging of Disciplines: New Directions in Pure Applied and Computational Mathematics, R. Ewing, K. Gross and C. Martin (eds.), pp 185–196. Springer-Verlag. USA. 1986.

[9] D. Wang and F. Zhao: *The Theory of Smale's Point Estimation and its Applications.* J. Comput. Appl. Math., **Vol. 60**, (1995) pp. 253–269.

[10] T. Yamamoto: *Historical development in convergence analysis for Newton's and Newton-like methods.* J. Comput. Appl. Math., **Vol. 124**, (2000) pp. 1–23.

[11] Zhen Shi-Ming: *Point estimates for Halley's iteration.* Acta Mathematicae Aplicate Sinica, **Vol. 14, No. 3**, (1991) pp. 381–383.

In: Computational Mathematics
Editor: Peter G. Chareton, pp. 213-260

ISBN 978-1-60876-271-2

Chapter 8

DISCRETE MAXIMUM PRINCIPLES FOR FEM SOLUTIONS OF NONLINEAR ELLIPTIC SYSTEMS

***János Karátson*[a*] *and Sergey Korotov*[b]**
[a] Department of Applied Analysis and Computational Mathematics,
Eötvös Loránd University; H–1518, Budapest, Pf. 120, Hungary
[b] Institute of Mathematics, Helsinki University of Technology
P.O. Box 1100, FI–02015 TKK, Finland

Abstract

The discrete maximum principle (DMP) is an important qualitative property of various discretized elliptic equations. Conditions that ensure the DMP have drawn much attention, including geometric properties for FEM discretizations. This chapter starts with a brief summary of some background on the DMP, including the algebraic case, and nonobtuseness or acuteness type conditions for FEM. When lower order terms are included in the operator, the DMP can be ensured for sufficiently fine mesh, under uniform acuteness or strict non-narrowness in the case of simplicial or rectangular FEM meshes, respectively. (Similar conditions also appear for prismatic FEM.) Our main interest is formed by nonlinear elliptic systems under standard linear or bilinear FEM discretizations. We first present our previous results on systems with second and zeroth order terms, then extend them to the case involving first order terms. The presentation includes a detailed exposition of the required theory, which needs a generalization of the usual underlying algebraic DMP and some Hilbert space background. The geometric properties of the FEM mesh are also discussed. In many applications the DMP implies (or reduces to) a natural requirement of nonnegativity for the approximations of the corresponding nonnegative physical quantities. Such applications are given to reaction-diffusion processes and diffusion-dominated transport systems, respectively.

1 Introduction

The maximum principle forms an important qualitative property of second order elliptic equations [40, 46]. Therefore its discrete analogues, the so-called discrete maxi-

*E-mail address: karatson@cs.elte.hu

mum principles (DMPs) have drawn much attention. Various DMPs, including geometric conditions on the computational meshes for FEM solutions, have been given e.g. in [4, 7, 8, 10, 11, 12, 19, 24, 33, 37, 38, 41, 47, 51, 53] for linear and [6, 25, 26, 27, 35] for nonlinear problems. For elliptic operators with only principal part, if the discretized operator L_h and the FEM solution u_h satisfy $L_h u_h \leq 0$, then the DMP has the simple form $\max_{\overline{\Omega}} u_h = \max_{\partial\Omega} u_h$. On the other hand, for operators with lower order terms as well, one has the weaker statement

$$\max_{\overline{\Omega}} u_h \leq \max\{0, \max_{\partial\Omega} u_h\}. \tag{1}$$

Also, in the latter case one can only provide the DMP for sufficiently fine mesh and needs stronger acuteness type conditions in the case of standard simplicial FEM meshes. Similar conditions on the shape and size of meshes also appear in the case when bilinear or prismatic finite elements are used [19].

The DMP has been extended to systems for the first time in [26]. The class of systems considered there has a coupling which is cooperative and weakly diagonally dominant, these conditions on the coupling also appear in the underlying continuous maximum principle [9, 16, 42, 43]. In the case of mixed boundary conditions and nonpositive right-hand sides, we have

$$\max_{k=1,\ldots,s} \max_{\overline{\Omega}} u_k^h \leq \max_{k=1,\ldots,s} \max\{0, \max_{\Gamma_D} u_k^h\} \tag{2}$$

where Γ_D is the Dirichlet boundary and k is the number of equations. The acuteness type conditions for simplicial FE meshes have also been suitably weakened in [26].

This chapter is devoted to the DMP for elliptic systems of general type. Its goal is twofold. First, after giving a proper background including algebraic properties and a suitable Hilbert space theory, we summarize our previous results on elliptic systems with second and zeroth order terms. Then, based on these, we develop various new results on systems which are regularly perturbed by first order terms, i.e. contain non-dominating convection type terms. Our general goal is to ensure (2). Further, in many applications the DMP reduces to the natural requirement of nonnegativity for the appropriate discrete quantities, hence this will be also addressed. Some applications are mentioned briefly to reaction-diffusion processes and transport systems, respectively.

We note that the DMP for a single nonsymmetric equation has been studied extensively in the last three decades, see e.g. the early papers [28, 44, 45], the monograph [23] and the references therein. Here a major issue is to provide a DMP for singularly perturbed (i.e. convection-dominated) problems, along with the construction of suitably stabilized Galerkin methods, see e.g. [13, 14, 29, 39, 48, 50]. The present chapter is devoted to regularly perturbed problems, with focus on the generalization of the standard DMP for Galerkin methods to systems; extension of these results to stabilized Galerkin methods may be the subject of further research.

2 Discrete Maximum Principles in Different Settings

2.1 Algebraic background and the 'matrix maximum principle'

Let us consider a system of equations of order $(k+m) \times (k+m)$:

$$\bar{\mathbf{A}}\bar{\mathbf{c}} = \bar{\mathbf{d}}, \tag{3}$$

where the matrix $\bar{\mathbf{A}}$ and the vectors $\bar{\mathbf{d}}$, $\bar{\mathbf{c}}$ have the following structure:

$$\bar{\mathbf{A}} = \begin{bmatrix} \mathbf{A} & \tilde{\mathbf{A}} \\ \mathbf{0} & \mathbf{I} \end{bmatrix}, \qquad \bar{\mathbf{d}} = \begin{bmatrix} \mathbf{d} \\ \tilde{\mathbf{d}} \end{bmatrix}, \qquad \bar{\mathbf{c}} = \begin{bmatrix} \mathbf{c} \\ \tilde{\mathbf{c}} \end{bmatrix} \tag{4}$$

where $\mathbf{I}$ is the $m \times m$ identity matrix and $\mathbf{0}$ is the $m \times k$ zero matrix. Then (3) becomes

$$\begin{bmatrix} \mathbf{A} & \tilde{\mathbf{A}} \\ \mathbf{0} & \mathbf{I} \end{bmatrix} \begin{bmatrix} \mathbf{c} \\ \tilde{\mathbf{c}} \end{bmatrix} = \begin{bmatrix} \mathbf{d} \\ \tilde{\mathbf{d}} \end{bmatrix}. \tag{5}$$

First we recall a basic definition in the study of DMP (cf. [52, p. 23]):

Definition 2.1. A square $k \times k$ matrix $\mathbf{A} = (a_{ij})_{i,j=1}^{k}$ is called *irreducibly diagonally dominant* if it satisfies the following conditions:

(i) $\mathbf{A}$ is irreducible, i.e., for any $i \neq j$ there exists a sequence of nonzero entries $\{a_{i,i_1}, a_{i_1,i_2}, \ldots, a_{i_s,j}\}$ of A, where $i, i_1, i_2, \ldots, i_s, j$ are distinct indices,

(ii) $\mathbf{A}$ is diagonally dominant, i.e., $|a_{ii}| \geq \sum\limits_{\substack{j=1\\ j\neq i}}^{k} |a_{ij}|$, $i = 1, \ldots, k$,

(iii) for at least one index $i_0 \in \{1, \ldots, k\}$ the above inequality is strict, i.e.,

$$|a_{i_0,i_0}| > \sum_{\substack{j=1\\ j\neq i_0}}^{k} |a_{i_0,j}|.$$

Following [11], we introduce

Definition 2.2. A $(k+m) \times (k+m)$ matrix $\bar{\mathbf{A}}$ with the structure (4) is said to be of *generalized nonnegative type* if the following properties hold:

(i) $a_{ii} > 0, \quad i = 1, \ldots, k$,

(ii) $a_{ij} \leq 0, \quad i = 1, \ldots, k,\ j = 1, \ldots, k+m \quad (i \neq j)$,

(iii) $\sum\limits_{j=1}^{k+m} a_{ij} \geq 0, \quad i = 1, \ldots, k$,

(iv) There exists an index $i_0 \in \{1, \ldots, k\}$ for which

$$\sum_{j=1}^{k} a_{i_0,j} > 0. \tag{6}$$

Remark 2.1. In the original definition in [11, p. 343], it is assumed instead of the above property (iv) that the principal block $\mathbf{A}$ is irreducibly diagonally dominant. However, if we assume that $\mathbf{A}$ is also irreducible, as will be done in Theorem 2.1, then its irreducibly diagonal dominance follows directly from Definition 2.2 under the given sign conditions on a_{ij}. We also note that a well-known theorem [52, p. 85] implies in this case that $\mathbf{A}^{-1} > 0$, i.e., the entries of the matrix $\mathbf{A}^{-1}$ are positive.

Many known results on various discrete maximum principles are based on the following theorem, considered as 'matrix maximum principle' (for a proof, see e.g. [11, Th. 3]).

Theorem 2.1. *Let $\bar{\mathbf{A}}$ be a $(k+m)\times(k+m)$ matrix with the structure (4), and assume that $\bar{\mathbf{A}}$ is of generalized nonnegative type in the sense of Definition 2.2, further, that $\mathbf{A}$ is irreducible.*

If the vector $\bar{\mathbf{c}} = (c_1,\ldots,c_{k+m})^T \in \mathbf{R}^{k+m}$ (where $(\,.\,)^T$ denotes the transposed) is such that $(\bar{\mathbf{A}}\bar{\mathbf{c}})_i \le 0$, $i = 1,\ldots,k$, then

$$\max_{i=1,\ldots,k+m} c_i \le \max\{0, \max_{i=k+1,\ldots,k+m} c_i\}. \tag{7}$$

The irreducibility of $\mathbf{A}$ is a technical condition which is sometimes difficult to check in applications, see e.g. [15, 20]. As shown in [26], it can be omitted from the assumptions if (iv) is suitably strengthened. This requires two definitions.

Definition 2.3. Let $\mathbf{A}$ be an arbitrary $k\times k$ matrix. The *irreducible blocks* of $\mathbf{A}$ are the matrices $\mathbf{A}^{(l)}$ $(l = 1,\ldots,q)$ defined as follows.

Let us call the indices $i, j \in \{1,\ldots,k\}$ *connectible* if there exists a sequence of nonzero entries $\{a_{i,i_1}, a_{i_1,i_2},\ldots,a_{i_s,j}\}$ of $\mathbf{A}$, where $i, i_1, i_2,\ldots,i_s, j \in \{1,\ldots,k\}$ are distinct indices. Further, let us call the indices i, j mutually connectible if both i, j and j, i are connectible in the above sense. (Clearly, mutual connectibility is an equivalence relation.) Let $N_1,\ldots,N_q$ be the equivalence classes, i.e. the maximal sets of mutually connectible indices. (Clearly, $\mathbf{A}$ is irreducible iff $q = 1$.) Letting $N_l = \{s_1^{(l)},\ldots,s_{k_l}^{(l)}\}$ for $l = 1,\ldots,q$, we have $k_1+\cdots+k_q = k$. Then we define for all $l = 1,\ldots,q$ the $k_l\times k_l$ matrix $\mathbf{A}^{(l)}$ by $\mathbf{A}^{(l)}_{pq} := a_{s_p^{(l)},s_q^{(l)}}$ $(p,q = 1,\ldots,k_l)$.

Remark 2.2. One may prove (cf. [1, Th. 4.2]) that by a proper permutation of indices, $\mathbf{A}$ becomes a block lower triangular matrix with the irreducible diagonal blocks $\mathbf{A}^{(l)}$.

Definition 2.4. A $(k+m)\times(k+m)$ matrix $\bar{\mathbf{A}}$ with the structure (4) is said to be of *generalized nonnegative type with irreducible blocks* if properties (i)-(iii) of Definition 2.2 hold, further, property (iv) therein is replaced by the following stronger one:

(iv') For each irreducible component of $\mathbf{A}$ there exists an index $i_0 = i_0(l) \in N_l = \{s_1^{(l)},\ldots,s_{k_l}^{(l)}\}$ for which $\sum\limits_{j=1}^{k} a_{i_0,j} > 0$.

Remark 2.3. Let assumptions (i)-(iii) hold in Definitions 2.2 or 2.4. Then for a given index $i_0 \in \{1,\ldots,k\}$, a sufficient condition for (6) to hold is that:

there exists an index $j_0 \in \{k+1,\ldots,k+m\}$ for which $a_{i_0,j_0} < 0$.

Namely, using also assumptions (ii) and (iii), respectively, we then have

$$\sum_{j=1}^{k} a_{i_0,j} > \sum_{j=1}^{k} a_{i_0,j} + a_{i_0,j_0} \geq \sum_{j=1}^{k} a_{i_0,j} + a_{i_0,j_0} + \sum_{\substack{j=k+1 \\ j\neq j_0}}^{k+m} a_{i_0,j} = \sum_{j=1}^{k+m} a_{i_0,j} \geq 0.$$

Theorem 2.2. *[[26]] Let* $\bar{\mathbf{A}}$ *be a* $(k+m)\times(k+m)$ *matrix with the structure* (4)*, and assume that* $\bar{\mathbf{A}}$ *is of generalized nonnegative type with irreducible blocks in the sense of Definition* 2.4.

If the vector $\bar{\mathbf{c}} = (c_1,\dots,c_{k+m})^T \in \mathbf{R}^{k+m}$ *is such that* $d_i \equiv (\bar{\mathbf{A}}\bar{\mathbf{c}})_i \leq 0,\ i=1,\dots,k,$ *then* (7) *holds.*

Consequently, in what follows, our main goal is to provide the stiffness matrix of the problems considered to be of generalized nonnegative type with irreducible blocks in the sense of Definition 2.4.

2.2 Some motivation for the DMP

2.2.1 Linear equations and continuous maximum principles

First we recall the *(continuous) maximum principle* (CMP) as it usually stands for linear second order elliptic problems. Let L denote the following linear operator, acting on smooth functions defined in a bounded domain Ω:

$$Lu \equiv -\operatorname{div}\Big(a(x)\,\nabla u\Big) + h(x)u, \tag{8}$$

where the coefficients $a \in C^1(\Omega)$ and $h \in C(\Omega)$ are such that $0 < \mu_0 \leq a(x) \leq \mu_1$ and $0 \leq h(x) \leq \mu_1$ with positive constants μ_0 and μ_1 independent of $x \in \Omega$. Further, we assume that $\Omega \subset \mathbf{R}^d, d = 2,3,\dots,$ has a piecewise smooth and Lipschitz continuous boundary $\partial\Omega$. The following basic result is found e.g. in [18, 40].

Theorem 2.3. *Let* $u \in C^2(\Omega)\cap C(\overline{\Omega})$ *be such that* $Lu \leq 0$ *in* Ω*, then*

$$\max_{\overline{\Omega}} u \leq \max\{0, \max_{\partial\Omega} u\}. \tag{9}$$

If, in addition, $h \equiv 0$*, then*

$$\max_{\overline{\Omega}} u = \max_{\partial\Omega} u. \tag{10}$$

Theorem 2.3 is also valid for more general differential operators, but we shall only use the operators in the form (8) in what follows. In the context of boundary value problems, we immediately obtain from Theorem 2.3 the following result:

Corollary 2.1. *Let* $u \in C^2(\Omega)\cap C(\overline{\Omega})$ *be a solution of the problem*

$$\begin{cases} Lu = f & \text{in } \Omega, \\ u = g & \text{on } \partial\Omega \end{cases} \tag{11}$$

where $g \in C(\partial\Omega)$. *If* $f \leq 0$ *in* Ω, *then*

$$\max_{\overline{\Omega}} u \leq \max\{0, \max_{\partial\Omega} g\}. \tag{12}$$

If, in addition, $h \equiv 0$, *then*

$$\max_{\overline{\Omega}} u = \max_{\partial\Omega} g. \tag{13}$$

The analogous (*continuous*) *minimum principles* can be immediately formulated by changing the sign condition (i.e., replacing u by $-u$). In this contribution we will be interested in the weak form (12) of the CMP.

Let us now consider a discretization of problem (11). The most widespread methods in this context are the finite element method (FEM) and finite difference method (FDM). Both of these discretizations normally lead to linear algebraic systems of the form (5), where the block decomposition corresponds to interior and boundary mesh points, respectively. For such discretizations, the goal is to ensure that the 'matrix maximum principle' (7) holds, i.e. to apply Theorem 2.1.

In this contribution we are mostly interested in FEM discretizations on simplices. Simplicial elements are most popular and present a basic special case of the FEM, because we can treat many complicated geometries with simplices. We emphasize here an important property related to FEM. Under standard assumptions, see later (66)-(67) (which hold e.g. for usual linear, bilinear or prismatic finite elements), statement (7) directly means that

$$\max_{\overline{\Omega}} u^h \leq \max\{0, \max_{\partial\Omega} g^h\} \tag{14}$$

for the discrete solution u^h of the boundary value problem (11). (Here g^h is the linear interpolant of g.) That is, the exact analogue of (12) is valid.

The main conditions that arise in this context are nonobtuseness (for problems with only principal part, i.e. $h \equiv 0$) or uniform acuteness conditions (for problems with lower order term) on the mesh. These conditions originate from the early papers [11, 12], see also [35]. Such geometric conditions will be discussed briefly in subsection 2.3.

When formulating discrete maximum principles, we will be interested in families of meshes and not in a given (single) mesh. Hence the following notion will be crucial for our study:

Definition 2.5. A set of FEM subspaces $\mathcal{V} = \{V_h\}_{h\to 0}$ is said to be a *family of FEM subspaces* if for any $\varepsilon > 0$ there exists $V_h \in \mathcal{V}$ with $h < \varepsilon$.

When the results are formulated in terms of simplicial FE meshes, one similarly defines the families of meshes $\mathcal{T} = \{\mathcal{T}_h\}_{h\to 0}$.

2.2.2 The DMP for a single nonlinear elliptic equation

The DMP for mixed nonlinear boundary value problems was first proved in [25]. Let us consider the problem

$$\begin{cases} -\operatorname{div}\Big(b(x,\nabla u)\,\nabla u\Big) + q(x,u) = f(x) & \text{in } \Omega, \\ b(x,\nabla u)\dfrac{\partial u}{\partial \nu} + s(x,u) = \gamma(x) & \text{on } \Gamma_N, \\ u = g(x) & \text{on } \Gamma_D, \end{cases} \tag{15}$$

where Ω is a bounded domain in $\mathbf{R}^d$, under the following

Assumptions 2.2.2.

(A1) Ω has a piecewise smooth and Lipschitz continuous boundary $\partial\Omega$; $\Gamma_N, \Gamma_D \subset \partial\Omega$ are measurable open sets, such that $\Gamma_N \cap \Gamma_D = \emptyset$ and $\overline{\Gamma}_N \cup \overline{\Gamma}_D = \partial\Omega$.

(A2) The scalar functions $b : \overline{\Omega} \times \mathbf{R}^d \to \mathbf{R}$, $q : \overline{\Omega} \times \mathbf{R} \to \mathbf{R}$ and $s : \overline{\Gamma}_N \times \mathbf{R} \to \mathbf{R}$ are continuously differentiable in their domains of definition. Further, $f \in L^2(\Omega)$, $\gamma \in L^2(\Gamma_N)$ and $g = g^*{}_{|\Gamma_D}$ with $g^* \in H^1(\Omega)$.

(A3) The function b satisfies

$$0 < \mu_0 \le b(x,\eta) \le \mu_1 \tag{16}$$

with positive constants μ_0 and μ_1 independent of (x,η), further, the diadic product matrix $\eta \cdot \frac{\partial b(x,\eta)}{\partial \eta}$ is symmetric positive semidefinite and bounded in any matrix norm by some positive constant μ_2 independent of (x,η).

(A4) Let $2 \le p_1$ if $d = 2$, or $2 \le p_1 \le \frac{2d}{d-2}$ if $d > 2$, further, let $2 \le p_2$ if $d = 2$, or $2 \le p_2 \le \frac{2d-2}{d-2}$ if $d > 2$. There exist functions $\alpha_1 \in L^{d/2}(\Omega)$, $\alpha_2 \in L^{d-1}(\Gamma_N)$ and a constant $\beta \ge 0$ such that for any $x \in \Omega$ (or $x \in \Gamma_N$, resp.) and $\xi \in \mathbf{R}$

$$0 \le \frac{\partial q(x,\xi)}{\partial \xi} \le \alpha_1(x) + \beta|\xi|^{p_1-2}, \qquad 0 \le \frac{\partial s(x,\xi)}{\partial \xi} \le \alpha_2(x) + \beta|\xi|^{p_2-2}. \tag{17}$$

(A5) Either $\Gamma_D \ne \emptyset$, or q increases strictly and at least linearly at ∞ in the sense that

$$q(x,\xi) \ge c_1|\xi| - c_2(x) \tag{18}$$

(with a constant $c_1 > 0$ and a function $c_2 \in L^1(\Omega)$) $\forall (x,\xi) \in \Omega \times \mathbf{R}$, or s increases strictly and at least linearly at ∞ in the same sense.

Theorem 2.4 ([25]). *Let* (A1)–(A5) *hold and let us consider a family of simplicial FEM meshes* $\mathcal{T} = \{\mathcal{T}_h\}_{h\to 0}$ *satisfying the following property: for any* $i = 1,\dots,n$, $j = 1,\dots,\bar{n}$ $(i \ne j)$, *the basis functions satisfy*

$$\nabla\phi_i \cdot \nabla\phi_j \le -\frac{\sigma_0}{h^2} < 0 \tag{19}$$

on supp $\phi_i \cap$ *supp* ϕ_j *with* $\sigma_0 > 0$ *independent of* i, j *and* h.

If the simplicial meshes $\mathcal{T}_h$ *are regular, i.e., there exist constants* $m_1, m_2 > 0$ *such that for any* $h > 0$ *and any simplex* $T_h \in \mathcal{T}_h$

$$m_1 h^d \leq meas(T_h) \leq m_2 h^d \tag{20}$$

(where $meas(T_h)$ *denotes the d-dimensional measure of* T_h*), then for sufficiently small h, the matrix* $\mathbf{A}(\overline{\mathbf{c}})$ *defined in* (46) *is of generalized nonnegative type in the sense of Definition 2.2, further,* $\mathbf{A}$ *is irreducible.*

Consequently, by Theorem 2.1, under the conditions of Theorem 2.4 and standard assumptions for the FEM mesh (see later (66)-(67)), if

$$f(x) - q(x,0) \leq 0, \; x \in \Omega, \qquad \text{and} \qquad \gamma(x) - s(x,0) \leq 0, \; x \in \Gamma_N. \tag{21}$$

then we have the DMP

$$\max_{\overline{\Omega}} u^h \leq \max\{0, \max_{\Gamma_D} g_h\}. \tag{22}$$

We note that for problems with only principal part, i.e. $q \equiv 0$ and $s \equiv 0$, it suffices to assume the weaker condition

$$\nabla\phi_i \cdot \nabla\phi_j \leq 0 \tag{23}$$

instead of (19), and we obtain the stronger DMP $\max_{\overline{\Omega}} u^h = \max_{\Gamma_D} g^h$.

Remark 2.4. It was also proved in [25] that, more generally, the above properties of the matrix $\mathbf{A}(\overline{\mathbf{c}})$ are also valid if the simplicial FE meshes $\mathcal{T}_h$ are only quasi-regular in the following sense: the left-hand side of (20) is replaced by

$$c_1 h^\gamma \leq meas(T_h), \tag{24}$$

where $\gamma \geq d$ satisfies

$$\begin{aligned} &2 \leq \gamma < 3 \quad \text{if } d = 2, \\ &3 \leq \gamma < \min\{\tfrac{12}{p_1 - 2},\, 5 - \tfrac{p_2}{2}\} \quad \text{if } d = 3, \\ &d \leq \gamma < \min\{\tfrac{4d}{(p_1-2)(d-2)},\, 3 + \tfrac{(4-p_2)(d-2)}{2}\} \quad \text{if } d > 3 \end{aligned} \tag{25}$$

with p_1, p_2 from assumption (A4) for problem (15).

2.3 Geometric properties to ensure the DMP

The values $\nabla\phi_i \cdot \nabla\phi_j$ are constant on each element, hence conditions (19) and (23) are not difficult to check, moreover, these conditions have a nice geometric interpretation. We briefly discuss some of these famous geometric properties, without the goal to give a detailed discussion. Some less strong assumptions will be discussed in subsection 3.4.

Conditions (19) and (23) have the following geometric meaning in view of well-known results. In order to satisfy condition (19) in the case of a simplicial mesh, it is sufficient if the employed mesh is uniformly acute, and similarly, condition (23) is satisfied if the employed mesh is nonobtuse [12, 35]. In the case of bilinear elements, condition (19) is equivalent to the so-called condition of non-narrow mesh, see [10]. The same issue for prismatic finite elements was recently treated in [19], where a convenient notion of (strictly) well-shaped prismatic partition is introduced.

We note that conditions (19) and (23) are sufficient but not necessary. For simplicial FEM, the DMP may still hold if some obtuse interior angles occur in the simplices of the meshes, i.e. if $\nabla\phi_i \cdot \nabla\phi_j$ is positive on each element. Namely, (19) was imposed to ensure the validity of the estimate

$$b_{ij}(\overline{\mathbf{c}}) = \int_{\Omega_{ij}} b(x, \nabla u^h)\, \nabla\phi_i \cdot \nabla\phi_j\, dx \leq -\frac{\sigma_1}{h^2} < 0 \tag{26}$$

with $\sigma_1 > 0$ independent of i, j and h, where $\Omega_{ij} = supp\ \phi_i \cap supp\ \phi_j$. However, using (16), we have in general

$$b_{ij}(\overline{\mathbf{c}}) \leq \mu_0 \sum_{K_l \in \mathcal{K}^-} \text{meas}\,(K_l)\, \nabla\phi_i \cdot \nabla\phi_j \,+\, \mu_1 \sum_{K_l \in \mathcal{K}^+} \text{meas}\,(K_l)\, \nabla\phi_i \cdot \nabla\phi_j\,, \tag{27}$$

with notations

$$\mathcal{K}^- = \{K \in \mathcal{T}_h : \nabla\phi_i \cdot \nabla\phi_j < 0 \text{ on } K\}, \quad \mathcal{K}^+ = \{K \in \mathcal{T}_h : \nabla\phi_i \cdot \nabla\phi_j \geq 0 \text{ on } K\}.$$

Then, it suffices to require that the expression in (27) is estimated above by $-\sigma_1/h^2$, which may allow the set $\mathcal{K}^+$ to be nonempty in certain situations. For linear problems, such weakened acute type conditions are given in e.g. in [33, 47].

One often wishes to solve the problem on finer meshes in order to obtain a more accurate approximation. These geometric conditions then need special attention. Namely, if we propose a global refinement of the initial mesh using some refinement technique (see e.g. [31]), then we must take care that the refined mesh preserves the desired acuteness (or nonobtuseness) property. Obviously, this is an easy task in the two-dimensional case, since, using the standard "2D red refinement" [31], we obtain a mesh consisting only of acute or nonobtuse triangular elements if the initial mesh had only acute or nonobtuse triangles, respectively. If we consider a tetrahedral mesh, the task is far from being trivial since in general it is not possible to refine any tetrahedron into eight subtetrahedra similar to it using "3D red refinement" (cf. [31]). A new technique, the so-called "3D yellow refinement" was developed in [30], which allows a global refinement of a nonobtuse tetrahedral mesh so that the resulting (conforming) mesh preserves the property of nonobtuseness. For local nonobtuse refinements (also in higher dimensions), see [2] and [32]. A construction of regular meshes, using a technique different from the red-refinement by midlines, is proposed in [34].

2.4 An algebraic DMP in Hilbert space

When dealing with elliptic systems, it is useful to state an algebraic (matrix) DMP in a Hilbert space setting in order to provide a clean line of thoughts. Namely, this setting will help an organized derivation of the corresponding results under the considered different conditions. The discussion below is based on [26], where it was applied to systems with second and zeroth order terms.

2.4.1 Formulation of the operator equation

Let H be a real Hilbert space and $H_0 \subset H$ a given subspace. We consider the following operator equation: for given vectors $\psi, g^* \in H$, find $u \in H$ such that

$$\langle A(u), v\rangle = \langle \psi, v\rangle \qquad (v \in H_0) \tag{28}$$

$$\text{and} \quad u - g^* \in H_0 \tag{29}$$

with an operator $A : H \to H$ satisfying the following conditions:

Assumptions 2.4.1.

(i) The operator $A : H \to H$ has the form

$$A(u) = B(u)u + N(u)u + R(u)u \tag{30}$$

where B, N and R are given operators mapping from H to $\mathcal{B}(H)$. (Here $\mathcal{B}(H)$ denotes the set of bounded linear operators in H.)

(ii) There exists a constant $m > 0$ such that

$$\Big\langle \Big(B(u) + N(u)\Big)v, v\Big\rangle \geq m\,\|v\|^2 \qquad (u \in H,\ v \in H_0). \tag{31}$$

(iii) There exist subsets of 'positive vectors' $D, P \subset H$ such that for any $u \in H$ and $v \in D$, we have

$$\langle R(u)w, v\rangle \geq 0 \tag{32}$$

provided that either $w \in P$ or $w = v \in D$.

(iv) There exists a continuous function $M_{NR} : \mathbf{R}^+ \to \mathbf{R}^+$ and another norm $|||.|||$ on H such that

$$\Big\langle \Big(N(u) + R(u)\Big)z, v\Big\rangle \leq M_{NR}(\|u\|)\, |||z|||\, |||v||| \qquad (u, z, v \in H). \tag{33}$$

In practice for PDE problems, g^* plays the role of boundary condition and H_0 will be the subspace corresponding to homogeneous boundary conditions, further, $B(u)$ is the principal part of A.

Assumptions 2.4.1 are not in general known to imply existence and uniqueness for (28)-(29). The following extra conditions already ensure well-posedness:

Assumptions 2.4.2.

(i) The operator A is Gateaux differentiable, further, A' is bihemicontinuous (i.e. mappings $(s,t) \mapsto A'(u + sk + tw)h$ are continuous from $\mathbf{R}^2$ to H).

(ii) There exists a continuous function $M_A : \mathbf{R}^+ \to \mathbf{R}^+$ such that

$$\langle A'(u)w, v\rangle \leq M_A(\|u\|)\, \|w\|\, \|v\| \qquad (u \in H,\ w, v \in H_0). \tag{34}$$

(iii) There exists a constant $m > 0$ such that

$$\langle A'(u)v, v\rangle \geq m\,\|v\|^2 \qquad (u \in H,\ v \in H_0). \tag{35}$$

Proposition 2.1. *If Assumptions* 2.4.1–2.4.2 *hold, then problem* (28)-(29) *is well-posed.*

The proof is based on uniform monotonicity and local Lipschitz continuity, see e.g. [17]. The proof of an equivalent formulation of Proposition 2.1 is given in [26].

2.4.2 Galerkin type discretization

Let $n_0 \leq n$ be positive integers and $\phi_1, \ldots, \phi_n \in H$ be given linearly independent vectors such that $\phi_1, \ldots, \phi_{n_0} \in H_0$. We consider the finite dimensional subspaces

$$V_h = \text{span}\{\phi_1, \ldots, \phi_n\} \subset H, \qquad V_h^0 = \text{span}\{\phi_1, \ldots, \phi_{n_0}\} \subset H_0 \tag{36}$$

with a real positive parameter $h > 0$. In practice, as is usual for FEM, h is inversely proportional to n, and one will consider a family of such subspaces in the sense of Definition 2.5.

We formulate here some connectivity type properties for these subspaces that we will need later. For this, certain pairs $\{\phi_i, \phi_j\} \in V_h \times V_h$ are called '*neighbouring basis vectors*', and then i, j are called '*neighbouring indices*'. The only requirement for the set of these pairs is that they satisfy Assumptions 2.4.3 below, given in terms of the *graph of neighbouring indices*, by which we mean the following. The corresponding indices $\{1, \ldots, n_0\}$ or $\{1, \ldots, n\}$, respectively, are represented as vertices of the graph, and the ith and jth vertices are connected by an edge iff i, j are neighbouring indices.

Assumptions 2.4.3. The set $\{1, \ldots, n\}$ can be partitioned into disjoint sets $S_1, \ldots, S_r$ such that for each $k = 1, \ldots, r$,

(i) both $S_k^0 := S_k \cap \{1, \ldots, n_0\}$ and $\tilde{S}_k := S_k \cap \{n_0 + 1, \ldots, n\}$ are nonempty;

(ii) the graph of all neighbouring indices in S_k^0 is connected;

(iii) the graph of all neighbouring indices in S_k is connected.

(In later PDE applications, these properties are meant to express that the supports of basis functions cover the domain, both its interior and the boundary.)

Now let $g^h = \sum\limits_{j=n_0+1}^{n} g_j \phi_j \in V_h$ be a given approximation of the component of g^* in $H \setminus H_0$. To find the Galerkin solution of (28)-(29) in V_h, we solve the following problem: find $u^h \in V_h$ such that

$$\langle A(u^h), v^h\rangle = \langle \psi, v^h\rangle \qquad (v^h \in V_h^0) \tag{37}$$

$$\text{and} \quad u^h - g^h \in V_h^0. \tag{38}$$

Using (30), we can rewrite (37) as

$$\langle B(u^h)u^h, v^h\rangle + \langle N(u^h)u^h, v^h\rangle + \langle R(u^h)u^h, v^h\rangle = \langle \psi, v^h\rangle \qquad (v^h \in V_h^0). \tag{39}$$

Let us now formulate the nonlinear algebraic system corresponding to (39). We set

$$u^h = \sum_{j=1}^{n} c_j \phi_j, \tag{40}$$

and look for the coefficients $c_1, \ldots, c_n$. For any $\bar{\mathbf{c}} = (c_1, \ldots, c_n)^T \in \mathbf{R}^n$, $i = 1, \ldots, n_0$ and $j = 1, \ldots, n$, we set

$$b_{ij}(\bar{\mathbf{c}}) := \langle B(u^h)\phi_j, \phi_i \rangle, \qquad n_{ij}(\bar{\mathbf{c}}) := \langle N(u^h)\phi_j, \phi_i \rangle, \qquad r_{ij}(\bar{\mathbf{c}}) := \langle R(u^h)\phi_j, \phi_i \rangle,$$

$$a_{ij}(\bar{\mathbf{c}}) := b_{ij}(\bar{\mathbf{c}}) + n_{ij}(\bar{\mathbf{c}}) + r_{ij}(\bar{\mathbf{c}}), \qquad d_i := \langle \psi, \phi_i \rangle. \tag{41}$$

Putting (40) and $v = \phi_i$ into (39), we obtain the $n_0 \times n$ system of algebraic equations

$$\sum_{j=1}^{n} a_{ij}(\bar{\mathbf{c}})\, c_j = d_i \qquad (i = 1, \ldots, n_0). \tag{42}$$

Using the notations

$$\mathbf{A}(\bar{\mathbf{c}}) := \{a_{ij}(\bar{\mathbf{c}})\},\ i, j = 1, \ldots, n_0, \qquad \tilde{\mathbf{A}}(\bar{\mathbf{c}}) := \{a_{ij}(\mathbf{c})\},\ i = 1, \ldots, n_0;\ j = n_0 + 1, \ldots, n,$$

$$\mathbf{d} := \{d_j\},\ \mathbf{c} := \{c_j\}, \quad j = 1, \ldots, n_0, \quad \text{and} \quad \tilde{\mathbf{c}} := \{c_j\}, \quad j = n_0 + 1, \ldots, n, \tag{43}$$

system (42) turns into

$$\mathbf{A}(\bar{\mathbf{c}})\mathbf{c} + \tilde{\mathbf{A}}(\bar{\mathbf{c}})\tilde{\mathbf{c}} = \mathbf{d}. \tag{44}$$

In order to obtain a system with a square matrix, we enlarge our system to an $n \times n$ one. Since $u^h - g^h \in V_h^0$, the coordinates c_i with $n_0 + 1 \le i \le n$ satisfy automatically $c_i = g_i$, i.e.,

$$\tilde{\mathbf{c}} = \tilde{\mathbf{g}} := \{g_j\}, \quad j = n_0 + 1, \ldots, n,$$

hence we can replace (44) by the equivalent system

$$\begin{bmatrix} \mathbf{A}(\bar{\mathbf{c}}) & \tilde{\mathbf{A}}(\bar{\mathbf{c}}) \\ 0 & \mathbf{I} \end{bmatrix} \begin{bmatrix} \mathbf{c} \\ \tilde{\mathbf{c}} \end{bmatrix} = \begin{bmatrix} \mathbf{d} \\ \tilde{\mathbf{g}} \end{bmatrix}. \tag{45}$$

Defining further

$$\bar{\mathbf{A}}(\bar{\mathbf{c}}) := \begin{bmatrix} \mathbf{A}(\bar{\mathbf{c}}) & \tilde{\mathbf{A}}(\bar{\mathbf{c}}) \\ 0 & \mathbf{I} \end{bmatrix}, \qquad \bar{\mathbf{c}} := \begin{bmatrix} \mathbf{c} \\ \tilde{\mathbf{c}} \end{bmatrix}, \tag{46}$$

we rewrite (44) as follows:

$$\bar{\mathbf{A}}(\bar{\mathbf{c}})\bar{\mathbf{c}} = \mathbf{d}. \tag{47}$$

2.4.3 Maximum principle for the abstract discretized problem

When formulating a discrete maximum principle for system (47), the notion of family of subspaces will be used in analogy of Definition 2.5. First we give sufficient conditions for the generalized nonnegativity of the matrix $\bar{\mathbf{A}}(\bar{\mathbf{c}})$.

Theorem 2.5. *Let Assumptions* 2.4.1 *and* 2.4.3 *hold. Let us consider the discretization of operator equation* (28)-(29) *in a family of subspaces* $\mathcal{V} = \{V_h\}_{h \to 0}$ *with bases as in* (36). *Let* $u^h \in V_h$ *be the solution of* (39) *and let the following properties hold:*

(a) *For all* $\phi_i \in V_h^0$ *and* $\phi_j \in V_h$, *one of the following holds: either*

$$\langle B(u^h)\phi_j, \phi_i \rangle = \langle N(u^h)\phi_j, \phi_i \rangle = 0 \quad \text{and} \quad \langle R(u^h)\phi_j, \phi_i \rangle \leq 0, \tag{48}$$

or

$$\langle B(u^h)\phi_j, \phi_i \rangle \leq -M_B(h) \tag{49}$$

with a proper function $M_B : \mathbf{R}^+ \to \mathbf{R}^+$ *(independent of* h, ϕ_i, ϕ_j*) such that, defining*

$$T(h) := \sup\{|||\phi_i||| : \phi_i \in V_h)\}, \tag{50}$$

we have

$$\lim_{h \to 0} \frac{M_B(h)}{T(h)^2} = +\infty. \tag{51}$$

(b) *If, in particular,* $\phi_i \in V_h^0$ *and* $\phi_j \in V_h$ *are neighbouring basis vectors (as defined for Assumptions* 2.4.3*), then* (49)-(51) *hold.*

(c) $M_{NR}(\|u^h\|)$ *is bounded as* $h \to 0$, *where* M_{NR} *is the function in Assumption* 2.4.1 (iv).

(d) *For all* $u \in H$ *and* $h > 0$, $\sum_{j=1}^{n} \phi_j \in ker B(u) \cap ker N(u)$.

(e) *For all* $h > 0$, $i = 1, \ldots, n$, *we have* $\phi_i \in D$ *and* $\sum_{j=1}^{n} \phi_j \in P$ *for the sets* D, P *introduced in Assumption* 2.4.1 (iii).

Then for sufficiently small h, the matrix $\bar{\mathbf{A}}(\bar{\mathbf{c}})$ *defined in* (46) *is of generalized nonnegative type with irreducible blocks in the sense of Definition* 2.4.

The proof is given in [26, Theorem 3.1] for the case $N(u) \equiv 0$. It is easy to see that the inclusion of $N(u)$ gives no difference in the proof, since $N(u)$ has been joined either to $B(u)$ or to $R(u)$ both in Assumptions 2.4.1 and in the appropriate conditions of Theorem 2.5. Hence in each step of the proof one has the same condition now with $N(u)$ as it was in [26, Theorem 3.1] without $N(u)$. (When applying Theorem 2.5 later, we will need $N(u)$ for the first order terms. Since it could not be joined either only to $B(u)$ or only to $R(u)$ above, we could not use [26, Theorem 3.1] formally in the original way.)

By Theorem 2.2, we immediately obtain the corresponding algebraic *discrete maximum principle:*

Corollary 2.2. *Let the assumptions of Theorem* 2.5 *hold. For sufficiently small h, if* $d_i \leq 0$ $(i = 1, \ldots, n_0)$ *in* (43) *and* $\bar{\mathbf{c}} = (c_1, \ldots, c_n)^T \in \mathbf{R}^n$ *is the solution of* (47), *then*

$$\max_{i=1,\ldots,n} c_i \leq \max\{0, \max_{i=n_0+1,\ldots,n} c_i\}. \tag{52}$$

Remark 2.5. Assumption (c) of Theorem 2.5 follows in particular if Assumptions 2.4.2 are added to Assumptions 2.4.1 as done in Proposition 2.1, provided that the functions $g^h \in V_h$ in (38) are bounded in H-norm as $h \to 0$. (In practice, the usual choices for g^h even produce $g^h \to g^*$ in H-norm.) In fact, in this case $\|u^h\|$ is bounded as $h \to 0$; then the continuity of M_{NR} yields that $M_{NR}(\|u^h\|)$ is bounded too. For more details see [26].

Remark 2.6. Assumptions 2.4.1. (iv) can be weakened such that one may allow different norms both for N and R and in the factors, i.e. (33) is replaced by

$$\langle N(u)z,v\rangle \leq M_N(\|u\|)\,|||z|||_{N_1}\,|||v|||_{N_2}\,, \tag{53}$$

$$\Big\langle R(u)w,v\Big\rangle \leq M_R(\|u\|)\,|||w|||_{R_1}\,|||v|||_{R_2} \tag{54}$$

(for all $u,w,v \in H$). Then Theorem 2.5 remains true if we appropriately replace (50) by

$$T(h) := \sup\Big\{\max\Big\{|||\phi_j|||_{N_1}\,|||\phi_i|||_{N_2}\,,\ |||\phi_j|||_{R_1}\,|||\phi_i|||_{R_2}\Big\} :\ \phi_i,\phi_j \in V_h\Big\}^{1/2}, \tag{55}$$

and require in assumption (c) that both $M_N(\|u^h\|)$ and $M_R(\|u^h\|)$ are bounded as $h \to 0$.

3 Discrete Maximum Principles for Elliptic Reaction-Diffusion Type Systems

We first study various types of nonlinear elliptic systems with second and zeroth order terms, quoting our results from [26]. The considered domain Ω and the diffusion coefficient functions b_k $(k = 1,\ldots,s)$ will satisfy common properties, formulated below:

Assumptions 3.0.

(i) $\Omega \subset \mathbf{R}^d$ is a bounded piecewise C^1 domain; Γ_D, Γ_N are disjoint open measurable subsets of $\partial\Omega$ such that $\partial\Omega = \overline{\Gamma}_D \cup \overline{\Gamma}_N$ and $\Gamma_D \neq \emptyset$.

(ii) (Ellipticity.) There exists $m > 0$ such that $b_k \geq m$ holds pointwise for all $k = 1,\ldots,s$.

3.1 Systems with nonlinear coefficients

3.1.1 Formulation of the problem

First we consider nonlinear elliptic systems of the form

$$\left.\begin{aligned} -\mathrm{div}\Big(b_k(x,u,\nabla u)\,\nabla u_k\Big) + \sum_{l=1}^{s} V_{kl}(x,u,\nabla u)\,u_l &= f_k(x) \quad \text{a.e. in } \Omega,\\ b_k(x,u,\nabla u)\frac{\partial u_k}{\partial \nu} &= \gamma_k(x) \quad \text{a.e. on } \Gamma_N,\\ u_k &= g_k(x) \quad \text{a.e. on } \Gamma_D \end{aligned}\right\} \quad (k = 1,\ldots,s) \tag{56}$$

with unknown function $u = (u_1,\ldots,u_s)^T$, under the following assumptions. Here ∇u denotes the $s \times d$ tensor with rows ∇u_k $(k = 1,\ldots,s)$, further, 'a.e.' means Lebesgue almost everywhere and inequalities for functions are understood a.e. pointwise for all possible arguments.

Assumptions 3.1.

(i) The domain Ω and the diffusion coefficients b_k satisfy Assumptions 3.0.

(ii) (Smoothness and boundedness.) For all $k,l = 1,\ldots,s$ we have $b_k \in (C^1 \cap L^\infty)(\Omega \times \mathbf{R}^s \times \mathbf{R}^{s\times d})$ and $V_{kl} \in L^\infty(\Omega \times \mathbf{R}^s \times \mathbf{R}^{s\times d})$.

(iii) (Cooperativity.) We have

$$V_{kl} \leq 0 \qquad (k,l = 1,\ldots,s,\ k \neq l). \tag{57}$$

(iv) (Weak diagonal dominance.) We have

$$\sum_{l=1}^{s} V_{kl} \geq 0 \qquad (k = 1,\ldots,s). \tag{58}$$

(v) For all $k = 1,\ldots,s$ we have $f_k \in L^2(\Omega)$, $\gamma_k \in L^2(\Gamma_N)$, $g_k = g^*_{k|\Gamma_D}$ with $g^*_k \in H^1(\Omega)$.

Remark 3.1. (i) Assumptions (57)-(58) imply

$$V_{kk} \geq 0 \qquad (k = 1,\ldots,s). \tag{59}$$

(ii) One may consider additional terms on the Neumann boundary, see Remark 3.4 later.

For the weak formulation of such problems, we define the Sobolev space

$$H^1_D(\Omega) := \{z \in H^1(\Omega) : z_{|\Gamma_D} = 0\}. \tag{60}$$

The weak formulation of problem (56) then reads as follows: find $u \in H^1(\Omega)^s$ such that

$$\langle A(u), v\rangle = \langle \psi, v\rangle \qquad (\forall v \in H^1_D(\Omega)^s) \tag{61}$$

$$\text{and} \quad u - g^* \in H^1_D(\Omega)^s, \tag{62}$$

where

$$\langle A(u), v\rangle = \int_\Omega \Big(\sum_{k=1}^{s} b_k(x,u,\nabla u)\, \nabla u_k \cdot \nabla v_k + \sum_{k,l=1}^{s} V_{kl}(x,u,\nabla u)\, u_l v_k \Big) \tag{63}$$

for given $u = (u_1,\ldots,u_s) \in H^1(\Omega)^s$ and $v = (v_1,\ldots,v_s) \in H^1_D(\Omega)^s$, further,

$$\langle \psi, v\rangle = \int_\Omega \sum_{k=1}^{s} f_k v_k + \int_{\Gamma_N} \sum_{k=1}^{s} \gamma_k v_k \tag{64}$$

for given $v = (v_1,\ldots,v_s) \in H^1_D(\Omega)^s$, and $g^* := (g^*_1,\ldots,g^*_s)$.

3.1.2 Finite element discretization

We define the finite element discretization of problem (56) in the following way. First, let $\bar{n}_0 \leq \bar{n}$ be positive integers and let us choose basis functions

$$\varphi_1,\ldots,\varphi_{\bar{n}_0} \in H^1_D(\Omega), \qquad \varphi_{\bar{n}_0+1},\ldots,\varphi_{\bar{n}} \in H^1(\Omega) \setminus H^1_D(\Omega), \tag{65}$$

which correspond to homogeneous and inhomogeneous boundary conditions on Γ_D, respectively. (For simplicity, we will refer to them as 'interior basis functions' and 'boundary basis functions', respectively, thus adopting the terminology of Dirichlet problems even in the general case.) These basis functions are assumed to be continuous and to satisfy

$$\varphi_p \geq 0 \quad (p = 1,\ldots,\bar{n}), \qquad \sum_{p=1}^{\bar{n}} \varphi_p \equiv 1, \tag{66}$$

further, that there exist node points $B_p \in \Omega$ $(p = 1,\ldots,\bar{n}_0)$ and $B_p \in \Gamma_D$ $(p = \bar{n}_0 + 1,\ldots,\bar{n})$ such that

$$\varphi_p(B_q) = \delta_{pq} \tag{67}$$

where δ_{pq} is the Kronecker symbol; and finally, there exists a constant $c > 0$ (independent of the basis functions) such that

$$\max |\nabla \varphi_t| \leq \frac{c}{diam(\operatorname{supp} \varphi_t)} \tag{68}$$

where supp denotes the support, i.e. the closure of the set where the function does not vanish. These conditions hold e.g. for standard linear, bilinear or prismatic finite elements. Finally, we assume that any two interior basis functions can be connected with a chain of interior basis functions with overlapping support. By its geometric meaning, this assumption obviously holds for any reasonable FE mesh.

We in fact need a basis in the corresponding product spaces, which we define by repeating the above functions in each of the s coordinates and setting zero in the other coordinates. That is, let $n_0 := s\bar{n}_0$ and $n := s\bar{n}$. First, for any $1 \leq i \leq n_0$,

$$\text{if } \; i = (k-1)\bar{n}_0 + p \quad \text{for some } 1 \leq k \leq s \text{ and } 1 \leq p \leq \bar{n}_0, \quad \text{then}$$

$$\phi_i := (0,\ldots,0,\varphi_p,0,\ldots,0) \qquad \text{where } \; \varphi_p \; \text{ stands at the } k\text{-th entry,} \tag{69}$$

that is, $(\phi_i)_m = \varphi_p$ if $m = k$ and $(\phi_i)_m = 0$ if $m \neq k$. From these, we let

$$V_h^0 := \operatorname{span}\{\phi_1,\ldots,\phi_{n_0}\} \subset H_D^1(\Omega)^s. \tag{70}$$

Similarly, for any $n_0 + 1 \leq i \leq n$, if

$$i = n_0 + (k-1)(\bar{n} - \bar{n}_0) + p - \bar{n}_0 \quad \text{for some } 1 \leq k \leq s \text{ and } \bar{n}_0 + 1 \leq p \leq \bar{n}, \text{ then}$$

$$\phi_i := (0,\ldots,0,\varphi_p,0,\ldots,0)^T \qquad \text{where } \; \varphi_p \; \text{ stands at the } k\text{-th entry,} \tag{71}$$

that is, $(\phi_i)_m = \varphi_p$ if $m = k$ and $(\phi_i)_m = 0$ if $m \neq k$. From (70) and these, we let

$$V_h := \operatorname{span}\{\phi_1,\ldots,\phi_n\} \subset H^1(\Omega)^s. \tag{72}$$

Using the above FEM subspaces, the finite element discretization of problem (56) leads to the task of finding $u^h \in V_h$ such that

$$\langle A(u^h), v^h \rangle = \langle \psi, v^h \rangle \qquad (\forall v^h \in V_h^0) \tag{73}$$

$$\text{and} \quad u^h - g^h \in V_h^0, \quad \text{i.e.,} \;\; u^h = g^h \text{ on } \Gamma_D \tag{74}$$

(where $g^h = \sum_{j=n_0+1}^{n} g_j\phi_j \in V_h$ is the approximation of g^* on Γ_D). Then, setting $u^h = \sum_{j=1}^{n} c_j\phi_j$ and $v = \phi_i$ $(i = 1,\ldots,n_0)$ in (61) (just as in (40)-(42)), we obtain the $n_0 \times n$ system of algebraic equations

$$\sum_{j=1}^{n} a_{ij}(\bar{\mathbf{c}})\, c_j = d_i \qquad (i = 1,\ldots,n_0), \tag{75}$$

where for any $\bar{\mathbf{c}} = (c_1,\ldots,c_n)^T \in \mathbf{R}^n$ and $i = 1,\ldots,n_0,\ \ j = 1,\ldots,n,$

$$a_{ij}(\bar{\mathbf{c}}) := \int_\Omega \Big(\sum_{k=1}^{s} b_k(x,u^h,\nabla u^h)\,(\nabla\phi_j)_k \cdot (\nabla\phi_i)_k + \sum_{k,l=1}^{s} V_{kl}(x,u^h,\nabla u^h)\,(\phi_j)_l\,(\phi_i)_k \Big) \tag{76}$$

$$\text{and} \qquad d_i := \int_\Omega \sum_{k=1}^{s} f_k(\phi_i)_k + \int_{\Gamma_N} \sum_{k=1}^{s} \gamma_k(\phi_i)_k\,. \tag{77}$$

In the same way as for (47), we enlarge system (75) to a square one by adding an identity block, and write it briefly as

$$\bar{\mathbf{A}}(\bar{\mathbf{c}})\bar{\mathbf{c}} = \mathbf{d}\,. \tag{78}$$

That is, for $i = 1,\ldots,n_0$ and $j = 1,\ldots,n$, the matrix $\bar{\mathbf{A}}(\bar{\mathbf{c}})$ has the entry $a_{ij}(\bar{\mathbf{c}})$ from (76).

In what follows, we will need notions of (patch-)regularity of the considered FE meshes, cf. [3].

Definition 3.1. Let $\Omega \subset \mathbf{R}^d$ and let us consider a family of FEM subspaces $\mathcal{V} = \{V_h\}_{h\to 0}$ constructed as above. Here $h > 0$ is the mesh parameter, proportional to the maximal diameter of the supports of the basis functions $\phi_1,\ldots,\phi_n$. The corresponding family of meshes will be called

(a) *regular from above* if there exists a constant $c_0 > 0$ such that for any $V_h \in \mathcal{V}$ and basis function $\varphi_p \in V_h$,

$$meas(\operatorname{supp}\varphi_p) \le c_0 h^d \tag{79}$$

(where *meas* denotes d-dimensional measure and supp denotes the support, i.e. the closure of the set where the function does not vanish);

(b) *regular* if there exist constants $c_1, c_2 > 0$ such that for any $V_h \in \mathcal{V}$ and basis function $\varphi_p \in V_h$,

$$c_1 h^d \le meas(\operatorname{supp}\varphi_p) \le c_2 h^d; \tag{80}$$

(c) *quasi-regular* if (80) is replaced by

$$c_1 h^\gamma \le meas(\operatorname{supp}\varphi_p) \le c_2 h^d \tag{81}$$

for some fixed constant

$$d \le \gamma < d+2. \tag{82}$$

3.1.3 Discrete maximum principle for systems with nonlinear coefficients

The theory of subsection 2.4 can be applied to derive a DMP for problem (56). The underlying operators have the following properties:

Lemma 3.1 ([26, Lemma 4.1]). *For any $u \in H^1(\Omega)^s$, let us define the operators $B(u)$ and $R(u)$ via*

$$\langle B(u)z, v\rangle = \int_\Omega \sum_{k=1}^{s} b_k(x,u,\nabla u)\,\nabla z_k \cdot \nabla v_k, \quad \langle R(u)z, v\rangle = \int_\Omega \sum_{k,l=1}^{s} V_{kl}(x,u,\nabla u)\,z_l v_k \tag{83}$$

($z \in H^1(\Omega)^s$, $v \in H^1_D(\Omega)^s$). Together with the operator A, defined in (63), the operators $B(u)$ and $R(u)$, together with $N(u) \equiv 0$, satisfy Assumptions 2.4.1 *in the spaces $H = H^1(\Omega)^s$ and $H_0 = H^1_D(\Omega)^s$, and with the new norm*

$$|||v|||^2 := \|v\|^2_{L^2(\Omega)^s} = \int_\Omega \sum_{k=1}^{s} v_k^2. \tag{84}$$

Now let us consider the finite element discretization for problem (56), developed in the previous subsection. One can then derive from Theorem 2.5 the following nonnegativity result for the stiffness matrix:

Theorem 3.1 ([26, Theorem 4.1]). *Let problem* (56) *satisfy Assumptions* 3.1. *Let us consider a family of finite element subspaces $\mathcal{V} = \{V_h\}_{h\to 0}$ satisfying the following property: there exists a real number γ satisfying*

$$d \le \gamma < d+2$$

(where d is the space dimension) such that for any $p = 1,\dots,\bar{n}_0$, $t = 1,\dots,\bar{n}$ $(p \neq t)$, if meas$(\operatorname{supp}\varphi_p \cap \operatorname{supp}\varphi_t) > 0$ then

$$\nabla\varphi_t \cdot \nabla\varphi_p \le 0 \;\; on\; \Omega \quad and \quad \int_\Omega \nabla\varphi_t \cdot \nabla\varphi_p \le -K_0\, h^{\gamma-2} \tag{85}$$

with some constant $K_0 > 0$ independent of p,t and h. Further, let the family of associated meshes be regular from above, according to Definition 3.1.

Then for sufficiently small h, the matrix $\bar{\mathbf{A}}(\bar{\mathbf{c}})$ defined in (76) *is of generalized nonnegative type with irreducible blocks in the sense of Definition* 2.4.

By (77) we have $d_i \le 0$ $(i = 1,\dots,n_0)$, hence Corollary 2.2 immediately yields

Corollary 3.1. *Let the assumptions of Theorem* 3.1 *hold and let $f_k \le 0$, $\gamma_k \le 0$ $(k = 1,\dots,s)$. For sufficiently small h, if $\bar{\mathbf{c}} = (c_1,\dots,c_n)^T \in \mathbf{R}^n$ is the solution of* (75) *with matrix $\bar{\mathbf{A}}(\bar{\mathbf{c}})$ defined in* (76), *then*

$$\max_{i=1,\dots,n} c_i \le \max\{0, \max_{i=n_0+1,\dots,n} c_i\}. \tag{86}$$

The meaning of (86) is as follows. Let us split the vector $\bar{\mathbf{c}} = (c_1,\dots,c_n)^T \in \mathbf{R}^n$ as in (46), i.e. $\bar{\mathbf{c}} = \begin{bmatrix}\mathbf{c}\\ \tilde{\mathbf{c}}\end{bmatrix}$, where $\mathbf{c} = (c_1,\dots,c_{n_0})^T$ and $\tilde{\mathbf{c}} = (c_{n_0+1},\dots,c_n)^T$. Following the notions introduced after (65), the vectors $\mathbf{c}$ and $\tilde{\mathbf{c}}$ contain the coefficients of the 'interior basis functions' and 'boundary basis functions', respectively. Then (86) states that the maximal coordinate is nonpositive or arises for a boundary basis function.

Our main interest is the meaning of Corollary 3.1 for the FEM solution $u^h = (u_1^h, \ldots, u_s^h)^T$ itself.

Theorem 3.2 ([26, Theorem 4.2]). *Let the basis functions satisfy* (66)-(67). *If* (86) *holds for the FEM solution* $u^h = (u_1^h, \ldots, u_s^h)^T$, *then* u^h *satisfies*

$$\max_{k=1,\ldots,s} \max_{\overline{\Omega}} u_k^h \leq \max_{k=1,\ldots,s} \max\{0, \max_{\Gamma_D} g_k^h\}. \tag{87}$$

Thus we obtain the *discrete maximum principle* for system (56):

Corollary 3.2. *Let the assumptions of Theorem* 3.1 *hold and let*

$$f_k \leq 0, \qquad \gamma_k \leq 0 \qquad (k = 1, \ldots, s).$$

Let the basis functions satisfy (66)-(67). *Then for sufficiently small* h, *if* $u^h = (u_1^h, \ldots, u_s^h)^T$ *is the FEM solution of system* (56), *then*

$$\max_{k=1,\ldots,s} \max_{\overline{\Omega}} u_k^h \leq \max_{k=1,\ldots,s} \max\{0, \max_{\Gamma_D} g_k^h\}. \tag{88}$$

Remark 3.2. (i) Let $f_k \leq 0$, $\gamma_k \leq 0$ for all k. The result (88) can be divided in two cases, both of which are remarkable: if at least one of the functions g_k^h has positive values on Γ_D then

$$\max_{k=1,\ldots,s} \max_{\overline{\Omega}} u_k^h = \max_{k=1,\ldots,s} \max_{\Gamma_D} g_k^h \tag{89}$$

(which can be called more directly a discrete maximum principle than (88)), and if $g_k \leq 0$ on Γ_D for all k, then we obtain the nonpositivity property

$$u_k^h \leq 0 \quad \text{on } \Omega \text{ for all } k. \tag{90}$$

(ii) Analogously, if $f_k \geq 0$, $\gamma_k \geq 0$ for all k, then (by reversing signs) we can derive the corresponding discrete minimum principles instead of (88) and (89), or the corresponding nonnegativity property instead of (90).

Remark 3.3. The key assumption for the meshes in the above results is property (85). A simple but stronger sufficient condition to satisfy (85) is (19), provided that the family of meshes is quasi-regular according to Definition 3.1. For simplicial FEM, assumption (19) corresponds to acute triangulations. Less strong assumptions to satisfy (85) will be discussed in subsection 3.4.

Remark 3.4. The results of this section may hold as well if there are additional terms $\sum_{l=1}^{s} \omega_{kl}(x, u, \nabla u)\, u_l$ on the Neumann boundary Γ_N, which we did not include for technical simplicity. Then ω_{kl} must satisfy similar properties as assumed for V_{kl} in (57)-(58).

3.2 Systems with general reaction terms of sublinear growth

In (56) both the principal and lower-order parts of the equations were given as containing products of coefficients with ∇u_k and u_l, respectively. Whereas this is widespread in real models for the principal part, the lower order terms are usually not given in such a coefficient form. Now we consider problems where the dependence on the lower order terms is given as general functions of x and u. In this section these functions are allowed to grow at most linearly, in which case one can reduce the problem to the previous one (56) directly. (Superlinear growth of q_k will be dealt with in the next section.)

Accordingly, let us now consider the system

$$\left.\begin{aligned} -\mathrm{div}\Big(b_k(x,u,\nabla u)\,\nabla u_k\Big) + q_k(x,u_1,\ldots,u_s) &= f_k(x) \quad \text{a.e. in } \Omega,\\ b_k(x,u,\nabla u)\tfrac{\partial u_k}{\partial \nu} &= \gamma_k(x) \quad \text{a.e. on } \Gamma_N,\\ u_k &= g_k(x) \quad \text{a.e. on } \Gamma_D \end{aligned}\right\} \quad (k=1,\ldots,s) \tag{91}$$

under the following assumptions:

Assumptions 3.2.

(i) The domain Ω and the diffusion coefficients b_k satisfy Assumptions 3.0.

(ii) (Smoothness and boundedness.) For all $k,l=1,\ldots,s$ we have $b_k \in (C^1 \cap L^\infty)(\Omega \times \mathbf{R}^s \times \mathbf{R}^{s\times d})$ and $q_k \in W^{1,\infty}(\Omega \times \mathbf{R}^s)$.

(iii) (Cooperativity.) We have

$$\frac{\partial q_k}{\partial \xi_l}(x,\xi) \le 0 \qquad (k,l=1,\ldots,s,\ k\ne l;\ x\in\Omega,\ \xi\in\mathbf{R}^s). \tag{92}$$

(iv) (Weak diagonal dominance for the Jacobians.) We have

$$\sum_{l=1}^{s} \frac{\partial q_k}{\partial \xi_l}(x,\xi) \ge 0 \qquad (k=1,\ldots,s;\ x\in\Omega,\ \xi\in\mathbf{R}^s). \tag{93}$$

(v) For all $k=1,\ldots,s$ we have $f_k \in L^2(\Omega)$, $\gamma_k \in L^2(\Gamma_N)$, $g_k = g^*_{k|\Gamma_D}$ with $g^* \in H^1(\Omega)$.

Remark 3.5. Similarly to (59), assumptions (92)-(93) now imply

$$\frac{\partial q_k}{\partial \xi_k}(x,\xi) \ge 0 \qquad (k=1,\ldots,s;\ x\in\Omega,\ \xi\in\mathbf{R}^s). \tag{94}$$

The basic idea to deal with problem (91) is to reduce it to (56) via suitably defined functions $V_{kl} : \Omega \times \mathbf{R}^s \to \mathbf{R}$. Namely, let

$$V_{kl}(x,\xi) := \int_0^1 \frac{\partial q_k}{\partial \xi_l}(x,t\xi)\,dt \qquad (k,l=1,\ldots,s;\ x\in\Omega,\ \xi\in\mathbf{R}^s). \tag{95}$$

Then the Newton-Leibniz formula yields

$$q_k(x,\xi) = q_k(x,0) + \sum_{l=1}^{s} V_{kl}(x,\xi)\,\xi_l \qquad (k=1,\ldots,s;\ x\in\Omega,\ \xi\in\mathbf{R}^s). \tag{96}$$

Defining

$$\hat{f}_k(x) := f_k(x) - q_k(x,0) \qquad (k=1,\ldots,s), \tag{97}$$

problem (91) then becomes

$$\left.\begin{aligned} -\mathrm{div}\left(b_k(x,u,\nabla u)\,\nabla u_k\right) + \sum_{l=1}^{s} V_{kl}(x,u)\,u_l &= \hat{f}_k(x) \quad \text{a.e. in } \Omega,\\ b_k(x,u,\nabla u)\frac{\partial u_k}{\partial \nu} &= \gamma_k(x) \quad \text{a.e. on } \Gamma_N,\\ u_k &= g_k(x) \quad \text{a.e. on } \Gamma_D \end{aligned}\right\} \quad (k=1,\ldots,s), \tag{98}$$

which is a special case of (56). Here the assumption $q_k \in W^{1,\infty}(\Omega\times\mathbf{R}^s)$ yields that $V_{kl} \in L^\infty(\Omega\times\mathbf{R}^s)$ $(k,l=1,\ldots,s)$. Clearly, assumptions (92) and (93) imply that the functions V_{kl} defined in (95) satisfy (57) and (58), respectively. The remaining items of Assumptions 3.1 and 3.2 coincide, therefore system (98) satisfies Assumptions 3.2.

Consequently, for a finite element discretization developed as in subsection 3.1.2, Theorem 3.2 yields the discrete maximum principle (87) for suitable discretizations of (98), provided $\hat{f}_k \le 0$ and $\gamma_k \le 0$ $(k=1,\ldots,s)$. For the original system (91), we thus obtain

Corollary 3.3. *Let problem* (91) *satisfy Assumptions* 3.2, *and let its FEM discretization satisfy the corresponding conditions of Theorem* 3.1. *If*

$$f_k \le q_k(x,0), \qquad \gamma_k \le 0 \qquad (k=1,\ldots,s)$$

and $u^h = (u_1^h,\ldots,u_s^h)^T$ *is the FEM solution of system* (91), *then for sufficiently small* h,

$$\max_{k=1,\ldots,s}\ \max_{\overline{\Omega}} u_k^h \le \max_{k=1,\ldots,s}\ \max\{0, \max_{\Gamma_D} g_k^h\}. \tag{99}$$

3.3 Systems with general reaction terms of superlinear growth

In the previous section we have required the functions q_k to grow at most linearly via the condition $q_k \in W^{1,\infty}(\Omega\times\mathbf{R}^s)$. However, this is a strong restriction and is not satisfied even by (nonlinear) polynomials of u_k that often arise in reaction-diffusion problems. In this section we extend the previous results to problems where the functions q_k may grow polynomially. This generalization, however, needs stronger assumptions in other parts of the problem, because we now need the monotonicity of the corresponding operator in the proof of the DMP. For this to hold, the row-diagonal dominance for the Jacobians in Assumption 3.2 (iv) must be strengthened to diagonal dominance w.r.t. both rows and columns. (In addition, the principal part must be more specific too, but this is not so much restrictive since in practice it is even linear.)

Accordingly, let us now consider the system

$$\left.\begin{aligned} -\mathrm{div}\left(b_k(x,\nabla u_k)\,\nabla u_k\right) + q_k(x,u_1,\dots,u_s) &= f_k(x) \quad \text{a.e. in } \Omega, \\ b_k(x,\nabla u_k)\frac{\partial u_k}{\partial \nu} &= \gamma_k(x) \quad \text{a.e. on } \Gamma_N, \\ u_k &= g_k(x) \quad \text{a.e. on } \Gamma_D \end{aligned}\right\} \quad (k=1,\dots,s) \tag{100}$$

under the following assumptions:

Assumptions 3.3.

(i) The domain Ω and the diffusion coefficients b_k satisfy Assumptions 3.0.

(ii) (Smoothness and growth.) For all $k,l=1,\dots,s$ we have $b_k \in (C^1 \cap L^\infty)(\Omega \times \mathbf{R}^d)$ and $q_k \in C^1(\Omega \times \mathbf{R}^s)$. Further, let

$$2 \le p < p^*, \quad \text{where } p^* := \tfrac{2d}{d-2} \text{ if } d \ge 3 \text{ and } p^* := +\infty \text{ if } d=2; \tag{101}$$

then there exist constants $\beta_1, \beta_2 \ge 0$ such that

$$\left|\frac{\partial q_k}{\partial \xi_l}(x,\xi)\right| \le \beta_1 + \beta_2 |\xi|^{p-2} \qquad (k,l=1,\dots,s;\ x\in\Omega,\ \xi \in \mathbf{R}^s). \tag{102}$$

(iii) (Ellipticity.) Defining $a_k(x,\eta) := b_k(x,\eta)\eta$ for all k, the Jacobian matrices $\frac{\partial}{\partial \eta} a_k(x,\eta)$ are uniformly spectrally bounded from both below and above.

(iv) (Cooperativity.) We have

$$\frac{\partial q_k}{\partial \xi_l}(x,\xi) \le 0 \qquad (k,l=1,\dots,s,\ k \ne l;\ x\in\Omega,\ \xi \in \mathbf{R}^s). \tag{103}$$

(v) (Weak diagonal dominance for the Jacobians w.r.t. rows and columns.) We have

$$\sum_{l=1}^{s} \frac{\partial q_k}{\partial \xi_l}(x,\xi) \ge 0, \qquad \sum_{l=1}^{s} \frac{\partial q_l}{\partial \xi_k}(x,\xi) \ge 0 \qquad (k=1,\dots,s;\ x\in\Omega,\ \xi \in \mathbf{R}^s). \tag{104}$$

(vi) For all $k=1,\dots,s$ we have $f_k \in L^2(\Omega)$, $\gamma_k \in L^2(\Gamma_N)$, $g_k = g^*_{k|\Gamma_D}$ with $g^* \in H^1(\Omega)$.

Remark 3.6. (i) Similarly to (59), assumptions (103)-(104) now imply

$$\frac{\partial q_k}{\partial \xi_k}(x,\xi) \ge 0 \qquad (k=1,\dots,s;\ x\in\Omega,\ \xi \in \mathbf{R}^s). \tag{105}$$

(ii) Similarly to Remark 3.4, one may include additional terms $s_k(x,u_1,\dots,u_s)$ on the Neumann boundary Γ_N, which we omit here for technical simplicity; then s_k must satisfy similar properties as assumed for q_k.

To handle system (100), we start as in the previous subsection by reducing it to a system with nonlinear coefficients: if the functions V_{kl} and $\hat{f}_k$ $(k,l=1,\dots,s)$ are defined as in (95) and (97), respectively, then (100) takes a form similar to (98):

$$\left.\begin{aligned} -\operatorname{div}\Big(b_k(x,\nabla u)\,\nabla u_k\Big) + \sum_{l=1}^{s} V_{kl}(x,u)\,u_l &= \hat{f}_k(x) \quad \text{a.e. in } \Omega, \\ b_k(x,u,\nabla u)\frac{\partial u_k}{\partial \nu} &= \gamma_k(x) \quad \text{a.e. on } \Gamma_N, \\ u_k &= g_k(x) \quad \text{a.e. on } \Gamma_D \end{aligned}\right\} \quad (k=1,\dots,s). \tag{106}$$

The difference compared to the previous subsection is the superlinear growth allowed in (102), which does not let us apply Theorem 3.2 directly as we did for system (91). Instead, we must reprove Theorem 3.1 under Assumptions 3.3.

First, when considering a finite element discretization developed as in subsection 3.1.2, we need a strengthened assumption for the quasi-regularity of the mesh.

Definition 3.2. Let $\Omega \subset \mathbf{R}^d$ and let us consider a family of FEM subspaces $\mathcal{V} = \{V_h\}_{h\to 0}$ constructed as in subsection 3.1.2. Here $h > 0$ is the mesh parameter, proportional to the maximal diameter of the supports of the basis functions $\phi_1,\dots,\phi_n$. The corresponding mesh will be called *quasi-regular* w.r.t. problem (100) if

$$c_1 h^{\gamma} \le meas(\operatorname{supp}\varphi_p) \le c_2 h^d, \tag{107}$$

where the positive real number γ satisfies

$$d \le \gamma < \gamma_d^*(p) := 2d - \frac{(d-2)p}{2} \tag{108}$$

with p from Assumption 3.3 (ii).

Remark 3.7. Assumption (108) makes sense for γ since by (101),

$$d < d + d(1 - \tfrac{p}{p^*}) = \gamma_d^*(p). \tag{109}$$

Note on the other hand that $\gamma_d^*(p) \le \gamma_d^*(2) = d+2$, which is in accordance with (82). Further, we have, in particular, in 2D: $\gamma_2^*(p) \equiv 4$ for all $2 \le p < \infty$, and in 3D: $\gamma_3^*(p) = 6 - (p/2)$ (where $2 \le p \le 6$, and accordingly $3 \le \gamma_3^*(p) \le 5$).

Next, as an analogue of Lemma 3.1, the following technical result holds for problem (100):

Lemma 3.2 ([26]). *Let Assumptions* 3.3 *hold. Analogously to* (83), *for any* $u \in H^1(\Omega)^s$ *let us define the operators* $B(u)$ *and* $R(u)$ *via*

$$\langle B(u)w, v\rangle = \int_{\Omega} \sum_{k=1}^{s} b_k(x,\nabla u)\,\nabla w_k \cdot \nabla v_k, \quad \langle R(u)w, v\rangle = \int_{\Omega} \sum_{k,l=1}^{s} V_{kl}(x,u)\, w_l\, v_k$$

($w \in H^1(\Omega)^s$, $v \in H^1_D(\Omega)^s$). *Together with* $A(u) := B(u)u + R(u)u$, *the operators* $B(u)$ *and* $R(u)$ *satisfy Assumptions* 2.4.1-2.4.2.

Then one can derive the desired nonnegativity result for the stiffness matrix, i.e. the analogue of Theorem 3.1 for system (100). Here the entries of $\bar{\mathbf{A}}(\bar{\mathbf{c}})$ are

$$a_{ij}(\bar{\mathbf{c}}) = \int_\Omega \Big(\sum_{k=1}^{s} b_k(x, \nabla u^h) \, (\nabla \phi_j)_k \cdot (\nabla \phi_i)_k + \sum_{k,l=1}^{s} V_{kl}(x, u^h) \, (\phi_j)_l \, (\phi_i)_k \Big), \tag{110}$$

where by (95),

$$V_{kl}(x, u^h(x)) = \int_0^1 \frac{\partial q_k}{\partial \xi_l}(x, t u^h(x)) \, dt \qquad (k, l = 1, \ldots, s; \; x \in \Omega). \tag{111}$$

Theorem 3.3 ([26]). *Let problem* (100) *satisfy Assumptions* 3.3. *Let us consider a family of finite element subspaces* V_h $(h \to 0)$ *satisfying the following property: there exists a real number* γ *satisfying* (108) *such that for any indices* $p = 1, \ldots, \bar{n}_0$, $t = 1, \ldots, \bar{n}$ $(p \neq t)$, *if* $meas(\operatorname{supp} \varphi_p \cap \operatorname{supp} \varphi_t) > 0$ *then*

$$\nabla \varphi_t \cdot \nabla \varphi_p \leq 0 \;\; \text{on } \Omega \quad \text{and} \quad \int_\Omega \nabla \varphi_t \cdot \nabla \varphi_p \leq -K_0 \, h^{\gamma - 2} \tag{112}$$

with some constant $K_0 > 0$ *independent of* p, t *and* h. *Further, let the family of meshes be regular from above, according to Definition* 3.1.

Then for sufficiently small h, *the matrix* $\bar{\mathbf{A}}(\bar{\mathbf{c}})$ *defined in* (110) *is of generalized nonnegative type with irreducible blocks in the sense of Definition* 2.4.

Similarly as in Corollary 3.3, using Theorem 3.3, Corollary 2.2 and Theorem 3.2, respectively, we obtain the *discrete maximum principle* for system (100):

Corollary 3.4. *Let problem* (100) *satisfy Assumptions* 3.3, *and let its FEM discretization satisfy the conditions of Theorem* 3.3. *If*

$$f_k \leq q_k(x, 0), \qquad \gamma_k \leq 0 \qquad (k = 1, \ldots, s)$$

then for sufficiently small h, *the FEM solution* $u^h = (u_1^h, \ldots, u_s^h)$ *of system* (100) *satisfies*

$$\max_{k=1,\ldots,s} \max_{\bar{\Omega}} u_k^h \leq \max_{k=1,\ldots,s} \max\{0, \max_{\Gamma_D} g_k^h\}. \tag{113}$$

Remark 3.8. As pointed out in Remark 3.2, the result (113) can be divided in two cases: a 'more direct' DMP (89) or the nonpositivity property (90). Further, if $f_k \geq q_k(x, 0)$, $\gamma_k \geq 0$ for all k, then (by reversing signs) one can derive the corresponding discrete minimum principle or nonnegativity property. We formulate the latter below for its practical importance.

Corollary 3.5. *Let problem* (100) *satisfy Assumptions* 3.3, *and let its FEM discretization satisfy the conditions of Theorem* 3.3. *If*

$$f_k \geq q_k(x, 0), \quad \gamma_k \geq 0, \quad g_k \geq 0 \qquad (k = 1, \ldots, s)$$

then for sufficiently small h, *the FEM solution* $u^h = (u_1^h, \ldots, u_s^h)^T$ *of system* (100) *satisfies*

$$u_k^h \geq 0 \quad \text{on } \Omega \qquad (k = 1, \ldots, s). \tag{114}$$

3.4 Sufficient conditions and their geometric meaning

The key assumption for the FEM subspaces V_h and the associated meshes in the above results has been the following property, see (85) in Theorem 3.1 and (112) in Theorem 3.3. There exists a real number γ satisfying (82) or (108), respectively, such that for any indices $p = 1,\ldots,\bar{n}_0$, $t = 1,\ldots,\bar{n}$ $(p \neq t)$, if $meas(\text{supp}\,\varphi_p \cap \text{supp}\,\varphi_t) > 0$ then

$$\nabla\varphi_t \cdot \nabla\varphi_p \leq 0 \text{ on } \Omega \quad \text{and} \tag{115}$$

$$\int_\Omega \nabla\varphi_t \cdot \nabla\varphi_p \leq -K_0\, h^{\gamma-2} \tag{116}$$

with some constant $K_0 > 0$ independent of p,t and h. (The family of meshes must also be regular from above as in (79), but that requirement obviously holds for the usual definition of the mesh parameter h as the maximal diameter of elements.)

A classical way to satisfy such conditions is a pointwise inequality like (19) together with suitable mesh regularity, see Remark 3.3. However, one can ensure (115)-(116) with less strong conditions as well. We summarize some possibilities below.

Proposition 3.1. *Let the family of FEM discretizations $\mathcal{V} = \{V_h\}_{h\to 0}$ satisfy either of the following conditions, where φ_t, φ_p are arbitrary basis functions such that $p = 1,\ldots,\bar{n}_0$, $t = 1,\ldots,\bar{n}$, $p \neq t$, we let*

$$\Omega_{pt} := \text{supp}\,\varphi_p \cap \text{supp}\,\varphi_t\,,$$

further, let

$$\sigma > 0 \quad \textit{and} \quad c_1, c_2, c_3 > 0$$

denote constants independent of the indices p,t and the mesh parameter h, and finally, d is the space dimension and γ satisfies (108).

(i) *Let the basis functions satisfy*

$$\nabla\varphi_t \cdot \nabla\varphi_p \leq -\frac{\sigma}{h^2} < 0 \quad \textit{on } \Omega_{pt}, \tag{117}$$

and the family of meshes be quasi-regular as in (107).

(ii) *Let there exist $0 < \varepsilon \leq \gamma - d$ such that the basis functions satisfy*

$$\nabla\varphi_t \cdot \nabla\varphi_p \leq -\frac{\sigma}{h^{2-\varepsilon}} < 0 \quad \textit{on } \Omega_{pt}, \tag{118}$$

but let the quasi-regularity (107) of the family of meshes be now strengthened to

$$c_1 h^{\gamma-\varepsilon} \leq meas(\text{supp}\,\varphi_p) \leq c_2 h^d\,. \tag{119}$$

(iii) *Let there exist subsets $\Omega_{pt}^+ \subset \Omega_{pt}$ for all p,t such that the basis functions satisfy*

$$\nabla\varphi_t \cdot \nabla\varphi_p \leq -\frac{\sigma}{h^2} < 0 \quad \textit{on } \Omega_{pt}^+ \quad \textit{and} \quad \nabla\varphi_t \cdot \nabla\varphi_p \leq 0 \textit{ on } \Omega_{pt} \setminus \Omega_{pt}^+ \tag{120}$$

and we have

$$\frac{meas(\Omega_{pt}^+)}{meas(\Omega_{pt})} \geq c_3 > 0\,, \tag{121}$$

further, let the family of meshes be quasi-regular as in (107).

Then (115)-(116) *holds.*

Proof is obvious. □

As discussed in subsection 2.3, conditions (115) and (117) have nice geometric interpretations for simplicial, bilinear and for prismatic finite elements, but these conditions are often restrictive. The weaker conditions (118) and (120) allow in theory easier refinement procedures as the property of (strict) acuteness is often hard to preserve in refinement procedures, e.g. by bisection algorithms [5, 34]. First, (118) may allow the acute mesh angles to deteriorate (i.e. tend to 90°) as $h \to 0$. Namely, if a family of simplicial meshes is regular then $|\nabla\varphi_t| = O(h^{-1})$ for all linear basis functions: hence, considering two basis functions φ_p, φ_t and letting α denote the angle of their gradients on a given simplex, the sufficient condition

$$\cos\alpha \leq -\sigma h^{\varepsilon} \tag{122}$$

(with some constant $\sigma > 0$ independent of h) implies

$$\nabla\varphi_t \cdot \nabla\varphi_p = |\nabla\varphi_t|\,|\nabla\varphi_p|\,\cos\alpha \leq -\frac{\sigma h^{\varepsilon}}{h^2},$$

i.e. (118) holds. Clearly, if $h \to 0$ then (122) allows $\cos\alpha \to 0$, i.e. $\alpha \to 90°$, for the angle of gradients, in which case the corresponding mesh angle also tends to 90 °. (In particular, for problem (56), when (108) coincides with $d \leq \gamma < d+2$ as in (82), then $\gamma - d$ can be chosen arbitrarily close to 2. Hence the exponent $2-\varepsilon$ in (118) can be arbitrarily close to 0, i.e. the decay of mesh angles to 90° may be fast as $h \to 0$.)

Second, (120) means that one can allow some right mesh angles, but each Ω_{pt}, which consists of a finite number of elements, must contain some elements with acute mesh angles and the measure of these must not asymptotically vanish.

4 Discrete Maximum Principles for Elliptic Systems Including First Order Terms

In this section we give various new results for elliptic systems including first order terms. We consider four types of systems, in which the diffusion and reaction terms are nonlinear: first the reaction terms are given as the unknown functions multiplied by nonlinear coefficients, later the reactions are general functions of x and the u_k. On the other hand, the first three types of problems involve linear convection terms and general Dirichlet data, whereas the last type contains nonlinear convection terms and homogeneous Dirichlet data. The allowed growth of the reaction terms is sublinear in two cases and superlinear in the other two cases. These differences require suitable modifications in the assumptions and the treatment.

4.1 Nonsymmetric systems with nonlinear reaction coefficients

First we consider nonlinear elliptic systems of the form

$$\left.\begin{aligned} -\mathrm{div}\left(b_k(x,u,\nabla u)\,\nabla u_k\right) + \mathbf{w}_k(x)\cdot\nabla u_k & \\ + \sum_{l=1}^{s} V_{kl}(x,u,\nabla u)\,u_l &= f_k(x) \quad \text{a.e. in } \Omega, \\ b_k(x,u,\nabla u)\frac{\partial u_k}{\partial \nu} &= \gamma_k(x) \quad \text{a.e. on } \Gamma_N, \\ u_k &= g_k(x) \quad \text{a.e. on } \Gamma_D \end{aligned}\right\} \tag{123}$$

$(k = 1,\ldots,s)$. The notations follow those of section 3.

Assumptions 4.1.

(i) The domain Ω and the diffusion coefficients b_k satisfy Assumptions 3.0.

(ii) (Smoothness and boundedness.) For all $k,l = 1,\ldots,s$ we have $b_k \in (C^1 \cap L^\infty)(\Omega \times \mathbf{R}^s \times \mathbf{R}^{s\times d})$, $\mathbf{w}_k \in W^{1,\infty}(\Omega)$ and $V_{kl} \in L^\infty(\Omega \times \mathbf{R}^s \times \mathbf{R}^{s\times d})$.

(iii) (Coercivity.) We have $\mathrm{div}\,\mathbf{w}_k \le 0$ on Ω and $\mathbf{w}_k \cdot \nu \ge 0$ on Γ_N $(k = 1,\ldots,s)$.

(iv) (Cooperativity.) We have

$$V_{kl} \le 0 \qquad (k,l = 1,\ldots,s,\ k \ne l). \tag{124}$$

(v) (Weak diagonal dominance.) We have

$$\sum_{l=1}^{s} V_{kl} \ge 0 \qquad (k = 1,\ldots,s). \tag{125}$$

(vi) For all $k = 1,\ldots,s$ we have $f_k \in L^2(\Omega)$, $\gamma_k \in L^2(\Gamma_N)$, $g_k = g^*_{k|\Gamma_D}$ with $g^*_k \in H^1(\Omega)$.

Remark 4.1. (i) Assumptions (124)-(125) imply (59).
(ii) One may consider additional terms on the Neumann boundary as in Remark 3.4.

For the weak formulation of problem (123), we suitable modify (61)–(62): find $u \in H^1(\Omega)^s$ such that

$$\langle A(u), v\rangle = \langle \psi, v\rangle \qquad (\forall v \in H^1_D(\Omega)^s) \tag{126}$$

$$\text{and} \quad u - g^* \in H^1_D(\Omega)^s, \tag{127}$$

where

$$\langle A(u), v\rangle = \int_\Omega \Big(\sum_{k=1}^{s} b_k(x,u,\nabla u)\,\nabla u_k\cdot\nabla v_k + \sum_{k=1}^{s} (\mathbf{w}_k(x)\cdot\nabla u_k)\,v_k + \sum_{k,l=1}^{s} V_{kl}(x,u,\nabla u)\,u_l v_k \Big) \tag{128}$$

for given $u = (u_1, \dots, u_s) \in H^1(\Omega)^s$ and $v = (v_1, \dots, v_s) \in H^1_D(\Omega)^s$, further,

$$\langle \psi, v \rangle = \int_\Omega \sum_{k=1}^s f_k v_k + \int_{\Gamma_N} \sum_{k=1}^s \gamma_k v_k \tag{129}$$

for given $v = (v_1, \dots, v_s) \in H^1_D(\Omega)^s$, and finally $g^* := (g_1^*, \dots, g_s^*)$.

The finite element discretization of problem (123) is defined in the same way as in subsection 3.1.2. Using the FEM subspaces (70) and (72), one seeks $u^h \in V_h$ such that

$$\langle A(u^h), v^h \rangle = \langle \psi, v^h \rangle \qquad (\forall v^h \in V_h^0) \tag{130}$$

$$\text{and} \quad u^h - g^h \in V_h^0, \quad \text{i.e.,} \ u^h = g^h \text{ on } \Gamma_D \tag{131}$$

(where $g^h = \sum_{j=n_0+1}^n g_j \phi_j \in V_h$ is the approximation of g^* on Γ_D). The only difference is that the entries of the stiffness matrix become, for any $\bar{\mathbf{c}} = (c_1, \dots, c_n)^T \in \mathbf{R}^n$ and $i = 1, \dots, n_0$, $j = 1, \dots, n$,

$$\begin{aligned} a_{ij}(\bar{\mathbf{c}}) := \int_\Omega \Bigg(& \sum_{k=1}^s b_k(x, u^h, \nabla u^h) \, (\nabla \phi_j)_k \cdot (\nabla \phi_i)_k + \sum_{k=1}^s \Big(\mathbf{w}_k(x) \cdot (\nabla \phi_j)_k \Big) (\phi_i)_k \\ & + \sum_{k,l=1}^s V_{kl}(x, u^h, \nabla u^h) \, (\phi_j)_l \, (\phi_i)_k \Bigg) \end{aligned} \tag{132}$$

instead of (76). With these, similarly to (75), we have the $n_0 \times n$ system of algebraic equations

$$\sum_{j=1}^n a_{ij}(\bar{\mathbf{c}}) \, c_j = d_i \qquad (i = 1, \dots, n_0). \tag{133}$$

Now our goal is to derive a DMP for problem (123) using Theorem 2.5. For this, we first define the underlying operators as in (83), for which Assumptions 2.4.1 must hold.

Lemma 4.1. *Let Assumptions* 4.1 *hold. For any* $u \in H^1(\Omega)^s$*, let us define the operators* $B(u)$*,* $N(u) \equiv N$ *and* $R(u)$ *via*

$$\begin{aligned} \langle B(u)z, v \rangle = \int_\Omega \sum_{k=1}^s b_k(x, u, \nabla u) \, \nabla z_k \cdot \nabla v_k, \qquad & \langle Nz, v \rangle = \int_\Omega \sum_{k=1}^s (\mathbf{w}_k(x) \cdot \nabla z_k) \, v_k, \\ \langle R(u)z, v \rangle = \int_\Omega \sum_{k,l=1}^s V_{kl}(x, u, \nabla u) \, z_l \, v_k & \end{aligned} \tag{134}$$

($z \in H^1(\Omega)^s$, $v \in H^1_D(\Omega)^s$)*. Together with the operator* A*, defined in* (128)*, the operators* $B(u)$*,* $N(u)$ *and* $R(u)$ *satisfy Assumptions* 2.4.1*, modified according to Remark* 2.6*, in the spaces* $H = H^1(\Omega)^s$ *and* $H_0 = H^1_D(\Omega)^s$*.*

Proof. By Lemma 3.1, we only need to prove those statements that do not concern only $B(u)$ or $R(u)$. We define

$$\|v\|^2 := \sum_{k=1}^s \Big(\int_\Omega |\nabla v_k|^2 + \int_{\Gamma_D} |v_k|^2 \Big) \tag{135}$$

on $H^1(\Omega)^s$, which is a norm since $\Gamma_D \neq \emptyset$. Then for $v \in H_D^1(\Omega)^s$ we have $\|v\|^2 = \sum_{k=1}^{s} \int_\Omega |\nabla v_k|^2$.

(i) It is obvious from (128) and (134) that $A(u) = B(u)u + N(u)u + R(u)u$.

(ii) By Lemma 3.1, we have $\langle B(u)v, v\rangle \geq m\|v\|^2$ $(u \in H^1(\Omega)^s,\ v \in H_D^1(\Omega)^s)$. Further, the assumptions and the divergence theorem imply for all $v_k \in H_D^1(\Omega)$ that

$$2\int_\Omega (\mathbf{w}_k \cdot \nabla v_k)v_k = -\int_\Omega (\operatorname{div}\mathbf{w}_k)\, v_k^2\, dx + \int_{\partial\Omega} (\mathbf{w}_k \cdot \nu)\, v_k^2\, d\sigma \geq 0. \tag{136}$$

Summing up for k and dividing by 2, we obtain that $\langle Nv, v\rangle \geq 0$. Hence (31) is valid.

(iii) This follows from Lemma 3.1, where P and D were defined as follows. Let $D \subset H^1(\Omega)^s$ consist of the functions that have only one nonzero coordinate that is nonnegative, i.e. $v \in D$ iff $v = (0,\ldots,0,g,0,\ldots,0)^T$ with g at the k-th entry for some $1 \leq k \leq s$ and $g \in H^1(\Omega)$, $g \geq 0$. Further, let $P \subset H^1(\Omega)^s$ consist of the functions that have identical nonnegative coordinates, i.e. $v \in P$ iff $v = (y,\ldots,y)$ for some $y \in H^1(\Omega), y \geq 0$.

(iv) By Lemma 3.1, we have for all $u, w, v \in H^1(\Omega)^s$

$$\langle R(u)z, v\rangle \leq M_R(\|u\|)\, |||z|||\, |||v|||,$$

for the new norm $|||v|||^2 = \|v\|^2_{L^2(\Omega)^s}$, i.e. (54) holds. In fact [26, Lemma 4.1], one has the constant function $M_R(r) \equiv s\tilde{V}$, where $\tilde{V} := \max_{k,l} \|V_{kl}\|_{L^\infty}$. For N, we have

$$\langle Nz, v\rangle \leq \|\mathbf{w}\|_{L^\infty(\Omega)^s} \|\nabla z\|_{L^2(\Omega)^s} \|v\|_{L^2(\Omega)^s} \leq \|\mathbf{w}\|_{L^\infty(\Omega)^s} \|z\|\, \|v\|_{L^2(\Omega)^s},$$

where $\|\mathbf{w}\|_{L^\infty(\Omega)^s} := \sup_{k,x} |\mathbf{w}_k(x)|$, i.e. (53) holds for the constant function $M_N(r) \equiv \|\mathbf{w}\|_{L^\infty(\Omega)^s}$ and the norms $|||z|||_{N_1} := \|z\|$, $|||v|||_{N_2} := \|v\|_{L^2(\Omega)^s}$. □

Now let us consider the finite element discretization for problem (56), developed as in subsection 3.1.2. First we need a strengthened assumption for the regularity of the mesh.

Definition 4.1. Let $\Omega \subset \mathbf{R}^d$ and let us consider a family of FEM subspaces $\mathcal{V} = \{V_h\}_{h\to 0}$ constructed as in subsection 3.1.2. Here $h > 0$ is the mesh parameter, proportional to the maximal diameter of the supports of the basis functions $\phi_1,\ldots,\phi_n$. The corresponding mesh will be called *quasi-regular* w.r.t. problem (123) if

$$c_1 h^\gamma \leq meas(\operatorname{supp}\varphi_p) \leq c_2 h^d, \tag{137}$$

where the positive real number γ satisfies

$$d \leq \gamma < \frac{d(d+2)}{d+1}. \tag{138}$$

One can then prove the desired nonnegativity result for the stiffness matrix:

Theorem 4.1. *Let problem* (123) *satisfy Assumptions* 4.1. *Let us consider a family of finite element subspaces* $\mathcal{V} = \{V_h\}_{h\to 0}$, *such that the corresponding family of meshes is quasi-regular according to Definition* 4.1, *further, for any* $p = 1,\ldots,\bar{n}_0$, $t = 1,\ldots,\bar{n}$ $(p \neq t)$, *if* $meas(\mathrm{supp}\,\varphi_p \cap \mathrm{supp}\,\varphi_t) > 0$ *then*

$$\nabla\varphi_t \cdot \nabla\varphi_p \leq 0 \;\; on\ \Omega \quad and \quad \int_\Omega \nabla\varphi_t \cdot \nabla\varphi_p \leq -K_0\, h^{\gamma-2} \tag{139}$$

where γ *is from* (138) *and* $K_0 > 0$ *is a constant independent of* p, t *and* h.

Then for sufficiently small h, *the matrix* $\bar{\mathbf{A}}(\bar{\mathbf{c}})$ *defined in* (132) *is of generalized nonnegative type with irreducible blocks in the sense of Definition* 2.4.

Proof. We wish to apply Theorem 2.5, modified according to Remark 2.6, in the spaces $H = H^1(\Omega)^s$ and $H_0 = H^1_D(\Omega)^s$. We use that the operators $B(u)$ and $R(u)$ satisfy the corresponding assumptions, since this was used in Theorem 3.1.

With the operator A defined in (128), our problem (126)-(127) coincides with (28)-(29). The FEM subspaces (70) and (72) fall into the class (36). Using the operators $B(u)$, $N(u) \equiv N$ and $R(u)$ in (134), the discrete problem (130)-(131) turns into the form (39) such that by Lemma 4.1, these operators satisfy Assumptions 2.4.1, modified according to Remark 2.6.

Now we follow the proof of Theorem 3.1, see [26, Theorem 4.1]. First we define neighbouring basis functions satisfying Assumptions 2.4.3. Let $\phi_i, \phi_j \in V_h$. Using definitions (69) and (71), assume that ϕ_i has φ_p at its k-th entry and ϕ_j has φ_t at its l-th entry. Then we call ϕ_i and ϕ_j neighbouring basis functions if $k = l$ and $meas(\mathrm{supp}\,\varphi_p \cap \mathrm{supp}\,\varphi_t) > 0$. Let $N := \{1,\ldots,n\}$ as before. For any $k = 1,\ldots,s$ let

$$S_k^0 := \{i \in N : \ i = (k-1)\bar{n}_0 + p \text{ for some } 1 \leq p \leq \bar{n}_0\},$$

$$\tilde{S}_k := \{i \in N : \ i = n_0 + (k-1)(\bar{n} - \bar{n}_0) + p - \bar{n}_0 \text{ for some } \bar{n}_0 + 1 \leq p \leq \bar{n}\},$$

$$S_k := S_k^0 \cup \tilde{S}_k\,,$$

i.e. by (69) and (71), the basis functions ϕ_i with index $i \in S_k$ have a nonzero coordinate φ_p for some p at the k-th entry, and in particular, $i \in S_k^0$ if this φ_p is an 'interior' basis function (i.e. $1 \leq p \leq \bar{n}_0$) and $i \in \tilde{S}_k$ if this φ_p is a 'boundary' basis function (i.e. $\bar{n}_0 + 1 \leq p \leq \bar{n}$). By [26, Theorem 4.1], these neighbouring basis functions satisfy Assumptions 2.4.3.

Our remaining task is to check assumptions (a)-(e) of Theorem 2.5.

(a) Let $\phi_i \in V_h^0$, $\phi_j \in V_h$, and let ϕ_i have φ_p at its k-th entry and ϕ_j have φ_t at its l-th entry. We must prove that either (48) or (49)-(51) holds.

If $k \neq l$, then from [26, Theorem 4.1] we have $\langle B(u^h)\phi_j, \phi_i\rangle = 0$ and $\langle R(u^h)\phi_j, \phi_i\rangle \leq 0$. Here ϕ_i and ϕ_j have no common nonzero coordinates, hence $\langle N\phi_j, \phi_i\rangle = 0$, i.e. (48) holds.

If $k = l$, then Assumption 3.0 (ii) and (85) yield

$$\langle B(u^h)\phi_j, \phi_i\rangle = \int_\Omega b_k(x, u^h, \nabla u^h)\, \nabla\varphi_t \cdot \nabla\varphi_p \leq m \int_{\Omega_{pt}} \nabla\varphi_t \cdot \nabla\varphi_p \tag{140}$$

where $\Omega_{pt} := \text{supp}\,\varphi_p \cap \text{supp}\,\varphi_t$. If $meas(\Omega_{pt}) = 0$ then $\langle B(u^h)\phi_j, \phi_i\rangle = \langle N\phi_j, \phi_i\rangle = 0$ and we have $\langle R(u^h)\phi_j, \phi_i\rangle \leq 0$ similarly as before, hence (48) holds again. If $meas(\Omega_{pt}) > 0$ then (85) and (140) imply

$$\langle B(u^h)\phi_j, \phi_i\rangle \leq -mK_0 h^{\gamma-2} \equiv -\hat{c}_1 h^{\gamma-2} =: -M_B(h) \tag{141}$$

and we must check (51). Let us estimate $T(h)^2$ from (55). As seen in the proof of Lemma 4.1, $|||z|||_{N_1} = \|z\|$ and $|||v|||_{N_2} = |||v|||_{R_1} = |||v|||_{R_2} = \|v\|_{L^2(\Omega)^s}$. Here $\|z\|$ denotes the $H^1(\Omega)$-norm, which we replace here by the equivalent norm $\|v\|^2_{H^1(\Omega)^s} = \sum_{k=1}^{s} \left(\int_\Omega |\nabla v_k|^2 + \int_\Omega |v_k|^2\right)$. Hence we must estimate

$$T(h)^2 = \sup\left\{\max\left\{\|\phi_i\|_{H^1(\Omega)^s}\|\phi_j\|_{L^2(\Omega)^s},\ \|\phi_i\|_{L^2(\Omega)^s}\|\phi_j\|_{L^2(\Omega)^s}\right\} :\ \phi_i, \phi_j \in V_h\right\}. \tag{142}$$

The L_2-norm of the basis functions satisfies the following estimate, where ϕ_j has φ_t at its l-th entry as before, and we use (137) and that (66) implies $\varphi_t \leq 1$:

$$\|\phi_j\|^2_{L^2(\Omega)^s} = \|\varphi_t\|^2_{L^2(\Omega)} \leq \int_{\text{supp}\,\varphi_t} 1 = meas(\text{supp}\,\varphi_t) \leq c_2 h^d \tag{143}$$

for all j. For the H^1-norm, let us first estimate the gradient term. Using the previous argument, (137) and (68), respectively,

$$\|\nabla\phi_j\|^2_{L^2(\Omega)^s} = \|\nabla\varphi_t\|^2_{L^2(\Omega)} = \int_{\text{supp}\,\varphi_t} |\nabla\varphi_t|^2 \leq \frac{meas(\text{supp}\,\varphi_t)}{diam^2(\text{supp}\,\varphi_t)}. \tag{144}$$

Here $diam(\text{supp}\,\varphi_t) \geq c_1 h^{\gamma/d}$ for some $c_1 > 0$ independent of h, since otherwise the l.h.s. of (137) would fail. With this and the r.h.s. of (137), we obtain

$$\|\nabla\phi_j\|^2_{L^2(\Omega)^s} \leq c_3 h^{d-\frac{2\gamma}{d}} \tag{145}$$

for some $c_3 > 0$ independent of h. Since this is larger (as $h \to 0$) than the L^2-norm estimate in (143), we also have

$$\|\phi_j\|^2_{H^1(\Omega)^s} \leq c_3 h^{d-\frac{2\gamma}{d}}, \tag{146}$$

and from (143) and (146) we obtain $\|\phi_i\|^2_{H^1(\Omega)^s}\|\phi_j\|^2_{L^2(\Omega)^s} \leq const. \cdot h^{2d-\frac{2\gamma}{d}}$ for all i, j. Also, in (142) the first term in the max is the greater than the second one, hence

$$T(h)^2 \leq \sup\left\{\|\phi_i\|_{H^1(\Omega)^s}\|\phi_j\|_{L^2(\Omega)^s} :\ \phi_i, \phi_j \in V_h\right\} \leq c_4 h^{d-\frac{\gamma}{d}}$$

for some $c_4 > 0$ independent of h. From this, using (141) and that (138) implies $\gamma + \frac{\gamma}{d} < d+2$, we obtain

$$\lim_{h\to 0} \frac{M_B(h)}{T(h)^2} \geq c_5 \lim_{h\to 0} h^{\gamma-2-d+\frac{\gamma}{d}} = +\infty. \tag{147}$$

(b) Let $\phi_i \in V_h^0$ and $\phi_j \in V_h$ be neighbouring basis vectors, i.e, as defined before in the proof, $k = l$ and $meas(\text{supp}\,\varphi_p \cap \text{supp}\,\varphi_t) > 0$. Then, as seen just above, we obtain (141) and (147), which coincide with (49)-(51).

(c) According to Remark 2.6, it is required here that $M_N(\|u^h\|)$ and $M_R(\|u^h\|)$ are bounded as $h \to 0$. Since we have the constant bounds $M_R(r) \equiv s\tilde{V}$ and $M_N(r) \equiv \|\mathbf{w}\|_{L^\infty(\Omega)^s}$, see part (iv) of the proof of Lemma 4.1, these are trivially bounded.

(d) For all $u \in H^1(\Omega)^s$ and $h > 0$, the definition of the functions ϕ_j and assumption (66) imply

$$\sum_{j=1}^{n} \phi_j = \begin{pmatrix} \sum\limits_{p=1}^{\bar{n}} \varphi_p \\ \sum\limits_{p=1}^{\bar{n}} \varphi_p \\ \cdots \\ \sum\limits_{p=1}^{\bar{n}} \varphi_p \end{pmatrix} = \begin{pmatrix} 1 \\ 1 \\ \cdots \\ 1 \end{pmatrix} =: \mathbf{1}. \tag{148}$$

Then by (134)

$$\langle B(u)(\sum_{j=1}^{n} \phi_j), v\rangle = \langle B(u)\mathbf{1}, v\rangle = \int_\Omega \sum_{k=1}^{s} b_k(x,u,\nabla u)\,\nabla 1 \cdot \nabla v_k = 0 \quad \text{and}$$
$$\langle N(\sum_{j=1}^{n} \phi_j), v\rangle = \langle N\mathbf{1}, v\rangle = \int_\Omega \sum_{k=1}^{s} (\mathbf{w}_k(x) \cdot \nabla 1)\, v_k = 0$$

for all $v \in H_D^1(\Omega)^s$, i.e. $\sum\limits_{j=1}^{n} \phi_j$ belongs to both $\ker B(u)$ and $\ker N$.

(e) This was proved in Theorem 3.1, see [26, Theorem 4.1]. □

Similarly to Corollary 3.2 before, we thus obtain

Corollary 4.1. *Let the assumptions of Theorem* 4.1 *hold and let*

$$f_k \le 0, \qquad \gamma_k \le 0 \qquad (k = 1, \ldots, s).$$

Let the basis functions satisfy (66)-(68). *Then for sufficiently small h, if* $u^h = (u_1^h, \ldots, u_s^h)^T$ *is the FEM solution of system* (123), *then*

$$\max_{k=1,\ldots,s} \max_{\overline{\Omega}} u_k^h \le \max_{k=1,\ldots,s} \max\{0, \max_{\Gamma_D} g_k^h\}. \tag{149}$$

4.2 Nonsymmetric systems with sublinear reaction terms

Now we consider problems where the dependence on the lower order terms is given as general functions of x and u, growing at most linearly, thus following subsection 3.2. One can then reduce the problem to the previous one (123) directly. Accordingly, let us consider the system

$$\left.\begin{aligned} -\mathrm{div}\left(b_k(x,u,\nabla u)\,\nabla u_k\right)+\mathbf{w}_k(x)\cdot\nabla u_k+q_k(x,u_1,\dots,u_s) &= f_k(x) \quad \text{a.e. in } \Omega,\\ b_k(x,u,\nabla u)\frac{\partial u_k}{\partial \nu} &= \gamma_k(x) \quad \text{a.e. on } \Gamma_N,\\ u_k &= g_k(x) \quad \text{a.e. on } \Gamma_D \end{aligned}\right\} \tag{150}$$

$(k=1,\dots,s)$.

Assumptions 4.2.

(i) The domain Ω and the diffusion coefficients b_k satisfy Assumptions 3.0.

(ii) (Smoothness and boundedness.) For all $k,l=1,\dots,s$ we have $b_k\in(C^1\cap L^\infty)(\Omega\times\mathbf{R}^s\times\mathbf{R}^{s\times d})$, $\mathbf{w}_k\in W^{1,\infty}(\Omega)$ and $q_k\in W^{1,\infty}(\Omega\times\mathbf{R}^s)$.

(iii) (Coercivity.) We have $\mathrm{div}\,\mathbf{w}_k\le 0$ on Ω and $\mathbf{w}_k\cdot\nu\ge 0$ on Γ_N $(k=1,\dots,s)$.

(iv) (Cooperativity.) We have

$$\frac{\partial q_k}{\partial \xi_l}(x,\xi)\le 0 \qquad (k,l=1,\dots,s,\ k\ne l;\ x\in\Omega,\ \xi\in\mathbf{R}^s). \tag{151}$$

(v) (Weak diagonal dominance for the Jacobians.) We have

$$\sum_{l=1}^{s}\frac{\partial q_k}{\partial \xi_l}(x,\xi)\ge 0 \qquad (k=1,\dots,s;\ x\in\Omega,\ \xi\in\mathbf{R}^s). \tag{152}$$

(vi) For all $k=1,\dots,s$ we have $f_k\in L^2(\Omega)$, $\gamma_k\in L^2(\Gamma_N)$, $g_k=g^*_{k|\Gamma_D}$ with $g^*\in H^1(\Omega)$.

The basic idea to deal with problem (150) is to reduce it to (123), similarly as in Subsection 3.2. Defining the functions $V_{kl}:\Omega\times\mathbf{R}^s\to\mathbf{R}$ and $\hat{f}_k$ as in (95) and (97), respectively, problem (150) becomes

$$\left.\begin{aligned} -\mathrm{div}\left(b_k(x,u,\nabla u)\,\nabla u_k\right)+\mathbf{w}_k(x)\cdot\nabla u_k+\sum_{l=1}^{s}V_{kl}(x,u,\nabla u)\,u_l &= \hat{f}_k(x) \quad \text{a.e. in } \Omega,\\ b_k(x,u,\nabla u)\frac{\partial u_k}{\partial \nu} &= \gamma_k(x) \quad \text{a.e. on } \Gamma_N,\\ u_k &= g_k(x) \quad \text{a.e. on } \Gamma_D \end{aligned}\right\} \tag{153}$$

$(k=1,\dots,s)$, which is a special case of (150).

Consequently, for a finite element discretization developed as in subsection 3.1.2, Corollary 4.1 discrete maximum principle (149) for suitable discretizations of (153), provided $\hat{f}_k\le 0$ and $\gamma_k\le 0$ $(k=1,\dots,s)$. For the original system (91), we thus obtain

Corollary 4.2. *Let problem* (150) *satisfy Assumptions* 4.2, *and let its FEM discretization satisfy the corresponding conditions of Theorem* 4.1. *If*

$$f_k\le q_k(x,0),\qquad \gamma_k\le 0\qquad (k=1,\dots,s)$$

and $u^h=(u_1^h,\dots,u_s^h)^T$ *is the FEM solution of system* (150), *then for sufficiently small h,*

$$\max_{k=1,\dots,s}\max_{\overline{\Omega}}u_k^h\le\max_{k=1,\dots,s}\max\{0,\max_{\Gamma_D}g_k^h\}. \tag{154}$$

4.3 Nonsymmetric systems with superlinear reaction terms

$$\left.\begin{aligned} -\mathrm{div}\left(b_k(x,\nabla u)\nabla u_k\right)+\mathbf{w}_k(x)\cdot\nabla u_k+q_k(x,u_1,\dots,u_s) &= f_k(x) \quad \text{a.e. in } \Omega,\\ b_k(x,\nabla u)\frac{\partial u_k}{\partial \nu} &= \gamma_k(x) \quad \text{a.e. on } \Gamma_N,\\ u_k &= g_k(x) \quad \text{a.e. on } \Gamma_D \end{aligned}\right\} \tag{155}$$

$(k=1,\dots,s)$.

Assumptions 4.3.

(i) The domain Ω and the diffusion coefficients b_k satisfy Assumptions 3.0.

(ii) (Smoothness and growth.) For all $k,l=1,\dots,s$ we have $b_k\in(C^1\cap L^\infty)(\Omega\times\mathbf{R}^d)$, $\mathbf{w}_k\in W^{1,\infty}(\Omega)$ and $q_k\in C^1(\Omega\times\mathbf{R}^s)$. Further, let

$$2\le p<p^*, \quad \text{where } p^*:=\tfrac{2d}{d-2} \text{ if } d\ge 3 \text{ and } p^*:=+\infty \text{ if } d=2; \tag{156}$$

then there exist constants $\beta_1,\beta_2\ge 0$ such that

$$\left|\frac{\partial q_k}{\partial \xi_l}(x,\xi)\right|\le \beta_1+\beta_2|\xi|^{p-2} \qquad (k,l=1,\dots,s;\ x\in\Omega,\ \xi\in\mathbf{R}^s). \tag{157}$$

(iii) (Ellipticity.) Defining $a_k(x,\eta):=b_k(x,\eta)\eta$ for all k, the Jacobian matrices $\frac{\partial}{\partial\eta}a_k(x,\eta)$ are uniformly spectrally bounded from both below and above.

(iv) (Coercivity.) We have $\mathrm{div}\,\mathbf{w}_k\le 0$ on Ω and $\mathbf{w}_k\cdot\nu\ge 0$ on Γ_N $(k=1,\dots,s)$.

(v) (Cooperativity.) We have

$$\frac{\partial q_k}{\partial \xi_l}(x,\xi)\le 0 \qquad (k,l=1,\dots,s,\ k\ne l;\ x\in\Omega,\ \xi\in\mathbf{R}^s). \tag{158}$$

(vi) (Weak diagonal dominance for the Jacobians w.r.t. rows and columns.) We have

$$\sum_{l=1}^s\frac{\partial q_k}{\partial \xi_l}(x,\xi)\ge 0, \qquad \sum_{l=1}^s\frac{\partial q_l}{\partial \xi_k}(x,\xi)\ge 0 \qquad (k=1,\dots,s;\ x\in\Omega,\ \xi\in\mathbf{R}^s). \tag{159}$$

(vii) For all $k=1,\dots,s$ we have $f_k\in L^2(\Omega)$, $\gamma_k\in L^2(\Gamma_N)$, $g_k=g^*_{k|\Gamma_D}$ with $g^*\in H^1(\Omega)$.

We note that Remark 3.6 is valid for the above system as well.

Now we proceed as in subsection 3.3. System (155) is first reduced to a system with nonlinear coefficients like (153). Owing to the superlinear growth allowed in (157), we must reprove Theorem 4.1 under Assumptions 4.3. For this, we first define the operators

$$\begin{aligned} \langle B(u)z,v\rangle &= \int_\Omega\sum_{k=1}^s b_k(x,\nabla u)\,\nabla z_k\cdot\nabla v_k, \qquad \langle Nz,v\rangle=\int_\Omega\sum_{k=1}^s(\mathbf{w}_k(x)\cdot\nabla z_k)\,v_k,\\ \langle R(u)z,v\rangle &= \int_\Omega\sum_{k,l=1}^s V_{kl}(x,u)\,z_l v_k \end{aligned} \tag{160}$$

$(z\in H^1(\Omega)^s,\ v\in H^1_D(\Omega)^s)$.

Lemma 4.2. *Let Assumptions* 4.3 *hold. For any $u \in H^1(\Omega)^s$, the above operators $B(u)$, $N(u) \equiv N$ and $R(u)$, together with the operator $A(u) = B(u)u + Nu + R(u)u$, satisfy Assumptions* 2.4.1, *modified according to Remark* 2.6, *and Assumptions* 2.4.2, *in the spaces $H = H^1(\Omega)^s$ and $H_0 = H^1_D(\Omega)^s$.*

Proof. This follows from Lemma 4.1 and Lemma 3.2, using the arguments of the former for $B(u)$ and N, and (under the polynomial growth) the arguments of the latter for $R(u)$. □

We recall the new norms for (53)–(54): we have

$$|||z|||_{N_1} = \|z\|_{H^1(\Omega)^s} \quad \text{and} \quad |||v|||_{N_2} = \|v\|_{L^2(\Omega)^s} \tag{161}$$

from Lemma 4.1, and

$$|||v|||^2_{R_1} = |||v|||^2_{R_2} = \|v\|^2_{L^{2q}(\Omega)^s} := \Big\| \sum_{k=1}^{s} v_k^2 \Big\|_{L^q(\Omega)} \qquad (v \in H^1(\Omega)^s) \tag{162}$$

from Lemma 3.2, see [26], where r is any fixed real number satisfying

$$\frac{d}{2+d-\gamma} < r \leq \frac{p^*}{p-2}. \tag{163}$$

and then q in (162) is chosen as

$$\frac{1}{r} + \frac{1}{q} = 1. \tag{164}$$

We also need to strengthen Definition 4.1 on the quasi-regularity of the mesh.

Definition 4.2. Let $\Omega \subset \mathbf{R}^d$ and let us consider a family of FEM subspaces $\mathcal{V} = \{V_h\}_{h\to 0}$ constructed as in subsection 3.1.2. The corresponding mesh will be called *quasi-regular* w.r.t. problem (155) if

$$c_1 h^\gamma \leq meas(\operatorname{supp} \varphi_p) \leq c_2 h^d, \tag{165}$$

where the positive real number γ satisfies

$$d \leq \gamma < \min\Big\{\gamma_d^*(p),\ \frac{d(d+2)}{d+1}\Big\} \tag{166}$$

with p from (157) and $\gamma_d^*(p)$ from (108).

Now we can derive the desired nonnegativity result for the stiffness matrix. Here the entries of $\overline{\mathbf{A}}(\overline{\mathbf{c}})$ are, for any $\overline{\mathbf{c}} = (c_1, \ldots, c_n)^T \in \mathbf{R}^n$ and $i = 1, \ldots, n_0$, $j = 1, \ldots, n$,

$$\begin{aligned} a_{ij}(\overline{\mathbf{c}}) := \int_\Omega \Bigg(& \sum_{k=1}^{s} b_k(x, \nabla u^h) (\nabla \phi_j)_k \cdot (\nabla \phi_i)_k + \sum_{k=1}^{s} \Big(\mathbf{w}_k(x) \cdot (\nabla \phi_j)_k\Big) (\phi_i)_k \\ & + \sum_{k,l=1}^{s} V_{kl}(x, u^h) (\phi_j)_l (\phi_i)_k \Bigg) \end{aligned} \tag{167}$$

where, as in (111),

$$V_{kl}(x, u^h(x)) = \int_0^1 \frac{\partial q_k}{\partial \xi_l}(x, t u^h(x))\, dt \qquad (k, l = 1, \ldots, s;\ x \in \Omega). \tag{168}$$

Theorem 4.2. *Let problem* (155) *satisfy Assumptions* 4.3. *Let us consider a family of finite element subspaces* $\mathcal{V} = \{V_h\}_{h\to 0}$, *such that the corresponding family of meshes is quasi-regular according to Definition* 4.2, *further, for any* $p = 1,\dots,\bar{n}_0$, $t = 1,\dots,\bar{n}$ $(p \neq t)$, *if* $meas(\operatorname{supp}\varphi_p \cap \operatorname{supp}\varphi_t) > 0$ *then* (139) *holds, where* γ *is from* (166) *and* $K_0 > 0$ *is a constant independent of* p,t *and* h.

Then for sufficiently small h, *the matrix* $\bar{\mathbf{A}}(\bar{\mathbf{c}})$ *defined in* (167) *is of generalized nonnegative type with irreducible blocks in the sense of Definition* 2.4.

Proof. The proof is similar to that of Theorem 4.1, combining it with the proof of Theorem 3.3 [26, Theorem 4.3]. We will only point out the differences.

First, owing to (161)–(162), instead of (142) we must estimate

$$T(h)^2 = \sup\Big\{\max\Big\{\|\phi_i\|_{H^1(\Omega)^s}\|\phi_j\|_{L^2(\Omega)^s},\ \|\phi_i\|_{L^{2q}(\Omega)^s}\|\phi_j\|_{L^{2q}(\Omega)^s}\Big\} : \ \phi_i,\phi_j \in V_h\Big\}. \quad (169)$$

We have for all i,j

$$\|\phi_i\|_{H^1(\Omega)^s}\|\phi_j\|_{L^2(\Omega)^s} \leq c_4 h^{d-\frac{\gamma}{d}}$$

from Theorem 4.1 and

$$\|\phi_i\|_{L^{2q}(\Omega)^s}\|\phi_j\|_{L^{2q}(\Omega)^s} \leq c_5\, h^{d/q}$$

from Theorem 3.3, hence

$$T(h)^2 \leq c_6 \max(h^{d/q}, h^{d-\frac{\gamma}{d}}).$$

Here in (166), γ has been chosen such that both $\gamma - 2 - \frac{d}{q} < 0$, see (143) in [26], and $\gamma - 2 - d + \frac{\gamma}{d} < 0$, see (147). Therefore

$$\lim_{h\to 0} \frac{M_B(h)}{T(h)^2} \geq c_7 \lim_{h\to 0} \min\{h^{\gamma-2-\frac{d}{q}},\ h^{\gamma-2-d+\frac{\gamma}{d}}\} = +\infty. \quad (170)$$

(Here c_4, c_5 etc. denote positive constants.)

It is left to verify assumption (c), i.e. that $M_N(\|u^h\|)$ and $M_R(\|u^h\|)$ are bounded as $h \to 0$. For the former, we have the constant bound $M_N(\|u^h\|) \equiv \|\mathbf{w}\|_{L^\infty(\Omega)^s}$, see part (iv) of the proof of Lemma 4.1. For the latter, the boundedness of $M_R(\|u^h\|)$ was proved in [26, Theorem 4.3].

The other parts of the proof coincide with that of Theorem 4.1. □

As before, we can derive the corresponding DMP:

Corollary 4.3. *Let problem* (155) *satisfy Assumptions* 4.3, *and let its FEM discretization satisfy the corresponding conditions of Theorem* 4.2. *If* $f_k \leq q_k(x,0)$ *and* $\gamma_k \leq 0$ $(k = 1,\dots,s)$, *then* (154) *holds.*

4.4 Nonsymmetric systems with nonlinear convection coefficients

Finally we study a system containing nonlinear convection terms. The required strengthening in the other assumptions is the strong uniform diagonal dominance (172)–(173) and the homogenity of the Dirichlet data. The applicability of these conditions will be illustrated in the example in subsection 5.3.

Let us consider the system

$$\left.\begin{aligned} -\mathrm{div}\left(b_k(x,\nabla u)\,\nabla u_k\right) + \mathbf{w}_k(x,u)\cdot\nabla u_k + q_k(x,u_1,\dots,u_s) &= f_k(x) \quad \text{a.e. in } \Omega,\\ b_k(x,\nabla u)\frac{\partial u_k}{\partial \nu} &= \gamma_k(x) \quad \text{a.e. on } \Gamma_N,\\ u_k &= 0 \quad \text{a.e. on } \Gamma_D \end{aligned}\right\} \tag{171}$$

$(k=1,\dots,s)$.

Assumptions 4.4.

(i) The domain Ω and the diffusion coefficients b_k satisfy Assumptions 3.0.

(ii) (Smoothness and growth.) For all $k,l=1,\dots,s$ we have $b_k \in (C^1 \cap L^\infty)(\Omega \times \mathbf{R}^d)$ and $q_k \in C^1(\Omega \times \mathbf{R}^s)$, further, (156) and (157) hold.

(iii) (Ellipticity.) Defining $a_k(x,\eta) := b_k(x,\eta)\eta$ for all k, the Jacobian matrices $\frac{\partial}{\partial \eta} a_k(x,\eta)$ are uniformly spectrally bounded from both below and above.

(iv) (Bounded convection term.) We have $\mathbf{w}_k \in L^\infty(\Omega \times \mathbf{R})$ $(k=1,\dots,s)$.

(v) (Cooperativity.) We have $\dfrac{\partial q_k}{\partial \xi_l}(x,\xi) \le 0 \quad (k,l=1,\dots,s,\ k \ne l;\ x \in \Omega,\ \xi \in \mathbf{R}^s)$.

(vi) (Uniform diagonal dominance for the Jacobians w.r.t. rows and columns.) There exists $\mu > 0$ such that

$$\sum_{l=1}^{s} \frac{\partial q_k}{\partial \xi_l}(x,\xi) \ge \mu, \qquad \sum_{l=1}^{s} \frac{\partial q_l}{\partial \xi_k}(x,\xi) \ge \mu \qquad (k=1,\dots,s;\ x \in \Omega,\ \xi \in \mathbf{R}^s). \tag{172}$$

Moreover,

$$\mu > \frac{\|\mathbf{w}\|^2_{L^\infty(\Omega)^s}}{4m} \tag{173}$$

where $\|\mathbf{w}\|_{L^\infty(\Omega)^s} := \sup\limits_{\substack{k=1,\dots,s\\(x,\xi)\in\Omega\times\mathbf{R}^s}} |\mathbf{w}_k(x,\xi)|$ and m is from Assumption 3.0 (ii).

(vii) For all $k=1,\dots,s$ we have $f_k \in L^2(\Omega)$ and $\gamma_k \in L^2(\Gamma_N)$.

We proceed similarly as in the previous subsection, system (171) is reduced to a system with nonlinear coefficients as before via the functions $V_{kl} : \Omega \times \mathbf{R}^s \to \mathbf{R}$ and $\hat{f}_k$ from (95) and (97), respectively. The difference is is the nonlinear convection term. Taking this into account, we must reprove Theorem 4.2 under Assumptions 4.4, but only those parts are addressed where the convection term is involved.

The operator corresponding to our problem is

$$\begin{aligned} \langle A(u), v\rangle = \int_\Omega \Big(&\sum_{k=1}^{s} b_k(x,\nabla u)\,\nabla u_k \cdot \nabla v_k + \sum_{k=1}^{s} (\mathbf{w}_k(x,u)\cdot\nabla u_k)\, v_k \\ &+ \sum_{k,l=1}^{s} V_{kl}(x,u)\, u_l\, v_k \Big) \end{aligned} \tag{174}$$

($u \in H^1(\Omega)^s$, $v \in H^1_D(\Omega)^s$). First we properly modify (160), where the main point is to compensate for the presence of the convection term in the positivity of the operator without a coercivity condition on $\mathbf{w}_k$. We define the operators

$$\begin{aligned}
\langle B(u)z, v\rangle &= \int_\Omega \sum_{k=1}^{s} \Big(b_k(x, \nabla u)\, \nabla z_k \cdot \nabla v_k + \mu z_k v_k \Big), \\
\langle N(u)z, v\rangle &= \int_\Omega \sum_{k=1}^{s} (\mathbf{w}_k(x,u) \cdot \nabla z_k)\, v_k \\
\langle R(u)z, v\rangle &= \int_\Omega \Big(\sum_{k,l=1}^{s} V_{kl}(x,u)\, z_l v_k - \mu \sum_{k=1}^{s} z_k v_k \Big)
\end{aligned} \tag{175}$$

($z \in H^1(\Omega)^s$, $v \in H^1_D(\Omega)^s$). We note that (95) and (172) yield

$$\sum_{l=1}^{s} V_{kl}(x,\xi) \geq \mu \qquad (k = 1,\ldots,s;\ x \in \Omega,\ \xi \in \mathbf{R}^s), \tag{176}$$

and hence, since $V_{kl}(x,\xi) \leq 0$ for $k \neq l$ by Assumption 4.4 (v), we also have

$$V_{kk}(x,\xi) \geq \mu \qquad (k = 1,\ldots,s;\ x \in \Omega,\ \xi \in \mathbf{R}^s). \tag{177}$$

Lemma 4.3. *Let Assumptions* 4.4 *hold. For any $u \in H^1(\Omega)^s$, the operators $B(u)$, $N(u)$ and $R(u)$, together with the operator $A(u)$ in* (174)*, satisfy Assumptions* 2.4.1*, modified according to Remark* 2.6*, in the spaces $H = H^1(\Omega)^s$ and $H_0 = H^1_D(\Omega)^s$.*

Proof. We must reprove those parts of Lemma 4.2 that involve the convection term or the modifications of $B(u)$ and $R(u)$ with the term containing μ.

(i) It is obvious from (174) and (175) that $A(u) = B(u)u + N(u)u + R(u)u$.

(ii) We must prove (31). Here for all $u \in H^1(\Omega)^s$ and $v \in H^1_D(\Omega)^s$,

$$\begin{aligned}
\Big\langle \Big(B(u)+N(u)\Big)v, v\Big\rangle &= \int_\Omega \sum_{k=1}^{s} \Big(b_k(x,\nabla u)\, |\nabla v_k|^2 + \mu v_k^2 \Big) + \int_\Omega \sum_{k=1}^{s} (\mathbf{w}_k(x,u) \cdot \nabla v_k)\, v_k \\
&\geq m \|\nabla v\|^2_{L^2(\Omega)^s} + \mu \|v\|^2_{L^2(\Omega)^s} - \omega \|\nabla v\|_{L^2(\Omega)^s} \|v\|_{L^2(\Omega)^s}
\end{aligned} \tag{178}$$

where $\omega := \|\mathbf{w}\|_{L^\infty(\Omega)^s}$. Using the basic inequality $xy \leq \frac{1}{2}\Big(\varepsilon x^2 + \frac{1}{\varepsilon} y^2\Big)$ ($\varepsilon > 0$, $x,y \in \mathbf{R}$) for the last two factors, we obtain

$$\Big\langle \Big(B(u)+N(u)\Big)v, v\Big\rangle \geq \Big(m - \frac{\omega\varepsilon}{2}\Big) \|\nabla v\|^2_{L^2(\Omega)^s} + \Big(\mu - \frac{\omega}{2\varepsilon}\Big) \|v\|^2_{L^2(\Omega)^s}.$$

Choosing $\varepsilon := \frac{\omega}{2\mu}$, we have

$$\Big\langle \Big(B(u)+N(u)\Big)v, v\Big\rangle \geq \hat{m}\, \|\nabla v\|^2_{L^2(\Omega)^s} \equiv \hat{m}\, \|v\|^2$$

where $\hat{m} := m - \frac{\omega^2}{4\mu} > 0$ by (173).

(iii) Let us consider the sets P and D, defined in paragraph (iii) of the proof of Lemma 4.1. That is, $v \in D$ iff $v = (0, \dots, 0, g, 0, \dots, 0)^T$ with g at the k-th entry for some $1 \le k \le s$ and $g \in H^1(\Omega)$, $g \ge 0$. Further, $v \in P$ iff $v = (y, \dots, y)$ for some $y \in H^1(\Omega)$, $y \ge 0$. We must prove that for any $u \in H^1(\Omega)^s$ and $v \in D$, we have

$$\langle R(u)z, v\rangle \ge 0 \tag{179}$$

provided that either $z \in P$ or $z = v \in D$.

If $z \in P$, then

$$\langle R(u)z, v\rangle = \int_\Omega \Big(\sum_{l=1}^{s} V_{kl}(x,u) - \mu\Big) yg \ge 0$$

by (176) and that $y, g \ge 0$. If $z = v \in D$, then by (177)

$$\langle R(u)v, v\rangle = \int_\Omega \Big(V_{kk}(x,u) - \mu\Big) g^2 \ge 0.$$

(iv) This follows in the same way as in Lemma 4.2. For $N(u)$, we can similarly factor out $\|\mathbf{w}\|_{L^\infty(\Omega)^s}$. For $R(u)$, the new norms can remain $|||v|||^2_{R_1} = |||v|||^2_{R_2} = \|v\|^2_{L^{2q}(\Omega)^s}$ as in (162), since the additional term in (175) can be bounded by the product L^2-norm $\|.\|_{L^2(\Omega)^s}$, which is (up to a constant factor) not larger than the norm $\|.\|^2_{L^{2q}(\Omega)^s}$ owing to the Sobolev inequality. □

Now we can derive the nonnegativity of the stiffness matrix. Here the entries of $\bar{\mathbf{A}}(\bar{\mathbf{c}})$ are, for any $\bar{\mathbf{c}} = (c_1, \dots, c_n)^T \in \mathbf{R}^n$ and $i = 1, \dots, n_0$, $j = 1, \dots, n$,

$$a_{ij}(\bar{\mathbf{c}}) := \int_\Omega \Bigg(\sum_{k=1}^{s} b_k(x, \nabla u^h)\,(\nabla\phi_j)_k \cdot (\nabla\phi_i)_k + \sum_{k=1}^{s} \Big(\mathbf{w}_k(x,u^h)\cdot(\nabla\phi_j)_k\Big)\,(\phi_i)_k + \sum_{k,l=1}^{s} V_{kl}(x,u^h)\,(\phi_j)_l\,(\phi_i)_k \Bigg) \tag{180}$$

where $V_{kl}(x,u^h)$ is as in (168).

Theorem 4.3. *Let problem* (171) *satisfy Assumptions* 4.4. *Let us consider a family of finite element subspaces* $\mathcal{V} = \{V_h\}_{h\to 0}$, *such that the corresponding family of meshes is quasi-regular according to Definition* 4.2, *further, for any* $p = 1, \dots, \bar{n}_0$, $t = 1, \dots, \bar{n}$ $(p \ne t)$, *if* $meas(\operatorname{supp}\varphi_p \cap \operatorname{supp}\varphi_t) > 0$ *then* (139) *holds, where* γ *is from* (166) *and* $K_0 > 0$ *is a constant independent of* p, t *and* h.

Then for sufficiently small h, the matrix $\bar{\mathbf{A}}(\bar{\mathbf{c}})$ *defined in* (180) *is of generalized nonnegative type with irreducible blocks in the sense of Definition* 2.4.

Proof. The proof is similar to that of Theorem 4.2, with a few differences. First, the proof for assumption (a) relies on (141), where by (175), now $\langle B(u^h)\phi_j, \phi_i\rangle$ contains the additional term $\int_\Omega \mu \sum_{k=1}^{s} (\phi_j)_k(\phi_i)_k$. However, this term is bounded by μs, hence altogether (141) is preserved with another constant instead of $\hat{c}_1$ and still tends to $-\infty$. In the other parts of the proof we only need the sum of $B(u)$ and $R(u)$, in which the additional terms vanish by definition.

Finally, Theorem 4.2 contains the boundedness of $M_R(\|u^h\|)$, see the end of its proof, which fact is quoted from [26, Theorem 4.3]. This part of the proof uses Assumptions 2.4.2, which have not yet been proved now. Assumptions 2.4.2 are used in [26, Theorem 4.3] to have uniform monotonicity of A in order to prove that

$$\langle A(u^h) - A(g^h), u^h - g^h \rangle \geq m \, \|u^h - g^h\|^2, \tag{181}$$

since this implies the boundedness of $\|u^h\|$ if we assume the boundedness of $\|g^h\|$ (as $h \to 0$). These properties are derived in [26, Remark 3.1]. Now we have $g^h = 0$ by the homogeneous Dirichlet data in (171), hence we only need (181) for the special case $g^h = 0$. Therefore, to prove our theorem, it suffices instead of Assumptions 2.4.2 to verify

$$\langle A(u^h), u^h \rangle \geq \tilde{m} \, \|u^h\|^2 \qquad (h > 0) \tag{182}$$

for some constant $\tilde{m} > 0$, independent of the FEM solution u^h of our problem.

Since $u^h = 0$ on Γ_D, we can substitute $u = v = u^h$ in (174):

$$\langle A(u^h), u^h \rangle = \int_\Omega \Big(\sum_{k=1}^{s} b_k(x, \nabla u^h) \, |\nabla u_k^h|^2 + \sum_{k=1}^{s} (\mathbf{w}_k(x, u^h) \cdot \nabla u_k^h) \, u_k^h \Big) + \int_\Omega \sum_{k,l=1}^{s} V_{kl}(x, u^h) \, u_l^h \, u_k^h \tag{183}$$

$$= \int_\Omega \Big(\sum_{k=1}^{s} b_k(x, \nabla u^h) \, |\nabla u_k^h|^2 + \mu |u_k^h|^2 + \sum_{k=1}^{s} (\mathbf{w}_k(x, u^h) \cdot \nabla u_k^h) \, u_k^h \Big) \tag{184}$$

$$+ \int_\Omega \sum_{k,l=1}^{s} \Big(V_{kl}(x, u^h) \, u_l^h \, u_k^h - \mu |u_k^h|^2 \Big). \tag{185}$$

We can estimate (184) in the same way as in (178), and obtain the lower bound $\hat{m} \, \|u^h\|^2$ where $\hat{m} := m - \frac{\omega^2}{4\mu} > 0$. For (185), note that (172) and (95) imply that μ is a lower uniform spectral bound for the matrices $V(x, \xi)$, i.e.

$$V(x, \xi) \, \zeta \cdot \zeta \equiv \sum_{k,l=1}^{s} V_{kl}(x, \xi) \, \zeta_l \, \zeta_k \geq \mu |\zeta|^2 \tag{186}$$

(for all $(x, \xi) \in \Omega \times \mathbf{R}^s$ and $\zeta \in \mathbf{R}^s$), which yields that the expression in (185) is nonnegative. Altogether, (182) holds with $\tilde{m} := \hat{m}$. □

As before, we can derive the corresponding DMP (154) under the conditions of Theorem 4.3. Since now $g = 0$, this becomes the discrete nonpositivity property $u_k^h \leq 0$. One can similarly obtain the discrete nonnegativity property, which is more noteworthy to formulate here:

Corollary 4.4. *Let problem* (171) *satisfy Assumptions* 4.4, *and let its FEM discretization satisfy the corresponding conditions of Theorem* 4.3. *If* $f_k \geq q_k(x, 0)$ *and* $\gamma_k \geq 0$ $(k = 1, \ldots, s)$, *then for sufficiently small h, the FEM solution* $u^h = (u_1^h, \ldots, u_s^h)^T$ *of system* (171) *satisfies*

$$u_k^h \geq 0 \quad on \ \Omega \qquad (k = 1, \ldots, s). \tag{187}$$

5 Some real-life examples

5.1 Reaction-diffusion systems in chemistry

The following result is quoted from [26]. The steady states of certain reaction-diffusion processes in chemistry are described by systems of the following form:

$$\left.\begin{aligned} -b_k\Delta u_k + P_k(x,u_1,\dots,u_s) &= f_k(x) \quad \text{in } \Omega, \\ b_k\frac{\partial u_k}{\partial \nu} &= \gamma_k(x) \quad \text{on } \Gamma_N, \\ u_k &= g_k(x) \quad \text{on } \Gamma_D \end{aligned}\right\} \qquad (k=1,\dots,s). \tag{188}$$

Here, for all k, the quantity u_k describes the concentration of the kth species, and P_k is a polynomial which characterizes the rate of the reactions involving the k-th species. A common way to describe such reactions is the so-called mass action type kinetics [21, 22], which implies that P_k has no constant term for any k, in other words, $P_k(x,0) \equiv 0$ on Ω for all k. Further, the reaction between different species is often proportional to the product of their concentration, in which case $P_k(x,u_1,\dots,u_s) = a_{kk}(x)u_k^\alpha + \sum_{k\neq l} a_{kl}(x)u_k u_l$. The function $f_k \geq 0$ describes a source independent of concentrations.

We consider system (188) under the following conditions, such that it becomes a special case of system (100). As pointed out later, such chemical models describe processes with cross-catalysis and strong autoinhibiton.

Assumptions 5.1.

(i) $\Omega \subset \mathbf{R}^d$ is a bounded piecewise C^1 domain, where $d=2$ or 3, and Γ_D, Γ_N are disjoint open measurable subsets of $\partial\Omega$ such that $\partial\Omega = \overline{\Gamma}_D \cup \overline{\Gamma}_N$.

(ii) (Smoothness and growth.) For all $k,l = 1,\dots,s$, the functions P_k are polynomials of arbitrary degree if $d=2$ and of degree at most 4 if $d=3$, further, $P_k(x,0) \equiv 0$ on Ω.

(iii) (Ellipticity.) $b_k > 0$ $(k=1,\dots,s)$ are given numbers.

(iv) (Cooperativity.) We have

$$\frac{\partial P_k}{\partial \xi_l}(x,\xi) \leq 0 \qquad (k,l=1,\dots,s,\ k\neq l;\ x\in\Omega,\ \xi\in\mathbf{R}^s). \tag{189}$$

(v) (Weak diagonal dominance for the Jacobians w.r.t. rows and columns.) We have

$$\sum_{l=1}^{s}\frac{\partial P_k}{\partial \xi_l}(x,\xi) \geq 0, \qquad \sum_{l=1}^{s}\frac{\partial P_l}{\partial \xi_k}(x,\xi) \geq 0 \qquad (k=1,\dots,s;\ x\in\Omega,\ \xi\in\mathbf{R}^s). \tag{190}$$

(vi) For all $k=1,\dots,s$ we have $f_k \in L^2(\Omega)$, $\gamma_k \in L^2(\Gamma_N)$, $g_k = g^*_{k|\Gamma_D}$ with $g^* \in H^1(\Omega)$.

Similarly to (94), assumptions (189)-(190) now imply

$$\frac{\partial P_k}{\partial \xi_k}(x,\xi) \geq 0 \qquad (k = 1,\dots,s;\ x \in \Omega,\ \xi \in \mathbf{R}^s). \tag{191}$$

Returning to the model described by system (188), the chemical meaning of the cooperativity (189) is cross-catalysis, whereas (191) means autoinhibiton. Cross-catalysis arises e.g. in gradient systems [49]. Condition (190) means that autoinhibition is strong enough to ensure both weak diagonal dominances.

By definition, the concentrations u_k are nonnegative, therefore a proper numerical model must produce such numerical solutions. We can use Corollary 3.5 to obtain the required property:

Corollary 5.1. *Let problem* (188) *satisfy Assumptions* 5.1, *and let its FEM discretization satisfy the conditions of Theorem* 3.3. *If*

$$f_k \geq 0, \quad \gamma_k \geq 0, \quad g_k \geq 0 \qquad (k = 1,\dots,s)$$

then for sufficiently small h, the FEM solution $u^h = (u_1^h,\dots,u_s^h)^T$ *of system* (188) *satisfies*

$$u_k^h \geq 0 \quad \text{on } \Omega \qquad (k = 1,\dots,s). \tag{192}$$

5.2 Linear elliptic systems

Maximum principles or nonnegativity preservation for linear elliptic systems have attracted great interest, as mentioned in the introduction. Hence it is worthwile to derive the corresponding DMPs from the previous results. Following [26], let us therefore consider linear elliptic systems of the form

$$\left.\begin{aligned} -\operatorname{div}(b_k(x)\nabla u_k) + \sum_{l=1}^{s} V_{kl}(x)\,u_l &= f_k(x) \quad \text{a.e. in } \Omega,\\ b_k(x)\frac{\partial u_k}{\partial \nu} &= \gamma_k(x) \quad \text{a.e. on } \Gamma_N,\\ u_k &= g_k(x) \quad \text{a.e. on } \Gamma_D \end{aligned}\right\} \quad (k = 1,\dots,s) \tag{193}$$

where for all $k,l = 1,\dots,s$ we have $b_k \in W^{1,\infty}(\Omega)$ and $V_{kl} \in L^\infty(\Omega)$.

Let Assumptions 3.1 hold (where in fact we do not need assumption (ii)). Then (193) is a special case of (56), hence Corollary 3.2 holds, as well as the analogous results mentioned in Remark 3.2. Here we formulate two of these that follow the most studied CMP results:

Corollary 5.2. *Let problem* (193) *satisfy Assumptions* 3.1, *let its FEM discretization satisfy the conditions of Theorem* 3.1 *and let h be sufficiently small. If* $u^h = (u_1^h,\dots,u_s^h)^T$ *is the FEM solution of system* (193), *then the following properties hold.*

(1) *If* $f_k \leq 0$, $\gamma_k \leq 0$ *(*$k = 1,\dots,s$*) and* $\max\limits_{k=1,\dots,s} \max\limits_{\Gamma_D} g_k^h > 0$, *then*

$$\max_{k=1,\dots,s} \max_{\overline{\Omega}} u_k^h = \max_{k=1,\dots,s} \max_{\Gamma_D} g_k^h. \tag{194}$$

(2) *If* $f_k \geq 0$, $\gamma_k \geq 0$ *and* $g_k \geq 0$ *(*$k = 1,\dots,s$*), then*

$$u_k^h \geq 0 \quad \text{on } \Omega \qquad (k = 1,\dots,s). \tag{195}$$

5.3 Nonsymmetric transport systems

The description of nonlinear transport processes for certain agents (pollutants), involving diffusion, convection and reaction, often leads to systems of the form

$$\left.\begin{aligned} -b_k\Delta u_k + \mathbf{w}_k(x,u)\cdot\nabla u_k + P_k(x,u_1,\dots,u_s) &= f_k(x) \quad \text{a.e. in } \Omega,\\ b_k\frac{\partial u_k}{\partial \nu} &= \gamma_k(x) \quad \text{a.e. on } \Gamma_N,\\ u_k &= 0 \qquad\ \ \text{a.e. on } \Gamma_D \end{aligned}\right\} \tag{196}$$

$(k=1,\dots,s)$. We consider diffusion-dominated processes, i.e. when the fixed numbers $b_k > 0$ are comparable to the magnitude of the coefficients $\mathbf{w}_k$. Here $u_k \geq 0$ are the concentrations of the agents. One expects any numerical solution method to reproduce the nonnegativity of the solution.

Assumptions 5.3.

(i) The numbers b_k and functions P_k, f_k and γ_k satisfy Assumptions 5.1.

(ii) We have $\mathbf{w}_k \in L^\infty(\Omega\times\mathbf{R})$ $(k=1,\dots,s)$.

(iii) There exists $\mu > 0$ such that

$$\sum_{l=1}^{s}\frac{\partial P_k}{\partial \xi_l}(x,\xi) \geq \mu, \qquad \sum_{l=1}^{s}\frac{\partial P_l}{\partial \xi_k}(x,\xi) \geq \mu \qquad (k=1,\dots,s;\ x\in\Omega,\ \xi\in\mathbf{R}^s). \tag{197}$$

Moreover,

$$\mu > \frac{\|\mathbf{w}\|^2_{L^\infty(\Omega)^s}}{4m} \tag{198}$$

where $\|\mathbf{w}\|_{L^\infty(\Omega)^s} := \sup_{\substack{k=1,\dots,s\\ (x,\xi)\in\Omega\times\mathbf{R}^s}} |\mathbf{w}_k(x,\xi)|$ and $m := \min_k b_k > 0$.

Systems of the form (196) typically arise from the time discretization of the time-dependent transport system

$$\frac{\partial u_k}{\partial t} - b_k\Delta u_k + \mathbf{w}_k(x,u)\cdot\nabla u_k + R_k(x,u_1,\dots,u_s) = g_k(x,t) \tag{199}$$

with the boundary conditions of (196) and an initial condition $u_k(x,0)=u_0(x)$ $(x\in\Omega)$. Here $\mathbf{w}_k(x,u)$ is the convective term, e.g. wind, and R_k is a polynomial which characterizes the rate of the reactions involving the k-th species, as in subsection 5.1. Here the R_k do not satisfy a condition like (197), this will come instead from the numerical process below.

The standard numerical solution first uses a time discretization, resulting in the following equations, where u_k^i denotes the solution on the ith time level t_i:

$$\frac{u_k^i - u_k^{i-1}}{\tau} - b_k\Delta u_k^i + \mathbf{w}_k(x,u^i)\cdot\nabla u_k^i + R_k(x,u_1^i,\dots,u_s^i) = g_k^i(x)\,.$$

Rearranging this as

$$-b_k\Delta u_k^i + \mathbf{w}_k(x,u^i)\cdot\nabla u_k^i + \Big(R_k(x,u_1^i,\dots,u_s^i) + \frac{1}{\tau}u_k^i\Big) = g_k^i(x) + \frac{1}{\tau}u_k^{i-1},$$

we obtain a system for the unknown function u_k^i in the form (196) with coefficients

$$P_k(x,\xi_1,\dots,\xi_s) := R_k(x,\xi_1,\dots,\xi_s) + \frac{1}{\tau}\xi_k \tag{200}$$

and $f_k(x) := g_k^i(x) + \frac{1}{\tau}u_k^{i-1}(x)$. Then the strong uniform diagonal dominance (197)–(198) can be ensured as follows. Assume that we have an estimate

$$\inf_{\substack{k=1,\dots,s\\(x,\xi)\in\Omega\times\mathbf{R}^s}} \sum_{l=1}^{s}\frac{\partial R_k}{\partial \xi_l}(x,\xi) \ge -\mu_0, \qquad \inf_{\substack{k=1,\dots,s\\(x,\xi)\in\Omega\times\mathbf{R}^s}} \sum_{l=1}^{s}\frac{\partial R_l}{\partial \xi_k}(x,\xi) \ge -\mu_0$$

for some $\mu_0 \ge 0$, and let μ be a number satisfying (198). Then we can choose the time-step τ to be small enough, namely, $\tau \le \frac{1}{\mu_0+\mu}$. In this case, using (200), we obtain

$$\sum_{l=1}^{s}\frac{\partial P_k}{\partial \xi_l}(x,\xi) \ge -\mu_0 + \frac{1}{\tau} \ge -\mu_0 + (\mu_0+\mu) = \mu,$$

and similarly for the other sum in (197).

Under the above conditions, system (196) is a special case of system (171), hence we can apply Corollary 4.4. Here, as mentioned in subsection 5.1, $P_k(x,0)\equiv 0$ on Ω for all k, further, we have homogeneous Dirichlet boundary conditions. Hence the result has the following form:

Corollary 5.3. *Let problem* (196) *satisfy Assumptions* 5.3, *and let its FEM discretization satisfy the corresponding conditions of Theorem* 4.3. *If* $f_k \ge 0$ *and* $\gamma_k \ge 0$ $(k=1,\dots,s)$, *then for sufficiently small h, the FEM solution* $u^h = (u_1^h,\dots,u_s^h)^T$ *of system* (196) *satisfies*

$$u_k^h \ge 0 \quad \text{on } \Omega \qquad (k=1,\dots,s). \tag{201}$$

Acknowledgments

The first author was supported by the Hungarian Research Grant OTKA No.K 67819 and by HAS under the Bolyai János Scholarship. The second author was supported by the projects no. 211512 and no. 124619 from the Academy of Finland.

References

[1] Axelsson, O., *Iterative Solution Methods,* Cambridge University Press, 1994.

[2] Brandts, J., Korotov, S., Křížek, M., Dissection of the path-simplex in $\mathbf{R^n}$ into n path-subsimplices, *Linear Algebra Appl.* 421 (2007), no. 2-3, 382–393.

[3] Brandts, J., Korotov, S., Křížek, M., On the equivalence of regularity criteria for triangular and tetrahedral finite element partitions, *Comput. Math. Appl.* 55 (2008), 2227–2233.

[4] Brandts, J., Korotov, S., Křížek, M., The discrete maximum principle for linear simplicial finite element approximations of a reaction-diffusion problem, *Linear Algebra Appl.* 429 (2008), 2344–2357.

[5] Brandts, J., Korotov, S., Křížek, M., Šolc, J., On nonobtuse simplicial partitions, *SIAM Rev.*, **51(2)** (2009), 317–335.

[6] Burman, E., Ern, A., Nonlinear diffusion and the discrete maximum principle for stabilized Galerkin approximations of the convection-diffusion-reaction equation, *Comput. Methods Appl. Mech. Engrg.* 191 (2002), 3833–3855.

[7] Burman, E., Ern, A., Discrete maximum principle for Galerkin approximations of the Laplace operator on arbitrary meshes, *C. R. Acad. Paris, Ser I* 338 (2004), 641–646.

[8] Burman, E., Ern, A., Stabilized Galerkin approximation of convection-diffusion-reaction equations: Discrete maximum principle and convergence, *Math. Comput.* 74 (2005), 1637-1652.

[9] Caristi, G., Mitidieri, E., Further results on maximum principles for noncooperative elliptic systems, *Nonlinear Anal.* 17 (1991), no. 6, 547–558.

[10] Christie, I., Hall, C., The maximum principle for bilinear elements, *Internat. J. Numer. Methods Engrg.* 20 (1984), 549–553.

[11] Ciarlet, P. G., Discrete maximum principle for finite-difference operators, *Aequationes Math.* 4 (1970), 338–352.

[12] Ciarlet, P. G., Raviart, P.-A., Maximum principle and uniform convergence for the finite element method, *Comput. Methods Appl. Mech. Engrg.* 2 (1973), 17–31.

[13] Codina, R., *A finite element formulation for the numerical solution of the convection-diffusion equation*, Monograph, 14. Centro Internacional de Métodos Numéricos en Ingeniería, Barcelona, 1993.

[14] Donea, J., Huerta, A., *Finite Element Methods for Flow Problems*, John Wiley and Sons, 2003.

[15] Draganescu, A., Dupont, T. F., Scott, L. R., Failure of the discrete maximum principle for an elliptic finite element problem, *Math. Comp.* 74 (2005), no. 249, 1–23.

[16] de Figueiredo, D. G., Mitidieri, E., Maximum principles for cooperative elliptic systems, *C. R. Acad. Sci. Paris Sér. I Math.* 310 (1990), no. 2, 49–52.

[17] Faragó, I., Karátson, J., *Numerical solution of nonlinear elliptic problems via preconditioning operators. Theory and applications.* Advances in Computation, Volume 11, NOVA Science Publishers, New York, 2002.

[18] Gilbarg, D., Trudinger, N. S., *Elliptic partial differential equations of second order* (2nd edition), Grundlehren der Mathematischen Wissenschaften 224, Springer, 1983.

[19] Hannukainen, A., Korotov, S., Vejchodský, T., Discrete maximum principles for FE solutions of the diffusion-reaction problem on prismatic meshes, *J. Comput. Appl. Math.* 226 (2009), 275–287.

[20] Hannukainen, A., Korotov, S., Vejchodský, T., On Weakening Conditions for Discrete Maximum Principles for Linear Finite Element Schemes, in: *Numerical Analysis and Applications*, eds. S. Margenov, L.G. Vulkov and J.Wasniewski, Lecture Notes Comp. Sci. No. 5434, pp. 297-304, Springer, 2009.

[21] Hárs, V., Tóth, J., On the inverse problem of reaction kinetics, In: *Qualitative Theory of Differential Equations* (Szeged, Hungary, 1979), Coll. Math. Soc. János Bolyai 30, ed. M. Farkas, North-Holland - János Bolyai Mathematical Society, Budapest, 1981, pp. 363-379.

[22] Horn, F., Jackson, R., General mass action kinetics, *Arch. Rat. Mech. Anal.* 47 (1972), 81–116.

[23] Ikeda, T., *Maximum principle in finite element models for convection-diffusion phenomena,* Lecture Notes in Numerical and Applied Analysis, 4. North-Holland Mathematics Studies, 76. Kinokuniya Book Store Co., Ltd., Tokyo, 1983.

[24] Ishihara, K., Strong and weak discrete maximum principles for matrices associated with elliptic problems, *Linear Algebra Appl.* 88/89 (1987), 431–448.

[25] Karátson, J., Korotov, S., Discrete maximum principles for finite element solutions of nonlinear elliptic problems with mixed boundary conditions, *Numer. Math.* 99 (2005), 669–698.

[26] Karátson, J., Korotov, S., An algebraic discrete maximum principle in Hilbert space with applications to nonlinear cooperative elliptic systems, *SIAM J. Numer. Anal.* 47 (2009), No. 4., pp. 2518-2549.

[27] Karátson J., Korotov, S., Křížek, M., On discrete maximum principles for nonlinear elliptic problems, *Math. Comput. Simul.* 76 (2007), 99–108.

[28] Kikuchi, F., Discrete maximum principle and artificial viscosity in finite element approximations to convective diffusion equations, Institute of Space and Aeronautical Science, University of Tokyo, Report no. 550, vol. 42, Sept. 1977, 153-166.

[29] Knobloch, P., Numerical solution of convection-diffusion equations using upwinding techniques satisfying the discrete maximum principle, *Proceedings of Czech-Japanese Seminar in Applied Mathematics* 2005, 69–76, COE Lect. Note, 3, Kyushu Univ. The 21 Century COE Program, Fukuoka, 2006.

[30] Korotov, S., Křížek, M., Acute type refinements of tetrahedral partitions of polyhedral domains, *SIAM J. Numer. Anal.* 39 (2001), 724–733.

[31] Korotov, S., Křížek, M., Tetrahedral partitions and their refinements, In: *Proc. Conf. Finite Element Methods: Three-dimensional Problems, Univ. of Jyväskylä, GAKUTO Internat. Ser. Math. Sci. Appl.*, vol. 15, Gakkotosho, Tokyo, 2001, 118–134.

[32] Korotov, S., Křížek, M., Global and local refinement techniques yielding nonobtuse tetrahedral partitions, *Comput. Math. Appl.* 50 (2005), 1105–1113.

[33] Korotov, S., Křížek, M., Neittaanmäki, P., Weakened acute type condition for tetrahedral triangulations and the discrete maximum principle, *Math. Comp.* 70 (2001), 107–119.

[34] Korotov, S., Křížek, M., Kropáč, A., Strong regularity of a family of face-to-face partitions generated by the longest-edge bisection algorithm, *Comp. Math. Math. Phys.* 9 (2008), 1687–1698.

[35] Křížek, M., Lin Qun, On diagonal dominance of stiffness matrices in 3D, *East-West J. Numer. Math.* 3 (1995), 59–69.

[36] Křížek, M., Neittaanmäki, P., *Mathematical and numerical modelling in electrical engineering: theory and applications*, Kluwer Academic Publishers, 1996.

[37] Kuzmin, D., On the design of algebraic flux correction schemes for quadratic finite elements, *J. Comput. Appl. Math.* 218 (2008), no. 1, 79–87.

[38] Kuzmin, D., Shashkov, M.J., Svyatskiy, D., A constrained finite element method satisfying the discrete maximum principle for anisotropic diffusion problems, *J. Comput. Phys.* 228 (2009), 3448–3463.

[39] Mizukami, A., Hughes, T. J. R., A Petrov-Galerkin finite element method for convection-dominated flows: an accurate upwinding technique for satisfying the maximum principle, *Comput. Methods Appl. Mech. Engrg.* 50 (1985), no. 2, 181–193.

[40] Ladyzhenskaya, O. A., Ural'tseva, N. N., *Linear and quasilinear elliptic equations*, Leon Ehrenpreis Academic Press, New York-London, 1968.

[41] Liska, R., Shashkov, M., Enforcing the discrete maximum principle for linear finite element solutions of second order elliptic problems, *Commun. Comput. Phys.* 3 (2008), no. 4, 852–877.

[42] López-Gómez, J., Molina-Meyer, M., The maximum principle for cooperative weakly coupled elliptic systems and some applications, *Diff. Int. Equations* 7 (1994), no. 2, 383–398.

[43] Mitidieri, E., Sweers, G., Weakly coupled elliptic systems and positivity, *Math. Nachr.* 173 (1995), 259–286.

[44] Ohmori, K., The discrete maximum principle for nonconforming finite element approximations to stationary convective diffusion equations, *Math. Rep. Toyama Univ.* 2 (1979), 33–52.

[45] Ohmori, K., Correction to: "The discrete maximum principle for nonconforming finite element approximations to stationary convective diffusion equations" [Math. Rep. Toyama Univ. 2 (1979), 33–52], *Math. Rep. Toyama Univ.* 4 (1981), 179–182.

[46] Protter, M. H., Weinberger, H. F., *Maximum principles in differential equations*, Springer-Verlag, New York, 1984.

[47] Ruas Santos, V., On the strong maximum principle for some piecewise linear finite element approximate problems of non-positive type, *J. Fac. Sci. Univ. Tokyo Sect. IA Math.* 29 (1982), 473–491.

[48] Stynes, M., Steady-state convection-diffusion problems, *Acta Numer.* 14 (2005), 445–508.

[49] Tóth, J., Gradient systems are cross-catalytic, *Reaction Kinetics and Catalysis Letters* 12 (3) (1979), 253–257.

[50] Vabishchevich, P. N., Samarskii, A. A., Monotone difference schemes for convection-diffusion problems on triangular grids, , *Comput. Math. Math. Phys.* 42 (2002), no. 9, 1317–1330

[51] Vejchodský, T., Šolín, P., Discrete maximum principle for higher-order finite elements in 1D, *Math. Comp.* 76 (2007), no. 260, 1833–1846.

[52] Varga, R., *Matrix iterative analysis*, Prentice Hall, New Jersey, 1962.

[53] Xu, J., Zikatanov, L., A monotone finite element scheme for convection-diffusion equations, *Math. Comp.* 68 (1999), 1429–1446.

In: Computational Mathematics
Editor: Peter G. Chareton, pp. 261-278

ISBN 978-1-60876-271-2

Chapter 9

NUMERICAL CONFORMAL MAPPINGS FOR WAVEGUIDES

Anders Andersson
International Centre for Mathematical Modelling, Växjö University
Växjö, Sweden and School of Engineering, Jönköping University
Jönköping, Sweden

Abstract

Acoustic or electro-magnetic scattering in a waveguide with varying direction and cross-section can, if the variations takes place in only one dimension at a time be reformulated as a two-dimensional scattering problem. By using the so-called Building Block Method, it is possible to construct the scattering properties of a combination of scatterers when the properties of each scatterer are known. Hence, variations in the waveguide geometry or in the boundary conditions can be treated one at a time. We consider in this work acoustic scattering, but the same techniques can be used for both electro-magnetic and some quantum scattering problems.

By suppressing the time dependence and by using the Building Block Method, the problem takes the form of the Helmholtz equation in a waveguide of infinite length and with smoothly varying geometry and boundary conditions. A conformal mapping is used to transform the problem into a corresponding problem in a straight horizontal channel, and by expanding the field in Fourier trigonometric series, the problem can be reformulated as an infinite-dimensional ordinary differential equation. From this, numerically solvable differential equations for the reflection and transmission operators are derived.

To be applicable in the Building Block Method, the numerical conformal mapping must be constructed such that the direction of the boundary curve can be controlled. At the channel ends, it is an indispensable requirement, that the two boundary curves are (at least) asymptotically parallel and straight. Furthermore, to achieve bounded operators in the differential equations, the boundary curves must satisfy different regularity conditions, depending on the properties of the boundary.

Several methods to accomplish such conformal mappings are presented. The Schwarz–Christoffel mapping, which is a natural starting point and for which also efficient numerical software exists, can be modified in different ways to round the polygon corners, and we show algorithms by which the parameter problem can be solved after such modifications. It is also possible to use the unmodified Schwarz–Christoffel mapping for regions with smooth boundary, by constructing an appropriate outer polygon to the considered region. Finally, we show how a so-called zipper algorithm can be used for waveguides.

1 Introduction

Conformal mappings are important tools in mathematics, physics and engineering. For boundary value problems, they offer simple variable substitutions, which means that many of the standard differential equations in mathematical physics are just mildly transformed, while the geometry in which the problem is stated is simplified significantly. The most striking example is the Laplace operator $\nabla_{xy}^2 = \partial^2/\partial x^2 + \partial^2/\partial y^2$, which by any conformal mapping $z = x + \mathrm{i}y \rightarrow w = u + \mathrm{i}v$ is transformed to a new Laplace operator multiplied by a scale factor, giving $\nabla_{xy}^2 = |dw/dz|^2 \nabla_{uv}^2$.

This paper is a summary of a series of articles and conference reports, all focusing on the same problem: to find numerically evaluable conformal mappings for waveguides with smooth boundaries. When consulting the extensive literature in the area, for example [13], [15], [24] or [26], there exist already a wealth of methods, Fourier series, integral equations, polynomial approximations just to mention a few, that accomplish this smoothness criterion.

However, as we shall see, the problems we are solving require also a good control of the boundary curve direction. The waveguides to which we apply our conformal mappings are infinitely long. Outside some bounded region, they are straight and have constant width . Any approximate conformal mapping must be able to preserve these conditions. It has therefore been natural to consider variants of the Schwarz–Christoffel mapping, which maps the upper half–plane to a polygonal region.

In this work, we present different ways to round the polygon corners in a Schwarz–Christoffel mapping. In an additional variant, we use a Schwarz–Christoffel mapping for an outer polygon to the considered region. Finally, we show how one of the so-called zipper algorithms, a conformal mapping technique that is not Schwarz–Christoffel-related, can be used for waveguides.

To give a picture of the context in which the conformal mappings are used, we give in Section 2 a rather detailed description of the wave scattering problems we are solving. In Section 3, the conformal mapping methods are presented, and in Section 4, a few concluding remarks are given.

2 Wave Scattering in Two-Dimensional Waveguides

2.1 The original problem

Waveguide scattering problems appear in similar settings in both acoustics, electro-magnetics and in some quantum applications. The waveguides are of course three-dimensional, but, if variations in geometry and boundary conditions take place in only one dimension at a time, it is often sufficient to solve a two-dimensional problem. We will here consider acoustical scattering in a two-dimensional waveguide, but similar methods can be used in electro-magnetics, see for example [7], or for scattering in quantum tubes, see [21].

The acoustic pressure $\tilde{p}$ at a time t and a point (x,y) is ruled by the wave equation

$$\nabla^2 \tilde{p}(x,y,t) = \frac{1}{c^2}\frac{\partial^2}{\partial t^2}\tilde{p}(x,y,t), \tag{1}$$

where c is the speed of sound in the medium. Assuming the time dependence $\mathrm{e}^{-\mathrm{i}\omega t}$ and letting $k = \omega/c$, Helmholtz equation

$$\nabla^2 p(x,y) + k^2 p(x,y) = 0 \tag{2}$$

follows from (1).

2.2 The Building Block Method

A waveguide with varying geometry and boundary properties can be seen as a combination of scatterers. The scattering properties of the waveguide can be constructed from known scattering properties of each scatterer by the so-called Building Block Method or Cascade Technique, see [23]. In this method, it is assumed that the wave field can be described as a sum of infinitely countable basis functions, that the scatterers are separated in space, and that the values of basis functions at two different points are related in a way that allows calculations of them at one point from known values at the other.

When applying this method, the waveguide is divided into a finite number of blocks, assuming that any two adjacent blocks, which each can have variations in geometry and/or boundary conditions, are joined by a straight part with constant boundary conditions and cross-section. Since this unvarying section can be infinitely short, this is not a real restriction. Each block is then dealt with separately. To avoid interaction from reflections at the waveguide ends when a single block is treated, the block is assumed to be of infinite length with asymptotically constant boundary conditions and cross-sections towards infinity at both ends.

2.3 Transforming the problem

When using the Building Block Method, it is therefore sufficient to consider the Helmholtz equation in a two-dimensional channel of infinite length that contains one or just a few simple variations, for example a bend, a change in width or a change in the boundary condition. The Building Block Method allows also the assumption that all such variations are smooth. To illustrate the method, we will also assume a Neumann boundary condition at one of the boundaries and a smoothly varying impedance condition at the other. (For a more general treatment that covers all symmetric waveguides, the case with a Dirichlet boundary condition at one boundary must also be solved, something that is done similarly to what is shown here.)

The above assumptions complete the Helmholtz equation (2) and lead to the boundary value problem

$$\begin{cases} \nabla^2 p(x,y) + k^2 p(x,y) = 0, & \\ \partial p/\partial n = 0 & \text{on the lower boundary,} \\ p = -\mathrm{i}Z_0 \partial p/\partial n & \text{on the upper boundary,} \end{cases} \tag{3}$$

where the acoustic impedance Z_0 is constant or smoothly varying on the upper boundary. In accordance with the Building Block Method, no variations in geometry or boundary condition are assumed to take place outside some bounded region in the xy-plane.

To transform this problem into a corresponding problem in a straight horizontal channel with constant cross-section 1 in the uv-plane, we use a conformal mapping $F : x+y\mathrm{i} \leftarrow u+v\mathrm{i}$, by which (3) becomes

$$\begin{cases} \nabla^2\Phi(u,v)+k^2\mu(u,v)\Phi(u,v)=0, \\ \dfrac{\partial\Phi}{\partial v}(u,0)=0, \\ \Phi(u,1)=-\mathrm{i}Z(u)\dfrac{\partial\Phi}{\partial v}(u,1). \end{cases} \tag{4}$$

Here, $\Phi(u,v)=p(x(u,v),y(u,v))$, $\mu(u,v)=|F'(u+\mathrm{i}v)|^2$ and $Z(u)=Z_0(u+\mathrm{i})/|F'(u+\mathrm{i})|$.

The Neumann condition on the lower boundary suggests Fourier cosine series over v as a functional basis for Φ, and we therefore set

$$\Phi(u,v)=\sum_{n=1}^{\infty}\Phi_n(u)\cos(v\lambda_n(u)). \tag{5}$$

The functions λ_n are determined by the boundary condition on the upper boundary from which it follows that

$$\cot(\lambda_n(u))=\mathrm{i}Z(u)\lambda_n(u). \tag{6}$$

By differentiating (5) and by setting

$$v\sin(v\lambda_n(u))=\sum_{m=1}^{\infty}\alpha_{mn}\cos(v\lambda_m(u)), \tag{7}$$

$$v^2\cos(v\lambda_n(u))=\sum_{m=1}^{\infty}\beta_{mn}\cos(v\lambda_m(u)) \tag{8}$$

and

$$\mu(u,v)\cos(v\lambda_n(u))=\sum_{m=1}^{\infty}\mu_{mn}(u)\cos(v\lambda_m(u)), \tag{9}$$

(4) can be expressed as an infinite-dimensional second-order ordinary differential equation

$$\vec{\Phi}''(u)-A(u)\vec{\Phi}'(u)-B^2(u)\vec{\Phi}(u)=0, \tag{10}$$

with an infinite vector $\vec{\Phi}=(\Phi_1\Phi_2,\ldots)^T$ and infinite matrices A and B^2 having elements

$$A_{mn}(u)=2\alpha_{mn}(u)\lambda_n'(u) \tag{11}$$

and

$$B^2_{mn}(u)=\alpha_{mn}(u)\lambda_n''(u)+\beta_{mn}(u)(\lambda_n'(u))^2-k^2\mu_{mn}(u)+\delta_{mn}(\lambda_n(u))^2. \tag{12}$$

2.4 Solving the resulting problem

Unfortunately, equation (10) is not numerically stable and it is therefore not possible to use a numerical solver on (10) to determine $\vec{\Phi}$ for all $u \in \mathbb{R}$, given its state at some $u = u_0$. This is due to the fact that Φ contains waves propagating in both directions, some of them decaying but others growing exponentially for an increasing u.

The idea behind the so-called RT method is to treat the waveguide piece, i.e., each building block in the Building Block Method, as a black box for which only reflection and transmission operators are determined. It turns out that numerically stable equations can be formulated for these scattering operators, even if the field inside the waveguide is not known.

Since we assume no variations in geometry or in the boundary conditions outside some bounded region, there exist real numbers u_1 and u_2 such that $Z(u)$ and $\mu(u,v)$ for each v are constant outside the interval $[u_1,u_2]$. Hence, $A=0$ and B constant outside this interval, and we set $B(u)=B_-$ for $u<u_1$ and $B(u)=B_+$ for $u>u_2$.

Let $\vec{\Phi}=\vec{\Phi}^+ + \vec{\Phi}^-$, a decomposition that can be well defined outside $[u_1,u_2]$, where $\vec{\Phi}^+$ and $\vec{\Phi}^-$ are waves propagating to the right and left respectively. Now, scattering operators R^+, R^-, T^+ and T^- can be defined such that

$$\begin{pmatrix}\vec{\Phi}^+(u_2)\\ \vec{\Phi}^-(u)\end{pmatrix} = \begin{pmatrix}T^+(u,u_2) & R^-(u,u_2)\\ R^+(u,u_2) & T^-(u,u_2)\end{pmatrix}\begin{pmatrix}\vec{\Phi}^+(u)\\ \vec{\Phi}^-(u_2)\end{pmatrix}. \tag{13}$$

Our aim is to determine $R^+(u_1,u_2)$, $R^-(u_1,u_2)$, $T^+(u_1,u_2)$ and $T^-(u_1,u_2)$. We can assume the existence of an operator C, varying with u, such that

$$\frac{\partial}{\partial u}\vec{\Phi}(u) = -C(u)\vec{\Phi}^+(u) + C(u)\vec{\Phi}^-(u). \tag{14}$$

From (10) and (14), the differential equation

$$\frac{\partial}{\partial u}\begin{pmatrix}\vec{\Phi}^+\\ \vec{\Phi}^-\end{pmatrix} = \begin{pmatrix}J & K\\ L & M\end{pmatrix}\begin{pmatrix}\vec{\Phi}^+\\ \vec{\Phi}^-\end{pmatrix}, \tag{15}$$

where

$$\begin{aligned} J &= (2C)^{-1}\left(-C' - B^2 + (A-C)C\right)\\ K &= (2C)^{-1}\left(C' - B^2 - (A-C)C\right)\\ L &= (2C)^{-1}\left(C' + B^2 - (A+C)C\right)\\ M &= (2C)^{-1}\left(-C' + B^2 + (A+C)C\right) \end{aligned} \tag{16}$$

can be derived. We have some freedom in the choice of C inside $[u_1,u_2]$, but it must be equal to B outside the interval. To achieve a C that varies smoothly over u, we can set $C(u) = B_- + f(u)(B_+ - B_-)$, where f is a regular real function such that $f(u)=0$ for $u<u_1$ and $f(u)=1$ for $u>u_2$.

It is now possible to deduce numerically stable Ricatti equations for the reflection and transmission operators, for details see [22] or [7]. The equations for R^+ and T^+ are

$$\frac{\partial R^+}{\partial u}(u,u_2) = L(u) + M(u)R^+(u,u_2) - R^+(u,u_2)J(u) - R^+(u,u_2)K(u)R^+(u,u_2) \tag{17}$$

and

$$\frac{\partial T^+}{\partial u}(u,u_2) = -T^+(u,u_2)J(u) - T^+(u,u_2)K(u)R^+(u,u_2), \tag{18}$$

and similar equations for R^- and T^- can also be stated.

The equations (17) and (18) can be solved numerically with an ODE solver if we treat them as matrix equations with truncated matrices in place of the operators A, B, C, J, K, L and M. To get a good accuracy in the uppermost leftmost $N_0 \times N_0$ parts of the R and T matrices, the matrices A, B, C, etc. must be of size $N \times N$ where N is considerably larger than N_0. It can be shown, see [22], that the $N_0 \times N_0$ parts of R and T converge to the correct values when $N \longrightarrow \infty$. However, for low frequencies it is often sufficient with a relatively small N. For example, in channels with slowly decreasing cross-sections like the ones given as examples in [7], results with good accuracy for $N_0 = 5$ are achieved when $N = 10$.

An important alternative to the RT method is given by introducing so called Dirichlet to Neumann operators, see for example [11, 12, 18, 19].

Let u_1 and u_2 be defined as above. Since $A = 0$ for $u < u_1$ and $u > u_2$, we note following [12], that

$$\begin{cases} \left(\frac{\partial}{\partial u} \pm B_-\right)\Phi^\pm(u) = 0, & u < u_1, \\ \left(\frac{\partial}{\partial u} \mp B_+\right)\Phi^\pm(u) = 0, & u > u_2. \end{cases} \tag{19}$$

If we assume no sources to the right, then $\Phi^-(u_2) = 0$ and

$$\frac{\partial \Phi}{\partial u}(u_2) = \frac{\partial \Phi^+}{\partial u}(u_2). \tag{20}$$

A Dirichlet-to-Neumann operator $\Lambda^+(u,u_2)$ can then be defined, such that in the transition region $\{u : u_1 \le u \le u_2\}$,

$$\left(\frac{\partial}{\partial u} + \Lambda^+(u,u_2)\right)\Phi(u) = 0. \tag{21}$$

It follows from (10) that $\Lambda^+(u,u_2)$ must satisfy the operator equation

$$\frac{\partial}{\partial u}\Lambda^+(u,u_2) = \left(\Lambda^+(u,u_2) + A(u)\right)\Lambda^+(u,u_2) - B^2(u), \tag{22}$$

and considering (10) for $u > u_2$ yields the initial condition $\Lambda^+(u_2,u_2) = B_+$. Seen as a matrix equation, (22) can be solved with an ODE solver if truncated matrices in place of A and B are used, after which it is possible to determine the first components of $\vec{\Phi}$ in the waveguide.

For a countable number of so-called trapped modes, singularities can appear in both the RT and the DtN method. Since this happens in different ways in the two methods, they complement each other well. We have not so far made any exhaustive studies of this phenomenon in our problem setting, but the techniques developed in [18] seem to be possible to use.

3 Conformal Mapping Methods

The Building Block Method allows us to assume that the two-dimensional channels in which our wave scattering problem takes place are bounded by smooth curves, and it is therefore (at least in theory) possible to construct a conformal mapping between the channel and a straight horizontal channel, such that the matrices A and B in (10) are bounded. This is accomplished by any conformal mapping having no singularities or zeros on the boundary.

However, as mentioned in the introduction, there is a second requirement, originating from the Building Block Method, that must be fulfilled. To avoid influences from the ends of each block, it is important that the block is an infinite channel in which the walls are parallel, at least asymptotically, when approaching infinity. The methods in this section are all constructed with the purpose to satisfy this criterion and at the same time preserve any smoothness requirements in the problem.

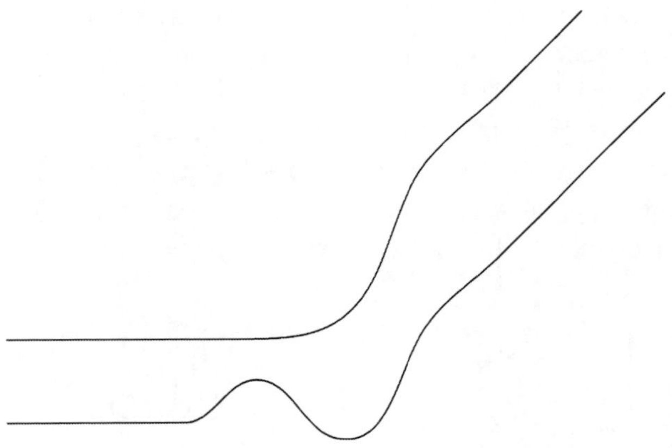

Figure 1: A channel with parallel walls at both ends

If the upper boundary condition in (3) is replaced with a Dirichlet boundary condition, the functions λ_n are constants, and hence in (10), A is zero and B is bounded given that μ is bounded on the boundary. It is therefore sufficient that the conformal mapping has a C^1 regularity on the boundary, something that is brought about by the methods described in Sections 3.1 and 3.4. For the general case, a C^3 regularity is needed. This is effectuated by the methods in Sections 3.2 and 3.3, which both produce C^∞ smooth boundary curves.

All the described methods construct approximate mappings from the upper half–plane to the channel under consideration. The function

$$w \longrightarrow e^{i\pi w} \tag{23}$$

maps the channel $\{w \in \mathbb{C} : 0 \leq \operatorname{Im} w \leq 1\}$ conformally on the upper half–plane, and it is used in combination with the Schwarz–Christoffel variants in Sections 3.1, 3.2 and 3.3. When using the geodesic algorithm described in Section 3.4, additional translations and scalings are needed.

3.1 Modified Schwarz–Christoffel mappings for polygons with rounded corners

Considering the parallel walls towards infinity, one's thoughts go naturally to the Schwarz–Christoffel mapping

$$f(w) = A\int_{w_0}^{w}\prod_{n=1}^{k}(\omega - w_k)^{\alpha_k - 1}d\omega + B, \tag{24}$$

which maps the upper half–plane or the unit circle to a polygonal region. The so-called prevertices $w_1,\ldots,w_n$ are mapped to polygon corners with inner angles $\alpha_1\pi,\ldots,\alpha_n\pi$ respectively. By this mapping, the direction of the boundary curve in the image region can be controlled perfectly, even if the polygon vertices are just approximately determined.

A few decades after the discovery of the Schwarz–Christoffel mapping, variants of the mapping where published in which one or several of the factors in (24) were replaced in order to achieve a "rounded corner". To accomplish this, $(\omega - w_k)^{\alpha_k-1}$ must be replaced with a factor having an argument that varies continuously over $\mathbb{R}$. The argument must also be equal to $(\alpha_k - 1)\pi$ to the left and 0 to the right of some interval $[a,b] \ni w_k$. A list of possible such "curve factors" was published already in 1915 by Leathem, see [17]. Other examples are found in [8] and [16].

One interesting example of a factor that rounds the corners in a Schwarz–Christoffel mapping is given by Henrici in [14]. It is used in [2], where algorithms for the use of curve factors in numerical implementations for arbitrary regions bounded by piecewise smooth curves are developed. If the factor $s_k(\omega) = (\omega - w_k)^{\alpha_k-1}$ in (24) is replaced with

$$h_k(\omega) = \begin{cases} \dfrac{1}{2}\left((\omega - w_k + \varepsilon_k)^{\alpha_k-1} + (\omega - w_k - \varepsilon_k)^{\alpha_k-1}\right), & \alpha_k > 1 \\ \dfrac{1}{2\alpha_k\varepsilon_k}\left((\omega - w_k + \varepsilon_k)^{\alpha_k} - (\omega - w_k - \varepsilon_k)^{\alpha_k}\right), & \alpha_k < 1 \end{cases}, \tag{25}$$

the corresponding vertex in the resulting region is rounded.

Since $\arg h_k(\omega) = \arg s_k(\omega)$ for $\omega \in \mathbb{R}\setminus[w_k - \varepsilon_k, w_k + \varepsilon_k]$, the direction of the boundary curve in the rounded polygon is the same as in the polygon produced by the corresponding Schwarz–Christoffel mapping. However, the replacement influences the length of the polygon sides, even far from the rounded corner. Therefore, the parameter problem must be solved again, i.e., the prevertices $w_1,\ldots,w_n$ and the multiplicative constant A must be re-determined.

In numerical implementations of the Schwarz–Christoffel mapping, see for example [25, 10, 9], this determination is done by placing the polygon vertices in correct positions or by fixing the distances between them. When the corners are rounded, there are no fixed positions for the vertices. In [2], tangent polygons are used instead.

On the boundary Γ of the model region Ω, a number of tangent points are placed in such a way that the tangents to Γ at the tangent points form a tangent polygon P_Γ, see Figure 2. This means that any point of inflexion must be included in the set of tangent point, and that the distance between two consequent tangent points must be small enough to allow the corresponding tangents to have a finite intersection on the correct side of Ω. A Schwarz–Christoffel mapping is constructed for P_Γ wherein the factors are replaced to round the corners. The resulting function maps the upper half–plane to a region bounded

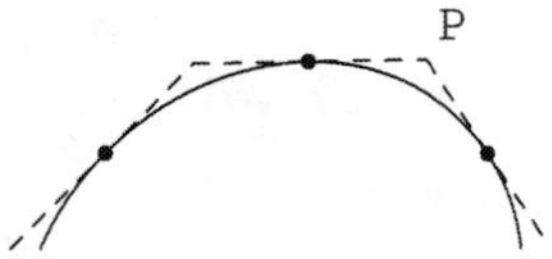

Figure 2: A part of a region with tangent points on the boundary and the corresponding tangent polygon.

by a curve C. Tangent points are put on the straight line parts of C, and a tangent polygon P_C is constructed. By comparing the side–lengths in the polygons P_Γ and P_C, equations are stated by which the parameters can be determined.

Note that the angular directions at corresponding tangent points on C and Γ are the same. It is therefore possible to construct an approximate conformal mapping in which the angular direction of the boundary curve is determined exactly wherever this is needed. Examples include channels with parallel straight line walls towards infinity and unrounded vertices with predefined inner angles in regions with piecewise smooth boundary.

The boundary of the resulting region has a regularity of order C^1; its angular direction varies continuously, but the second and third derivative of the mapping function has singularities in the points $w_k \pm \varepsilon_k$.

A conformal mapping, constructed with this method is shown in Figure 3.

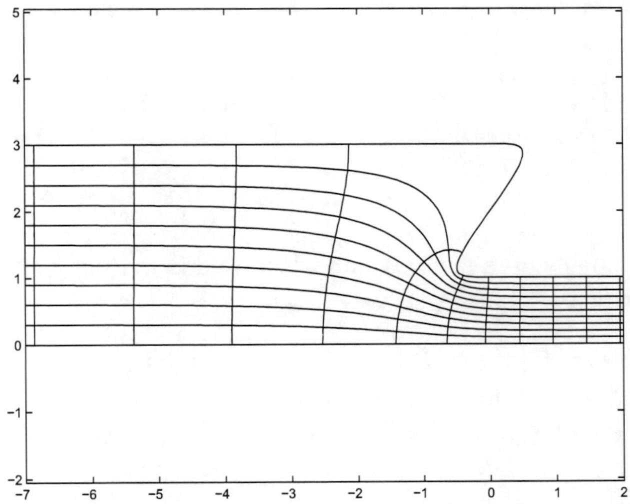

Figure 3: Conformal mapping for a channel using rounded corners with the technique described in Section 3.1. In the construction of tangent polygons, 25 tangent points are used on the upper boundary.

3.2 Approximate curve factors

If the requisite on the boundary curve direction is alleviated from constant pre-determined values outside some circle to asymptotically reached values towards infinity, this opens up for new variants of the Schwarz–Christoffel mapping. The factors $(\omega - w_k)^{\alpha_k - 1}$ can be replaced with factors with arguments that are just asymptotically equal to $(\alpha_k - 1)\pi$ and 0 when a real ω approaches negative and positive infinity, respectively.

One example is constructed from the function

$$z = \sqrt{w^2 - c^2}, \quad c > 0, \tag{26}$$

which, if the square root branch is chosen such that

$$\arg z = (\arg(w+c) + \arg(w-c))/2, \tag{27}$$

maps the upper half–plane on itself minus a slit on the imaginary axis of height c. Adding a small imaginary amount $b\mathrm{i}$ to w and removing the same amount from z, will result in a function that maps the real axis on a smooth curve in the upper half–plane that asymptotically coincides with the real axis for large negative or positive w values. The interior of the upper half–plane is mapped conformally on the region above this curve.

Hence, replacing $(\omega - w_k)^{\alpha_k - 1}$ in a Schwarz–Christoffel mapping for a polygon P with

$$\left(\sqrt{(\omega + b\mathrm{i} - w_k)^2 - c^2} - b\mathrm{i}\right)^{\alpha_k - 1} \tag{28}$$

means that the kth vertex in P is replaced with a C^∞ smooth rounded corner. Similarly, approximate curve factors giving C^∞ smoothness can be constructed from existing curve factors of the type described in Section 3.1 which give C^1 boundary curves. A major result is the following:

> Assume that C is a curve factor which if it is used in a Schwarz–Christoffel-like mapping rounds a corner with a C^m boundary curve, $m \geq 1$. Then,
>
> $$C(w + b\mathrm{i}) - b\mathrm{i} \tag{29}$$
>
> is an approximate curve factor which when used in the same mapping rounds the corner with a C^∞ smooth curve.

An example is shown in Figure 4. The picture looks similar to Figure 3, and it certainly is, but there is one important difference. The factors of type (25) in the conformal mapping in Figure 3 are all modified using (29) with $b = 10^{-5}$, and the boundary in Figure 4 is therefore smooth.

Naturally, the replacement of a Schwarz–Christoffel factor with an approximate curve factor yields effects on the resulting "polygon" beyond the rounded corner. There are no more straight lines or no more exact pre-determined angles at the remaining vertices. However, if the difference between the curve factor and the corresponding Schwarz–Christoffel factor tends to zero rapidly when $\omega \to \infty$, as is the case with the factor in (28) if sufficiently small b and c are used, these effects can be small enough to be negligible. In channels or other unbounded regions, walls towards infinity are asymptotically straight. Furthermore, the methods for determination of parameters described in section 3.1 can with small changes be applied, even when approximate curve factors are used to round the corners.

A theoretical framework which includes many types of curve factors, and several examples is given in [4].

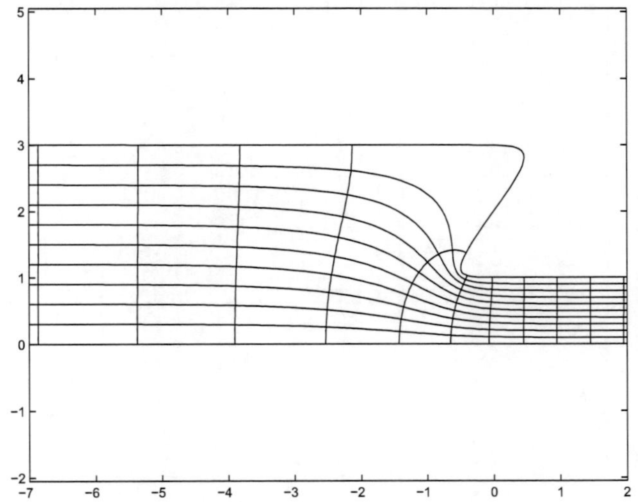

Figure 4: A conformal mapping with approximate curve factors constructed from the factors in the mapping in Figure 3, using (29).

A Schwarz–Christoffel-like mapping from the upper half–plane to a channel with parallel straight walls at both ends has the form

$$w \to A \int_{w_0}^{w} \frac{\prod_k C_k(\omega)}{\omega - a}\, d\omega. \tag{30}$$

It maps the real number a to infinity at one of the channel ends, while infinity is mapped to the other infinite channel end. The channel width at both ends is easily calculated, see [4], which means that the multiplicative constant A and one of the factors C_k can be determined analytically, while the constant a can be set in advance.

An example is given in Figure 5, where factors of type (28) is used at the two vertices on the upper boundary. To avoid influences from the approximate curve factors on the lower boundary, the constructed function maps the upper half–plane on a region with four rounded corners that is symmetric about the real axis. What we see in Figure 5 is the image of the second quadrant.

To achieve similar curvature about the convex and concave vertex in Figure 5, the parameters b and c in (28) must be substantially larger in the factor corresponding to the concave vertex. However, to large parameters in that factor may yield notable effects to the left of the convex vertex, something that restricts how much a single vertex can be rounded. If a suitable mapping is impossible to construct with few factors, there is always the alternative to use additional factors, i.e., to use a polygon with more vertices.

3.3 The Outer Polygon Method

Similar results as in Section 3.2 are achieved, using a simple idea that we call "The Outer Polygon method", see [3]. In this method, an unmodified Schwarz–Christoffel mapping is used, and it can therefore take advantage of existing efficient Schwarz–Christoffel software, for example the Schwarz–Christoffel toolbox for MATLAB [9].

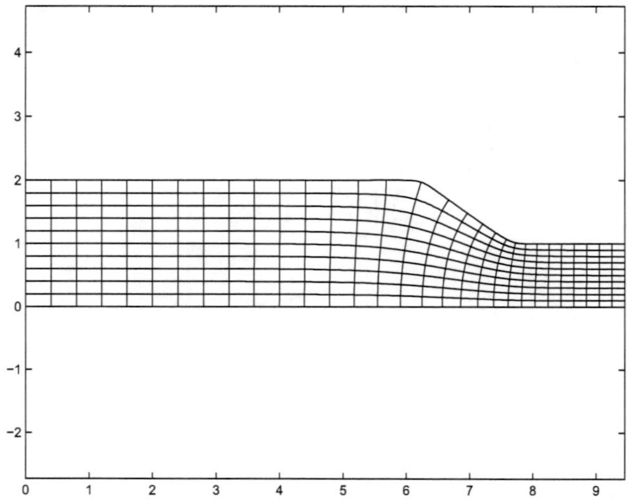

Figure 5: A conformal mapping for a channel using approximate curve factors of type (28) at the two rounded vertices on the upper boundary.

Let f be a Schwarz–Christoffel mapping from the unit disk to a polygon P, and let for $\varepsilon \in]0,1[$ $D_\varepsilon = \{w \in \mathbb{C} : |w| \leq 1-\varepsilon\}$. Then, $f(D_\varepsilon)$ is a region, somewhat similar to P if ε is small, but bounded by a smooth curve.

Assume now that Ω is a region bounded by a smooth curve for which a conformal mapping shall be constructed, and let ε be a small positive number. The aim of the Outer Polygon Method is to construct a polygon $P_\Omega^\circ \supset \Omega$ such that the Schwarz–Christoffel mapping f from the unit disk to P_Ω° maps D_ε approximately on Ω.

The vertices of the outer polygon P_Ω° can be determined by methods similar to those described in Section 3.1. Tangent polygons to Ω and $f(D_\varepsilon)$ are compared, giving equations for determination of the outer polygon vertices.

Many unbounded regions can be treated in a similar way. For a region with one infinite boundary point, it is natural to use a Schwarz–Christoffel mapping from the upper half–plane and $D_\varepsilon = \{w \in \mathbb{C}; \mathrm{Im}(w) \geq \varepsilon\}$. A channel with parallel walls at both ends, the example that has been the motivation for all the methods described here, can be handled using a mapping from a channel $\{w \in \mathbb{C} : 0 \leq \mathrm{Im}(w) \leq 1\}$ to an outer polygon, using $D_\varepsilon = \{w \in \mathbb{C} : \varepsilon_1 \leq \mathrm{Im}(w) \leq 1-\varepsilon_2\}$ or equivalently, a Schwarz–Christoffel mapping from the upper half–plane where D_ε is the region above two rays from the point on the real axis that is mapped to one of the infinite channel ends.

It is easy to prove that a required accuracy can be achieved just by using a tangent polygon to Ω as an outer polygon if the tangent polygon has sufficiently many sides and ε is sufficiently small. However, it is also proved in [3] that an outer polygon with fewer sides can always be found if ε is small.

A very small ε means a polygon-like result with large variations in curvature and in the derivatives of the mapping function on the boundary. This can have crucial relevance in an application like the one described in Section 2.1. A further investigation on how the size of ε affects the curvature as well as limits for ε for a bounded curvature is given in [5].

A simple example is shown in Figure 6. Given is a channel that makes a 135° bend and

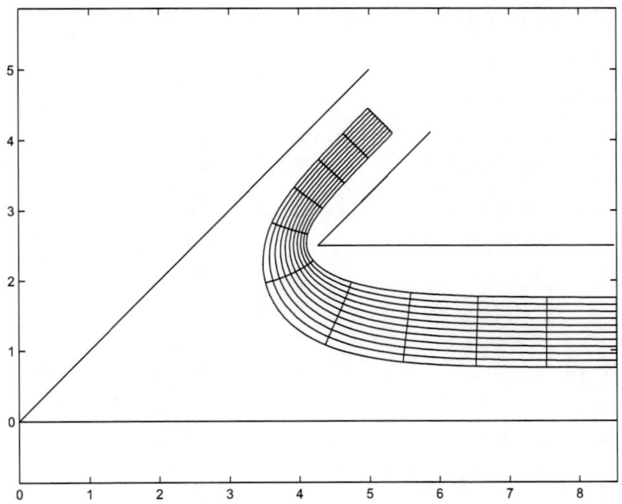

Figure 6: A conformal mapping of a bending channel with increasing width. The picture shows also the outer polygon used in the mapping.

at the same time widens from 1 to 2. The Schwarz–Christoffel mapping from the upper half–plane to the outer polygon in the figure is

$$s(w) = A \int_1^w \frac{(\omega - 1)^{3/4}}{(\omega - a)(\omega + 1)^{3/4}} \, d\omega, \tag{31}$$

where the constants $A = 5/(2\pi)$ and $a = -(2^{4/3} - 1)/(2^{4/3} + 1)$ are chosen such that the image of the region above the rays from a with directions $3\pi/10$ and $7\pi/10$ has the prescribed widths at both ends.

Other examples are given in [3].

3.4 Using the geodesic algorithm for channels

In [20], Marshall and Rohde present three algorithms for numerical conformal mappings. Their common idea is that the boundary of region Ω is cut in a large number of small pieces, which one by one are mapped to the real axis, until finally the whole boundary is mapped to the real axis and Ω to the upper half–plane. These mappings are all reversible and together, they constitute a conformal mapping between Ω and the upper half–plane.

One of the algorithms, the so-called geodesic algorithm, approximates the boundary of Ω with a chain of circular arcs, connected in such a way that the approximated boundary curve gets C^1 regularity. The algorithm is competitive in both accuracy and efficiency at least when Ω is bounded and has a boundary, simple enough to be approximated well without letting the number of vertices grow uncontrolled.

Considering the waveguide problems that motivate the conformal mappings in this work, the natural question is: Is it possible to use the geodesic algorithm to construct a mapping for a two-dimensional channel with parallel walls at both ends?

The answer is affirmative, but a sequence of mappings is required before the geodesic algorithm can be applied. Assume that Ω is an infinite channel bounded below by a straight line and above by a smooth curve which is parallel to the lower boundary at both ends. First, the inverse of a simple Schwarz–Christoffel or a Schwarz–Christoffel-like mapping s for a region exterior to Ω is used to transform the channel to a region $s^{-1}(\Omega)$, bounded below by a curve that coincides with the real axis, except at an interval $[a,b]$ where it has an indentation in the upper half–plane. Note that the Schwarz–Christoffel-like mapping can be chosen such that the boundary of $s^{-1}(\Omega)$ is (C^1) smooth everywhere, even at a and b. The mapping

$$J(w) = \frac{w^2 - ab}{2w - a - b} \tag{32}$$

maps the upper half–plane minus a semi-circle between a and b on the upper half–plane, and with an appropriate choice of square root branch, $J^{-1} \circ s^{-1}(\Omega)$ is a region in the upper half–plane, again bounded by the real axis and a smooth curve γ between a and b, but with γ perpendicular to the real axis at a and b.

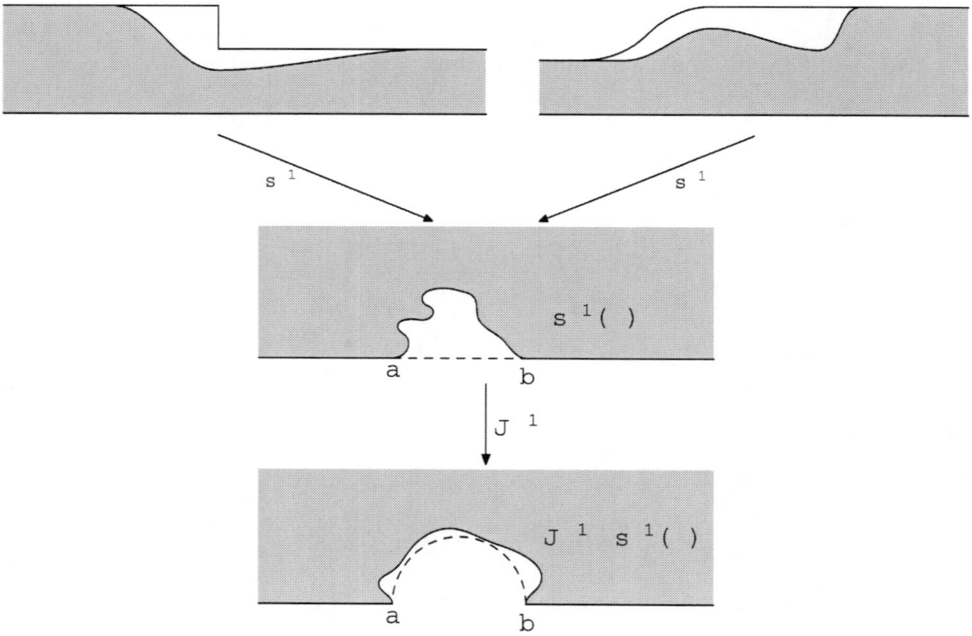

Figure 7: Mapping a channel to a region on which the geodesic algorithm can be applied.

This is a situation in which Marshall's geodesic algorithm can be applied. Translate the region so that $a = 0$. Put control points $z_0 = a, z_1, \ldots, z_n, z_{n+1} = b$ on γ. Assuming that the curve between z_0 and z_1 is a circular arc, the mapping

$$f_{z_1}(z) = \sqrt{\frac{z^2}{(1 - z/b)^2} + c^2}, \tag{33}$$

where $b = |z_1|^2 / \mathrm{Re}(z_1)$ and $c = |z_1|^2 / \mathrm{Im}(z_1)$, maps this circular arc to the real axis. Furthermore, $f(z_1) = 0$, $f(z_2), \ldots, f(z_n)$ are, with the right square root branch, in the upper half–plane, while the images of all points on the real axis including z_0 and z_{n+1} are real.

This can be repeated recursively until finally Ω is mapped to the upper half–plane. All the mappings in the geodesic algorithm have easily evaluable inverses, and it therefore offers a very fast conformal mapping between the channel Ω and the upper half–plane. The most time-consuming part is in fact the calculation of s^{-1}, the inverse of the Schwarz–Christoffel mapping, especially if a modified mapping is used.

The channel in Figure 1 is treated similarly. The inverse Schwarz–Christoffel-like mapping maps the channel on a half–plane with two indentations that are treated one by one with the geodesic algorithm.

To get an high accuracy in the mapping, i.e., a good correspondence between the boundary of Ω and the boundary of the region produced by the constructed mapping, many control points on γ can be used. However, the number of control points cannot be enhanced unrestrictedly. The reason is that for each use of f in (33), the distance to the following control point increases. Hence, a too large number of control points can cause numerical problems. This is especially troublesome in a situation like the one in Figure 1, where two indentations must be handled by the geodesic algorithm.

A more detailed description of the method is given in [1], and an example of a resulting conformal mapping i shown in Figure 8.

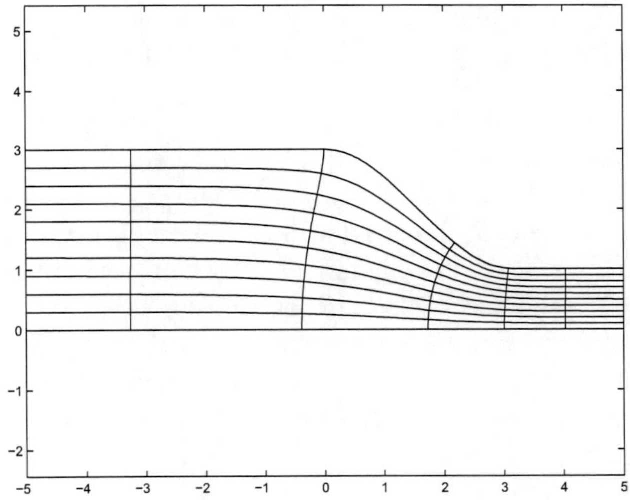

Figure 8: A conformal mapping of a channel constructed using the geodesic algorithm.

4 Conclusion

When applying the Building Block Method in the modelling of wave scattering in waveguides, it is important that a global conformal mapping for one of the blocks keeps a good control of the boundary curve direction, especially towards infinity at the channel ends. It is also important that the regularity of the boundary curve can be controlled.

We have shown several ways to construct such conformal mappings. Most of them are variants of the Schwarz–Christoffel mapping, where different techniques to get rounded polygon corners are used. For these modified Schwarz–Christoffel mappings, we have presented algorithms for the solution of the parameter problem. The accuracy of such mappings depends mainly on the number of polygon vertices, and to achieve a good correspondence between the boundary curves in the original channel and the one produced by the mapping, it might be necessary to use polygons with many vertices. A possible drawback with these methods is that the resulting Schwarz–Christoffel-like mappings with complicated multiple-factor integrands could be computationally slow.

Apart from the desire of a simple and fast mapping function, there are other things to consider. In an application like the ones in [6] or [7], it is important that the smoothness properties of the boundary are preserved. This means that huge variations in the first, second and third derivative of the mapping function on the boundary should be avoided. Often, this is easier to achieve with a Schwarz–Christoffel-like function containing fewer but more extensively modified factors.

The accuracy in the geodesic algorithm depends on the number of control points on the curved part of the boundary. Increasing this number has comparatively small effects on the speed of the mapping function. Instead, there is another problem. Intermediary results can grow very large which risks to bring about negative effects on the accuracy.

A competitive alternative is to use modified Schwarz–Christoffel mappings with approximate curve factors of the type described in Section 3.2. These factors result in C^∞ smooth boundary curves, and one has also, by manipulating the parameters, some possibilities to adopt the factors for particular geometries. Hence, a Schwarz–Christoffel-like mapping with good accuracy can be constructed with comparatively few factors.

The main drawback is of course that the influences on the boundary curve direction from an approximate curve factor is not limited to a small region about the vertex. Sometimes, undesired influences can be dealt with using symmetry properties in a similar way as in the example shown in Figure 5. In other situations, these effects yields restrictions on the parameters in the factors, which can diminish the degree of freedom substantially.

However, efficient and accessible Schwarz–Christoffel software, which gives good starting approximations when solving the parameter problem, together with the possibility to mix different types of curve factors, means that variants of the Schwarz–Christoffel mapping are good alternatives as conformal mappings both for waveguides and many other geometries.

References

[1] Anders Andersson. Using a zipper algorithm to find a conformal map for a channel with smooth boundary. In *AIP Conference Proceedings: 2nd Conference on Mathematical Modeling of Wave Phenomena*, volume 834, pages 3–12. AIP, 2006.

[2] Anders Andersson. A modified Schwarz–Christoffel mapping for regions with piecewise smooth boundary. *J. Comput. Appl. Math.*, 213:56–70, 2008.

[3] Anders Andersson. Schwarz–Christoffel mappings for nonpolygonal regions. *SIAM Journal on Scientific Computing*, 31(1):94–111, 2008.

[4] Anders Andersson. Modified Schwarz–Christoffel mappings using approximate curve factors. Technical Report 09001, MSI, Växjö University, Växjö, Sweden, 2009.

[5] Anders Andersson. On the curvature of some inner curves in a Schwarz–Christoffel mapping. In H.G.W. Begehr, A.O. Çelebi, and R.P. Gilbert, editors, *Further Progress in Analysis: Proceedings of the 6th International ISAAC Congress* , pages 281–290. World Scientific, 2009.

[6] Anders Andersson and Börje Nilsson. Acoustic transmission in ducts of various shapes with an impedance condition. In *AIP Conference Proceedings: International Conference on Numerical Analysis and Applied Mathematics* , volume 1048, pages 33–36. AIP, 2008.

[7] Anders Andersson and Börje Nilsson. Electro-magnetic scattering in variously shaped waveguides with an impedance condition. In *AIP Conference Proceedings: 3rd Conference on Mathematical Modeling of Wave Phenomena* , volume 1106, pages 36–45. AIP, 2009.

[8] George F. Carrier, Max Krook, and Carl E. Pearson. *Functions of a complex variable* , volume 49 of *Classics in Applied Mathematics* . Society for Industrial and Applied Mathematics (SIAM), Philadelphia, PA, 2005. Theory and technique, Reprint of the 1966 original.

[9] Tobin A. Driscoll. Schwarz–Christoffel toolbox for Matlab. Available from +http://www.math.udel.edu/ driscoll/SC/+, 1994-2003.

[10] Tobin A. Driscoll and Lloyd N. Trefethen. *Schwarz-Christoffel mapping*, volume 8 of *Cambridge Monographs on Applied and Computational Mathematics* . Cambridge University Press, Cambridge, 2002.

[11] L. Fishman. One-way wave propagation methods in direct and inverse scalar wave propagation modeling. *Radio Science*, 28:865–876, September 1993.

[12] L. Fishman, A. K. Gautesen, and Z. Sun. *New Perspectives on Problems in Classical and Quantum Physics: A Festschrift in Honor of Herbert Überall*, pages 75–97. CRC Press, 1998.

[13] Peter Henrici. *Applied and computational complex analysis. Vol. 3* . Pure and Applied Mathematics (New York). John Wiley & Sons Inc., New York, 1986. Discrete Fourier analysis—Cauchy integrals—construction of conformal maps—univalent functions, A Wiley-Interscience Publication.

[14] Peter Henrici. *Applied and computational complex analysis. Vol. 1* . Wiley Classics Library. John Wiley & Sons Inc., New York, 1988. Power series—integration—conformal mapping—location of zeros, Reprint of the 1974 original, A Wiley-Interscience Publication.

[15] Prem K. Kythe. *Computational conformal mapping* . Birkhäuser Boston Inc., Boston, MA, 1998.

[16] M. A. Lawrentjew and B. W. Schabat. *Methoden der komplexen Funktionentheorie*. Übersetzung und wissenschaftliche Redaktion von Udo Pirl, Reiner Kühnau und

Lothar Wolfersdorf. Mathematik für Naturwissenschaft und Technik, Band 13. VEB Deutscher Verlag der Wissenschaften, Berlin, 1967.

[17] J. G. Leathem. Some applications of conformal transformation to problems in hydrodynamics. *Royal Society of London Philosophical Transactions Series A* , 215:439–487, 1915.

[18] Ya Yan Lu. Exact one-way methods for acoustic waveguides. *Math. Comput. Simulation*, 50(5-6):377–391, 1999.

[19] Ya Yan Lu. A fourth-order Magnus scheme for Helmholtz equation. *J. Comput. Appl. Math.*, 173(2):247–258, 2005.

[20] Donald E. Marshall and Steffen Rohde. Convergence of a variant of the zipper algorithm for conformal mapping. *SIAM J. Numer. Anal.*, 45(6):2577–2609 (electronic), 2007.

[21] Börje Nilsson. Scattering in quantum tubes, 2001.

[22] Börje Nilsson. Acoustic transmission in curved ducts with various cross-sections. *Proc. R. Soc. Lond. A*, 458:1555–1574, 2002.

[23] Börje Nilsson and Olle Brander. The propagation of sound in cylindrical ducts with mean flow and bulk-reacting lining. IV. Several interacting discontinuities. *IMA J. Appl. Math.*, 27(3):263–289, 1981.

[24] Roland Schinzinger and Patricio A. A. Laura. *Conformal mapping: methods and applications*. Elsevier Science Publishers B.V., Amsterdam, 1991.

[25] Lloyd N. Trefethen. Numerical computation of the Schwarz-Christoffel transformation. *SIAM J. Sci. Statist. Comput.*, 1(1):82–102, 1980.

[26] Lloyd N. Trefethen, editor. *Numerical conformal mapping*. North-Holland Publishing Co., Amsterdam, 1986. Reprint of J. Comput. Appl. Math. **14** (1986), no. 1-2.

In: Computational Mathematics
Editor: Peter G. Chareton, pp. 279-312

ISBN 978-1-60876-271-2

Chapter 10

Computational Study of the 3D Affine Transformation

Béla Paláncz[1]***, Piroska Zaletnyik,***[2∗] ***Joseph L. Awange***[3]***,***
and Robert H. Lewis[4]

[1] Department of Photogrammetry and Geoinformatics,
Budapest University of Technology and Economics, H-1521, Pf. 91. Hungary
[2] Department of Geodesy and Surveying, and Research Group of Physical
Geodesy and Geodynamics of the Hungarian Academy of Sciences,
Budapest University of Technology and Economics, H-1521, Pf. 91. Hungary
[3] Western Australian Centre for Geodesy, Department of Spatial Sciences,
Division of Science and Engineering,Curtin University of Technology,
GPO Box U1987, Perth,WA 6845, Australia
[4] Department of Mathematics, Fordham University,
Bronx, NY 10458, USA

1 Introduction

Three-dimensional datum transformations play a central role in contemporary Euclidean point positioning. In precise positioning with global positioning system (GPS), coordinates given in the World Geodetic System 1984 (WGS84) often have to be transformed into local geodetic coordinate system. The transformation between the traditional terrestrial coordinate system and the satellite observations derived network is a difficult task due to the heterogeneity of the data.

The most frequently used datum transformation method is the 7-parameter similarity transformation, also called Helmert or $C_7(3,3)$ transformation. In geodesy, to solve the 3D Helmert transformation problem, there are different types of methods. The traditional numerical methods use linearization and linear regression subsequently, which needs initial values, i.e. Hooijberg (2008). Awange and Grafarend (2005) solve the 3-point problem in symbolic form with reduced Groebner basis.

∗E-mail address: zaletnyikp@hotmail.com

Due to the distortions between the traditional terrestrial and the GPS derived networks, the 7-parameter similarity transformations in some cases may not offer satisfactory precision. For example transforming GPS local coordinates to the local Hungarian system with global similarity transformation gives 0.5 m maximal residuals, see, e.g., Papp and Szucs (2005). To reduce the remaining residuals, other transformation models with more parameters can be used.

The 9-parameter affine transformation is not only a logical extension but even a generalization of the 7-parameter similarity model (Figure 1). This transformation is the modification of the Helmert $C_7(3,3)$ transformation, where 3 different scales are used in the corresponding coordinate axes instead of one scale factor. In case of the 3 scale parameters being equal, the model reverts back to similarity transformation.

The estimation of the 9-parameter model parameters was achieved by Spath (2004) using numerical minimization technique of the residuum vector as well as by Papp and Szucs (2005) using linearized least squares method. Watson (2006) pointed out that the Gauss-Newton method or its variants can be easily implemented for the nine parameter problem using separation of variables and iteration with respect to the rotation parameters alone, while other parameters can be calculated via simple linear least square solution. The method he suggested, is analogous to other methods for separated least square problems, which goes back at least to Golub and Pereyra (1973).

The 9-parameter affine transformation is also included in some coordinate-transformation softwares developed by the request of GPS users (see e.g. Mathes 2002 and Frohlich and Broker 2003).

For determining the 9 parameter of the 3D affine transformation we need minimum 3 points with known coordinates (X_i, Y_i, Z_i) in both systems. This is the so called 3-point problem. However in geodesy usually we have $N > 3$ known points.

The N-point problem is basically an overdetermined problem and because of the size of real world problems, it is not proper for symbolic solution, although some new research results have been published recently, for example Szanto(2008).

Awange and Grafarend (2005) employed the symbolic solution of the 3-point problem with Gauss-Jacobi combinatoric algorithm with complicated weighting technique to solve N-point Helmert problem. This method does not need an initial guess, but has two important disadvantages in case of real world problems: (1) increasing number of data points (N) causes computational run-away, i.e. in case of $N = 100$ we need 161 700 evaluations of the 3-point models; (2) real data are not well-behaved, different 3 points give very different parameter values and their proper weighting is quite difficult as will be illustrated in the last section on this study. They also suggested another, somewhat new and effective fully numerical method, the Procrustean transformation, which also alleviates the need for linearization and starting values, see, e.g., Awange and Grafarend (2005).

In our study the solution of the overdetermined model of the parameter estimation problem of the 3D affine transformation was investigated via two numerical methods, with global minimization using genetic algorithm and with a local method, the Newton-Raphson method with deflation employing singular value decomposition to compute pseudoinverse of the Jacobian. This latest can be considered as the modern version of the linearization followed by linear regression method labeled as traditional method above, see Quoc-Nam Tran (1994) and Zhao (2007). The overdetermined model consists of $3N$ equations.

In order to reduce the size and make the system square, the overdetermined model can be transformed into a determined one by employing symbolic evaluation of the least square objective and its symbolic derivation providing the necessary condition of its minimum. However, in this case one will get a higher order and more complex polynomial system than the original overdetermined one, but in that case we always have a system of 9 equations irrespectively of the number of the data points. The symbolic solution of this determined model proved to be hopeless because of its complexity, the solution even with coefficient in form of rational numbers seems to be very difficult (Lewis 2007). To solve the determined model in numerical way Newton-Raphson method with Krylov iteration was employed.

The initial value for the two local methods, Newton-Raphson with deflation for the overdetermined model and Newton-Raphson with Krylov iteration for the determined model can be computed via a symbolic-numeric method. This symbolic-numeric method employs explicit analytical expressions developed by computer algebra technique via Dixon resultant as well as reduced Groebner basis for solving 3-point problem. This solution can be used as initial value for a Newton-Raphson-Krylov numerical method to solve the N-point problem in a form of determined system of 9 polynomials developed by symbolic computation, as well as for the inital value for Newton - Raphson with deflation for the overdetermined model. As an alternative method, homotopy solution of the 3-point problem has been also investigated. Criteria for the proper selection of the 3 points from the N ones, is also given.

In addition an extension of the general Procrustes algorithm also has been investigated, which is a fully numerical method and in case of Helmert transformation model employing a single scale parameter does not require iteration at all. However in case of 3D affine transformation, one needs initial guess for the 3 different scale parameters, which can be estimated effectively from the measurements data using a modified version of the method proposed by Albertz and Kreiling (1975).

Numerical illustrations are presented with real world geodetic data representing 1138 points of the Hungarian Datum, the OGPSH network and also with 81 points.

In addition other methods were also investigated, like general polynomial solver based on numerical Groebner basis, linear homotopy continuation method.

Numerical computations were carried out with *Mathematica* and *Fermat* computer algebra systems on hp workstation xw 4100 with XP operation system, 3 GHz P4 Intel processor and 1 GB RAM. The details can be found in *MathSource* of Wolfram Research Inc.

2 Definition of the 3 - Point Problem

The 9-parameter 3D affine transformation is one possible generalization of the 7-parameter Helmert transformation, using three different scale (s_1, s_2, s_3) parameters instead of a single one.

$$\begin{pmatrix} x_i \\ y_i \\ z_i \end{pmatrix} = WR \begin{pmatrix} X_i \\ Y_i \\ Z_i \end{pmatrix} + \begin{pmatrix} X_0 \\ Y_0 \\ Z_0 \end{pmatrix}, \quad (1)$$

where W is the scale matrix.

In this case the scale factors can be modeled by a diagonal matrix

$$W = \begin{pmatrix} s_1 & 0 & 0 \\ 0 & s_2 & 0 \\ 0 & 0 & s_3 \end{pmatrix} \tag{2}$$

and X_0, Y_0, Z_0 are the translation parameters and R the rotation matrix.

The rotation matrix in general is given by using 3 axial rotation angles (α, β, γ - Cardan angles) (Figure 1).

$$R = R_1(\alpha)\, R_2(\beta)\, R_3(\gamma) \tag{3}$$

with

$$R_1(\alpha) = \begin{pmatrix} 1 & 0 & 0 \\ 0 & \cos\alpha & \sin\alpha \\ 0 & -\sin\alpha & \cos\alpha \end{pmatrix}$$

$$R_2(\beta) = \begin{pmatrix} \cos\beta & 0 & -\sin\beta \\ 0 & 1 & 0 \\ \sin\beta & 0 & \cos\beta \end{pmatrix}$$

$$R_3(\gamma) = \begin{pmatrix} \cos\gamma & \sin\gamma & 0 \\ -\sin\gamma & \cos\gamma & 0 \\ 0 & 0 & 1 \end{pmatrix}$$

leading to

$$R(\alpha,\beta,\gamma) = \begin{pmatrix} \cos\beta\cos\gamma & \cos\beta\sin\gamma & -\sin\beta \\ \sin\alpha\sin\beta\cos\gamma - \cos\alpha\sin\gamma & \sin\alpha\sin\beta\sin\gamma + \cos\alpha\cos\gamma & \sin\alpha\cos\beta \\ \cos\alpha\sin\beta\cos\gamma + \sin\alpha\sin\gamma & \cos\alpha\sin\beta\sin\gamma - \sin\alpha\cos\gamma & \cos\alpha\cos\beta \end{pmatrix}. \tag{4}$$

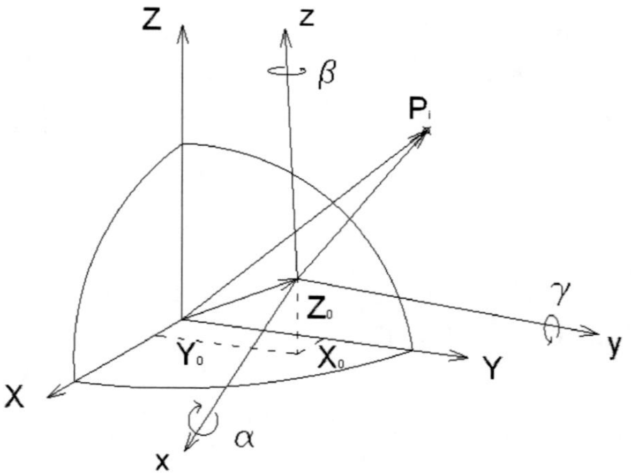

Figure 1: Translation and rotations

In the traditional 7 or 9 parameter transformation solution, the single 3 × 3 rotation matrix is simplified from three separate rotation matrices by assuming that each axial rotation is differentially small (typically less than five arc seconds for most geodetic networks), thus permitting binomial series expansions of the sine and cosine terms for radian measure (Papp and Szucs 2005).

The rotation matrix can be expressed with the skew-symmetric matrix (S) also (see Awange and Grafarend 2003), and this facilitates the symbolic-numeric solution of the problem without using simplifications.

$$S = \begin{pmatrix} 0 & -c & b \\ c & 0 & -a \\ -b & a & 0 \end{pmatrix} \tag{5}$$

The rotation matrix is

$$R = (I_3 - S)^{-1}(I_3 + S), \tag{6}$$

where I_3 is a 3×3 identity matrix.

The rotation matrix with parameters a, b and c is

$$R = \begin{pmatrix} \frac{1+a^2-b^2-c^2}{1+a^2+b^2+c^2} & \frac{2ab-2c}{1+a^2+b^2+c^2} & \frac{2(b+ac)}{1+a^2+b^2+c^2} \\ \frac{2(ab+c)}{1+a^2+b^2+c^2} & \frac{1-a^2+b^2-c^2}{1+a^2+b^2+c^2} & -\frac{2(a-bc)}{1+a^2+b^2+c^2} \\ \frac{2(-b+ac)}{1+a^2+b^2+c^2} & \frac{2(a+bc)}{1+a^2+b^2+c^2} & \frac{1-a^2-b^2+c^2}{1+a^2+b^2+c^2} \end{pmatrix}, \tag{7}$$

for which the next restriction is true:

$$RR^T = \begin{pmatrix} 1 & 0 & 0 \\ 0 & 1 & 0 \\ 0 & 0 & 1 \end{pmatrix}.$$

The axial rotation angles (Cardan angles) can be obtained from the rotation matrix (R) through (Awange and Grafarend 2005):

$$\begin{gathered} \tan\alpha = \frac{r_{23}}{r_{33}} \Rightarrow \alpha = \arctan\frac{r_{23}}{r_{33}} \\ \tan\beta = \frac{-r_{31}}{\sqrt{r_{11}^2 + r_{12}^2}} \Rightarrow \beta = \arctan\frac{-r_{31}}{\sqrt{r_{11}^2 + r_{12}^2}} \text{ or } \beta = -\arcsin r_{13} \\ \tan\gamma = \frac{r_{12}}{r_{11}} \Rightarrow \gamma = \arctan\frac{r_{12}}{r_{11}}. \end{gathered} \tag{8}$$

For the determination of the 9 parameters (a, b, c, X_0, Y_0, Z_0, s_1, s_2, s_3) of the 3D affine transformation we need 3 non-collinear points with known coordinates in both coordinate systems. In further, instead of the scale parameters (s_1, s_2, s_3), we will use $(\sigma_1, \sigma_2, \sigma_3)$ to get more simple equations.

Let us call $\sigma_i = 1/s_i$ and introduce $\Omega = W^{-1}$

$$\Omega = \begin{pmatrix} \sigma_1 & 0 & 0 \\ 0 & \sigma_2 & 0 \\ 0 & 0 & \sigma_3 \end{pmatrix}. \tag{9}$$

Expressing the rotation matrix with the skew-symmetric matrix and using the inverse of the scale matrix (Ω) the nonlinear system ($f_i = 0$) to be solved for determining the 9 transformation parameters leads to:

$$
\begin{aligned}
f_1 =&\ -X_1 + cY_1 - bZ_1 + x_1\sigma_1 - X_0\sigma_1 + cy_1\sigma_2 - cY_0\sigma_2 - bz_1\sigma_3 + bZ_0\sigma_3 \\
f_2 =&\ -cX_1 - Y_1 + aZ_1 - cx_1\sigma_1 + cX_0\sigma_1 + y_1\sigma_2 - Y_0\sigma_2 + az_1\sigma_3 - aZ_0\sigma_3 \\
f_3 =&\ bX_1 - aY_1 - Z_1 + bx_1\sigma_1 - bX_0\sigma_1 - ay_1\sigma_2 + aY_0\sigma_2 + z_1\sigma_3 - Z_0\sigma_3 \\
f_4 =&\ -X_2 + cY_2 - bZ_2 + x_2\sigma_1 - X_0\sigma_1 + cy_2\sigma_2 - cY_0\sigma_2 - bz_2\sigma_3 + bZ_0\sigma_3 \\
f_5 =&\ -cX_2 - Y_2 + aZ_2 - cx_2\sigma_1 + cX_0\sigma_1 + y_2\sigma_2 - Y_0\sigma_2 + az_2\sigma_3 - aZ_0\sigma_3 \\
f_6 =&\ bX_2 - aY_2 - Z_2 + bx_2\sigma_1 - bX_0\sigma_1 - ay_2\sigma_2 + aY_0\sigma_2 + z_2\sigma_3 - Z_0\sigma_3 \\
f_7 =&\ -X_3 + cY_3 - bZ_3 + x_3\sigma_1 - X_0\sigma_1 + cy_3\sigma_2 - cY_0\sigma_2 - bz_3\sigma_3 + bZ_0\sigma_3 \\
f_8 =&\ -cX_3 - Y_3 + aZ_3 - cx_3\sigma_1 + cX_0\sigma_1 + y_3\sigma_2 - Y_0\sigma_2 + az_3\sigma_3 - aZ_0\sigma_3 \\
f_9 =&\ bX_3 - aY_3 - Z_3 + bx_3\sigma_1 - bX_0\sigma_1 - ay_3\sigma_2 + aY_0\sigma_2 + z_3\sigma_3 - Z_0\sigma_3\ .
\end{aligned}
\tag{10}
$$

Similarly to $C_7(3,3)$ problem (Awange and Grafarend 2005) we can reduce the equation system by differencing the equations. In this way the translation parameters (X_0, Y_0, Z_0) can be eliminated. The elimination of the translation parameters can be done according to the next substractions

$$
\begin{aligned}
r_1 &= f_1 - f_7 \\
r_2 &= f_4 - f_7 \\
r_3 &= f_2 - f_8 \\
r_4 &= f_5 - f_8 \\
r_5 &= f_3 - f_9 \\
r_6 &= f_6 - f_9.
\end{aligned}
\tag{11}
$$

In the remaining 6 equations ($g_i = 0$) there are only 6 unknown parameters (a, b, c, σ_1, σ_2, σ_3). This system can be written in a simplified form, if we introduce some new variables, let us call them relative coordinates instead of the original ones, see Awange and Grafarend (2003). Then our equation system is

$$
\begin{aligned}
g_1 &= -X_{13} + cY_{13} - bZ_{13} + x_{13}\sigma_1 + cy_{13}\sigma_2 - bz_{13}\sigma_3 \\
g_2 &= -X_{23} + cY_{23} - bZ_{23} + x_{23}\sigma_1 + cy_{23}\sigma_2 - bz_{23}\sigma_3 \\
g_3 &= -cX_{13} - Y_{13} + aZ_{13} - cx_{13}\sigma_1 + y_{13}\sigma_2 + az_{13}\sigma_3 \\
g_4 &= -cX_{23} - Y_{23} + aZ_{23} - cx_{23}\sigma_1 + y_{23}\sigma_2 + az_{23}\sigma_3 \\
g_5 &= +bX_{13} - aY_{13} - Z_{13} + bx_{13}\sigma_1 - ay_{13}\sigma_2 + z_{13}\sigma_3 \\
g_6 &= +bX_{23} - aY_{23} - Z_{23} + bx_{23}\sigma_1 - ay_{23}\sigma_2 + z_{23}\sigma_3,
\end{aligned}
\tag{12}
$$

where the relative coordinates are

$$\begin{aligned}
&x_{12}=x_1-x_2,\ x_{13}=x_1-x_3,\ x_{23}=x_2-x_3\\
&y_{12}=y_1-y_2,\ y_{13}=y_1-y_3,\ y_{23}=y_2-y_3,\\
&z_{12}=z_1-z_2,\ z_{13}=z_1-z_3,\ z_{23}=z_2-z_3,\\
&X_{12}=X_1-X_2,\ X_{13}=X_1-X_3,\ X_{23}=X_2-X_3,\\
&Y_{12}=Y_1-Y_2,\ Y_{13}=Y_1-Y_3,\ Y_{23}=Y_2-Y_3,\\
&Z_{12}=Z_1-Z_2,\ Z_{13}=Z_1-Z_3,\ Z_{23}=Z_2-Z_3.
\end{aligned} \tag{13}$$

3 Numerical Solutions

Here we shall consider three global methods working without guess of initial values:

a) a general solver for polynomial systems based on the application of numerical Groebner basis and eigensystem method, built in the *Mathematica* system as a function **NSolve**

b) Global minimization based on genetic algorithm, built in *Mathematica* as **NMinimize**

c) A linear homotopy method implemented in *Mathematica* as a function **LinearHomotopyFR**, and using the built in *Mathematica* function **FindRoot**, employing Newton-Raphson method, see Palancz (2008a).

Let us consider the numerical values of 3 Hungarian points form the 81 data points in the system of ETRS89 $(x_1, y_1, z_1, \ldots, z_3)$ and in the local Hungarian system HD72 (Hungarian Datum 1972) $(X_1, Y_1, Z_1, \ldots, Z_3)$ (*Table 1*). Then our equations are, see Palancz et al (2008a)

$$\begin{aligned}
g_1&=-215777+140851\,b-141039\,c+215777\,\sigma_1-141039\,c\,\sigma_2+140850\,b\,\sigma_3,\\
g_2&=-191326+61552.7\,b-334829\,c+191325\,\sigma_1-334829\,c\,\sigma_2+61552.1\,b\,\sigma_3,\\
g_3&=+141039-140851\,a-215777\,c-215777\,c\,\sigma_1-141039\,\sigma_2-140850\,a\,\sigma_3,\\
g_4&=+334829-61552.7\,a-191326\,c-191325\,c\,\sigma_1-334829\,\sigma_2-61552.1\,a\,\sigma_3,\\
g_5&=+140851+141039\,a+215777\,b+215777\,b\,\sigma_1+141039\,a\,\sigma_2-140850\,\sigma_3,\\
g_6&=+61552.7+334829\,a+191326\,b+191325\,b\,\sigma_1+334829\,a\,\sigma_2-61552.1\,\sigma_3.
\end{aligned} \tag{14}$$

Table 1: Coordinates of 3 Hungarian points in ETRS89 and in HD72

	X [m]	Y [m]	Z [m]	x [m]	y [m]	z [m]
1	4171409.677	1470823.777	4580140.907	4171352.311	1470893.887	4580150.178
2	4146957.889	1277033.850	4659439.264	4146901.301	1277104.509	4659448.287
3	3955632.880	1611863.197	4720991.316	3955575.649	1611933.124	4721000.952

Table 2: Results of the general solver

a	b	c
15.248	2.857	3.082
1078855	-0.3499	5.3366
-0.0655	202158	-0.1873
825375.943	293148	970906
-3.3119	$-3.4112\cdot 10^{-6}$	2.8155
$-1.2115\cdot 10^{-6}$	1.1763	-0.3551
$9.2691\cdot 10^{-7}$	$-4.9466\cdot 10^{-6}$	$-3.2439\cdot 10^{-7}$
0.3019	-0.8501	$-1.0299\cdot 10^{-6}$
σ_1	σ_2	σ_3
-1.000005408	-0.9999978437	0.99998929
1.000005408	-0.9999978437	-0.99998929
-1.000005408	0.9999978437	-0.99998929
-1.000005408	-0.9999978437	-0.99998929
1.000005408	-0.9999978437	0.99998929
-1.000005408	0.9999978437	0.99998929
1.000005408	0.9999978437	0.99998929
1.000005408	0.9999978437	-0.99998929

3.1 General polynomial solver based on numerical Groebner basis and eigensystem method

The results provided by the general solver are summarized in *Table 3.1*.

Employing general solver, we should select from the eight solutions the only one, which provides positive value for all scale variables, $\sigma_i > 0, i = 1, 2, 3$.

3.2 Global minimization

In order to use directly the least squares method, we can define the objective function as the sum of the square of the errors, which should be minimized,

$$\Delta(a,b,c,\sigma_1,\sigma_2,\sigma_3) = \sum_{i=1}^{6} g_i^2(a,b,c,\sigma_1,\sigma_2,\sigma_3). \tag{15}$$

The minimization was carried out with **NMinimize** employing genetic algorithm. Here we get a single solution, see *Table 3.2*.

3.3 Homotopy Solution

We can employ the convex linear or *Bézout* homotopy, see Drexler(1977), Garcia and Zangwill (1979). The homotopy function is

$$H(x,\beta) = \gamma\,(1-\beta)\,G(x) + \beta\,F(x). \tag{16}$$

Table 3: Results of the global minimization

a	$9.269080333 \cdot 10^{-7}$
b	$-4.946616459 \cdot 10^{-6}$
c	$-3.243922350 \cdot 10^{-7}$
σ_1	1.000005408
σ_2	0.9999978437
σ_3	0.99998929

The homotopy function will transform $G(x) = 0 \implies F(x) = 0$ smoothly, while $H(x,0) \implies H(x,1)$. The original system $F(x)$ is called target system and $G(x)$ is called as start system. We call $x(\beta)$ the homotopy path, the location of the solution of $H(x,\beta) = 0$ for every $\beta \in [0,1]$, with other words the function of the location of the roots of the homotopy function, depending on β.

A proper start system is

$$G(x_0) = 0. \tag{17}$$

The $x(\beta)$ function start with $x(0) = x_0$ and at $\beta = 1$, we get the solution of system $F(x)$, namely $x(1) = x_v$ and $F(x_v) = 0$.

An appropriate start system as well as the corresponding initial values can be generated in the following way, Hazaveh et al. (2003),

Let $f_i(x_1,\ldots,x_n)$, $i = 1,\ldots,n$ be a system of n polynomials. We are interested in the common zeros of the system, namely

$$f = (f_1(x),\ldots,f_n(x)) = 0. \tag{18}$$

Let d_j denote the degree of the jth polynomial - that is the degree of the highest order term in the equation. Then the start system is,

$$G_j(x) = e^{i\phi_j}\left(x_j{}^{d_j} - \left(e^{i\theta_j}\right)^{d_j}\right) = 0, \qquad j = 1,\ldots,n, \tag{19}$$

where ϕ_j and θ_j are random real numbers in the interval $[0,\ 2\pi]$. The equation above has the obvious particular solution $x_j = e^{i\theta_j}$ and the complete set of the starting solutions for $j = 1,\ldots,n$ is given by

$$x_{0j} = e^{\left(i\theta_j + \frac{2\pi i k}{d_j}\right)}, \qquad k = 0,1,\ldots,d_j - 1. \tag{20}$$

Mathematica function was written to generate such a system in general, see Palancz (2008a). The start systems and their initial values can be computed with this function. We have 8 different values and 8 solutions accordingly, see *Table 3.1.* The start system belonging to the only acceptable solution is

$$\begin{aligned}
G_1 &= (-0.665071+0.746781\,i)\,((0.0686677-0.99764\,i)+a)\\
G_2 &= (0.885355-0.464915\,i)\,((0.211415+0.977396\,i)+b)\\
G_3 &= (-0.55332-0.832968\,i)\,((0.995053+0.0993405\,i)+c)\\
G_4 &= (0.385686+0.92263\,i)\,((0.270671+0.962672\,i)+\sigma_1)\\
G_5 &= (0.267103+0.963668\,i)\,((-0.544854-0.838531\,i)+\sigma_2)\\
G_6 &= (0.245239+0.969463\,i)\,((-0.998031-0.0627203\,i)+\sigma_3)
\end{aligned} \tag{21}$$

and the corresponding initial values

$$\begin{aligned}
x_0 = \{&-0.0686677+0.99764\,i,\ -0.211415-0.977396\,i,\\
&-0.995053-0.0993405\,i,\ -0.270671-0.962672\,i,\\
&0.544854+0.838531\,i,\ 0.998031+0.0627203\,i\}.
\end{aligned} \tag{22}$$

Now, the start system is $G = \{G_j, \quad j = 1,\ldots,6\}$ and the target system is $F = \{g_j, \quad j = 1,\ldots,6\}$. The homotopy function then

$$H(a,b,c,\sigma_1,\sigma_2,\sigma_3,\beta) = \gamma\,\beta\,F + (1-\beta)\,G, \tag{23}$$

with the initial conditions

$$H_0 = H(0,x_0). \tag{24}$$

Applying

$$\gamma = \{1,1,1,1,1,1\} \tag{25}$$

we have got the proper real solution which is the same as the solution in *Table 3.2*.

This is a general method to generate the start system and its solution. However in our case a more simple method is also possible. Let us consider our start system as the linear part of the nonlinear system g_i,

$$\begin{aligned}
G_1 &= -X_{13} + cY_{13} - bZ_{13} + x_{13}\sigma_1\\
G_2 &= -X_{23} + cY_{23} - bZ_{23} + x_{23}\sigma_1\\
G_3 &= -cX_{13} - Y_{13} + aZ_{13} + y_{13}\sigma_2\\
G_4 &= -cX_{23} - Y_{23} + aZ_{23} + y_{23}\sigma_2\\
G_5 &= bX_{13} - aY_{13} - Z_{13} + z_{13}\sigma_3\\
G_6 &= bX_{23} - aY_{23} - Z_{23} + z_{23}\sigma_3.
\end{aligned} \tag{26}$$

The start values can be easily computed by solving this linear system

$$G(x_0) = 0, \tag{27}$$

where $x_0 = (1.85381 \times 10^{-6}, -9.89319 \times 10^{-6}, -6.48787 \times 10^{-7}, 1.00001, 0.999998, 0.999989)$.

Employing this start system and initial values in the homotopy function, the results are again the same, but the computation is somewhat faster than in the first case, (0.031sec vs. 0.047 sec).

Table 4: Comparing Global Numerical Methods for solving 3-point problem

Method	Time of computation (sec)	Solution
Numerical Groebner basis with eigensystem method	0.110	many
Global minimization via genetic algorithm	1.172	one
Linear homotopy method	0.031 (0.047)	many

In *Table 3.3* we summarized the computation times of the different methods.

Remark: Although all of the three methods gave the same numerical result, homotopy solution is roughly 2-3 times faster than the general solver and 30- 40 times faster than global minimization.

4 Symbolic Solutions

Symbolic solution of the system g_i means the reduction of the multivariate polynomial system via computer algebra to a single univariate polynomial and computing its roots. Then the other unknowns can be computed backwards similarly to Gaussian - elimination in case of linear systems. Different methods exist to carry out such reduction, see Awange and Grafarend (2005). Here we employed the Dixon resultant method.

4.1 Dixon's Resultant: Basic Concepts

Consider a system of polynomials

$$\mathcal{F} = \{f_0, f_1, \ldots, f_d\}, \tag{28}$$

where

$$f_i = \sum_{\alpha \in \mathcal{A}_i} c_{i,\alpha} \mathbf{x}^{\alpha} \tag{29}$$

and

$$\mathbf{x}^{\alpha} = x_1^{\alpha_1} x_2^{\alpha_2} \cdots x_1^{\alpha_d} \tag{30}$$

for each $i = 0, \ldots, d$, and with the polynomial coefficients $c_{i,\alpha}$, which are also referred as parameters of the system.

The goal is to derive a condition on parameters $c_{i,\alpha}$ such that the polynomial system $\mathcal{F} = 0$ has a solution. We can consider this problem as the elimination of variables from the polynomial system. Elimination theory tells us that such a condition exists for large family of polynomial systems, and is called the resultant of the polynomial system. Since the number of equations is more than the number of variables, in general, for arbitrary values of $c_{i,\alpha}$ the polynomial system does not have any solution.

If there is only one such polynomial representing this family, then its defining equation is called the *resultant*.

One way to compute the resultant of a given polynomial system is to construct a matrix with property that whenever the polynomial system has a solution, such a matrix has a deficient rank, thereby implying that the determinant of any maximal minor is a multiple of the resultant.

A simple way to construct a resultant matrix is to use the dialytic method, i.e. multiply each polynomial with a finite set of monomials, and rewrite the resulting system in matrix form. We call such matrix the dialytic matrix. This alone, however, does not guarantee that a matrix so constructed is a resultant matrix.

Note that such matrices are usually quite sparse: matrix entries are either zero or coefficients of the polynomials in the original system. Good examples of resultant dialytic matrices are for univariate case, and as well as Newton sparse matrices for the multivariate case; they all differ only in the selection of multiplier monomial sets.

In contrast to dialytic matrices, the Dixon matrix is dense since its entries are combinations of the coefficients of the polynomials in the original system. It has the advantage of being an order of magnitude smaller in comparison to a dialytic matrix, which is important as the computation of the determinant of a matrix symbolic entries is sensitive to its size. The Dixon matrix is constructed through the computation of the Dixon polynomial, which is expressed in matrix form.

Let $\pi_i(\mathbf{x}^{\alpha}) = \sigma_1^{\alpha_1} \cdots \sigma_i^{\alpha_i} x_{i+1}^{\alpha_{i+1}} \cdots x_d^{\alpha_d}$ where $i \in \{0, 1, \ldots, d\}$ and the σ_i 's are new variables; $\pi_0(\mathbf{x}^{\alpha}) = \mathbf{x}^{\alpha}\pi_i$ is extended to polynomials in a natural way as

$$\pi_i(f(x_1, \ldots, x_d)) = f(\sigma_1, \ldots \sigma_i, x_{i+1}, \ldots, x_d). \tag{31}$$

Given a polynomial system $\mathcal{F} = \{f_0, f_1, \ldots, f_d\}$ define its *Dixon polynomial* as

$$\begin{aligned} &\theta(f_0, f_1, \ldots, f_d) = \delta(x_1, \ldots, x_d, \sigma_1, \ldots, \sigma_d) \\ &= \prod_{i=1}^{d} \frac{1}{\sigma_i - x_i} \quad \mathrm{Det} \begin{pmatrix} \pi_0(f_0) & \pi_0(f_1) & \cdots & \pi_0(f_d) \\ \pi_1(f_0) & \pi_1(f_1) & \cdots & \pi_1(f_d) \\ \vdots & & \ddots & \vdots \\ \pi_d(f_0) & \pi_d(f_1) & \cdots & \pi_d(f_d) \end{pmatrix}. \end{aligned} \tag{32}$$

The order in which original variables in $\mathbf{x}$ are replaced by new variables in σ is significant in the sense that Dixon polynomial computed using two different variable orderings may be different.

A Dixon polynomial $\delta(\mathbf{x}, \boldsymbol{\sigma})$ can be written in bilinear form as

$$\delta(x_1, \ldots, x_d, \sigma_1, \ldots, \sigma_d) = \boldsymbol{\Xi\Theta}\mathbf{X}^T, \tag{33}$$

where $\Xi = (\sigma^{\beta_1}, \ldots, \sigma^{\beta_k})$ and $\mathbf{X} = (x^{\alpha_1}, \ldots, x^{\alpha_s})$ are row vectors. The $k \times s$ matrix Θ is called the *Dixon matrix*.

4.2 Construction of Dixon Resultant

Cayley's formulation of Bézout's method

Let us recall here the Cayley's formulation Cayle (1865) of Bézout's method for solving two polynomial equations in univariate case. However that method is actually due to Euler (e.g., Salmon 1859).

Let us consider two univariate polynomial $f(x)$ and $g(x)$, and let

$$\deg = \max(\text{degree}(f), \text{degree}(g)) \tag{34}$$

and let σ be an auxiliary variable. The quantity

$$\delta(x,\sigma) = \frac{1}{x-\sigma}\det\begin{pmatrix} f(x) & g(x) \\ f(\sigma) & g(\sigma) \end{pmatrix} = \frac{f(x)g(\sigma) - f(\sigma)g(x)}{x-\sigma} \tag{35}$$

is a symmetric polynomial in x and σ of $deg-1$ which is called *Dixon polynomial* of f and g. Every common zero of f and g is a zero of $\delta(x,\sigma)$ for all values of σ.

Example

Let us consider two univariate polynomials with parameter π,

$$f(x) = (x-1)(x+3)(x-4) = 12 - 11x - 2x^2 + x^3$$
$$g(x) = (x-\pi)(x+4) = -4\pi + 4x - \pi x + x^2.$$

Then the *Dixon polynomial* is

$$\delta = \frac{1}{x-\sigma}\det\begin{pmatrix} f(x) & g(x) \\ f(\sigma) & g(\sigma) \end{pmatrix} =$$
$$-48 + 56\pi - 12x + 8\pi x - 4\pi x^2 - 12\sigma + 8\pi\sigma + 3x\sigma$$
$$-2\pi x\sigma + 4x^2\sigma - \pi x^2\sigma - 4\pi\sigma^2 + 4x\sigma^2 - \pi x\sigma^2 + x^2\sigma^2.$$

Hence at a common zero, each coefficient of σ^i in δ is vanishing

$$\sigma^2 : -4\pi + 4x - \pi x + x^2 = 0$$
$$\sigma^1 : -12 + 8\pi + 3x - 2\pi x + 4x^2 - \pi x^2 = 0$$
$$\sigma^0 : -48 + 56\pi - 12x + 8\pi x - 4\pi x^2 = 0.$$

It is a homogeneous system in variables x^0, x^1 and x^2.

$$\begin{pmatrix} -4\pi & 4-\pi & 1 \\ -12+8\pi & 3-2\pi & 4-\pi \\ -48+56\pi & -12+8\pi & -4\pi \end{pmatrix} \begin{pmatrix} x^0 \\ x^1 \\ x^2 \end{pmatrix} = \begin{pmatrix} 0 \\ 0 \\ 0 \end{pmatrix}$$

This system has non-trivial solutions if and only if its determinant, D is zero. D is called the *Dixon resultant* of f and g. The matrix of the system M is the *Dixon matrix*,

$$M = \begin{pmatrix} -4\pi & 4-\pi & 1 \\ -12+8\pi & 3-2\pi & 4-\pi \\ -48+56\pi & -12+8\pi & -4\pi \end{pmatrix}$$

and its determinant is

$$D = -480 + 440\pi + 80\pi^2 - 40\pi^3.$$

The polynomials have common zeros if D is vanishing. Really, solving the equation $D = 0$ we get

$$\pi_1 = -3, \quad \pi_2 = 1 \quad \text{and} \quad \pi_3 = 4.$$

Dixon's generalization of the Cayley-Bézout's method

Dixon generalized Cayley's approach to Bézout's method to systems of three polynomials equations in three unknowns, Dixon (1908). Let

$$\begin{aligned} f(x,y,z) &= 0 \\ g(x,y,z) &= 0 \\ h(x,y,z) &= 0. \end{aligned} \tag{36}$$

Now the Dixon polynomial is defined by

$$\delta(x,y,z,\sigma,\xi) = \frac{1}{(x-\sigma)(y-\xi)} \det \begin{pmatrix} f(x,y,z) & g(x,y,z) & h(x,y,z) \\ f(\sigma,y,z) & g(\sigma,y,z) & h(\sigma,y,z) \\ f(\sigma,\xi,z) & g(\sigma,\xi,z) & h(\sigma,\xi,z) \end{pmatrix} \tag{37}$$

Example

Let us consider three three-variate polynomials

$$\begin{aligned} f &= x^2+y^2-1 \\ g &= x^2+z^2-1 \\ h &= y^2+z^2-1. \end{aligned}$$

The *Dixon polynomial* is

$$\begin{aligned} \delta(x,y,z,\sigma,\xi) &= \frac{1}{(x-\sigma)(y-\xi)} \det \begin{pmatrix} x^2+y^2-1 & x^2+z^2-1 & y^2+z^2-1 \\ \sigma^2+y^2-1 & \sigma^2+z^2-1 & y^2+z^2-1 \\ \sigma^2+\xi^2-1 & \sigma^2+z^2-1 & \xi^2+z^2-1 \end{pmatrix} \\ &= yz-2x^2yz+y\xi-2x^2y\xi+z\sigma-2x^2z\sigma+\xi\sigma-2x^2\xi\sigma. \end{aligned}$$

Now we eliminate the variables y and z, so considering that

$$\delta(x,y,z,\sigma,\xi) = \sigma\xi\left(1-2x^2\right)+\xi y\left(1-2x^2\right)+\sigma z\left(1-2x^2\right)+yz\left(1-2x^2\right)$$

and system of equations is

$$\begin{array}{lcccc} . & y^0z^0 & y^1z^0 & y^0z^1 & y^1z^1 \\ \sigma^0\xi^0: & 0 & 0 & 0 & 1-2x^2 \\ \sigma^1\xi^0: & 0 & 0 & 1-2x^2 & 0 \\ \sigma^0\xi^1: & 0 & 1-2x^2 & 0 & 0 \\ \sigma^1\xi^1: & 1-2x^2 & 0 & 0 & 0. \end{array}$$

Consequently the *Dixon matrix*

$$M = \begin{pmatrix} 0 & 0 & 0 & 1-2x^2 \\ 0 & 0 & 1-2x^2 & 0 \\ 0 & 1-2x^2 & 0 & 0 \\ 1-2x^2 & 0 & 0 & 0 \end{pmatrix}$$

and its determinant, the *Dixon resultant*

$$D = 1 - 8x^2 + 24x^4 - 32x^6 + 16x^8.$$

Dixon proved that for three 2 degree generic polynomials, the vanishing of D is a necessary condition for the existence of a common zero. Furthermore, D is not identically zero. Dixon's method and proofs easily generalize to a system of $n+1$ generic $n$degree polynomials in n unknowns. Recall, that a polynomial is generic if all its coefficients are independent parameters, unrelated to each other. A polynomial in n variables is $n$degree if all powers to the maximum of each variable appear in it, see Kapur, Saxena and Yang (1994) for more details.

4.3 Improved Dixon resultant - Kapur, Saxena and Yang method

First we used the improved Dixon resultant method (Dixon - KSY) suggested by Kapur, Saxena and Yang (1994) as well as Nakos and Williams (2002). Employing this method, using pairwise-elimination step by step we got a univariate polynomial of degree 29 for σ_1, see Zaletnyik et al (2007) and Zaletnyik (2008). The size of the coefficients are very big, therefore we present only its numerical form computed from the coordinate values of the 3 points,

$$\begin{aligned} p(\sigma_1) = &-1.727698196 \times 10^6 - 9.74498664 \times 10^6\sigma_1 - 3.866075363 \times 10^7\sigma_1^2 \\ &- 9.20725651 \times 10^7\sigma_1^3 - 2.506853275 \times 10^7\sigma_1^4 + 9.83323298 \times 10^8\sigma_1^5 \\ &+ 2.687098281 \times 10^9\sigma_1^6 - 1.677570696 \times 10^9\sigma_1^7 - 1.852770679 \times 10^{10}\sigma_1^8 \\ &- 3.405605447 \times 10^{10}\sigma_1^9 - 2.433848603 \times 10^{10}\sigma_1^{10} + 9.37949292 \times 10^9\sigma_1^{11} \\ &+ 4.051192059 \times 10^{10}\sigma_1^{12} + 4.479618094 \times 10^{10}\sigma_1^{13} + 1.859916305 \times 10^{10}\sigma_1^{14} \\ &- 1.442601078 \times 10^{10}\sigma_1^{15} - 2.481001170 \times 10^{10}\sigma_1^{16} - 1.193451221 \times 10^{10}\sigma_1^{17} \\ &+ 2.779876803 \times 10^9\sigma_1^{18} + 7.50398359 \times 10^9\sigma_1^{19} + 5.15047763 \times 10^9\sigma_1^{20} \\ &+ 1.271679920 \times 10^9\sigma_1^{21} - 1.042156062 \times 10^9\sigma_1^{22} - 1.366950032 \times 10^9\sigma_1^{23} \\ &- 8.36299106 \times 10^8\sigma_1^{24} - 3.502473078 \times 10^8\sigma_1^{25} - 1.067078887 \times 10^8\sigma_1^{26} \\ &- 2.193636502 \times 10^7\sigma_1^{27} - 2.150540534 \times 10^6\sigma_1^{28} - 5.52314307\sigma_1^{29}. \end{aligned} \tag{38}$$

Fortunately we do not need to find all of the roots of this polynomial, because for the value of σ_1 a very good estimation can be given as $\sigma_1 = 1/s_1$, where s_1 (the first scale parameter) can be estimated dividing the sum of distances from the center of gravity in both systems, see Albertz and Kreiling (1975).

The center of gravity in the two systems (x_s, y_s, z_s) and (X_s, Y_s, Z_s) can be computed as follows,

$$\begin{aligned} x_s &= \frac{\sum_{i=1}^3 x_i}{3}, \quad y_s = \frac{\sum_{i=1}^3 y_i}{3}, \quad z_s = \frac{\sum_{i=1}^3 z_i}{3}, \\ X_s &= \frac{\sum_{i=1}^3 X_i}{3}, \quad Y_s = \frac{\sum_{i=1}^3 Y_i}{3}, \quad Z_s = \frac{\sum_{i=1}^3 Z_i}{3}. \end{aligned} \tag{39}$$

The estimated scale parameter according to Albertz and Kreimlig in case of the Helmert similarity transformation

$$s_{est} = \frac{\sum_{i=1}^{3} \sqrt{(x_i - x_s)^2 + (y_i - y_s)^2 + (z_i - z_s)^2}}{\sum_{i=1}^{3} \sqrt{(X_i - X_s)^2 + (Y_i - Y_s)^2 + (Z_i - Z_s)^2}}. \quad (40)$$

The estimated scale parameters according to the modified Albertz-Kreimlig expression for the 9 parameter affine transformation are see Zaletnyik (2008)

$$s_{1,est} = \frac{\sum_{i=1}^{3} \sqrt{(x_i - x_s)^2}}{\sum_{i=1}^{3} \sqrt{(X_i - X_s)^2}},$$
$$s_{2,est} = \frac{\sum_{i=1}^{3} \sqrt{(y_i - y_s)^2}}{\sum_{i=1}^{3} \sqrt{(Y_i - Y_s)^2}}, \quad (41)$$
$$s_{3,est} = \frac{\sum_{i=1}^{3} \sqrt{(z_i - z_s)^2}}{\sum_{i=1}^{3} \sqrt{(Z_i - Z_s)^2}}.$$

Then the estimated $\sigma_{i,est}$

$$\sigma_{i,est} = \frac{1}{s_{i,est}}. \quad (42)$$

In our example the estimated value for σ_1 is $\sigma_{1,est} = 1.000001248$.

This value can be employed as a good starting value for finding the the proper root of $p(\sigma_1) = 0$ employing Newton-Raphson method. The result is $\sigma_1 = 1.000005408167503$.

4.4 Heuristic methods to accelerate the Dixon resultant

The basic idea of the Dixon method is to construct a square matrix M whose determinant D is a multiple of the resultant. Usually M is not unique, it is obtained as a maximal minor in a larger matrix we shall call M^+, and there are usually many maximal minors - any one of which will do. The entries in M are polynomials in parameters. The factors of D that are not the resultant are called the *spurious factors*, and their product is sometimes the *spurious factor*.

The naive way to proceed is to compute D, factor it, and separate the spurious factor from the actual resultant. But there are problems. On one hand the determinant may be so large as to be impractical or even impossible to compute, even though the resultant is relatively small, the spurious factor is huge. On the second hand, the determinant may be so large that factoring it is impractical.

Lewis developed three heuristic methods to overcome these problems, Lewis(2002) and Lewis(2004). The first may be used on any polynomial system. It uses known factors of D to compute other factors. The second also may be used on any polynomial system and it discovers factors of D so that the complete determinant is never produced. The third applies only when the resultant appears as a factor of D in certain exponential patterns. These methods were discovered by experimentation and may apply to other resultant formulation, such as the Macaulay.

4.5 Early discovery of factors: the EDF method

This method is to exploit the observed fact that D has many factors. In other words, we try to turn the existence of spurious factors to our advantage. By elementary row and column manipulations (Gaussian elimination) we discover probable factors of D and pluck them out of $M_0 \equiv M$. Any denominators that form in the matrix are plucked out. This produces a smaller matrix M_1 still with polynomial entries, and list of discovered numerators and denominators. Here is a very simple example.

Example

Given initially

$$M_0 = \begin{pmatrix} 9 & 2 \\ 4 & 4 \end{pmatrix} \qquad \text{numerators : ---} \qquad \text{denominators : ---}$$

We factor a 2 out of the second column, then a 2 from the second row. Thus

$$M_0 = \begin{pmatrix} 9 & 1 \\ 2 & 1 \end{pmatrix} \qquad \text{numerators} : 2,2 \qquad \text{denominators : ---}$$

Note that $9 \times 4 - 2 \times 4 = 2 \times 2 \times (9 \times 1 - 2 \times 1)$.
We change the second row by subtracting 2/9 of the first

$$M_0 = \begin{pmatrix} 9 & 1 \\ 0 & 7/9 \end{pmatrix} \qquad \text{numerators} : 2,2 \qquad \text{denominators : ---}$$

We pull out the denominator 9 from the second row, and factor out 9 from the first column:

$$M_0 = \begin{pmatrix} 1 & 1 \\ 0 & 7 \end{pmatrix} \qquad \text{numerators} : 2,2,9 \quad \text{denominators} : 9$$

Note that $9 \times 7/9 - 1 \times 0 = \frac{2\times2\times9}{9} \times (1 \times 7 - 1 \times 0)$.
We "clean up" or consolidate by dividing out the common factor of 9 from the numerator and denominator lists; any one that occurs may be erased and the list compacted since the first column is canonically simple, we are finished with one step the algorithm, and have produced a smaller M_1

$$M_1 = (7) \qquad \text{numerators} : 2,2 \qquad \text{denominators : ---}$$

The algorithm terminates by pulling out the 7:

$$\text{numerators} : 2,2,7 \quad \text{denominators : ---}$$

As expected (since the original matrix contained all integers) the denominator list is empty. The product of all the entries in the numerator list is the determinant, but we never needed to deal with any number larger than 9.

4.6 Application of the EDF method

As explained in the previous section, the univariate polynomial for σ_1 can also be computed by employing the accelerated Dixon resultant by the Early Discovery Factors (Dixon - EDF) algorithm, which was suggested and implemented in the computer algebra system *Fermat* by Lewis and Bridgett (2003) as well as Lewis (2008). Using this method one can get the result in the following form

$$\prod_{i=1}^{5} \varphi_i(\sigma_1)^{K_i} \tag{43}$$

where $\varphi_i(\sigma_1)$ are irreducible polynomials with low degree, but their powers, K_i are very big positive integer numbers, so expanding this expression would result into millions of terms!

Consequently, we shall consider $K_i = 1$, for $i = 1,\ldots,5$, namely

$$\prod_{i=1}^{5} \varphi_i(\sigma_1) \tag{44}$$

as the Dixon resultant. These polynomials are as follows

$$\begin{aligned}
\varphi_1 =& y_{13}\, z_{23} - y_{23}\, z_{13};\\
\varphi_2 =& x_{13}^2\, y_{23}\, z_{23}\, \sigma_1^2 - x_{13}\, x_{23}\, y_{13}\, z_{23}\, \sigma_1^2 - x_{13}\, x_{23}\, y_{23}\, z_{13}\, \sigma_1^2 + x_{23}^2\, y_{13}\, z_{13}\, \sigma_1^2\\
& - Z_{13}^2\, y_{23}\, z_{23} - Y_{13}^2\, y_{23}\, z_{23} - X_{13}^2\, y_{23}\, z_{23} + Z_{13}\, Z_{23}\, y_{13}\, z_{23}\\
& + Y_{13}\, Y_{23}\, y_{13}\, z_{23} + X_{13}\, X_{23}\, y_{13}\, z_{23} + Z_{13}\, Z_{23}\, y_{23}\, z_{13} + Y_{13}\, Y_{23}\, y_{23}\, z_{13}\\
& + X_{13}\, X_{23}\, y_{23}\, z_{13} - Z_{23}^2\, y_{13}\, z_{13} - Y_{23}^2\, y_{13}\, z_{13} - X_{23}^2\, y_{13}\, z_{13};\\
\varphi_3 =& x_{13}\, y_{23}\, \sigma_1 - x_{23}\, y_{13}\, \sigma_1 + X_{13}\, y_{23} - X_{23}\, y_{13};\\
\varphi_4 =& Z_{13}\, x_{13}\, x_{23}\, z_{23}\, \sigma_1^2 - Z_{23}\, x_{13}^2\, z_{23}\, \sigma_1^2 - Z_{13}\, x_{23}^2\, z_{13}\, \sigma_1^2 + Z_{23}\, x_{13}\, x_{23}\, z_{13}\, \sigma_1^2\\
& + X_{13}\, Z_{13}\, x_{23}\, z_{23}\, \sigma_1 - 2\, X_{13}\, Z_{23}\, x_{13}\, z_{23}\, \sigma_1 + X_{23}\, Z_{13}\, x_{13}\, z_{23}\, \sigma_1\\
& + X_{13}\, Z_{23}\, x_{23}\, z_{13}\, \sigma_1 - 2\, X_{23}\, Z_{13}\, x_{23}\, z_{13}\, \sigma_1 + X_{23}\, Z_{23}\, x_{13}\, z_{13}\, \sigma_1\\
& - X_{13}^2\, Z_{23}\, z_{23} + X_{13}\, X_{23}\, Z_{13}\, z_{23} + X_{13}\, X_{23}\, Z_{23}\, z_{13} - X_{23}^2\, Z_{13}\, z_{13};\\
\varphi_5 =& Z_{13}\, z_{23} - Z_{23}\, z_{13};
\end{aligned} \tag{45}$$

The factor polynomial providing proper root has degree of two, therefore its solution can be expressed in analytical form (only one of the roots is correct, the positive σ_1), namely

$$\begin{aligned}
\sigma_1 =& \sqrt{}\,(X_{23}^2 y_{13} z_{13} + y_{13} Y_{23}^2 z_{13} + X_{13}^2 y_{23} z_{23} + Y_{13}^2 y_{23} z_{23} + y_{23} Z_{13}^2 z_{23}\\
& - X_{13} X_{23}(y_{23} z_{13} + y_{13} z_{23}) - Y_{13} Y_{23}(y_{23} z_{13} + y_{13} z_{23}) - y_{23} z_{13} Z_{13} Z_{23} -\\
& y_{13} Z_{13} z_{23} Z_{23} + y_{13} z_{13} Z_{23}^2)\\
& / \left(\sqrt{(x_{23} y_{13} - x_{13} y_{23})(x_{23} z_{13} - x_{13} z_{23})}\right)
\end{aligned} \tag{46}$$

Similarly we can get simple explicit forms for σ_2 and σ_3 as

$$\begin{aligned}
\sigma_2 =& \sqrt{}\,(-X_{13}^2 x_{23} z_{23} + X_{13} X_{23}(x_{23} z_{13} + x_{13} z_{23})\\
& + x_{23}\left(Y_{13} Y_{23} z_{13} - Y_{13}^2 z_{23} - Z_{13}^2 z_{23} + z_{13} Z_{13} Z_{23}\right)\\
& - x_{13}\left(X_{23}^2 z_{13} + Y_{23}^2 z_{13} - Y_{13} Y_{23} z_{23} - Z_{13} z_{23} Z_{23} + z_{13} Z_{23}^2\right))\\
& / \left(\sqrt{-(x_{23} y_{13} - x_{13} y_{23})(-y_{23} z_{13} + y_{13} z_{23})}\right)
\end{aligned} \tag{47}$$

and

$$\begin{aligned}\sigma_3 =& \sqrt{}\,(X_{13}^2 x_{23} y_{23} - X_{13} X_{23} (x_{23} y_{13} + x_{13} y_{23})\\ &+ x_{23}\left(Y_{13}^2 y_{23} - y_{13} Y_{13} Y_{23} + y_{23} Z_{13}^2 - y_{13} Z_{13} Z_{23}\right)\\ &+ x_{13}\left(X_{23}^2 y_{13} - Y_{13} y_{23} Y_{23} + y_{13} Y_{23}^2 - y_{23} Z_{13} Z_{23} + y_{13} Z_{23}^2\right))\\ &/\left(\sqrt{(x_{23} z_{13} - x_{13} z_{23})(y_{23} z_{13} - y_{13} z_{23})}\right)\end{aligned} \tag{48}$$

Using the numerical values the results are the same as computed by numerical methods, see *Table 3.2*.

Remark: The result of Dixon-EDF method is not only faster and more elegant but also a bit more precise than that of the Dixon- KSY method. Although, one should check the solutions of all polynomials with degree 1 and 2 in order to choose the proper result! The expressions above for σ_i are already the properly selected solutions.

The further advantage of this elimination technique is that it can give result also for the other parameters (a, b and c) as function of the coordinates, directly.

4.7 Application of Reduced Groebner basis

Lichtblau (2007) showed that the same result for σ_i, $i = 1, 2, 3$ can be achieved by using reduced Groebner basis built in *Mathematica* with parameter *MonomialOrder* $\rightarrow$ *EliminationOrder*. Unfortunately this technique failed to get the other parameters a, b and c with neither Buchberger nor Groebner walk. The computation time of the analytical form of the scale parameters is less than that of required by the Dixon-EDF but the necessary storage space is bigger.

4.8 Computation of other parameters

The other parameters can be computed using computer algebra elimination via Dixon resultant or reduced Groebner basis. The parameters of the skew-symmetric matrix can be computed, too. Let us express a from equations g_5 and g_6 as

$$\begin{aligned}a =& (X_{23} Z_{13} - X_{13} Z_{23} + (-X_{23} z_{13} + X_{13} z_{23})\,\sigma_3\\ &+ \sigma_1\,(x_{23} Z_{13} - x_{13} Z_{23} + (-x_{23} z_{13} + x_{13} z_{23})\sigma_3))\\ &/\,(-X_{23} Y_{13} + X_{13} Y_{23} + (-X_{23} y_{13} + X_{13} y_{23})\sigma_2\\ &+ \sigma_1\,(-x_{23} Y_{13} + x_{13} Y_{23} + (-x_{23} y_{13} + x_{13} y_{23})\sigma_2))\end{aligned} \tag{49}$$

The parameter b is given by one of the two equations, let say from g_5,

$$b = \frac{aY_{13} + Z_{13} + ay_{13}\sigma_2 - z_{13}\sigma_3}{X_{13} + x_{13}\sigma_1} \tag{50}$$

and parameter c from g_1

$$c = \frac{X_{13} + bZ_{13} - x_{13}\sigma_1 + bz_{13}\sigma_3}{Y_{13} + y_{13}\sigma_2} \tag{51}$$

The translation parameters can be similarly computed, but now from the original system of equations, f_i. From f_1,f_2 and f_3, we get

$$X_0 = \frac{1}{(1+a^2+b^2+c^2)\sigma_1}\left((-1-a^2+b^2+c^2)X_1+(-2ab+2c)Y_1 - 2bZ_1 - 2acZ_1 + x_1\sigma_1 + a^2x_1\sigma_1 + b^2x_1\sigma_1 + c^2x_1\sigma_1\right) \tag{52}$$

and

$$Y_0 = \frac{1}{(1+a^2+b^2+c^2)\sigma_2}\left(-2(ab+c)X_1+(-1+a^2-b^2+c^2)Y_1 + 2aZ_1 - 2bcZ_1 + y_1\sigma_2 + a^2y_1\sigma_2 + b^2y_1\sigma_2 + c^2y_1\sigma_2\right) \tag{53}$$

The third translation parameter can be computed from f_1

$$Z_0 = \frac{X_1 - cY_1 + bZ_1 - x_1\sigma_1 + X_0\sigma_1 - cy_1\sigma_2 + cY_0\sigma_2 + bz_1\sigma_3}{b\sigma_3} \tag{54}$$

Let us summarize our results in case of the 3-point problem in *Table 5.*

Table 5: Results in case of 3-point problem

Method	Running Time (sec)
Numerical Groebner Basis & eigensystem method	0.11
Global minimization with genetic algorithm	1.17
Linear Homotopy	0.03
Analytic solution computed via computer algebra	~ 0.00

As we have seen before the Dixon-KSY method required estimation for the initial value as well as Newton iteration, while the other two symbolic solutions (Dixon-EDF and reduced Groebner basis) provide the same result, a fully analytic solution requiring neither initial condition nor iteration, therefore the computation times for both methods are practically zero.

Considering *Table 5* it is clear that the best choice to solve 3-point problem is to use the analytic formulas developed by the help of the Dixon-EDF method or reduced Groebner basis.

The computer algebra method, namely the accelerated Dixon resultant with the technique of Early Discovery Factors as well as the reduced Groebner basis, provides a very simple, elegant symbolic solution for the 3-point problem. Although, numerical methods without initial guess values, like linear homotopy and numerical Groebner basis with eigensystem method are also very efficient. The main advantages of the symbolic solution originated from its iteration-free feature, are the very short- practically zero -computation time and the independence on value of the actual numerical data.

The details of the *Mathematica* computations can be found in Palancz et al (2008a).

5 Definition of the N-Point Problem

For the determination of the 9 parameters of the transformation $(a,b,c,X_0,Y_0,Z_0,s_1,s_2,s_3)$ we need at least 3 non-collinear points with known coordinates in both coordinate systems. In further, instead of the scale parameters (s_1,s_2,s_3), we will use again $(\sigma_1,\sigma_2,\sigma_3)$ to get more simple equations. However in case of $N>3$ points in both systems, expressing the rotation matrix with the skew-symmetric matrix and using the inverse of the scale matrix (Ω), the nonlinear system to be solved for $(f_i=0)$ determining the 9 parameters leads to a system of 3 N polynomial equations. For example in one of our numerical examples, we have $N=1138$ points. It means we have 3414 equations and 9 unknown parameters, so our system is an overdetermined multivariate polynomial system,

$$
\begin{aligned}
e_1 &= -X_1+cY_1-bZ_1+x_1\sigma_1-X_0\sigma_1+cy_1\sigma_2-cY_0\sigma_2-bz_1\sigma_3+bZ_0\sigma_3\\
e_2 &= -cX_1-Y_1+aZ_1-cx_1\sigma_1+cX_0\sigma_1+y_1\sigma_2-Y_0\sigma_2+az_1\sigma_3-aZ_0\sigma_3\\
e_3 &= bX_1-aY_1-Z_1+bx_1\sigma_1-bX_0\sigma_1-ay_1\sigma_2+aY_0\sigma_2+z_1\sigma_3-Z_0\sigma_3\\
&\vdots\\
e_{3N-2} &= -X_N+cY_N-bZ_N+x_N\sigma_1-X_0\sigma_1+cy_N\sigma_2-cY_0\sigma_2-bz_N\sigma_3+bZ_0\sigma_3\\
e_{3N-1} &= -cX_N-Y_N+aZ_N-cx_N\sigma_1+cX_0\sigma_1+y_N\sigma_2-Y_0\sigma_2+az_N\sigma_3-aZ_0\sigma_3\\
e_{3N} &= bX_N-aY_N-Z_N+bx_N\sigma_1-bX_0\sigma_1-ay_N\sigma_2+aY_0\sigma_2+z_N\sigma_3-Z_0\sigma_3
\end{aligned}
\tag{55}
$$

6 Solution of the Overdetermined Model

To solve such a system in symbolic form is a difficult problem, however there are some general results in this field like Giusti and Schost (1999), Mourrain (2002), Szanto (2008) etc. Sometimes, if the number of equations is small the Gauss-Jacobi combinatoric solution can be realistic. In that case one consider every triplets $\binom{N}{3}$ from the N points and use the solution of the 3-point for each. Then the weighted sum of these solutions can provide the solution of the overdetermined system, see for example Awange and Grafarend (2005). In this study we applied numerical solutions. First as a reference of the solution we employed again direct global minimization.

6.1 Direct Numerical Solution via Global Minimization

The objective function can be constructed by considering the sum of the square residual of the equations above, namely

$$\Delta(a,b,c,X_0,Y_0,Z_0,\sigma_1,\sigma_2,\sigma_3)=\sum_{i=1}^{3N}e_i^2 \tag{56}$$

which should be minimized. To carry out global minimization, again genetic algorithm was employed.

6.2 Newton-Raphson with Deflation

This method solves the numerical Jacobi matrix in every steps with singular value decomposition method (SVD), therefore it can solve overdetermined system in least squares sense.

As it is known every A matrix $m \times n$, $m \geq n$ can be decomposed as

$$A = U\Sigma V^T \tag{57}$$

where $(.)^T$ denotes the transposed matrix and U is $m \times n$ matrix, V is $n \times n$ matrix satisfying

$$U^T U = V^T V = V V^T = I_n \tag{58}$$

and $\Sigma = < \sigma_1, \ldots, \sigma_n >$ a diagonal matrix.

These $\sigma_i' s$, $\sigma_1 \geq \sigma_2 \geq, \ldots, \sigma_n \geq 0$ are the square roots of the non negative eigenvalues of $A^T A$ and are called as the *singular values* of matrix A.

As it is well - known from linear algebra, singular value decomposition is a technique to compute *pseudoinverse* for singular or ill-conditioned matrix of linear systems. In addition this method provides least square solution for overdetermined system and minimal norm solution in case of undetermined system.

The pseudoinverse of a matrix A $(m \times n)$ is a matrix A^+ $(n \times m)$ satisfying

$$AA^+A = A, \quad A^+AA^+ = A^+, \quad \left(A^+A\right)^* = A^+A, \quad \left(AA^+\right)^* = AA^+ \tag{59}$$

where $(.)^*$ denotes the conjugate transpose of the matrix.

Always exists a unique A^+ which can be computed using SVD:

a) If $m \geq n$ and $A = U\Sigma V^T$ then

$$A^+ = V\Sigma^{-1}U^T \tag{60}$$

where $\Sigma^{-1} = < 1/\sigma_1, \ldots, 1/\sigma_n >$

b) If $m < n$ then compute the $\left(A^T\right)^+$, pseudoinverse of A^T and then

$$A^+ = \left(\left(A^T\right)^+\right)^T, \tag{61}$$

The idea of using pseudoinverse in order to generalize the Newton method is not new but has been suggested in Haselgrove (1961), Ben-Israel (1966), Fletcher (1970), Ben-Israel and Greville (1974), Quoc-Nam Tran (1994) etc.

It means that in the iteration formula, the pseudoinverse of the Jacobian matrix will be employed,

$$x_{i+1} = x_i - J^+ \left(x_i\right) f\left(x_i\right) \tag{62}$$

In *Mathematica* pseudoinverse can be computed in symbolic as well as in numeric form. Let us see an example, in Sommese and Wampler (2005).

Example

Let us consider a simple monomial system,

$$f_1 = x^2$$
$$f_2 = xy$$
$$f_3 = y^2$$

This system is a "monomial ideal" and trivial for computer algebra, while its Groebner basis is $\{y^2, xy, x^2\}$. Now the Jacobian of the system is

$$J = \begin{pmatrix} 2x & 0 \\ y & x \\ 0 & 2y \end{pmatrix}$$

its pseudoinverse

$$J^+(x,y) = \begin{pmatrix} \frac{x^3+4xy^2}{2(x^4+4x^2y^2+y^4)} & \frac{y^3}{x^4+4x^2y^2+y^4} & -\frac{xy^2}{2(x^4+4x^2y^2+y^4)} \\ -\frac{x^2y}{2(x^4+4x^2y^2+y^4)} & \frac{x^3}{x^4+4x^2y^2+y^4} & \frac{4x^2y+y^3}{2(x^4+4x^2y^2+y^4)} \end{pmatrix}$$

and at the point $x_i, y_i = (1., 1.)$

$$J^+(x,y) = \begin{pmatrix} 2. & 0 \\ 1. & 1. \\ 0 & 2. \end{pmatrix}$$

Then the new values x_{i+1}, y_{i+1} in the next iteration step are

$$\begin{pmatrix} x_{i+1} \\ y_{i+1} \end{pmatrix} = \begin{pmatrix} x_i \\ y_i \end{pmatrix} - J^+(x_i, y_i) \begin{pmatrix} f_1(x_i, y_i) \\ f_2(x_i, y_i) \\ f_3(x_i, y_i) \end{pmatrix}$$

which means

$$\begin{pmatrix} 0.5 \\ 0.5 \end{pmatrix} = \begin{pmatrix} 1 \\ 1 \end{pmatrix} - \begin{pmatrix} 0.416667 & 0.166667 & -0.0833333 \\ -0.0833333 & 0.166667 & 0.416667 \end{pmatrix} \begin{pmatrix} 1 \\ 1 \\ 1 \end{pmatrix}$$

This method was implemented in *Mathematica*, see Palancz (2008b).

This method as a local method requires initial guess values. Selecting 3 from N points, these initial values can be computed with the symbolic formulas developed for the 3-point problem in *Section 4*. We use now the numerical values computed on the basis of the first three points of the data set of 1138 points employing the analytic expressions developed by computer algebra method via Dixon EDF, see *Table 6.2*

This algorithm is faster with about two magnitudes than global minimization (3.03 sec vs. 130.91 sec) and it is quite robust, see the considerable deviations between the result and the initial guess values in *Table 6.2*.

Table 6: Result of the N-point problem in case of $N = 1138$

	Initial values computed from the first three point	Solution of the N-point problem
a	$-6.991 \cdot 10^{-6}$	$-8.389 \cdot 10^{-6}$
b	$+8.021 \cdot 10^{-6}$	$+1.142 \cdot 10^{-4}$
c	$+1.596 \cdot 10^{-6}$	$-3.482 \cdot 10^{-5}$
σ_1	0.9999741620	0.9997302444
σ_2	0.9999995727	1.0001887097
σ_3	1.0000271383	1.0002452426
X_0	−209.460	−2298.589
Y_0	−1.391	+526.509
Z_0	+248.532	+2143.716

6.3 Extended general Procrustes algorithm

The Procrustean approach for solving least square problem appeared nearly 25 years ago (see, e.g., Gower 1984) is very fast method, which needs neither initial starting value nor iteration. Procrustes being a technique of matching one configuration into another and producing a measure of match, seeks the isotropic dilation and the rigid translation, reflection and rotation needed to best match one configuration to another, Cox and Cox (1994).

The Procrustes problem is concerned with fitting a configuration $\mathcal{B}$ into $\mathcal{A}$ as close as possible. The most simple Procrustes case is one in which both configurations have the same dimensionality and the same number of points, which can be brought into a 1 - 1 correspondence by substantive considerations, Borg and Groenen (1997). Let us consider the case where both $\mathcal{A}$ and $\mathcal{B}$ matrices are of the same dimension. The *partial Procrustes* problem is then formulated as

$$\mathcal{A} = \mathcal{B}\, T. \tag{63}$$

The rotation matrix T is then solved by measuring the distances between corresponding points in both configurations, square these values, and add them to obtain the sum of squares $\|\mathcal{A} - \mathcal{B}\, T\|$ which is then minimized. One proceeds via Frobenius norm as follows

$$\|\mathcal{A} - \mathcal{B}\, T\| = \sqrt{tr\,((\mathcal{A}^T - T^T \mathcal{B}^T)\,(\mathcal{A} - \mathcal{B}\, T))} \rightarrow \min_{T} \tag{64}$$

where

$$T^T T = I. \tag{65}$$

This partial Procrustes problem can be extended for additional translation and scaling, beside the rotation, Awange and Grafarend (2005)

$$\mathcal{A} = s\, \mathcal{B}\, T + \mathcal{D}. \tag{66}$$

Then the objective function to be minimized

$$\|\mathcal{A} - s\,(\mathcal{B}\, T + \mathcal{D})\|_W = \sqrt{tr\,((\mathcal{A}^T - s\,(T^T \mathcal{B}^T + \mathcal{D}))\; W\; (\mathcal{A} - s\,(\mathcal{B}\, T + \mathcal{D})))} \rightarrow \min_{T,s,\mathcal{D}} \tag{67}$$

where W is a weight matrix, s is the scaling parameter and $\mathcal{D}$ is translation vector. This Procrustes problem is called as *general Procrustes* problem.

These Procrustes problems arise in applications related to, e.g. rigid body movement and psychometrics, factor analysis, multivariate analysis, multidimensional scaling and global positioning system, see for example, Gower and Dijksterhuis (2004), and Meridith (1977).

The solution of the general Procrustes problem provides the solution of the parameter estimation problem of the Helmert transformation with 7 parameter, see Awange and Grafarend (2005).

Recently, an extention of general Procrustes method for 3D affine 9-parameter transformation has been published, see Awange, Bae and Claessens (2008).

This method is very fast, and the computation effort for iteration to improve the S scale matrix is negligible. The main problem with this method is that after shifting the two systems into the origin of the coordinate system, the rotation matrix is computed via SVD. It means that the physical rotation as well as the deformation (distortion) caused by different scaling in the different directions of the 3 different principal axes will be also involved in the computation of the rotation matrix. This can be approximately allowed only when these scale factors do not differ from each other considerably. In addition, the necessary condition for getting optimal scale matrix is not restricted for diagonal matrix, and the off-diagonal elements are simply deleted in every iteration step, which will not ensure real global minimum for the trace of the error matrix.

A modification of this extension of the general Procrustes method can cure these problems, but there is a price for it! This method start with an initially guessed diagonal S scale matrix (which can be calculated with the modified Albertz-Kreimlig expression (see *Section 4.3*) or with the general Procrustes method described in Awange, Bae and Claessens 2008), and first using the inverse of this matrix to eliminate the deformation caused by scaling, and only after the elimination of this "distortion" will be SVD applied to compute the rotation matrix itself. It goes without saying that in this way, the computation of the rotation matrix should be repeated in iterative way in order to decrease the error. That is why that this method is precise and correct for any ratio of scale parameters, but takes longer time to achieve the results, than this or other methods, see Palancz et al (2008b).

7 The Determined Model

Determined model can be achieved by employing symbolic evaluation of the least square objective and its symbolic derivation of the overdetermined model providing the necessary condition of its minimum. However, in this case one will get a higher order and more complex polynomial system than the original overdetermined one. In symbolic form most of these equations have many thousands of terms, therefore it is useful to collect terms corresponding to the same multivariate expression via computer algebra, see Zaletnyik (2008). Here, just as an illustration, let us see the first equation resulted from the derivation of the objective with respect to the variable a,

$$\begin{aligned}
Eq_a = & - bNX_0Y_0\sigma_1\sigma_2 + aNY_0^2\sigma_2^2 - cNX_0Z_0\sigma_1\sigma_3 + aNZ_0^2\sigma_3^2 + bY_0\sigma_1\sigma_2\sum_{i=1}^{N} x_i \\
& + cZ_0\sigma_1\sigma_3\sum_{i=1}^{N} x_i + bY_0\sigma_2\sum_{i=1}^{N} X_i + cZ_0\sigma_3\sum_{i=1}^{N} X_i + aY_0\sigma_2^2\sum_{i=1}^{N}(-2y_i) \\
& + bX_0\sigma_1\sigma_2\sum_{i=1}^{N} y_i + b\sigma_1\sigma_2\sum_{i=1}^{N}(-x_iy_i) + b\sigma_2\sum_{i=1}^{N}(-X_iy_i) + a\sigma_2^2\sum_{i=1}^{N} y_i^2 \\
& + aY_0\sigma_2\sum_{i=1}^{N}(-2Y_i) + bX_0\sigma_1\sum_{i=1}^{N} Y_i + Z_0\sigma_3\sum_{i=1}^{N} 2Y_i + b\sigma_1\sum_{i=1}^{N}(-x_iY_i) \\
& + b\sum_{i=1}^{N}(-X_iY_i) + a\sigma_2\sum_{i=1}^{N} 2y_iY_i + aZ_0\sigma_3^2\sum_{i=1}^{N}(-2z_i) + cX_0\sigma_1\sigma_3\sum_{i=1}^{N} z_i \\
& + c\sigma_1\sigma_3\sum_{i=1}^{N}(-x_iz_i) + c\sigma_3\sum_{i=1}^{N}(-X_iz_i) + \sigma_3\sum_{i=1}^{N}(-2Y_iz_i) + a\sigma_3^2\sum_{i=1}^{N} z_i^2 \\
& + Y_0\sigma_2\sum_{i=1}^{N}(-2Z_i) + aZ_0\sigma_3\sum_{i=1}^{N}(-2Z_i) + cX_0\sigma_1\sum_{i=1}^{N} Z_i + c\sigma_1\sum_{i=1}^{N}(-x_iZ_i) \\
& + c\sum_{i=1}^{N}(-X_iZ_i) + \sigma_2\sum_{i=1}^{N} 2y_iZ_i + a\sigma_3\sum_{i=1}^{N} 2z_iZ_i + a\sum_{i=1}^{N}\left(Y_i^2 + Z_i^2\right) = 0.
\end{aligned} \tag{68}$$

In this way we get 9 polynomial equations $(Eq_a, Eq_b, \ldots, Eq_{\sigma_3})$ (see Palancz et al 2008c) with the 9 unknown parameters $(a, b, c, X_0, Y_0, Z_0, \sigma_1, \sigma_2, \sigma_3)$.

8 Numerical Solution of the Determined Model

This model can be solved only in numerical way. First, we considered two methods which do not need initial values. Global minimization with genetic algorithm was unsuccessful, while it failed to converge to the requested accuracy or precision within 200 iterations and the general polynomial solver also failed to provide results after 500 seconds. Therefore local method, namely Newton-Raphson method with Krylov iteration was employed. Krylov iteration can improve the inverse of the Jacobian and in this way stabilize the solution even when Jacobian has high condition number, see Stoer and Bulirsh (2006), while the traditional Newton - Raphson method failed because singular Jacobian was encountered. This Newton - Raphson method with Krylov iteration has been proved to be very robust as well as fast too, see Zaletnyik (2008) and Palancz et al (2008c). The initial values in *Table 8* were calculated from the symbolic solution of the 3-point problem for the first three points of the 1138 data points.

Let us summarize our results in case of the N- point problem in *Table 8*.

9 Complexity Study of the Algorithms

In order to get more information about the algorithms considered as numerical methods for solving N-point problem, the computations have been carried out for different N values,

Table 7: Results in case of N-point Problem

Model	Method	Initial guess value	Time (sec)
overdetermined	Global minimization with genetic algorithm	no	130.91
overdetermined	Newton method with Deflation using SVD	yes*	3.03
determined	Global minimization with genetic algorithm	no	failed
determined	Numerical Groebner Basis and eigensystem method	no	failed
determined	Newton Method with Krylov iteration	yes*	0.75

* initial starting values are based on the result of the 3-point problem, taken the first three points of the 1138 points data set.

too. We considered here only the two Newton methods, namely Newton - Raphson method with deflation for overdetermined system and the Newton - Raphson with Krylov iteration for the determined model.

This implies that for N-point problem probably the best choice is employing the determined model with Newton - Raphson method with Krylov iteration, if the size of the problem is big, $N > 200$. While for smaller problem the Newton - Raphson method with deflation using overdetermined system can be the proper choice.

The initial values for this method can be computed from the 3-point model, but one need to be cautious and not to do it blindly. In the last section, we shall discuss how one can properly select 3 points from the N ones, in order compute good initial values ensuring fast convergence of the Newton method employed for the solution of the N-point problem.

From *Table 8.* it is clear that Newton -Raphson method with Krylov iteration for the determined system has zero complexity as it was expected.

10 The Proper Selection of the 3 Points for Initial Guess Values

Although the Newton - Raphson Method with Krylov iteration is very robust for the initial values, it is important to select properly the 3 points from the N points, which are chosen to calculate symbolically the initial guess values for the N-point problem. Let us choose two different triplets from the data set Hungarian Datum of 81 points [1] (*Figure 2*) and calculate with the symbolic solution of the parameters of the coordinate transformation $(X_0, Y_0, Z_0, a, b, c, \sigma_1, \sigma_2, \sigma_3)$.

The results for the two triplets are quite different (see *Table 10.*)

[1]We used here a smaller data set in order to be able to visualize the points clearly.

Table 8: Results of the complexity study

Number of points	Time in case A (sec)	Time in case B (sec)
200	0.47	0.78
400	1.67	0.81
600	1.89	0.69
800	2.50	0.73
1000	2.78	0.73
1138	3.03	0.75

where
case A - stands for the Newton - Raphson method with deflation for overdetermined system and case B - stands for the Newton - Raphson method with Krylov iteration for the determined system.

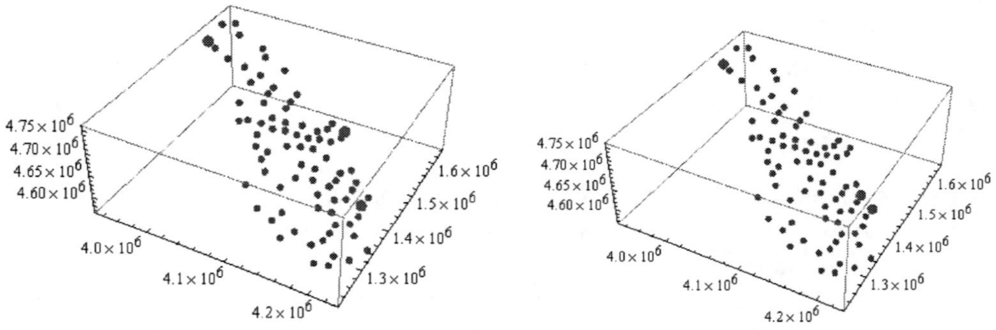

Figure 2: Two chosen triplets from the 81 points in the three dimensional Cartesian coordinate system (X_i, Y_i, Z_i)

Using these parameters as initial values for the Newton-Rapshon method to solve the N-point solution (here 81 points), in the case of the first triplet the method converges rapidly in 4 iteration steps, but in the second case the method is not converging, not even after 100 iteration steps (see Figure 3).

There is correlation between the geometry of the chosen triplet and the goodness of the calculated initial values. According to our numerical calculations we get the best initial values when the geometry of the triplet is similar to a equilateral triangle, and the worse when the geometry of the three points is nearly on a line. A geometrical index can be introduced to represent the geometry of the selected 3 points to avoid the solutions which provide disadvantageous starting values. This geometrical index is the sine of the minimum angle in the triangle, it's maximal value is $\frac{\sqrt{3}}{2}$ when the triangle is an equilateral triangle, and around zero when the three points are nearly collinear. In the earlier examples the geometrical index in the first case, which gave good initial values, was 0.529 and in the second case was 0.002.

Table 9: Calculated coordinate transformation parameters from two different triplets in case Hungarian Datum 81 points

	1st triplet	2nd triplet
X_0	-77.523	-496.192
Y_0	+90.366	+124.797
Z_0	+25.151	+543.061
a	$+1.520 \cdot 10^{-6}$	$+5.470 \cdot 10^{-6}$
b	$+1.526 \cdot 10^{-6}$	$+27.692 \cdot 10^{-6}$
c	$-0.543 \cdot 10^{-6}$	$-2.415 \cdot 10^{-6}$
σ_1	0.999998865	0.999957317
σ_2	1.000001182	0.999989413
σ_3	1.000001642	1.000069623

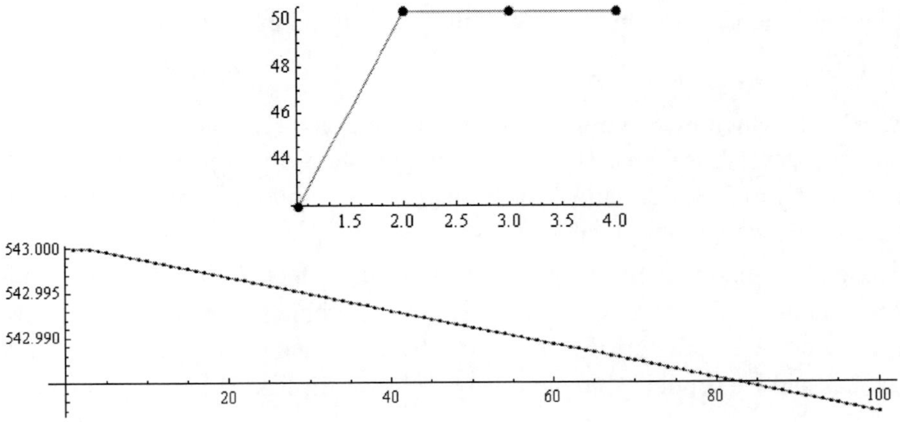

Figure 3: Iteration steps for the Z_0 parameter values with the Newton-Raphson method using two different initial guess values (vertical axis is for Z_0 and the horizontal axis is for the number of iteration steps)

To check the correlation between the geometry and the goodness of the initial values we calculated the transformation parameters for all 3 points combinations from the 81 points, so we examined $\binom{81}{3} = 85320$ combinations.

Let check the Z_0 values for all combinations. The real Z_0 value calculated with Newton-Rapshon method from the 81 points is 50.342 m, but the values calculated from the different triplets can be very different from this, for example the maximum value for Z_0 was 24 679 629 m! The geometrical index of this extreme triplet was 0.003, so these 3 points were almost collinear.

In *Figure 4* the calculated Z_0 values are represented in function of the geometrical index for all of the 85320 combinations. The figure is only a window of the whole graphics, for the better representation $|Z_0| > 1000$ values are not represented (the minimum Z_0 is -1 798 501 and the maximum is 24 679 629). In the figure the adjusted value of Z_0 based on the 81 data points is also represented by a line.

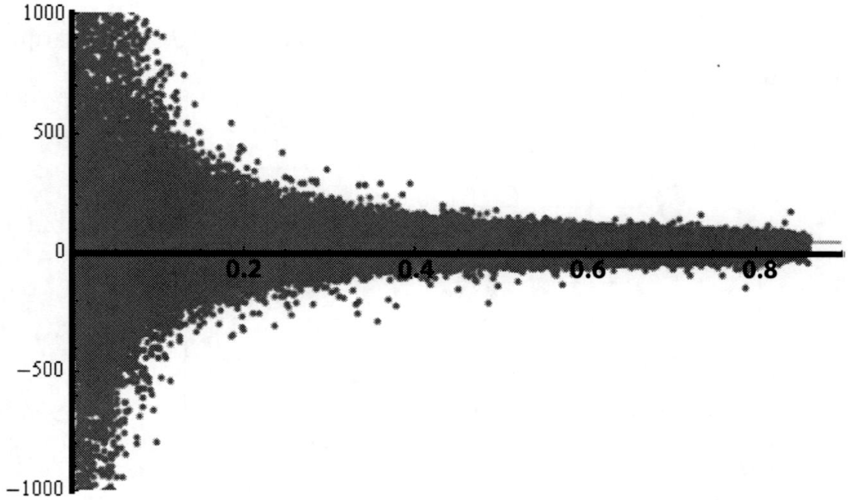

Figure 4: Values of Z_0 in function of the geometrical index for all combinations

The triplets which give extremely different solutions from the adjusted value all have geometrical indexes less then 0.1. In this range the round-off error is dominating, while for higher value of this index the improper fitting of the measurement data to the 9 parameter model causes the deviations in value of Z_0.

In general the more similar is the geometry of the selected three points to a equilateral triangle the better is the initial value for the N-point problem. Accordingly to these results it is very important to examine the geometry of the selected three points for calculating the initial values and we should avoid the nearly collinear triplets, Zaletnyik (2008).

11 Conclusions

According to the results of our computational study, up to now the most effective method to compute the parameters of 3D affine transformation model is based on a symbolic-numeric algorithm. This algorithm computes the parameters of the 3-point problem employing analytical expression developed by computer algebra (Dixon resultant or reduced Groebner basis), then using these values as initial values for a Newton - Raphson method with Krylov iteration to solve the N-point problem of the determined model developed from the overdetermined model via computer algebra. This method is fast, robust and has very low complexity. Criteria for selecting an appropriate triplet from data points is also given.

However, for relatively small system, $N < 200$ data points, the solution of the original overdetermined system via Newton - Raphson method with deflation can be more efficient.

Although, in case of the Helmert transformation (7 parameters), the general Procrustes method is very efficient, its application for 3D affine transformation needs a time consuming iteration process if the scale factors are strongly different, therefore up to now it seems not a good choice.

Acknowledgments

The authors are indebted to Professor Daniel Lichtblau (Wolfram Research) for his valuable remarks and suggestions, especially calling up their attention to the proper implementation of **NSolve** as well as to the application of the reduced Groebner basis.

References

[1] Albertz, J., W. Kreiling, 1975: *Photogrammetric Guide.* Herbert Wichmann Verlag, Karlsruhe, 58-60.

[2] Awange, J.L, and E. W. Grafarend, 2003: Closed form solution of the overdetermined nonlinear 7 parameter datum transformation. *Allgemeine Vermessungs-Nachrichten (AVN)*, **110**, 130-148.

[3] Awange, J.L., and E. W. Grafarend, 2005: *Solving Algebraic Computational Problems in Geodesy and Geoinformatics.* Springer, Berlin, 333 pp.

[4] Awange, J. L., K. Bae, and S. J. Claessens, 2008: Procrustean solution of the 9-parameter transformation problem. *Earth, Planets and Space*, **60**, 529-537

[5] Ben-Israel, A., 1966: A Newton-Raphson Method for the solution of systems of equations. *J.Math. Anal. Applic.*, **15**, 243-252.

[6] Ben-Israel, A., and T. N. E. Greville, 1974: *Generalized Inverses: Theory and Applications*, Wiley and Sons, New York, 395 pp.

[7] Borg, I., and P. Groenen, 1997: *Modern multidimensional scaling*. Springer, New York, 496 pp.

[8] Cayley, A., 1865: On the theory of elimination. *Cambridge and Dublin Mathematical Journal*, **3**, 210-270.

[9] Cox, T.F., and M. A. A. Cox, 1994: *Multidimensional scaling*. Chapman & Hall, 213 pp.

[10] Dixon, A. L., 1908: The eliminant of three quantics in two independent variables. *Proc. London Math.Soc.*, **6**, 468-478.

[11] Drexler, F.J., 1977: Eine Methode zur Berechnung sämtlicher Lösungen von Polynomgleichungssystemen. *Numer. Math.*, **29**, 45-58.

[12] Fletcher, R., 1970: Generalized inverses for nonlinear equations and optimization. In Rabinowitz, P. ed. *Numerical Methods for Nonlinear Algebraic Equations*, Gordon and Breach, London, 75-85.

[13] Frohlich, H., and G. Broker, 2003: *Trafox version 2.1. - 3d-kartesische Helmert-Transformation.* [online] Available from: http://www.koordinatentransformation.de/data/trafox.pdf [Accessed 27 August 2008]

[14] Garcia, C.B., and W. I. Zangwill, 1979: Determining all solutions to certain systems of nonlinear equations. *Math. Operations Res.*, **4**, 1-14.

[15] Giusti, M., and E. Schost, 1999: Solving Some Overdetermined Polynomial Systems. *ISSAC 99 Vancouver*, British Columbia, Canada.

[16] Golub, G. H., and V. Pereyra, 1973: The differentiation of pseudo-inverses and nonlinear least squares problems whose variables separate. *SIAM J. Num. Anal.*, **10**, 413-432.

[17] Grafarend, E. W., and B. Schaffrin, 1993: *Ausgleichungsrechnung in Linearen Modellen*, B.I. Wissenschaftsverlag, Mannheim, 483 pp.

[18] Gower, J.C., 1984: Multivariate Analysis: Ordination, Multidimensional Scaling and Allied Topics. *Handbook of Applicable Mathematics*, E. H. Lloyd Ed., **VI**: Statistics, 727-781.

[19] Gower, J.C., and G. B. Dijksterhuis, 2004: *Procrustes Problems*. Oxford University Press, 233 pp.

[20] Haselgrove, C.B., 1961: The solution of nonlinear equations and of differential equations with two-point boundary conditions. *Computing J.*, **4**, 255-259.

[21] Hazaveh, K., D. J. Jeffrey, G. J. Reid, S. M. Watt, and A. D. Wittkopf, 2003: *An exploration of homotopy solving in Maple*. [online] http://www.apmaths.uwo.ca/~djeffrey/Offprints/ascm2003.pdf [Accessed 27 August 2008]

[22] Hooijberg, M., 2008: *Geometrical Geodesy*, Springer, Berlin, 440 pp.

[23] Kapur, D., T. Saxena, and L. Yang, 1994: Algebraic and geometric reasoning using Dixon resultants. In *ACM ISSAC 94*, International Symposium on Symbolic and Algebraic Computation, Oxford, England, July 1994, 99-107.

[24] Lewis, R.H., 2002: Using the Dixon resultant on big problems. in *CBMS Conference*, Texas A&M University, Available from: http://www.math.tamu.edu/conferences/cbms/abs.html [Accessed 27 August 2008]

[25] Lewis, R.H., and S. Bridgett, 2003: Conic tangency equations and Apollonius problems in biochemistry and pharmacology. *Mathematics and Computers in Simulation*, **61**, 101-114.

[26] Lewis, R.H., 2004: Exploiting symmetry in a polynomial system with the Dixon resultant. in *Int. Conf. on Appl. of Computer Algebra (ACA)*, Lamar University, Texas, July 2004.

[27] Lewis, R.H., 2007: Private communication.

[28] Lewis, R. H., 2008: Heuristics to Accelerate the Dixon Resultant. *Mathematics and Computers in Simulation*, **77**, Issue 4, 400-407.

[29] Lichtblau, D., 2007: Private communication.

[30] Mathes, A. 2002: EasyTrans Pro-Edition, Professionelle Koordinatentransformation für Navigation, *Vermessung und GIS*, ISBN 978-3-87907-367-2, CD-ROM mit Benutzerhandbuch

[31] Meridith, W., 1977: On weighted Procrustes and Hyperplane fitting in factor analytic rotation. *Psyhometrika*, **42**, 491-522.

[32] Mourrain, B., 2002: *Symbolic-numeric methods for solving polynomial equations and applications.* [online] Available from: http://mate.dm.uba.ar/~visita16/cimpa/notes/Mourrain.pdf [Accessed 27 August 2008]

[33] Nakos, G., and R. Williams, 2002: A fast algorithm implemented in Mathematica provides one-step elimination of a block of unknowns from a system of polynomial equations, Wolfram Library Archive, *MathSource*, Available from: http://library.wolfram.com/infocenter/MathSource/2597/ [Accessed 27 August 2008]

[34] Palancz, B., 2008a: Introduction to Linear Homotopy, Wolfram Library Archive, *MathSource*, Available from: http://library.wolfram.com/infocenter/MathSource/7119/ [Accessed 27 August 2008]

[35] Palancz, B., 2008b: Extended Newton-Raphson's Method, Wolfram Library Archive, *MathSource*, Available from: http://library.wolfram.com/infocenter/MathSource/7118/ [Accessed 27 August 2008]

[36] Palancz, B., R. H. Lewis, P. Zaletnyik, and J. L. Awange, 2008a: Computational Study of the 3D Affine Transformation, Part I. 3-point problem,Wolfram Library Archive, *MathSource*, Available from: http://library.wolfram.com/infocenter/MathSource/7090/ [Accessed 27 August 2008]

[37] Palancz, B., P. Zaletnyik, and J. L. Awange, 2008b: Extension of Procrustes algorithm for 3D affine coordinate transformation, Wolfram Library Archive, *MathSource*, Available from: http://library.wolfram.com/infocenter/MathSource/7171/ [Accessed 27 August 2008]

[38] Palancz, B., P. Zaletnyik, R. H. Lewis, and J. L. Awange, 2008c: Computational Study of the 3D Affine Transformation, Part II. N-point problem,Wolfram Library Archive, *MathSource*, Available from: http://library.wolfram.com/infocenter/MathSource/7121/ [Accessed 27 August 2008]

[39] Papp, E., and L. Szucs, 2005: Transformation Methods of the Traditional and Satellite Based Networks (in Hungarian with English abstract), *Geomatikai Kozlemenyek* VIII. 85-92.

[40] Quoc-Nam Tran, 1994: Extended Newton's method for finding the roots of an arbitrary system of nonlinear equations. In Hamza, M.H. ed. *Proc. 12th IASTED Int. Conf.on Applied Informatics*, IASTED, Anaheim, CA.

[41] Salmon, G., 1859: *Lessons: Introductory to Modern Higher Algebra.* Hodges and Smith, Dublin, [5th edition, June 1986, Chelsea Pub Co]

[42] Sommese, A. J., C. W. Wampler, 2005: *The Numerical Solution of Systems of Polynomials arising in Engineering and Sciences*, World Scientific Publishing Company, 401 pp.

[43] Spath, H., 2004: A numerical method for determining the spatial Helmert transformation in case of different scale factors, *Zeitschrift für Geodäsie, Geoinformation und Landmanagement*, **129**, 255-257.

[44] Stoer, J., and R. Bulirsch, 2005: *Numerische Mathematik 2*. 5. Auflage, Springer, Berlin, 394 pp.

[45] Szanto, A., 2008: Solving over-determined systems by the subresultant method (with an appendix by Marc Chardin), *J.of Symbolic Computation*, **43**, 46-74.

[46] Watson, G.A., 2006: Computing Helmert transformations, *Journal of Computational and Applied Mathematics*, **197**, 387-395.

[47] Zaletnyik, P., B. Palancz, J. L. Awange, and E. W. Grafarend, 2007: Application of CAS to geodesy: a live approach. IUGG XXIV General Assembly, "Earth: our changing planet" Perugia, Italy 2-13. *IAG Symposia*, **133**, Springer Verlag Berlin, Heidelberg, New York., (in press)

[48] Zaletnyik, P., 2008: *Application of Computer Algebra and Neural Networks to Solve Coordinate Transformation Problems*. PhD. Thesis at Department of Geodesy and Surveying [in Hungarian], Budapest University of Technology and Economics, Hungary, 115 pp.

[49] Zhao, 2007: *Newton's Method with Deflation for Isolated Singularities of Polynomial Systems*, MSc. Thesis. University of Illinois at Chicago, USA, 94 pp. Available from: http://www.math.uic.edu/~jan/Students/ailing_thesis.pdf [Accessed 27 August 2008]

In: Computational Mathematics
Editor: Peter G. Chareton, pp. 313-351
ISBN 978-1-60876-271-2

Chapter 11

DISTANCES BASED ON NEIGHBORHOOD SEQUENCES IN THE TRIANGULAR GRID

Benedek Nagy*
Faculty of Informatics, University of Debrecen,
4032, Egyetem tér 1., Debrecen, 4010 PO Box 12. Debrecen, Hungary

This work is dedicated to the memory of Attila Kuba

Abstract

In computers the discrete plane/space is used since this world is digital. The discrete space is usually defined by a grid. There are three regular grids in two dimensions: the square grid, the hexagonal grid and the triangular grid. The square grid is the most used due to the simplicity of the Cartesian coordinate frame. There are two types of neighbors defined naturally [52, 53]: the city-block and the chessboard neighborhoods. The hexagonal grid is the simplest one, since there is only one natural neighborhood among the hexagons of the grid. The triangular grid is a little bit sophisticated (there are three types of natural neighborhoods), however it has some nice and interesting properties. In the digital space the usage of the usual (Euclidean) distance may lead to some strange phenomena [25]. Instead, path-based, so-called digital distances can be used based on the neighborhood structure of the grid. Digital distances are frequently used in computerized applications of geometry, e.g., in image processing, in computer graphics. There are two main approaches to define digital distances: distances based on neighborhood sequences in which the used types of neighbors is varied along a path; and weighted distances, where various types of steps have various weights (lengths). In the present chapter distances based on neighborhood sequences on the triangular grid are detailed. First, an effective coordinate system is presented to the grid. By this system the grid can handle as easily, as the square grid by the Cartesian frame. After the definitions of neighborhood sequences, paths and distances, some results are detailed: a greedy algorithm that provides a shortest path, formula to compute the distance from a point to another point defined by a given neighborhood sequence. Interesting properties of these distances, such as non-metrical distances are shown (triangular inequality and symmetry may be violated). A necessary and sufficient condition to define metrical distances is proved. Some details on digital circles based on these distances are also presented. Finally some further directions of research and open problems close the chapter.

*E-mail address: nbenedek@inf.unideb.hu

1 Introduction

The digital geometry has important applications in image processing and computer graphics. In digital geometry the spaces we work in consist of discrete points with integer coordinates. In the square (cubic, hypercubic) grids, two points are neighbors if their coordinate difference values are at most one. In the n dimensional grid there are n kinds of neighborhood relations according to the number of differences. In the hexagonal grid only one neighborhood criterion is widely used, while in the case of the triangular grid there are three kinds of neighbors. We define the distance of two points as the number of steps in a shortest path (it is possible that several shortest paths of the same path-length exist), where by a step we mean a movement from a point to one of its neighbor points. This distance function depends not only on the points, but also the given neighborhood criterion. By varying the neighborhood relations in a path, we get the concept of neighborhood sequences. In this chapter we analyze the distance functions based on neighborhood sequences in the triangular grid.

In the next subsection we present a short bibliographic introduction as a survey.

1.1 A brief history of digital geometry

In this part we show some references on grids in digital geometry. The history of neighborhood sequences is also presented.

The birth of digital geometry is connected to [52] and [53]. In these papers Rosenfeld and Pfaltz defined two basic neighborhood relations in the square grid. They gave two types of motions in the two dimensional square grid. The cityblock motion allows horizontal and vertical movements only, while with the chessboard motion the diagonal directions are also permitted. So in this grid we have two kinds of distances, based on these motions. Figure 1 shows a point together with those points, which have distance 1 from it. Both cityblock and chessboard distances are shown. Moreover, the theory of neighborhood sequences started with the paper [53]. The authors recommended the alternate use of the two possible motions for distance measuring. This so-called octagonal distance obtains a better approximation of the Euclidean distance than the ones using only one kind of steps. We come back to the history of neighborhood sequences later.

The higher dimensional square grids were also investigated ([4, 58]). There are some survey papers about the digital metrics of square grids. Short summaries of these kinds of examinations can be found in [26, 29] and in [51]. In the case of the square grid, each coordinate of a point is independent of the others. For n dimensions we use n coordinates. In the n dimensional cubic grid, the structure of the nodes is isomorphic to the structure of the n dimensional cubes. In Figure 1, both options are shown in two dimensions, because we will use both kinds of grid.

We note here, that another branch of the mathematics works on these grids: the 'Geometry of Numbers'. The crystallography has also close relation to digital geometry. The concept 'lattice' is used with the same meaning with which we use 'grid' here. So let us see where and how these grids are used in the field of digital geometry.

Parallel to the square grid the hexagonal and the triangular grid were also investigated in digital geometry. A connection among the cubic, hexagonal and triangular grids can be

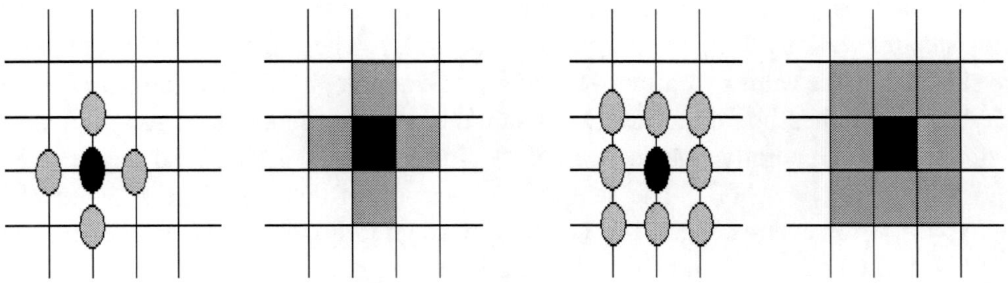

Figure 1: 'Cityblock' and 'chessboard' neighborhood relations in the square grid of nodes and of regions

found in [24, 35, 38, 59], and we will explain it.

We can say that the hexagonal grid is popular. There is a natural neighborhood relation in this grid and it is used almost every time. One of the first papers about the hexagonal plane is [28], in which the name 'rhombic array' was introduced. In pattern recognition the idea of using a non-cartesian coordinate system can be found, for example, in [3, 5, 14] and in [18]. There are some fairly old papers on this topic; hence, the hexagonal grid is also well described.

The hexagonal grid is used in many applications and in practice as well, because it is very natural and simple. The distance function based on the neighborhood relation of the hexagonal grid can be found in [27].

The triangular grid is the third "basic grid". The three kinds of neighborhood criteria of the triangular grid can be found in [14], where thinning algorithms are shown in the three basic grids.

The triangular grid is a valid concurrent plane to the square one in digital geometry. The triangular grids play an increasing role in geometric modeling; many 3D-scanners produce triangulations. These grids are generally not regular, but at high enough resolution they are close to regular ones. The human retina is often modeled by a Delauney triangulation. Many algorithms of computer graphics are also given for the triangular grid ([17, 54]). Therefore we can say that the triangular grid is also one of the most important grids in digital geometry, in digital image processing and in the theory of cellular neural networks also (see [50]).

Now we will consider the theory of neighborhood sequences.

Formally, in digital geometry we use a discrete space, i.e., points can have only integer coordinates. In the square cases we do not have more restrictions for the values of coordinates. Two different points in $\mathbb{Z}^m$ are k-neighbors, $(k, m \in \mathbb{N}, k \leq m)$, if their corresponding coordinate values are equal up to at most k exceptions, and the difference of the exceptional values are at most 1. After fixing k, we may define the distance of two points as the number of steps of a/the shortest path between these points, where a step means moving from a point to one of its k-neighbors (see [10, 12]). It is easy to check that by this definition we get a metric on $\mathbb{Z}^m$, for each $k \in \{1, 2, \ldots, m\}$, and that these metrics are different for the separate values of k.

To obtain these metrics we fixed k in the beginning, in other words, we used the same k in each step for walking from a point p to a point q in $\mathbb{Z}^m$. The situation is more complicated if we can change the value of k after every step. For example, by allowing arbitrary mixtures of cityblock and chessboard motions we obtain the concept of neighborhood sequences in two dimensions. Generally, a sequence $(b(i))_{i=1}^{\infty}$ is called a neighborhood sequence in $\mathbb{Z}^m$ if $b(i) \in \{1,\dots,m\}$ $(i \in \mathbb{N})$. The sequence is periodic if there is some $l \in \mathbb{N}$ such that $b(i+l) = b(i)$ for every $i \in \mathbb{N}$. The concept of neighborhood sequences was introduced in [9] by Das, Chakrabarti and Chatterji, and in more general way (not connected to any specific neighborhood relation) in [60, 61] by Yamashita and his co-authors.

With the help of a neighborhood sequence $(b(i))_{i=1}^{\infty}$ we may define the distance of $p,q \in \mathbb{Z}^m$ in the following way. We take the length of a shortest path from p to q, but at the i-th step now we may move from a point to another if and only if they are $b(i)$-neighbors. Certainly this notion is a generalization of the original one, as we may choose $b(i) = k$ for each $i \in \mathbb{N}$, with any $k \in \{1,\dots,m\}$.

As we mentioned, the neighborhood sequence $(b(i))_{i=1}^{\infty}$ with $b(i) = k$ $(i \in \mathbb{N})$ generates a metric on $\mathbb{Z}^m$ $(m \in \mathbb{N})$ for any $1 \leq k \leq m$. However, it is easy to find neighborhood sequences, such that the distances with respect to these sequences do not provide metrics on $\mathbb{Z}^m$. In [34] there is a nice characterization of the neighborhood sequences, for which the above defined distance functions provide a metric on $\mathbb{Z}^m$.

The main advantage of neighborhood sequences over the classical distances, using only one neighborhood criterion at each step, is that they provide more flexibility in moving in space. Making use of this property, Das and Chatterji [2, 8, 11, 13] were able to determine distance functions that provide a good approximation of the Euclidean distance in the square grid. The authors in [9, 13] analyzed the geometric properties of the octagons occupied by a neighborhood sequence during "spreading" on the 2D plane. This is another aspect of the analysis of neighborhood sequences, examining how the occupied areas develop step by step. It is obvious that in the square grid using only one type of neighborhood in each step, we get squares (in the case of chessboard) or diamonds (in the case of cityblock). Using both kinds of steps we obtain octagons as digital circles. Some results about these wave-front sets were described by Das *et al.* (see [13]). In the three dimensional digital space Danielsson described the digital spheres ([7]). In [30, 34, 20, 46] the vertices of the n dimensional digital hyperspheres were computed.

The concept of neighborhood sequences is also introduced for the triangular grid [31, 33, 32]. An algorithm to find a shortest path by using arbitrary neighborhood sequence can be found in [31, 33] and the detailed analysis of the algorithm in [33]. We showed the strange property that the distance based on neighborhood sequences in the triangular grid may not be symmetric. In [32] there is a necessary and sufficient condition for neighborhood sequences generating symmetric and metrical distances, while the condition of the triangular inequality is detailed in [43]. In [37, 42] a characterization and analysis of some properties of digital circles is given while in [21, 45] we approximated the Euclidean circles by using neighborhood sequences in the triangular grid as well. It turned out that the digital circles of the triangular grid are better approximations of the Euclidean circles than the ones in the square grid.

In Section 2 we show some basic definitions. We present some formal and informal preliminaries as well. In Section 3 we write about the shortest path problem, and we present an algorithm which generates a minimal path. In Section 4 we give a necessary and sufficient condition for a distance based on a neighborhood sequence to be a metric. Some interesting properties are also discussed. In Section 5 the embedding of the triangular grid into the cubic grid is shown, and based on this fact a formula to calculate the distance is presented. In Section 6 we describe the digital circles of the triangular grid. Finally, Section 7 concludes the chapter.

2 Basic Definitions and Notations

First we introduce a notation. In this chapter we will use the function $sgn(x)$ for $x \in \mathbb{R}$:

$$sgn(x) = \begin{cases} +1, & \text{if } x > 0 \\ 0, & \text{if } x = 0 \\ -1, & \text{if } x < 0. \end{cases}$$

The concept of distance plays an important role in this chapter. We use the name *"distance"* for any function which is defined on the square of the Universe (V). Usually we use points (or regions) with integer coordinates and therefore the elements of the Universe are vectors with integer values.

Definition 2.1. A function $d: V \times V \to \mathbb{R}$ is called a metric on V if it satisfies the following conditions:

1. $\forall p,q \in V$: $d(p,q) \geq 0$, and $d(p,q) = 0$ if and only if $p = q$ (positive definiteness),
2. $\forall p,q \in V$: $d(p,q) = d(q,p)$ (symmetry),
3. $\forall p,q,r \in V$: $d(p,q) + d(q,r) \geq d(p,r)$ (triangular inequality).

We note that usually the metric properties are defined as above, but the non-negative property is a consequence of the other properties.

Using same-size regular triangles there is a way for tiling the plane. The grid of the triangular areas (the so-called triangular grid) is isomorphic to the grid of hexagonal nodes (tiling by same-sized regular hexagons and addressing the nodes as points). In this chapter we consider these grids, and we mostly use the triangular regions.

We present a suitable method to formulate the concept of neighborhoods and neighborhood sequences in the triangular grid.

We can use more types of neighbors in an arbitrary grid. Informally, we assume that two geometrical objects (of the same dimension n as the space where we are) are neighbors of each other if there is at least one point which is on the border of both. Two geometrical objects are neighbors of type m (or shortly, m-neighbors) if we can make at most m steps from one to the other in such a way that in each step we step through an ($n-1$ dimensional) border line of two objects.

The neighborhood relations in triangular grid is based on the widely used relations (see [14]), we use three types of neighbors as Figure 2 shows.

Each triangle (not considering the original one) has three 1-neighbors, nine 2-neighbors (the 1-neighbors, and six more 2-neighbors), and twelve 3-neighbors (nine 2-neighbors, and three more 3-neighbors).

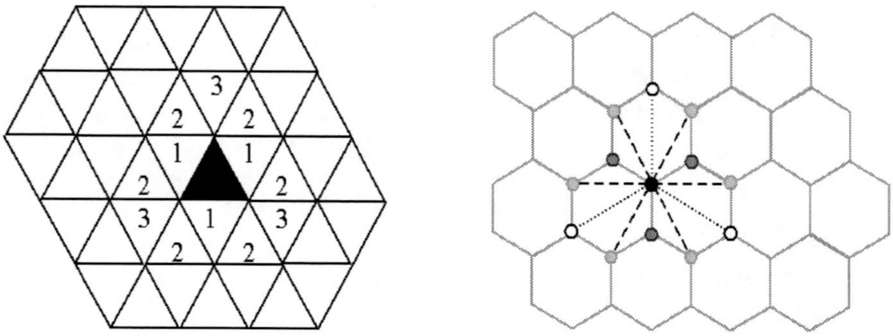

Figure 2: Types of neighbors in the triangular grid of areas and in the hexagonal grid of nodes

Analogously, we can define the relation *"m-neighborhood"* among the nodes in (planar) graphs. Two nodes are neighbors if they are on the border of the same region. They are m-neighbors if they are neighbors and the shortest path between them includes at most m edges.

Sometimes we use the concept of strict m-neighbors when two objects are m-neighbors, but they are not $(m-1)$-neighbors.

It is obvious that these m-neighborhood relations are reflexive and symmetric relations. Moreover they have the following inclusion properties. All $(m-1)$-neighbors of an object are also its m-neighbors.

In Figure 2, for hexagonal grid of nodes we use the dark grey points to represent the 1-neighbors. With these points the light grey ones are the 2-neighbors, and together with them the white points are the 3-neighbors. (Only the 1-neighbors are directly connected by a side, the 2 and 3-neighbors are at the positions of diagonals, respectively.)

These relations are reflexive (i.e., the pixel marked dark triangle is a 1-, 2-, and 3-neighbor of itself). In addition, all 1-neighbors of a pixel are its 2-neighbors and all 2-neighbors are 3-neighbors, as well (i.e., increasing and inclusion properties).

Using this definition of m-neighborhood we have to pay attention to the coordinatization of the grid. The aim is to assign the coordinate values to the objects in such a way that we can redefine the neighborhood criteria in natural way by using coordinate values.

The triangular grid has triangular symmetry, therefore three coordinates are recommended to analyze this system ([36]). In [39] some useful geometrical properties of the grid and the coordinate-system is presented.

To describe mathematically the triangular grid we need an appropriate coordinate system. The next procedure shows how we will order the coordinate values to the triangles.

Procedure 2.1. Choose a point for the origin, whose coordinate values are $(0,0,0)$. Take the three lines through the centre of the origin triangle, which are orthogonal to its sides. Fix these lines as the coordinate axes x,y and z, as Figure 3 shows. We assign the coordinate values to the points inductively. Let the coordinate values of a triangle p be known. Consider a triangle q which has not coordinate values yet and has a common side with p. This common side is orthogonal to one of the coordinate axes. According to the direction

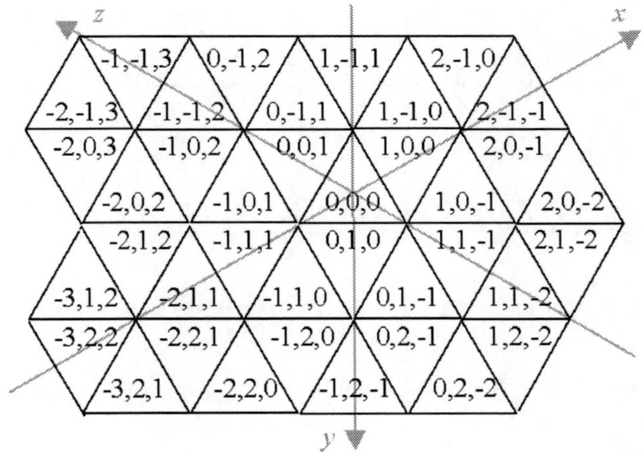

Figure 3: Coordinate values on the triangular grid

of this axis, we increase or decrease the corresponding coordinate value of p by 1 to get the corresponding coordinate of q. The other two values of p and q are equal.

Figure 3 shows a part of the triangular grid with the associated coordinate values.
Due to the coordinate values we can redefine the neighborhood relations.

Definition 2.2. The points p and q of the triangular grid are m-neighbors ($m = 1,2,3$), if the following two conditions hold:

1. $|p(i) - q(i)| \leqslant 1$, for $i = 1,2,3$,
2. $|p(1) - q(1)| + |p(2) - q(2)| + |p(3) - q(3)| \leqslant m$.

Remember that we use the term strict m-neighbors if the second condition is equality.

Remark 2.1. It is easy to check that the formal definition above with the presented coordinate values (Figure 3) gives the neighborhood relations shown in Figure 2.

We adopt some definitions from the literature mentioned earlier.
According to the possible types of neighbors, we can define the so-called neighborhood sequences.

Definition 2.3. The infinite sequence $B = (b(i))_{i=1}^{\infty}$, – in which the values $b(i) \in \mathbb{N}$ are possible types of neighborhood criteria in the digital space that is used – is called a neighborhood sequence (or abbreviated n.s.). If for some $l \in \mathbb{N}$, $b(i) = b(i+l)$ holds for every $i \in \mathbb{N}$, then B is called periodic (with period l). In periodic cases we will use the abbreviation $B = (b(1),\ldots,b(l))$.

In 1984 the name of "neighborhood sequences" appeared [60]. In 1987, Das *et al.* [9] used the theory of neighborhood sequences in arbitrary finite dimensional square grids.

Definition 2.4. Let p and q be two points and $B = (b(i))_{i=1}^{\infty}$ a n.s. A finite point sequence $\Pi(p,q;B)$ of the form $p = p_0, p_1, \ldots, p_m = q$, where p_{i-1}, p_i are $b(i)$-neighbor points for

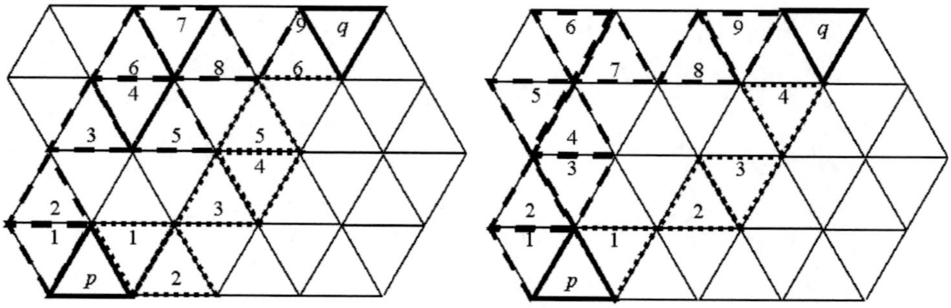

Figure 4: Paths of different lengths from p to q using B_1 and B_2

$1 \leq i \leq m$, is called a B-path from p to q. We write $m = |\Pi(p,q;B)|$ for the length of the path. Let us denote by $\Pi^*(p,q;B)$ a shortest path from p to q, and set $d(p,q;B) = |\Pi^*(p,q;B)|$. We call $d(p,q;B)$ the B-distance of p and q.

Note that the shortest path problem in our case looks more like a graph-theoretical problem (see [23]) than the problem in Euclidean space. There can be several shortest paths opposite to the Euclidean case when it is always only one.

In the triangular grid we can define several types of distances according to the neighborhood criteria used. Since there are three types of neighborhoods in this grid, we will use the neighborhood sequences containing numbers of the set $\{1,2,3\}$.

Now, we show some paths defined by neighborhood sequences in the triangular grid.

In Figure 4 there are some paths given between the points $p = (-3,2,1)$ and $q = (2,-1,0)$ by the help of $B_1 = (1,1,2)$ and $B_2 = (1,3,1,2,2,2,2,2,2,2,2,2\ldots)$. As we can see, there are paths with different lengths between the points. On the left-hand side of Figure 4 we show two paths with the neighborhood sequence B_1, one of them has length 10 and the other has length 7. The points of the paths are represented as dashed and dotted shapes. The numbers in the triangles refer to the steps of the given path. Similarly, in the right-hand side of the figure, we show other paths, using B_2, of length 5 (it is the shortest path) and 10 between the same points.

We can define the concept "lanes" for geometric descriptions.

Definition 2.5. The points having the same value as x, y, or z-coordinate, form a lane.

Remark 2.2. Each lane is orthogonal to one of the coordinate axis. For the points of a lane a coordinate value is fixed. The other two values changes by ± 1.

Remark 2.3. The points and their coordinate values are assigned by a one-to-one mapping. We can see this in the following way using the concept of lanes. Let us fix two coordinate values. They define two non-parallel lanes, whose intersection contains two points. The third coordinate of each of them should have a value such that the sum of the coordinates are 0 and 1, respectively. Moreover the points of the triangular plane are exactly the points with three integer coordinate values with sum of coordinate value 0 or 1.

There are two types of points according to the values of the sum of its coordinates. To distinguish them we define the parity of the points.

Figure 5: Examples for lanes in the triangular grid

Definition 2.6. If the sum of the coordinate values of the point p is 0, then we call the point p even. If the sum is 1, then the point p has odd parity.

In our figures the even and odd triangles are of the shape $\triangle$ and ∇, respectively. The points of the grid have the following properties.

Remark 2.4. If two points are 1-neighbors, then there exist two lanes containing both of them, and their parities are different. If two points are strict 2-neighbors then only one lane contains both of them and their parities are the same. If two points are strict 3-neighbors then their parities are different and no lane contains both of them.

We can define distance between points and lanes, which depends only on these objects and independent of the neighborhood sequences. The distance of a point and a lane is the minimal number of parallel lanes which we have to go through to reach the point from the lane. We can compute this distance in an iterative way.

Procedure 2.2. The distance between the lane A and the point p is zero if p is on the lane A. The next parallel lanes of the lane which contains the point p have distance 1 from p. Let the distance of the lane A' and p be d. Then the next parallel lane with A', – which does not have distance $d-1$ from the point p – has distance $d+1$ from p.

The distance between a lane and a point, with respect to the distance of points, has the following property.

Remark 2.5. If the distance between the lane A and the point p is d, then starting from p we can reach some points of A in the d-th step by using the constant neighborhood sequence (2).

Definition 2.7. The difference $w_{p,q} = (w(1), w(2), w(3))$ of two points p and q is defined by: $w(i) = q(i) - p(i)$. If $w(1) + w(2) + w(3) = 0$, then the parity of w is even, else it is odd. Let $v_{p,q}$ be the sorted difference of the points, i.e., we order the values of $w_{p,q}$ by non-decreasing way by their absolute values. Formally: let h_i $(i = 1, 2, 3)$ be a permutation

Figure 6: An example for a parallelogram between two points

of $(1,2,3)$, such that $|w(h_1)| \geqslant |w(h_2)| \geqslant |w(h_3)|$, and $sgn(w(h_1)) \neq sgn(w(h_2))$. Then $v_{p,q} = (w(h_1), w(h_2), w(h_3))$. In obvious cases we omit the indices of $w_{p,q}$ and $v_{p,q}$.

The distance using the neighborhood sequence (1) is a special one, for using it we introduce the abbreviation as follows.

Notion 2.1. The distance of any two points with the neighborhood sequence (1) (i.e., in which every element is 1) is called 1-distance.

One can connect any triangle to any other one by the neighborhood sequence (1), using one or two lanes. In the latter case, the angle of the directions of the motion on these lanes can be chosen to be $\frac{2\pi}{3}$. (In one of these lanes that coordinate value remains constant, which has the smallest absolute value in the difference w of the points. In the other one that coordinate value is fixed which has the second largest absolute value in w.) Generally, we obtain a parallelogram, see Figure 6.

3 The Shortest Paths

In the triangular grid, we have some difficulties in changing the coordinate values. Such difficulties do not occur in case of the hexagonal and square grids. In the triangular grid, when moving from a point to one of its neighbors, we have to take care of the parity of these points. Namely, we have to change the coordinates of a point in such a way that the sum of the coordinate values must be 0 or 1 (see Remark 2.3).

Now we give a greedy algorithm which solves the problem of constructing a shortest path between two given points ([31, 33]). We prove that the algorithm is correct, i.e., it finds a shortest path from the first point to the other one.

Algorithm 3.1.
Input: two points p, q; a neighborhood sequence B.
step 1. Let w be the difference of p and q, and let $x_0 = p$, $\Pi = (x_0)$ and $j = 0$.

step 2. While $w \neq (0,0,0)$ do
step 3. Let $j = j+1$. Let h_i $(i = 1,2,3)$ be a permutation of $(1,2,3)$, such that $|w(h_1)| \geqslant |w(h_2)| \geqslant |w(h_3)|$ and $sgn(w(h_1)) \neq sgn(w(h_2))$.
step 4. If $b(j) = 1$,
then if x_{j-1} is even/odd,
change by 1 the positive/negative one from $w(h_1)$ and $w(h_2)$, respectively: $w(h_i) = sgn(w(h_i))|w(h_i) - 1|$, where $i = 1$ or 2;
go to step 8.
step 5. If $b(j) = 2$,
then let $w(h_1) = sgn(w(h_1))|w(h_1) - 1|$ and $w(h_2) = sgn(w(h_2))|w(h_2) - 1|$;
go to step 8.
step 6. If the parity of x_{j-1} is even
then if w has two coordinates with positive values,
then let $w(i) = sgn(w(i))|w(i) - 1|$ $(i = 1,2,3)$,
else let $w(h_1) = sgn(w(h_1))|w(h_1) - 1|$ and $w(h_2) = sgn(w(h_2))|w(h_2) - 1|$.
step 7. If the parity of x_{j-1} is odd
then if w has two negative coordinate values,
then let $w(i) = sgn(w(i))|w(i) - 1|$ $(i = 1,2,3)$,
else let $w(h_1) = sgn(w(h_1))|w(h_1) - 1|$ and $w(h_2) = sgn(w(h_2))|w(h_2) - 1|$.
step 8. Let $x_j(i) = q(i) - w(i)$ $(i = 1,2,3)$.
step 9. Concatenate x_j to the path Π.
step 10. End while.
step 11. Output: one of the shortest paths Π from p to q, using B and the length of the path is j. End.

Now we give a detailed description of the algorithm.

In the first step we initialize the algorithm: p is the starting point, $p = x_0$ is the first element of the path Π which contains x_0 only, w is the difference between the last point of Π and q. The length of the path j starts from 0.

In step 2 we check whether we have finished or not. If yes, we go to Step 11, where the output values are given, and the algorithm terminates.

In step 3 we increase the length of the path j, and order the elements of the difference w. First, suppose that among these elements, there exists one, with largest absolute value. In this case, this element has opposite sign from the others, or some of the others are equal to zero. If two elements have the same absolute value, and the third one has smaller absolute value, then the two elements with larger absolute value have opposite signs. Then the third element must be 0 or ± 1, because of the restriction of the sum of the coordinates of w. Hence the permutation satisfies our conditions. If all three elements have the same absolute value, then this value must be 1, and their sum is ± 1. Hence the permutation can be made in every case.

As we mentioned and showed in Figure 6, we can connect points x_j and q by two lanes and we have a parallelogram. If we can move to a 1- or a 2-neighbor of x_j then we make this step on the lane which goes through x_j and is closer to q.

We use step 4 if we move to a 1-neighbor. In this case we can move from x_{j-1} to a point of different parity. We decrease one of the absolute values of the first two elements of the permutation of w by 1. If x_{j-1} is even, then the sum of the elements of w is 1, if q is odd, and the sum of the elements of w is 0, if q is even. Since w has a non-zero element, $w(h_1)$ or $w(h_2)$ must be positive, as well. So we can change this positive value. If x_{j-1} is odd and q is even, then the sum of the elements of w is -1; the sum of the elements of w is 0, if q is odd. Since w has a non-zero element, $w(h_1)$ or $w(h_2)$ must be negative, as well. So we can change this negative value. Thus, at this step we get closer to q in a coordinate by 1.

Let us consider step 5. If we can move to a 2-neighbor of x_{j-1}, we have two cases. First, if there is only one non-zero element of w, then it must be ± 1, so we change this element to 0. In this case only this element is changed since $sgn(0) = 0$. In the other case, the first two elements of the permutation have opposite signs, hence we move from x_{j-1} to a point of the same parity, by changing these two values.

In step 4 and 5 we step in the lane which has smaller distance from the destination point. We can decrease the higher absolute values in w, but if we can move to a 3-neighbor then we step in the lane in which we can decrease the higher absolute values in w, and if possible we step to the next parallel lane which is closer to the end-point. So if we move to a 3-neighbor of x_j, then if possible we step to the point which is in the intersection of two lanes which are closer to q than the previous ones. (This point is on the next lanes as x_j.) If this is impossible, then we move to a 2-neighbor of x_j. We describe these cases below.

We use step 6 or 7 if $b(j) = 3$, so we move to a 3-neighbor of x_{j-1}. If the parity of x_{j-1} is even (Step 6), then we may step to an odd point, and change all of its coordinates by 1, if w contains two positive and one negative values. If x_{j-1} has odd parity, and w contains two negative and one positive values, then we step to an even point, by changing every coordinate value by 1. If w does not let us to do so, we can step only to a 2-neighbor, like at Step 5.

In step 8 and 9 the algorithm calculates the coordinates of x_j, by using the values of w, and adds x_j to the path Π (w is the difference of q and x_j).

Step 10 guarantees the repetition of the procedure, starting from step 2.

Before we present some examples how the algorithm works in practice, we prove that it is correct.

Theorem 3.1. *Algorithm* 3.1 *provides a shortest path.*

Proof. Let $p = y_0, y_1, \ldots y_m = q$ be a B-path, and for $i = 0, \ldots, m$ put $v_i = (q(1) - y_i(1), q(2) - y_i(2), q(3) - y_i(3))$ and $h_i = |y_i(1) - q(1)| + |y_i(2) - q(2)| + |y_i(3) - q(3)|$. Similarly, for the path provided by the algorithm ($p = x_0, x_1, \ldots x_n = q$), let $w_i = (q(1) - x_i(1), q(2) - x_i(2), q(3) - x_i(3))$ and $g_i = |x_i(1) - q(1)| + |x_i(2) - q(2)| + |x_i(3) - q(3)|$ ($i = 1, \ldots, n$).

We show that $g_i \leqslant h_i$ for all $i \leqslant \min(m,n)$. We use induction. For $i = 0$ we have $g_0 = h_0$, so our assumption holds. Suppose that $g_i \leqslant h_i$. We prove that $g_{i+1} \leqslant h_{i+1}$. We distinguish three cases according to the values of $b(i)$.

If $b(i) = 1$, then $g_{i+1} = g_i - 1$, and $h_{i+1} \geqslant h_i - 1$. Hence, by $g_i \leqslant h_i$, we have $g_{i+1} \leqslant h_{i+1}$.

If $b(i) = 2$, then $h_{i+1} \geqslant h_i - 2$, and if $g_i \geqslant 2$ then $g_{i+1} = g_i - 2$. Hence, using $g_i \leqslant h_i$, we get $g_{i+1} \leqslant h_{i+1}$. If $g_i = 1$, then $g_{i+1} = 0$, and $g_{i+1} \leqslant h_{i+1}$ holds.

Finally, let $b(i) = 3$. Then $g_{i+1} = g_i - 3$ yields $g_{i+1} \leqslant h_{i+1}$, as $h_{i+1} \geqslant h_i - 3$. If $g_{i+1} = 0$, then we also have $g_{i+1} \leqslant h_{i+1}$. Otherwise, $g_{i+1} = g_i - 2$, and we have two possibilities: the parity of x_j is even, and w_j contains two negative values, or the parity of x_j is odd, and w_j

contains two positive values. If $h_i > g_i$, then $g_{i+1} \leqslant h_{i+1}$, since $h_{i+1} \geqslant h_i - 3$. If $h_i = g_i$, then the parity of w_i is the same, as the parity of v_i. We have the following cases. If v_i has the same number of positive and negative elements as w_i, then y_{i+1} can differ from y_i in at most two coordinates, and we still have the inequality $g_{i+1} \leqslant h_{i+1}$. Otherwise y_{i+1} differs from y_i in all coordinates, and we have $g_{i+1} \leqslant h_{i+1}$ again, since the difference of some coordinate of y_{i+1} and q must grow. If $h_i = g_i$ then v_i cannot contain the same number of positive elements as the number of negative elements of w_i, and vice versa. This is because the number of the positive and negative elements in w_i is the same as in the difference between q and p (until one or more of them become zero). If v_i contains an element c, which has opposite sign in w_i, then h_i must be greater than the sum, where c is replaced by 0. Since we assumed that $g_i = h_i$, it is a contradiction.

We have $g_i \leqslant h_i$ for all $i \leqslant \min(m,n)$, and the sequence g_i strictly monotonously goes to zero. This implies that $n \leqslant m$. So the algorithm stops after finitely many steps, and it provides a shortest path from p to q. □

The following remark is a consequence of the work of the algorithm.

Remark 3.1. The distance of any two points p and q, with respect to a neighborhood sequence B, depends on the difference and the parity of the points, and on the neighborhood sequence only.

Algorithm 3.1 is a greedy algorithm, since at every step it changes as many coordinate values as possible to get closer to the end point.

Let us analyze the complexity of Algorithm 3.1. It is clear that we need only memory to store the point where we are (x_i), and what is the following element of the neighborhood sequence. So if we can write the output path and we can read the sequence while the algorithm is running, then we need only constant memory.

What is the time-complexity of our algorithm?

It is easy to show that there is a constant upper bound for the time that an iteration takes. (In Steps 3-9 there is ordering of three elements, evaluation of conditions and changing values.) So an iteration takes maximum c time. In the worst case (with neighborhood sequence (1)) we must make $|w(1)| + |w(2)| + |w(3)|$ steps, where w is the difference of the start and end points.

So our algorithm terminates at most after $c(|w(1)| + |w(2)| + |w(3)|)$ time, which is linear in the different of the coordinate values of the starting and ending points. So we can say that our algorithm is efficient.

The next examples show how the algorithm works in practice.

Table 1: Construction of a shortest path to Example 3.1

w	j	$w(h_1)$	$w(h_2)$	$w(h_3)$	$b(j)$	x_j
$(-3,4,-2)$	0	–	–	–	–	$(3,-4,2)$
$(-3,4,-2)$	1	y:4	$x:-3$	$z:-2$	2	$(2,-3,2)$
$(-2,3,-2)$	2	y:3	$x:-2$	$z:-2$	3	$(1,-2,1)$
$(-1,2,-1)$	3	y:2	$x:-1$	$z:-1$	1	$(1,-1,1)$
$(-1,1,-1)$	4	$x:-1$	y:1	$z:-1$	3	$(0,0,0)$

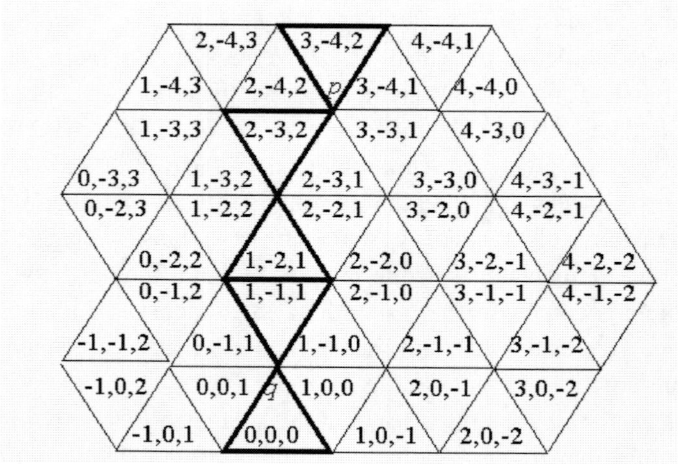

Figure 7: The shortest path in Example 3.1

Example 3.1. Let $p = (3,-4,2)$ and $q = (0,0,0)$ be two points, and $B = (2,3,1,3)$ a neighborhood sequence. Table 1 shows how the values change during the algorithm. The notation used in Table 1 is the same as used at Algorithm 3.1. The first row of the table contains the initial values of the algorithm. Every row contains values obtained after moving to the next point. The presented shortest path is in Figure 7.

Further in this subsection, we study distance functions with strange properties.

Table 2: The shortest path from r to s in Example 3.2

w	j	$w(h_1)$	$w(h_2)$	$w(h_3)$	$b(j)$	x_j
$(-1,2,-1)$	0	–	–	–	–	$(1,-2,1)$
$(-1,2,-1)$	1	y:2	x:-1	z:-1	1	$(1,-1,1)$
$(0,1,-1)$	2	x:-1	y:1	z:-1	3	(0,0,0)

We illustrate that our algorithm works also in the case, when the distance based on a neighborhood sequence is not symmetric and/or does not meet the triangular inequality.

Table 3: The shortest path from s to r in Example 3.2

w	j	$w(h_1)$	$w(h_2)$	$w(h_3)$	$b(j)$	x_j
$(1,-2,1)$	0	–	–	–	–	(0,0,0)
$(1,-2,1)$	1	y:-2	x:1	z:1	1	(1,0,0)
$(0,-1,0)$	2	y:-2	z:1	x:0	3	$(1,-1,1)$
$(0,0,-1)$	3	y:-1	x:0	z:0	2	(1,-2,1)

Example 3.2. Let $r = (1,-2,1)$ and $s = (0,0,0)$ be two points, and $B = (1,3,2)$ a neighborhood sequence. First, we calculate $d(r,s;B)$ by using Algorithm 3.1. By Table 2 we get $d(r,s;B) = 2$. Now let us calculate $d(s,r;B)$. The result is in Table 3: $d(s,r;B) = 3$. As $d(r,s;B) \neq d(s,r;B)$, this distance function is not symmetric.

Table 4: Calculating a shortest path from $(0,0,0)$ to $(0,1,-1)$ by $B=(2,1,1)$

w	j	$w(h_1)$	$w(h_2)$	$w(h_3)$	$b(j)$	x_j
(0,1,−1)	0	–	–	–	–	(0,0,0)
(0,1,−1)	1	y:1	z:−1	x:0	2	(0,1,−1)

Table 5: Calculating a shortest path from $(0,1,-1)$ to $(0,2,-2)$ by $B=(2,1,1)$

w	j	$w(h_1)$	$w(h_2)$	$w(h_3)$	$b(j)$	x_j
(0,1,−1)	0	–	–	–	–	(0,1,−1)
(0,1,−1)	1	y:1	z:−1	x:0	2	(0,2,−2)

Example 3.3. Let $r=(0,0,0)$, $s=(0,1,-1)$ and $t=(0,2,-2)$ be three points, and $B=(2,1,1)$ a neighborhood sequence. The calculation of $d(r,s;B)$ is in Table 4. So $d(r,s;B)=1$. In Table 5 we present the determination of $d(s,t;B)=1$. The calculation of $d(r,t;B)$ is presented in Table 6. As we can see, $d(r,t;B)=3$, and $d(r,t;B)>d(r,s;B)+d(s,t;B)$. Thus the triangular inequality does not hold.

As we can see in the previous examples, we can find neighborhood sequences which do not generate metrics.

Example 3.4. The distance function generated by the neighborhood sequence $B=(3,1)$ is non-symmetric and non-triangular.

In the hexagonal and square cases the distance functions must be symmetric, because by the symmetry of the grid we can interchange the start- and endpoints. However in the triangular plane this does not work in every case because of the parity of the points. The digital distance based on the number of steps in the hexagonal grid (of areas) is a metric ([27]). In the square grid the triangular inequality may not hold, in [34] there is a necessary and sufficient condition for metric B-distances. The non-symmetric distance is a new phenomenon in the digital geometry (with symmetric neighborhood structure), it appears in the triangular grid with neighborhood sequences. In the next section we are going to give a condition for the neighborhood sequence to generate a metric distance in the triangular grid.

4 Condition for Metric Distances

In view of the previous examples, we have the following natural question: knowing B, how can we decide whether the distance function defined by B is a metric on the triangular grid. We give the answer in this section (based on [32, 43]).

Lemma 4.1. *The* 1*-distance of p and q is given by* $d(p,q;(1))=|w(1)|+|w(2)|+|w(3)|$.

Proof. The statement is an easy consequence of the definitions. □

We have seen by the work of Algorithm 3.1 that it is possible that it is less efficient to use a 3-step as a strict 3-step than a 2-step (even if the endpoint is not reached yet). As is shown in the previous section, we can step by 1 or 2 in all possible cases in the shortest path

Table 6: Calculating a shortest path from $(0,0,0)$ to $(0,2,-2)$ by $B=(2,1,1)$

w	j	$w(h_1)$	$w(h_2)$	$w(h_3)$	$b(j)$	x_j
(0,2,−2)	0	–	–	–	–	(0,0,0)
(0,2,−2)	1	y:2	z:−2	x:0	2	(0,1,−1)
(0,1,−1)	2	y:1	z:−1	x:0	1	(0,2,−1)
(0,0,−1)	3	z:−1	x:0	y:0	1	(0,2,−2)

(expect if we need only a 1-step to reach the endpoint and we have a 2 in the neighborhood sequence).

Now we introduce the concept of *minimal equivalent neighborhood sequence*, which will be helpful when deriving conditions for metricity.

Definition 4.1. Let B and B' be two neighborhood sequences. B' is the minimal equivalent neighborhood sequence of B, if the following conditions hold:

1. $d(p,q;B)=d(p,q;B')$ for all points p,q; and
2. for each neighborhood sequence B_1, if $d(p,q;B)=d(p,q;B_1)$ for all points p, q, then $b'(i)\leqslant b_1(i)$ for all i.

Lemma 4.2. *The minimal equivalent neighborhood sequence B' of B is uniquely determined, and it is given by*

$b'(i)=2$, *if* $b(i)=3$ *and there is a* $j<i$ *such that* $b(j)=3$ *and* $\sum_{k=j+1}^{i-1} b'(k)$ *is even, where* $j=\max\{l|l<i \text{ and } b'(l)=3\}$.
else $b'(i)=b(i)$.

Proof. We start from the beginning of the neighborhood sequence B, and we give the element $b'(i)$ using $b(i)$ and the previous part of the sequence B' using Algorithm 3.1.

It is obvious, that if $b(i)<3$ then $b'(i)=b(i)$, because we use these steps as strict 1- or 2-steps.

The question is when we may use the strict 3-steps and when it is impossible (to move farther by a strict 3-step than a 2-step).

There are two cases possible for strict 3-steps:

Let $b(i)=3$, and let $p=p_0,p_1,\dots p_n=q$ be a minimal path from p to q produced by Algorithm 3.1.

We can change all coordinate values in the i-th step in the following two cases: (according to the parity and the direction)

- p_{i-1} is odd and we need to decrease two coordinates, and increase only one value to go to q the difference of q and p_{i-1} has two negative and a positive value,
- p_{i-1} is even and we need to increase two coordinate values, and decrease only one of them: the difference of q and p_{i-1} has two positive and a negative value.

Therefore in a given path produced by the algorithm at most one of the two cases occurs. In this way strict 3-steps are possible, if

this $b(i) = 3$ is the first element 3 of the neighborhood sequence B; or

$b(i) = 3$ is not the first element 3, but $\sum_{k=j+1}^{i-1} b'(k)$ is odd, where $j = \max\{l | l < i, b'(l) = 3\}$.

Indeed, if this subsum is even then we are in a point which has the same parity as the point after the previous 3-step, hence we cannot use a strict 3-step in the i-th step.

By the above argument one can easily see that B' is uniquely determined. Thus the lemma is proved. □

Corollary 4.1. *Let B' be the minimal equivalent neighborhood sequence of an arbitrary neighborhood sequence B. Let $j \in \mathbb{N}$ and let $r = \sum_{i=1}^{j} b'(i)$. There are points p and q, such that their 1-distance is r and B'-distance is j, i.e., we can use all elements 3 of the neighborhood sequence B' to strict 3-steps.*

However, it is possible (depending the coordinate values of p and q) that, we use the first element 3 as value 2 in the shortest path according to Algorithm 3.1. For this case we will use the concept of *reduced minimal equivalent neighborhood sequence*.

Definition 4.2. The reduced minimal equivalent neighborhood sequence B'' of B is given by:

- $b''(k) = 2$, where k is the index of the first element 3 of B;
- $b''(i) = b'(i)$, for all other value of i, where $b'(i)$ are the correspondent elements of the minimal equivalent neighborhood sequence B' of B.

Now we are turning to analyze the metrical properties of the distances.

Lemma 4.3. *The distance based on a neighborhood sequence $B = (b(i))_{i \in \mathbb{N}}$, is non-symmetric if and only if $b(i) = 3$ for an $i \in \mathbb{N}$, and one of the following conditions is true with $i = \min\{l | b(l) = 3\}$:*

- $\sum_{k=1}^{i-1} b(k)$ *is odd; or*
- *there is a j such that $b(j) = 1$, and $i < j$.*

Proof. First assume that the first condition holds. We give an example when this distance is not symmetric.

Let i be the index of the first element 3 of B, and l is the index of the previous element 1, i.e., $l < i, b(l) = 1$, and there is no h, such that $l < h < i$ and $b(h) = 1$. There must be such an index l, because $\sum_{k=1}^{i-1} b(k)$ is odd, and this part of B does not contain any elements 3, it contains only elements 1 and 2.

Let $x = \frac{\sum_{k=1}^{l-1} b(k)}{2}$ and $y = \frac{\sum_{k=l+1}^{i-1} b(k)}{2} = i - l - 1$. These sums must be even by our assumptions and the choices of the indexes i and l. Let $p = (0,0,0)$ and

$q = (-1, -x-y-1, x+y+2)$. We can calculate the distances between these points by using Algorithm 3.1. We get $d(p,q;B) = i$, because a shortest path goes through the following points: before the l-th step we are in $(0,-x,x)$, after this we are in $(0,-x,x+1)$. Before the i-th step we are in $(0,-x-y,x+y+1)$ (we arrive here from p on the lane where the first coordinate value is 0), and after the i-th step we are in $(-1,-x-y-1,x+y+2) = q$ (we take a strict 3-step). But $d(q,p;B) = i+1$, because a shortest path goes through the following points: before the l-th step we are in $(-1,-y-1,y+2)$, after this we are in $(-1,-y,y+2)$. Before the i-th step we are in $(-1,0,2)$ (we go from q to $(-1,0,2)$ on the lane where the first coordinate value is -1), and after the i-th step we are in $(0,0,1)$, so we need one more step to reach p (because we cannot use a strict 3-step).

Now we assume that the first condition is false, but the second is true. We provide a counterexample again.

Let i be the same as above and $b(j) = 1$, such that $i < j$, and j is minimal with these properties. Then let $x = \frac{\sum_{k=1}^{i-1} b(k)}{2}$ (it is an integer because the first condition is false). Let B' be the minimal equivalent sequence of B. It follows from Lemma 4.2, that $b'(k) = 2$ for all $i < k < j$, hence let $y = \frac{\sum_{k=i+1}^{j-1} c(k)}{2} = j-i-1$. Let $p = (0,0,0)$ and $q = (1,-x-y-2,x+y+1)$. Then a shortest path from p to q: We start from $p = (0,0,0)$, before the i-th step we go to $(0,-x,x)$ on the lane where the first coordinate value is zero, after this step we are in $(1,-x-1,x+1)$ (in this step we can go to another lane with a strict 3-step). After the i-th step we move on the lane where the first coordinate is still 1, and before the j-th step we arrive the point $(1,-x-y-1,x+y+1)$, and after the j-th step $(1,-x-y-2,x+y+1) = q$, hence this distance is j. From q to p with B: we started from $q = (1,-x-y-2,x+y+1)$, before the i-th step $(1,-y-2,y+1)$, after this $(1,-y-1,y)$ (in this step we cannot go to another lane because the parity and the direction, hence we cannot move a strict 3-step, so we use the lane where the first coordinate is 1 to go away). Before the j-th step we are in $(1,-1,0)$, and after the j-th step $(1,0,0)$, and we need one more step to reach p, hence this distance is $j+1$.

Finally we show that if both conditions are false then the B-distance is symmetric. Let p and q be two arbitrary points. We prove that $d(p,q;B) = d(q,p;B)$.

If B does not contain the element 3, then we use Algorithm 3.1 to construct the shortest paths. It is easy to see that we have no problem if we use only 1- and 2-neighbors (we can move strict 2-steps and strict 1-steps from each point to any direction), hence $d(p,q;B) = d(q,p;B)$.

If there is an element 3 in B, then we must use the minimal equivalent sequence B' instead of B. We know from the false conditions and from Lemma 4.2 that the subsum before the first element 3 is even, and after this we have only elements 2 in B'. Let i be the index of the 3 in B'. Put $x = \frac{\sum_{k=1}^{i-1} b(k)}{2}$. This sum is even because the first condition of the theorem is false. If $d(p,q;(1)) < 2x+1$, then we have the same case as if B does not contain any elements 3. If $d(p,q;(1)) = 2x+1$ or $d(p,q;(1)) = 2x+2$, then we step only to a 2-neighbor at the last step, so both distances are i. If $d(p,q;(1)) > 2x+2$ then we either can use the strict 3 in the i-th step, or not. Let us see first the case when we can use this strict 3-step. Before the i-th step we are in a point with the same parity as our starting point.

As we can see in the proof of Lemma 4.2, we can use the strict 3-step if

- we go from an odd point and we need to decrease two coordinates, and increase only one value to go to q
- we go from an even point and we need to increase two coordinate values, and decrease only one of them.

It is obvious that if p and q have different parities then we can use this 3-step in the same way (strict 3-steps in both paths or in none of them). If the parity of p and q are the same, then we can use this strict 3-step only in one way. After this strict 3-step we are in a point of opposite parity than p. Hence in the last step we must step only to a 1-neighbor to reach the endpoint, but we may step to a 2-neighbor, so now we lose what we won at the strict 3-step. So in this case we have $d(p,q;B) = d(q,p;B)$ again. □

And now, we are dealing with the triangular inequality.

Lemma 4.4. *Let B be a neighborhood sequence, which does not contain the element 3. For the distance based on this B, the triangular inequality does not hold if and only if there are such i and j that* $\sum_{k=1}^{i} b(k) > \sum_{k=j+1}^{j+i} b(k)$.

Proof. Assume that for the neighborhood sequence B there are i and j such that $\sum_{k=1}^{i} b(k) > \sum_{k=j+1}^{j+i} b(k)$ holds for them. We will construct points p,q and r such that the triangular inequality does not hold for their B-distances. For the sake of simplicity let $p = (0,0,0)$. Let $q = \left(\left\lceil \frac{\sum_{k=1}^{j} b(k)}{2} \right\rceil, -\left\lfloor \frac{\sum_{k=1}^{j} b(k)}{2} \right\rfloor, 0 \right)$, where we used the upper and lower integer of the fraction. (It is clear, that q is a point of the grid: if the sum $\sum_{k=1}^{j} b(k)$ is even, then q is even and if the sum odd, then q is odd. Moreover $q(1) + |q(2)| = \sum_{k=1}^{j} b(k)$. Now, similarly let $r = \left(\left\lceil \frac{\sum_{k=1}^{j} b(k) + \sum_{k=1}^{i} b(k)}{2} \right\rceil, -\left\lfloor \frac{\sum_{k=1}^{j} b(k) + \sum_{k=1}^{i} b(k)}{2} \right\rfloor, 0 \right)$. One can move strict 2-steps and strict 1-steps from each point to any directions, therefore in a shortest path the changes of the coordinate values are the same as the given elements of the neighborhood sequence B (but the last step in which it is possible that we need a 1-step to reach the endpoint independently of the element of the neighborhood sequence). Therefore $d(p,q;B) = j$ and $d(q,r;B) = i$. Let us determine $d(p,r;B)$. It is easy to show that we need to change $\sum_{k=1}^{j} b(k) + \sum_{k=1}^{i} b(k)$ coordinate values together in this shortest path. Since this value is greater than $\sum_{k=i}^{j+i} b(k)$ we have $d(p,r;B) > i+j$. This proves that the triangular inequality does not hold.

Let us prove the other direction. Since we can move strict 2-steps and strict 1-steps from each point to any directions the B-distance of two points p and r is the minimal number of the elements of B from the beginning such that their sum is not less than $|p(1)-r(1)|+|p(2)-r(2)|+|p(3)-r(3)|$. Now, let us assume that for the neighborhood sequence B the condition $\sum_{k=1}^{i} b(k) \leq \sum_{k=j+1}^{j+i} b(k)$ holds for every i, j. Let the points p, q and r are arbitrary. Let us construct paths between the points. There is a path from p to r which is the concatenation of the/a shortest B-path from p to q and the/a shortest B_*-path from q to r using the elements of B starting with the $(d(p,q;B)+1)$-st element (i.e., $B_* = (b(i))_{i=d(p,q;B)+1}^{\infty}$). Then we have a B-path from p to r. Its length is not less than $d(p,r;B)$. So $d(p,r;B) \leq d(p,q;B) + d(q,r;B_*)$. But we know that $\sum_{k=1}^{i} b(k) \leq \sum_{k=j+1}^{j+i} b(k)$ also holds for the special case $j = d(p,q;B)+1$. Because the B_*-distance of two points q and r is the minimal number of the elements of B from the j-th such that their sum is not less than $|q(1)-r(1)|+|q(2)-r(2)|+|q(3)-r(3)|$ we have the following fact: $d(q,r;B) \geq d(q,r;B_*)$. Therefore $d(p,r;B) \leq d(p,q;B) + d(q,r;B)$. Since the points were arbitrary the triangular inequality must hold for this B-distance.

We finished the proof. □

For arbitrary B, the following statement holds.

Lemma 4.5. *The triangular inequality holds for the B-distance, if and only if*

$$\sum_{k=1}^{i} \min(2, b(k)) \leq \sum_{k=j+1}^{j+i} \min(2, b(k))$$

for every i, j and

$$S(i) \leq \sum_{k=j+1}^{j+i} b'(k)$$

for all pairs of i and j, where $b'(k)$ is the k-th element of the minimal equivalent sequence of B and $S(i) = \sum_{k=1}^{i} b'(k)$ sums the elements $b'(k)$ of the minimal equivalent neighborhood sequence if $\sum_{k=1}^{j} b'(k)$ is even and $S(i) = \sum_{k=1}^{i} b''(k)$ sums the elements $b''(k)$ of the reduced minimal equivalent sequence when $\sum_{k=1}^{j} b'(k)$ is odd.

Proof. First, assume that the first condition does not hold for B and we will prove that the triangular inequality does not hold for this case. Consider the neighborhood sequence $B^{(2)} = (b^{(2)}(i))_{i\in\mathbb{N}} = (\min(2, b(i)))_{i\in\mathbb{N}}$ constructed from B substituting all elements 3 by 2. (This, so-called 2-limited sequence of B is used in the first condition of the lemma.)
Then using Lemma 4.4, there are three points, for instance,

$$p = (0,0,0), \quad q = \left(\left\lceil \frac{\sum_{k=1}^{j} b^{(2)}(k)}{2} \right\rceil, -\left\lfloor \frac{\sum_{k=1}^{j} b^{(2)}(k)}{2} \right\rfloor, 0 \right)$$

$$\text{and } r = \left(\left\lceil \frac{\sum\limits_{k=1}^{j} b^{(2)}(k) + \sum\limits_{k=1}^{i} b^{(2)}(k)}{2} \right\rceil, - \left\lfloor \frac{\sum\limits_{k=1}^{j} b^{(2)}(k) + \sum\limits_{k=1}^{i} b^{(2)}(k)}{2} \right\rfloor, 0 \right),$$

such that the triangular inequality does not hold for them using $B^{(2)}$. But the third coordinate is 0 for all of them. Therefore using elements 3 instead of any elements 2 in the neighborhood sequence for each pair of these points there is an identical shortest path with the shortest path using $B^{(2)}$. (There is useless to change the third value in a path.) It implies that the B-distances are the same as the $B^{(2)}$-distances for these points. Therefore the triangular inequality does not true for this B-distance.

Now assume that the second condition does not hold for B and we will prove that there are three points such that the triangular inequality does not hold for them. Let the value of i and j are such that $\sum\limits_{k=1}^{i} b''(k) > \sum\limits_{k=j+1}^{j+i} b'(k)$ holds (with minimal equivalent sequence B' and reduced minimal equivalent sequence B'' of B). Now we construct the coordinate values of the points. First the coordinate values of the points p and q are constructed using the minimal equivalent neighborhood sequence of B (by the following procedure):

1. Let $p = q = (0,0,0)$.
2. For $k = j-1$ downto $k = 1$
   2.1. case of $b'(k)$
   $b'(k) = 3$: if p is even then
      let $p = (p(1)+1, p(2)+1, p(3)-1)$
      if p is odd then let $q = (0,0,1)$ {it can be only for the first element 3 of B}
   $b'(k) = 2$: let $p = (p(1)+1, p(2), p(3)-1)$
   $b'(k)$=1 : if p is even then let $p = (p(1)+1, p(2), p(3))$
      else let $p = (p(1), p(2), p(3)-1)$.

It is trivial by the construction that $d(p,q;B) = j-1$.

Now we construct the point r:

3. Let $r = q$. {it is (0,0,0) or (0,0,1)}
4. For $k = 1$ to $k = i$
   4.1. case of $b'(k)$
   $b'(k) = 3$: if r is odd then let $r = (r(1)-1, r(2)-1, r(3)+1)$
      else let $r = (r(1)-1, r(2), r(3)+1)$
   $b'(k) = 2$: let $r = (r(1)-1, r(2), r(3)+1)$
   $b'(k) = 1$: if r is even then let $r = (r(1), r(2), r(3)+1)$
      else let $r = (r(1)-1, r(2), r(3))$.

By the construction we have $d(q,r;B) = i$. Now $d(p,q;B)+d(q,r;B) = i+j-1$. Let us calculate $d(p,r;B)$. It is clear that q is in a shortest path from p to r. We reach q exactly at the $(j-1)$-st step. We know that $\sum\limits_{k=1}^{i} b''(k) > \sum\limits_{k=j+1}^{j+i} b'(k)$ which means that after reached the point q, i steps cannot be enough to reach r. Therefore $d(p,r;B) > i+j-1$; the triangular inequality does not hold.

And now we will prove the other direction. Let the conditions $\sum\limits_{k=1}^{i} b''(k) \leq \sum\limits_{k=j+1}^{j+i} b'(k)$

and $\sum_{k=1}^{i} \min(2, b(k)) \leq \sum_{k=j+1}^{j+i} \min(2, b(k))$ be fulfilled by B. Let p, q and r be three points. Then there is path from p to r which is the concatenation of the/a shortest B-path from p to q and the/a shortest B_*-path from q to r using the elements of B starting with the $(d(p,q;B)+1)$-st element. ($B_* = (b(i))_{i=d(p,q;B)+1}^{\infty}$) Then we have a B-path from p to r. Its length is not less than $d(p,r;B)$. So $d(p,r;B) \leq d(p,q;B) + d(q,r;B_*)$. But we know that both $\sum_{k=1}^{i} b''(k) \leq \sum_{k=j+1}^{j+i} b'(k)$ and $\sum_{k=1}^{i} \min(2, b(k)) \leq \sum_{k=j+1}^{j+i} \min(2, b(k))$ hold for the special case $j = d(p,q;B)+1$ and for all i. These two conditions imply that $d(q,r;B) \geq d(q,r;B_*)$. Therefore $d(p,r;B) \leq d(p,q;B) + d(q,r;B)$. Since the points were arbitrary the triangular inequality must hold for this B-distance. The theorem is proved. □

Now we are in the position to answer the question whether a neighborhood sequence generates a metric or not.

Theorem 4.1. *Let B be a neighborhood sequence. The distance function based on B is a metric if and only if the following conditions hold:*

- *if* $b(j) = 3$ *and* $b(i) = 1$ *then* $i < j$,
- *if B contains* 3, *then* $\sum_{b(k)=1} b(k)$ *is even,*
- $\sum_{k=1}^{i} b(k) \leqslant \sum_{k=j+1}^{j+i} b(k)$, *when* $i + j < l$, *where* l *is the index of the first element* 3 *in B,* (*if* 3 *is not in B then this condition must hold for all* i , j).

Proof. The theorem is a consequence of Lemmas 4.3 and 4.5. □

Note that by the first two conditions of the theorem, if there is an element 3 of B, then the number of the elements 1 must be finite and even: all of them must be before the (first) element 3.

According to the previous theorem we can exactly give the neighborhood sequences, which provide metrics.

Corollary 4.2. *B generates a metric if and only if B has one of the following forms:*

a) $b(1) = 1$, *B does not contain* 3, *and* $\sum_{k=1}^{i} b(k) \leqslant \sum_{k=j+1}^{j+i} b(k)$ *for all* $i, j \in \mathbb{N}$.

b) $b(i) > 1$ *for all* $i \in \mathbb{N}$.

c) $b(1) = 1$, *if* l *is the minimal index such that* $b(l) = 3$ *then* $\sum_{k=1}^{l-1} b(k)$ *is even,* $\sum_{k=1}^{i} b(k) \leqslant \sum_{k=j+1}^{j+i} b(k)$ *if* $i + j < l$ *and* $b(k) > 1$ *for* $k > l$.

The condition c) is a mixture of a) and b). The first part of B has the same property as in case a), and the other part beginning by $b(l)$ *has the same property as in case b).*

Neighborhood sequences, which are both periodic and give metric distances, are of special importance and interest.

Corollary 4.3. *A periodic B generates a metric if and only if one of the following conditions holds:*

there are only elements 1 and 2, and $\sum_{k=1}^{i} b(k) \leqslant \sum_{k=j+1}^{j+i} b(k)$ *for all* $i, j \in \mathbb{N}$, *or*

there are only elements 2 *and* 3 *in B.*

We will calculate the B-distance of any two points with a given neighborhood sequence in the next section.

5 Computing the Distance

First in this section a relationship between the triangular grid and the cubic grid is established ([35, 38]).

Let us see how the triangular plane can be embedded into $\mathbb{Z}^3$, i.e., what points in $\mathbb{Z}^3$ are used to represent the points of the triangular grid. Figure 8 shows the points of the grid of hexagonal nodes in the cubic grid. They form two parallel planes, according to the parities of points.

The lanes of this grid are also parallel planes with a plane including two coordinate axes in $\mathbb{Z}^3$ intersected with the two planes representing the triangular grid.

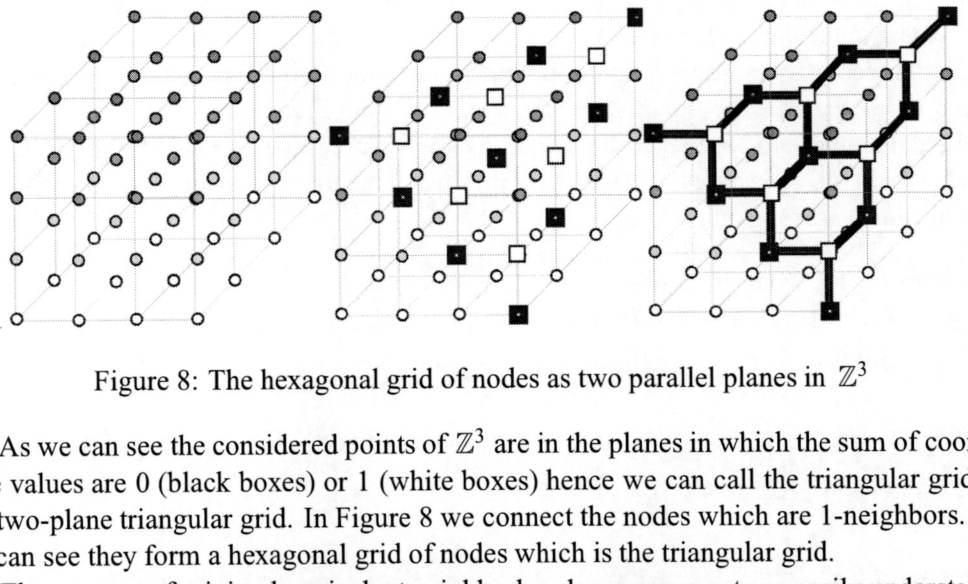

Figure 8: The hexagonal grid of nodes as two parallel planes in $\mathbb{Z}^3$

As we can see the considered points of $\mathbb{Z}^3$ are in the planes in which the sum of coordinate values are 0 (black boxes) or 1 (white boxes) hence we can call the triangular grid as the two-plane triangular grid. In Figure 8 we connect the nodes which are 1-neighbors. As we can see they form a hexagonal grid of nodes which is the triangular grid.

The concept of minimal equivalent neighborhood sequences gets an easily understandable explanation. To build a shortest path between two points of the cubic grid we do not care about the 'parity' of the points. We can use the points in our path with any kind of coordinate sum. Unlike in the previous case in triangular grid we cannot step to a point which is not included in the two given planes of the grid. Using the minimal equivalent neighborhood sequence instead of the original one we reduce the number of the possible wrong steps to 1, i.e., we have to care only the first occurrence of the element 3.

In the next part of the section we will use this mapping from the triangular grid into the cubic grid to determine the distance of points.

Now we recall the formula to compute the distance in cubic grid ([41, 44, 46]).

Proposition 5.1. *Let B be a neighborhood sequence and p,q be two points in the three dimensional space. Then*

$$d(p,q;B) = \max\{v(1), d_2, d_3\},$$

where v is the absolute difference vector of p and q, i.e., containing the values $|p(1) - q(1)|, |p(2) - q(2)|, |p(3) - q(3)|$ in a non-increasing order,

$$d_2 = \max\left\{ i \,\middle|\, v(1) + v(2) > \sum_{j=1}^{i-1} b^{(2)}(j) \right\}, \text{ and}$$

$$d_3 = \max\left\{ i \,\middle|\, v(1) + v(2) + v(3) > \sum_{j=1}^{i-1} b(j) \right\},$$

where the neighborhood sequence $B^{(2)} = (b^{(2)}(i))_{i\in\mathbb{N}} = (\min(2, b(i)))_{i\in\mathbb{N}}$ is the 2-limited sequence of B.

Further in this section we present formulae for the distances in the triangular grid ([44]).

The concept of the minimal equivalent neighborhood sequences shows that it is possible that both theoretically and practically we cannot change all the three coordinate values to step closer to the endpoint even if there is an element 3 in the neighborhood sequence.

In this section we use the previous mapping and modify the formula of Proposition 5.1 to our case. We will use the minimal equivalent neighborhood sequence B' instead of the original neighborhood sequence B. Recall that, as we used in the proof of Lemma 4.3, in a shortest path, in some cases, it is worth modifying all the three coordinate values at 3-steps, and in some cases it is not.

Corollary 5.1. *Going from p to q it is better to use a 3-step as a strict 3-step by the first element 3 of the neighborhood sequence B (let its index be k), if and only if one of the following holds:*

1. *p is even, $\sum_{i=1}^{k-1} b(i)$ is even and we need to decrease two coordinates, and increase only one value to go to q;*
2. *p is odd, $\sum_{i=1}^{k-1} b(i)$ is odd and we need to decrease two coordinates, and increase only one value to go to q;*
3. *p is odd, $\sum_{i=1}^{k-1} b(i)$ is even and we need to increase two coordinate values, and decrease only one of them to direction to q;*
4. *p is even, $\sum_{i=1}^{k-1} b(i)$ is odd and we need to increase two coordinate values, and decrease only one of them.*

From the concept of minimal equivalent neighborhood sequence it is obvious that by building a shortest path from the point p to q, it is worth modifying all coordinate values only when there is an element 3 in the minimal equivalent neighborhood sequence of B. Therefore we must use it instead of B in our calculation. Moreover it is possible (as we already showed) that one uses the first element 3 as value 2 in a shortest path (in the cases that are not listed above in Corollary 5.1). In these cases we will use the reduced minimal equivalent neighborhood sequences in the formula for distance. Notice that the 2-limited sequence of B, B' and B'' are the same ($B^{(2)} = (\min(2, b(i)))_{i\in\mathbb{N}}$).

We introduce some notations using the sorted difference v of the points (Definition 2.7):

$$d_2 = \max\left(i \,\middle|\, |v(1)|+|v(2)| > \sum_{j=1}^{i-1} b^{(2)}(j)\right),$$

$$d_3' = \max\left(i \,\middle|\, |v(1)|+|v(2)|+|v(3)| > \sum_{j=1}^{i-1} b'(j)\right),$$

$$d_3'' = \max\left(i \,\middle|\, |v(1)|+|v(2)|+|v(3)| > \sum_{j=1}^{i-1} b''(j)\right).$$

Finally we can state our theorem using the abbreviations above.

Theorem 5.1. *The distance from a point p to q with a given neighborhood sequence B can be calculated in the following way:*

$$d' = \max\left(|v(1)|, d_2, d_3'\right)$$

and

$$d'' = \max\left(|v(1)|, d_2, d_3''\right).$$

The first equation gives the result $(d(p,q;B) = d')$ if we can modify all the 3 coordinate values by step of the first element 3 (let its index be k). We must use the second equation in other cases (and get $d(p,q;B) = d''$).

So we must use the value d'' in the following case: if the distance $d' \geq k$ and one of the followings hold:

- *p is even, $\sum_{i=1}^{k-1} b(i)$ is even, and the difference $w_{p,q}$ has two negative and a positive value;*
- *p is odd, $\sum_{i=1}^{k-1} b(i)$ is odd and the difference of q and p has two negative and a positive value;*
- *p is odd, $\sum_{i=1}^{k-1} b(i)$ is even and the difference of q and p has two positive and a negative value;*
- *p is even, $\sum_{i=1}^{k-1} b(i)$ is odd and the difference of q and p has two positive and a negative value.*

Proof. Using our previous notations and lemmas, with the shortest path algorithm (Algorithm 3.1) and Corollary 5.1, it is consequence of Proposition 5.1. □

The previous theorem is one of our most important results. In the triangular grid the difference of any two points is 'balanced', i.e., $w(1)+w(2)+w(3) \in \{-1,0,1\}$. The differences of the coordinate values are distributed among the three values. If the two points are in a common lane, then we use the 2-limited neighborhood sequence $B^{(2)}$, i.e., the sequence in which all values 3 are replaced by values 2. Elsewhere the minimal or the restricted minimal equivalent neighborhood sequence can be used according to the parity and direction of the points.

Now we show an example.

Example 5.1. Let $r = (1,-2,1)$ and $s = (0,0,0)$ be two points, and $B = (3,1,1)$ a neighborhood sequence. The minimal equivalent neighborhood sequence of B is: $B' = (3,1,1,2,1,1,2,1,1,\ldots)$ with repeating part $(2,1,1)$.

First we will calculate the value of $d(r,s;B)$.

The value of v is $(2,-1,-1)$ in this case. Try to use the first form of Theorem 5.1.
$d_2 = \max\left(i \,\middle|\, |2|+|-1| > \sum_{j=1}^{i-1} b^{(2)}(j)\right) = 2$ and $d_3' = \max\left(i \,\middle|\, 2+1+1 > \sum_{j=1}^{i-1} b'(j)\right) = 2$, therefore $d' = \max(|2|,2,2) = 2$. This is greater then $k = 1$, and the parity of r is even, $\sum_{i=1}^{k-1} b(i) = 0$ is even and the difference w is even, moreover it has two negative and a positive value, therefore we must use the second formula, i.e., the neighborhood sequence $B'' = (2,1,1)$: $d_2 = 2$ and $d_3'' = 3$, $(4 > 2+1)$. Therefore $d'' = 3$ and so $d(r,s;B) = 3$.

Now let us calculate $d(s,r;B)$. Then $v_{s,r} = (-2,1,1)$. Using the first form: $d' = \max(2,2,2) = 2$. It is greater than $k = 1$, but the other conditions fail, so the result: $d(s,r;B) = 2$. Thus this distance function is not symmetric (as it was also proved in Lemma 4.3).

As we mentioned before, the non-symmetric distance functions are exotic in the field of digital geometry, since they do not exist in the square and hexagonal cases using symmetric neighborhood system (see also in [40] where a detailed comparison can be found for these grids).

Looking at the formulae for calculating distance we state the following important property.

Theorem 5.2. *A distance $d(p,q;B) > k$ depends on the order of the first k elements of B if and only if there is a permutation of these elements such that using it as the initial part of the neighborhood sequence the distance function is not symmetric.*

Proof. Let B be a neighborhood sequence for which the distance is not symmetric. So let p and q be two points that $d(p,q;B) \neq d(q,p;B)$. We can assume that $d(p,q;B) = k < d(q,p;B) = l$. If we reorder the first k elements of B in inverse order then we get that $d(q,p;B_*) = k$. (We have a shortest path from q to p which is the symmetric pair of the original shortest path between p and q.) Let us prove the other direction. Let us assume that the distance depends on the ordering. One can check (and it can be found in [21, 45]) that interchanging an element $b(i) = 2$ with any other element $b(j)$ (where $i, j < k$) does not cause dependency, i.e., the distance remains the same. Dependency occurs only if there are $b(i) = 3$ and $b(j) = 1$. But in these cases using Lemma 4.3 we can have a non-symmetric distance with these elements. □

6 Digital Circles

In this section[1] we present some other properties in which the triangular grid differs from the square grid. We analyze the changing and development of wave-fronts, and give an illustrated description of the digital circles with neighborhood sequences in triangular grid ([37]). At the end of the section we show a characterization of digital circles.

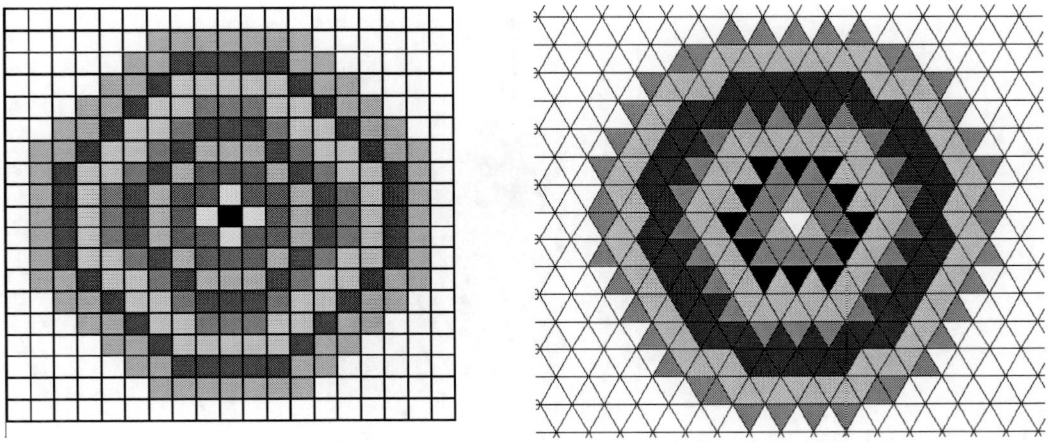

(a) Circles in square grid (b) Circles in triangular grid

Figure 9: Examples for growing digital circles up to radius 8

In this section we investigate the way a neighborhood sequence spreads in the digital space starting from a point of the triangular grid. This spreading is translation-invariant among the points of the same parity and it is central-symmetric concerning points with different parities. So, for simplicity we may choose the origin o as the starting point.

Definition 6.1. Let B be a neighborhood sequence in the triangular grid. For $k \in \mathbb{N}$, let

$$C_k^B = \{p \,|\, d(o,p;B) \leq k\}.$$

C_k^B is the region (digital circle) occupied by B after k steps.

In square grid we use the

$$O_k^B = \{q \,|\, d(o,q;B) \leq k\}$$

notations for the occupied regions where q is a point in the square grid ($\mathbb{Z}^2$), and in this case $1 \leq b(i) \leq 2$ for all i, and k is a natural number. (In obvious cases we reduce the notion to C_k and O_k.)

[1] A large part of this section including figures is reprinted from [37] with permission from Elsevier.

In Figure 9 there are some examples of growing digital circles.

In the following we summarize some simple observations about the digital circles. We underline some properties which are different for the digital circles in square grid and in triangular grid.

In the square grid the region O_k occupied by k steps of a neighborhood sequence B is independent of the ordering of the first k element of B.

Contrary to the case of square grid, it is possible for a neighborhood sequence B and for a $k \in \mathbb{N}$, that the region C_k does depend on the order of the first k elements of B. This property is based on the fact shown in Theorem 5.2. We will show an example. The regions $C_2^{(1,3)}$ and $C_2^{(3,1)}$ differ as Figure 10 shows in the margins of row 3.

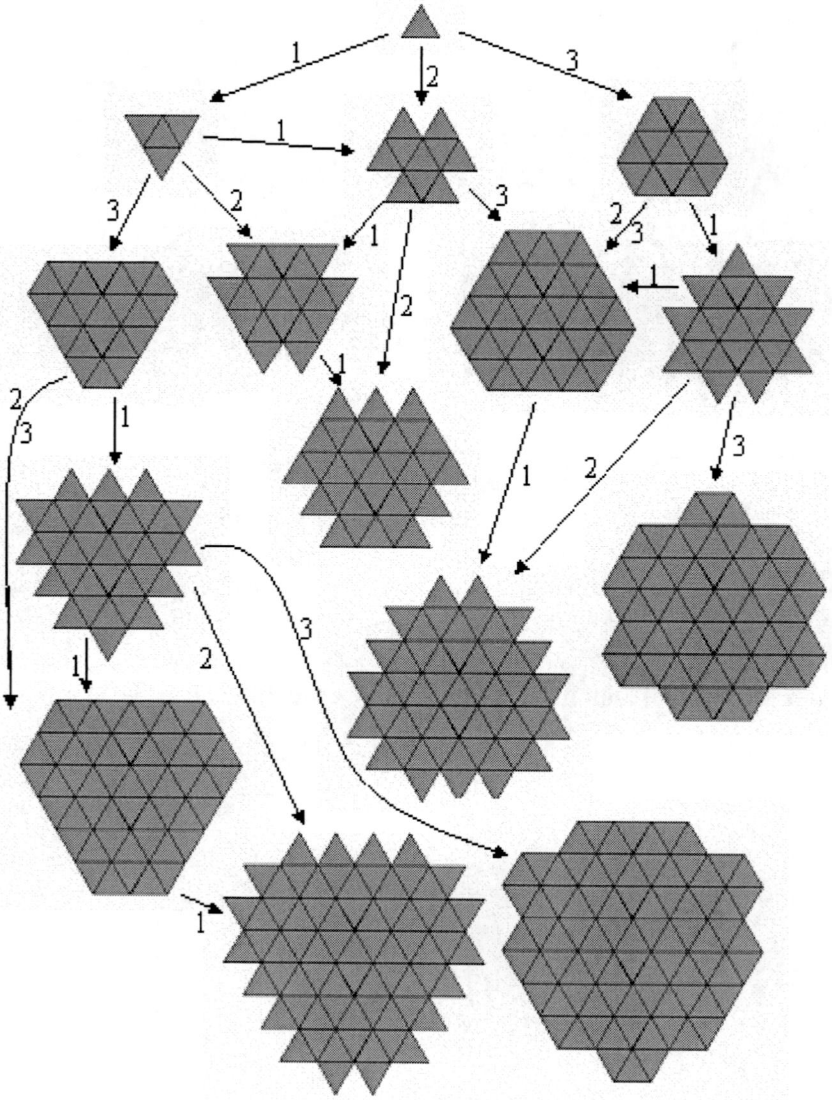

Figure 10: Basic digital circles and their hierarchy

Remark 6.1. For any neighborhood sequence B, the sequence of regions $(C_k)_{k=1}^{\infty}$ is a strictly monotone increasing sequence. That is, $k > l$ implies $C_k \supsetneq C_l$.

In the square grid it is impossible for $k \neq l$ and any two neighborhood sequence B_1 and B_2 that $O_k^{B_1} = O_l^{B_2}$. This statement follows from the fact, that for any n.s. B the point $p = (0,k) \in O_l$ if and only if $l \geq k$ (when $o = (0,0)$).

Lemma 6.1. *Contrary to the square grid, in the triangular grid there are neighborhood sequences B_1, B_2 and $k, l \in \mathbb{N}$ such that $C_k^{B_1} = C_l^{B_2}$ with $k \neq l$.*

Proof. We present an example. Let $B_1 = (1)$ and $B_2 = (2)$ then $C_2^{(1)} = C_1^{(2)}$ (see Figure 10). □

In the square grid the digital circles with the same radius form a well ordered set by inclusion. Formally, for all pairs of B_1, B_2 one of the following relations hold:

$$O_r^{B_1} \subseteq O_r^{B_2} \text{ or } O_r^{B_1} \supset O_r^{B_2}.$$

Moreover

$$O_r^{(1)} \subseteq O_r^{B_1} \subseteq O_r^{(2)}.$$

(The circle $O_r^{B_1}$ is 'larger' than $O_r^{B_2}$ if and only if the number of 2's are greater among the first r elements in B_1 than B_2.)

Contrary, in the triangular grid the following property holds.

Lemma 6.2. *In the triangular grid there are neighborhood sequences B_1, B_2 and $r \in \mathbb{N}$ such that $C_r^{B_1} \nsubseteq C_r^{B_2}$ and $C_r^{B_1} \nsupseteq C_r^{B_2}$.*

Proof. We present an example. Let $B_1 = (1,3)$ and $B_2 = (3,1)$. The $p = (2,-1,-1) \in C_2^{(1,3)}$, but $p \notin C_2^{(3,1)}$. The $q = (1,-2,1) \notin C_2^{(1,3)}$, but $q \in C_2^{(3,1)}$. □

We will use the concept of minimal equivalent neighborhood sequences in this section as well. In the triangular grid, for certain neighborhood sequences, it can happen that a 3-step is equivalent to a 2-step for our investigations, i.e., the digital circles in triangular grid have the following property.

Remark 6.2. According to the definition of the minimal equivalent neighborhood sequence (Definition 4.1), the same digital circles are obtained by B and B': $C_k^{B'} = C_k^{B}$ for any $k \in \mathbb{N}$.

In Figure 10 we present some simple digital circles obtained in a few steps (small radii). In the next part we analyze the wavefronts.

Table 7: State transition table of edges by taking a step

original edge type	after a 1-step	after a 2-step	after a 3-step
'sawtooth'	'smooth'	'sawtooth'	'smooth'
'hilly'	'hilly'	'hilly'	'hilly'
'smooth'	'sawtooth'	'smooth'	'smooth'

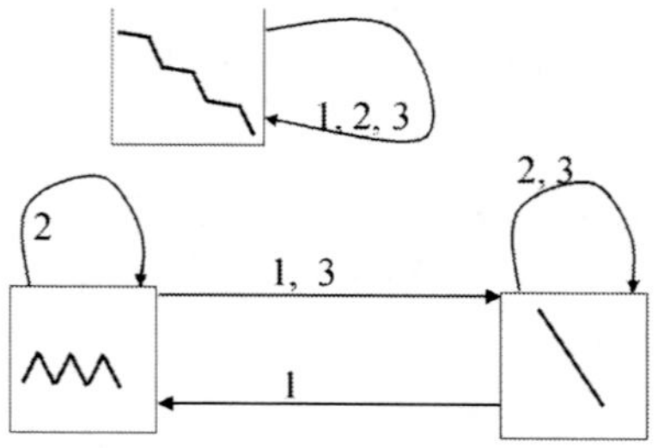

Figure 11: State transition diagram of types of the edges of digital circles

We present the types and the development of wavefronts in the triangular plane. In [13] Das and Chatterji showed that for every neighborhood sequence B, (O_k) is always an octagon. In square grid we have two kinds of sides: 'smooth' and 'stair'-types as shown in Figure 9. The vertical and horizontal edges are 'smooth', the other four edges are 'stair'-type. In the next cases the octagon is degenerate, i.e., it is a square with only one type edges. With the neighborhood sequence $B_1 = (1)$ we get only four 'stair'-type edges, while using $B_2 = (2)$ we get a square with only 'smooth' edges. In case we use both 1-step and 2-step our result is a non-degenerated octagon.

In the triangular plane we have three kinds of possible "limit lines" (edges). These are the 'smooth', the 'hilly' and the 'sawtooth'. They change to each other by using a step from C_k to C_{k+1}. The diagram of the changing of types of edges is given by the Table 7 and by Figure 11.

Table 8: State transition table of corners by taking a step

original corner type	edges after a 1-step	edges after a 2-step	edges after a 3-step
'smooth'-'smooth' (1)	'sawtooth'-'sawtooth' (3)	'smooth'-'smooth' (1)	'smooth'-'smooth' (1)
'smooth'-'sawtooth' (2)	'smooth'-'sawtooth' (2)	'smooth'-'sawtooth' (2)	'smooth'-'smooth' (1)
'sawtooth'-'sawtooth' (3)	'smooth'-'smooth' (1)	'sawtooth'-'sawtooth' (3)	'smooth'-'hilly'-'smooth' (4,4)
'smooth'-'hilly' (4)	'sawtooth'-'hilly' (5)	'smooth'-'hilly' (4)	'smooth'-'hilly' (4)
'sawtooth'-'hilly' (5)	'smooth'-'hilly' (4)	'sawtooth'-'hilly' (5)	'smooth'-'hilly' (4)
'hilly'-'hilly' (type 6)	'hilly'-'hilly' (type 7)	'smooth'-'hilly'-'smooth' (4,4)	'smooth'-'hilly'-'smooth' (4,4)
'hilly'-'hilly' (type 7)	'smooth'-'hilly'-'smooth' (4,4)	'hilly'-'sawtooth'-'hilly' (5,5)	'hilly'-'hilly' (type 6)

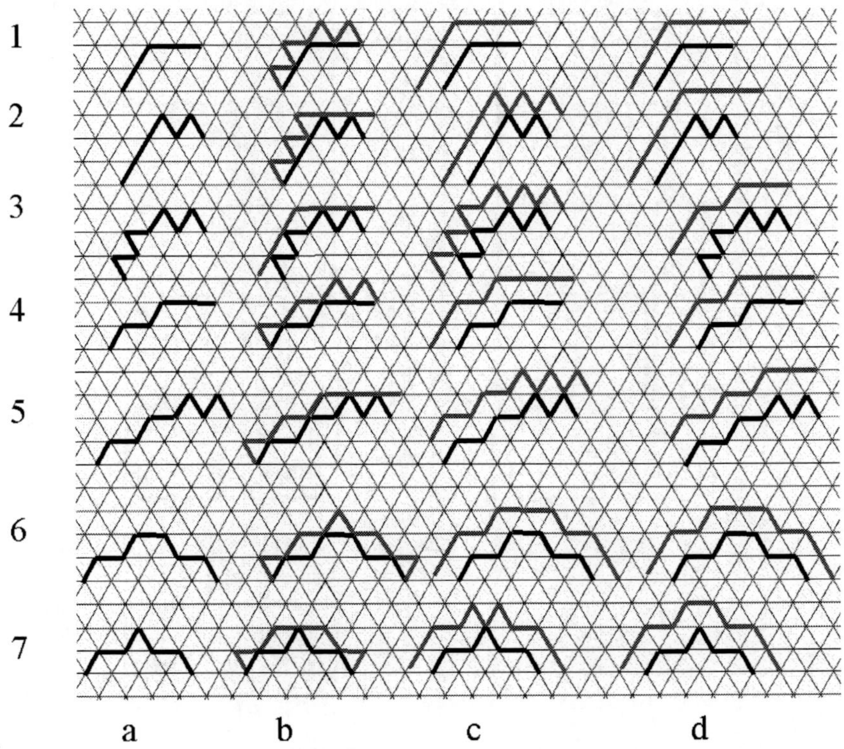

Figure 12: Changing the corners after a step (a) original edges (b) after a 1-step (c) after a 2-step (d) after a 3-step

In the next statements we summarize our experiences.

Proposition 6.1.

- *After a 3-step the 'smooth' and 'sawtooth' edges go to 'smooth' lines.*
- *The 2-steps do not change the type of the edges.*
- *The 'hilly' edge cannot change into another type of edge.*

In Figure 10 we showed the basic digital circles and here we analyzed the edges. In the next part we analyze how the possible corners, i.e., the connections of the possible type of edges change in growing steps. In Table 8 we show what kind of corners occur in different digital circles. Figure 12 shows all of the cases of changing corners by a step in upward direction.

Table 9 gathers the types of corners which occur in basic digital circles (we refer to the basic circles of Figure 10 as the name of possible steps (signs on the edges of the graph) to get them starting with the origin-triangle). We use here all digital circles occurring in Figure 10, for which the figure does not contain all the three possible growing steps.

Based on Table 8 we summarize how the vertices change via the growing procedure.

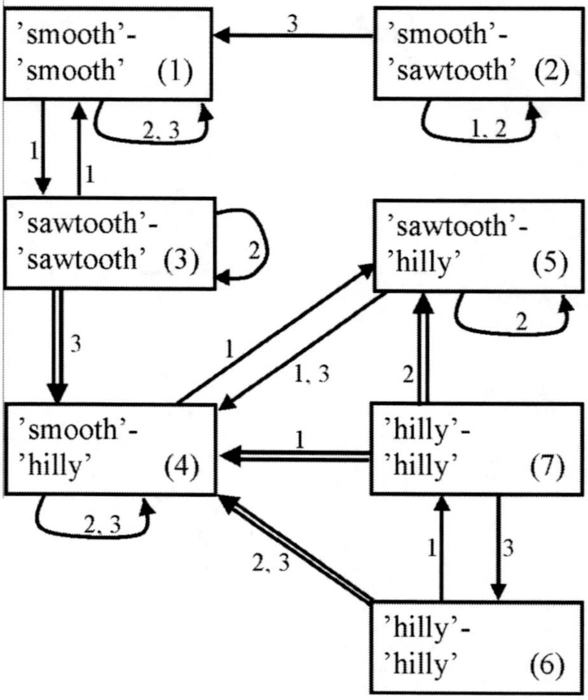

Figure 13: State transition diagram of corners by growing digital circles

Proposition 6.2.

- *There are two types of corners between 'hilly' edges, as we used type* 6 *and* 7. (*Their difference can be seen in Figure* 12.)
- *The following corners can start a new type of edge, as we denote in the table by two values: between two 'sawtooth' edges with a* 3*-step we get a new 'hilly' edge between the 'smooth' ones; between two 'hilly' edges a new 'smooth' one appears* (*in case* 6 *with* 2*-step or* 3*-step*) *and we get a 'smooth' and a 'sawtooth' one using* 1*-step and* 2*-step, respectively, in case* 7.

We draw the state transition diagram of the corners in Figure 13. The double arrows mean that using these transitions, two corners of the same type are obtained.

As we can see, our edge-types and corner-types are in closed sets, i.e., we cannot step out from the above used sets by the growing steps. One can check also, that all kinds of vertices and edges occur in digital circles.

Using the basic digital circles and our growing tables we get all possible digital circles of the triangular grid.

In the remaining part of this section – based on our previous experience – we characterize the digital circles with neighborhood sequences in the triangular plane.

In the following, based on Lemma 4.2 and Remark 6.2 we will use the minimal equivalent neighborhood sequence B' instead of the original sequence B. In Table 10 and in Figure 14 we show the types of the possible digital circles C_k. In the figure we can see the state transition diagram for them.

Table 9: Corner-types of basic digital circles in triangular grid

name and sign of corner-type	'smooth'-'smooth' (1)	'smooth'-'sawtooth' (2)	'sawtooth'-'sawtooth' (3)	'smooth'-'hilly' (4)	'saw-tooth'-'hilly' (5)	'hilly'-'hilly' (type 6)	'hilly'-'hilly' (type 7)
occurrence in basic circles generated by steps ...	23=32 132=1311	21=12 22=112= =121=211 =1111	231=321= 312=3111= =1131 1321= 1312=13111	313 1313	-	313 1313	-

Lemma 6.3. *Table 10 contains the digital circles for all possible neighborhood sequences.*

Proof. It is evident, that – using for steps the equivalent neighborhood sequence B' instead of B (using Remark 6.2) – all initial parts of all neighborhood sequences occur in the second column. □

Table 10: The possible types of digital circles

A) triangle	basic: 0 (only the starting triangle) or by a 1-step
B) hexagon – six 'smooth' edges	only one 3-step and the others are 1-steps and 2-steps; the sum after the 3-step is even
C) hexagon – three 'smooth' and three 'saw-tooth' edges	with only 1-steps and 2-steps (without any 3-step)
D) hexagon – six 'sawtooth' edges	only one 3-step and the others are 1-steps and 2-steps; the sum after the 3-step is odd
E) enneagon – six 'hilly' and three 'smooth'	only odd steps: 1-steps and 3-steps by turns (with minimum 2 3-steps); without any 2-step or repetition (double 1-step) with 3-step at last
F) enneagon – six 'hilly' and three 'sawtooth'	only 1-steps and 3-steps by turns (minimum 2 3-steps) with 1-step at last
G) dodecagon – six 'hilly' and six 'smooth'	at least (a 2-step or repeated 1-steps) and at least two 3-steps and after the last 3-step the sum is even
H) dodecagon – six 'hilly' and six 'sawtooth'	at least (a 2-step or repeated 1-steps) and at least two 3-steps and after the last 3-step the sum is odd

Proposition 6.3. *All basic digital circles in Figure* 10 *occur in Table* 10, *and their types are correct.*

Theorem 6.1. *Table* 10 *contains all possible digital circles, and they are in the correct places, respectively.*

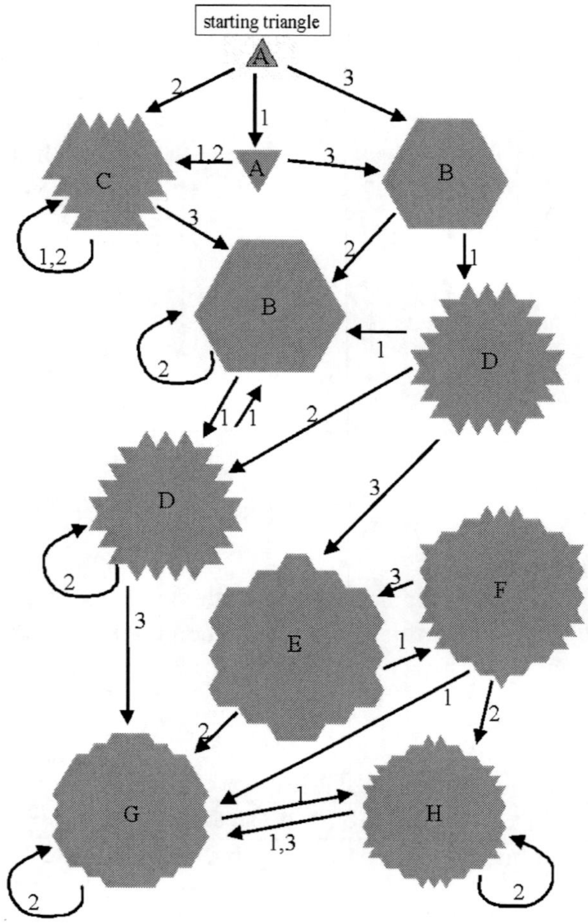

Figure 14: State transition of types of digital circles

Proof. By using Lemma 6.3 we know that all possible initial parts for the possible neighborhood sequences are in Table 10. Therefore we need to prove only the statement that, for all rows of the table, the given digital circles are correct. Our proof is by induction. From Proposition 6.3 we know this fact for the basic digital circles. Now we suppose for each digital circle that it is in the correct row in Table 10. Our induction steps are based on state transitions of the wave-fronts (Table 8 and 7 for the corners and edges respectively). Using these facts we get the state transition diagram that we show in Figure 14. Thus the theorem is proved. □

In Figure 14 we used the minimal equivalent neighborhood sequences to represent all neighborhood sequences. For this reason, one cannot see arrows representing 3-steps from the types of digital circles for which the sum of the elements after the last used 3-step must be even. For example types B, E and G are in this position. In that case, if the next element of the n.s. B is 3, then we get the same result as we get with element 2. (Therefore we would have used both values 2 and 3 on the arrows representing 2-steps if we had used the original n.s. B.)

Now, the coordinate values of the corners are of interest, they describe these polygons. Since the triangles are addressed by coordinate-triplets some of the corners will be addressed by a pair of points (at 'smooth'-'smooth' and at 'sawtooth'-'sawtooth' corners). Moreover these points are exactly those points of the grid which have the same B-distance from the origin (centre of the circle) as the other points on the edges, but they have maximal Euclidean distance from the origin among the points on the edges. Now we show how the coordinate values of these points can be determined. They can be computed by an incremental algorithm based on Tables 8 and 9 and on Figure 13. The coordinate-values of these points of the first basic circles are given in Table 11. (Remember that the same circles can be generated by the minimal equivalent n.s. of B as by B, therefore B' can be used in this computation also.)

Table 11: Coordinate values of corners of circles obtained by at most 1 step

origin	a 1-step	a 2-step	a 3-step
(0,0,0)	(1,0,0)	(1,-1,0)	(1,-1,1)
	(0,1,0)	(1,0,-1)	(1,-1,0)
	(0,0,1)	(0,1,-1)	(1,0,-1)
		(-1,1,0)	(1,1,-1)
		(-1,0,1)	(0,1,-1)
		(0,-1,1)	(-1,1,0)
			(-1,1,1)
			(-1,0,1)
			(0,-1,1)

Now, let us see how the corners develop in terms of coordinate values:

– by a 1-step a positive value is increasing by 1 (if it was an even point), or a non-positive value is decreasing by 1. (in the case when there are two positive or two negative values among the coordinates, the one with larger absolute value will be changed; if they have the same value, then both possibilities are applied and the number of corner points is increasing)

– by a 2-step a positive value is increasing by 1, and a non-positive value is decreasing by 1. (in the case when there are two positive or two negative values among the coordinates, the one with larger absolute value will be changed; if they have the same value, then both possibilities are applied, and so the number of corner points is increased)

– by a 3-step two non-negative values are increasing and a non-positive value is decreasing (if it was an even point), or the two non-positive is decreasing and a non-negative is increasing. (and in the case when there was a value 0 among the coordinate values, some new corner points must also be added in the way as a 2-step develops from the given points)

Observe, that all corners can be obtained by a permutation of the coordinate values of at most two corners. (One can see a detailed algorithm in [42].)

In this section we were growing regions in the triangular grid using 3 kinds of neighboring relations in various n.s. In image processing, region growing is often used method for analyzing pictures (find a connected region etc., see [1, 19, 6, 15, 22, 55]). In practice, starting from a point of an image, a variation of our method can be used. We unite only those new points of the wave-front set to our region which satisfy another desired property.

In Cellular Neural Networks (CNN-UM structures) and in several image processes there are effective algorithms which are based on morphological procedures of waves on binary pictures. It could be interesting further research to analyze these algorithms in the triangular grid.

7 Conclusion

In this chapter we dealt with digital geometry, especially digital distances based on neighborhood sequences on the triangular grid.

After describing the grid by three coordinate values, a greedy algorithm was presented to solve the shortest path problem. Formulae to compute the distances were also presented based on the fact that the triangular grid is a special subspace of the cubic grid. Several properties of the distances and of the digital circles were detailed, such as metrical properties. Non-symmetric distances and distances that do not meet the triangular inequality were characterized. Some interesting properties are presented. The distances based on neighborhood sequences, in some cases, as we proved, depend on the order of the first elements of the applied n.s. Moreover there are digital circles that can be obtained by various neighborhood sequences and with various radii.

There are two alternative ways to define digital distances. The present chapter was about distances based on neighborhood sequences. The other way is to define various weights for the various neighbors and then define weighted distances ([16]). This direction is open for the triangular grid. Moreover a mixture of these approaches, the weighted distances based on neighborhood sequences unites the advantages of both approaches (see [57] for results on the square grid). These extensions will make the triangular grid more applicable.

We note here that the triangular grids has extensions on higher dimensions, there are also some results on these (diamond, face-centered cubic, body-centered cubic) grids ([47, 48, 49, 56]). The presented theory can be applied in digital image processing, in computer graphics, in (computer and network) architectures and in computing models (neural-networks, cellular automata). There are also some possible applications in physics (simulations, random walk, etc.) and in crystallography.

References

[1] M. R. Anderberg: Cluster Analysis for Application, Academic Press, NY, 1973.

[2] M. Aswatha Kumar, J. Mukherjee, B. N. Chatterji, and P. P. Das: A geometric approach to obtain best octagonal distances, *Ninth Scandinavian Conf. Image Process.* (1995), pp. 491–498.

[3] S. B. M. Bell, F. C. Holroyd and D. C. Mason: A digital geometry for hexagonal pixels, *Image and vision comp.* **7** (1989) 194–204.

[4] G. Borgefors: Distance transformations in arbitrary dimensions, *Comput. Vision Graphics Image Process.* **27** (1984), 321-345.

[5] G. Borgefors: Distance transformations on hexagonal grids, *Pattern Recognition Lett.* **9** (1989) 97–105.

[6] C. H. Chen, L. F. Pau and P. S. P. Wang (editors): Handbook of pattern recognition & computer vision, (Handbooks in Science and Technology), Academic Press, Inc., Orlando, FL, 1986.

[7] P. E. Danielsson: 3D octagonal metrics, *Eighth Scandinavian Conf. Image Process.* (1993), pp. 727–736.

[8] P. P. Das: Best simple octagonal distances in digital geometry, *J. Approx. Theory* **68** (1992), 155–174.

[9] P. P. Das, P. P. Chakrabarti and B. N. Chatterji: Distance functions in digital geometry, *Inform. Sci.* **42** (1987), 113–136.

[10] P. P. Das, P. P. Chakrabarti and B. N. Chatterji: Generalised distances in digital geometry, *Inform. Sci.* **42** (1987), 51–67.

[11] P. P. Das and B. N. Chatterji: Estimation of errors between Euclidean and m-neighbor distance, *Inform. Sci.* **48** (1989), 1–26.

[12] P. P. Das and B. N. Chatterji: Hyperspheres in digital geometry, *Inform. Sci.* **50** (1990), 73–91.

[13] P. P. Das and B. N. Chatterji: Octagonal distances for digital pictures, *Inform. Sci.* **50** (1990), 123–150.

[14] E. S. Deutsch: Thinning algorithms on rectangular, hexagonal and triangular arrays, *Communications of the ACM*, **15** No.3 (1972) 827–837.

[15] B. Everitt: Cluster Analysis, Heinemann Educational Books Ltd, London, 1973.

[16] C. Fouard, R. Strand, and G. Borgefors: Weighted distance transforms generalized to modules and their computation on point lattices. *Pattern Recognition*, **40** (2007), 2453-2474.

[17] H. Freeman: Algorithm for Generating a Digital Straight Line on a Triangular Grid, *IEEE Transactions on Computers*, **C-28** (1979), 150–152.

[18] M. J. E. Golay: Hexagonal Parallel Pattern Transformations, *IEEE Transactions on Computers*, **C-18** (1969), 733–740.

[19] R. C. Gonzalez and R. E. Woods: Digital image processing. Addison-Wesley, Reading, MA, 1992.

[20] A. Hajdu: Geometry of Neighbourhood Sequences, *Pattern Recognition Lett.*, **24** (2003), 2597–2606.

[21] A. Hajdu and B. Nagy: Approximating the Euclidean circle using neighbourhood sequences, *KEPAF conference*, **3** (2002), 260–271.

[22] A. Hajdu, B. Nagy and Z. Zörgő: Indexing and segmenting colour images using neighbourhood sequences, *ICIP'03, IEEE International Conference on Image Processing*, Barcelona, Spain, (2003) I 957–960.

[23] F. Harary, R. A. Melter and I. Tomescu: Digital metrics: a graph theoretical approach, *Pattern Recognition Lett.* **2** (1984), 159–163.

[24] I. Her: Geometric transformations on the hexagonal grid, *IEEE Transaction on Image Processing* **4** No. 9 (1995), 1213–1221.

[25] R. Klette and A. Rosenfeld: Digital geometry. Geometric methods for digital picture analysis, Morgan Kaufmann Publishers, San Francisco, CA; Elsevier Science B.V., Amsterdam, 2004.

[26] T. Y. Kong and A. Rosenfeld: Digital topology: Introduction and survey, *Computer Vision, Graphics, and Image Processing* **48** (1987), 357–393.

[27] E. Luczak and A. Rosenfeld: Distance on a Hexagonal Grid, *IEEE Transaction on Computers* (1976), 532–533.

[28] B. H. McCormick: The Illinois pattern recognition computer-ILLIAC III, *IEEE Trans. Electronic Computers* **EC-12** (1963), 791–813.

[29] R. A. Melter: A Survey on Digital Metrics, *Contemp. Math.* **119** (1991), 95–106.

[30] B. Nagy: Distance functions based on neighbourhood sequences, 5th *International Conference on Applied Informatics*, Eger-Noszvaj, (2001), 183–190.

[31] B. Nagy: Finding shortest path with neighbourhood sequences on triangular grids, *ITI - ISPA'01, 2nd International Symposium on Image and Signal Processing and Analysis*, Pula, Croatia (2001), 55–60.

[32] B. Nagy: Metrics Based on Neighbourhood Sequences in Triangular Grids, *Pure Math. Appl.*, **13** (2002), 259–274.

[33] B. Nagy: Shortest Path in Triangular Grids with Neighbourhood Sequences, *Journal of Comp. and Inf. Techn.*, **11** (2003), 111–122.

[34] B. Nagy: Distance functions based on neighbourhood sequences, *Publ. Math. Debrecen* **63** (2003), 483–493.

[35] B. Nagy: A Family of Triangular Grids in Digital Geometry, *ISPA'03, 3rd International Symposium on Image and Signal Processing and Analysis* , Rome, Italy, (2003), 101–106.

[36] B. Nagy: A symmetric coordinate frame for hexagonal networks, *IS-TCS'04, ACM Conference on Theoretical Computer Science - Information Society*, Ljubljana, Slovenia, (2004), 193–196.

[37] B. Nagy: Characterization of Digital Circles in Triangular Grid, *Pattern Recognition Lett.*, **25** (2004), 1231–1242.

[38] B. Nagy: Generalised Triangular Grids in Digital Geometry, *Acta Math. Acad. Paed. Nyíregy.*, **20** (2004), 63–78.

[39] B. Nagy: Transformations of the triangular grid, *GRAFGEO, Third Hungarian Conference on Computer Graphics and Geometry*, Budapest, Hungary, (2005) 155–162.

[40] B. Nagy: A Comparison among Distances Based on Neighborhood Sequences in Regular Grids, *SCIA 2005, 14th Scandinavian Conference on Image Analysis*, Joensuu, Finnland, *Lecture Notes in Computer Science, LNCS* **3540** (2005), 1027–1036.

[41] B. Nagy: Metric and non-metric distances on $\mathbb{Z}^n$ by generalized neighbourhood sequences, *ISPA 2005, 4th International Symposium on Image and Signal Processing and Analysis*, Zagreb, Croatia, (2005), 215–220.

[42] B. Nagy: Geometry of neighborhood sequences in hexagonal grid, *DGCI 2006, Discrete Geometry for Computer Imagery*, Szeged, Hungary, *Lecture Notes in Computer Science, LNCS* **4245** (2006), 53–64.

[43] B. Nagy: Nonmetrical Distances on the Hexagonal Grid Using Neighborhood Sequences, *Pattern Recognition and Image Analysis* **17** (2007), 183–190.

[44] B. Nagy: Distances with Neighbourhood Sequences in Cubic and Triangular Grids, *Pattern Recognition Letters* **28** (2007), 99–109.

[45] B. Nagy: Optimal Neighborhood Sequences on the Hexagonal Grid (2007), *ISPA 2007, 5th International Symposium on Image and Signal Processing and Analysis* , Istanbul, Turkey, (2007), 310–315.

[46] B. Nagy: Distance with Generalised Neighbourhood Sequences in *n*D and ∞D, *Discrete Applied Mathematics* **156** (2008), 2344-2351.

[47] B. Nagy, and Strand, R.: Approximating Euclidean Distance Using Distances based on Neighbourhood Sequences in Non-Standard Three-Dimensional Grids, *IWCIA 2006, 11th International Workshop on Combinatorial Image Analysis* , Berlin, Germany, *Lecture Notes in Computer Science, LNCS* **4040** (2006), 89–100.

[48] B. Nagy, and R. Strand: Neighborhood Sequences in the Diamond Grid, in: *Image Analysis – From Theory to Applications, Research Publishing* , editors: Reneta P. Barneva, Valentin E. Brimkov, Singapore, Chennai (2008), 187–195.

[49] B. Nagy, and R. Strand: A Connection between $\mathbb{Z}^n$ and Generalized Triangular Grids, *ISVC 2008, 4th International Symposium on Visual Computing* , Las Vegas, Nevada, USA, *Lecture Notes in Computer Science, LNCS* **5359** (2008), 1157–1166.

[50] A. Radványi: On the rectangular grid representation of general CNN networks, *Int. J. Circ. Theor. Appl.* **30** (2002), 181–193.

[51] A. Rosenfeld, and R. A. Melter: Digital Geometry, *The mathematical intelligencer*, **11** No 3 (1989), 69–72.

[52] A. Rosenfeld and J. L. Pfaltz: Sequential operations in digital picture processing. *Journal of the ACM*, **13** No 4 (1966), 471-494.

[53] A. Rosenfeld and J. L. Pfaltz: Distance functions on digital pictures, *Pattern Recognition* **1** (1968), 33–61.

[54] K. Shimizu: Algorithm for Generating a Digital Circle on a Triangular Grid, *CGIP* **15** (1981), 401–402.

[55] M. Sonka, V. Hlavac and R. Boyle: Image processing, analysis, and machine vision. Brooks/Cole Publishing Company, Pacific Grove, CA, 1999.

[56] R. Strand, and B. Nagy: Distances Based on Neighbourhood Sequences in Non-Standard Three-Dimensional Grids, *Discrete Applied Mathematics* **155** (2007), 548–557.

[57] R. Strand: Weighted distances based on neighbourhood sequences, *Pattern Recognition Letters*, **28** (2007), 2029-2036.

[58] K. Voss: Discrete images, objects, and functions in $\mathbb{Z}^n$. Springer-Verlag, Berlin, 1991.

[59] C. A. Wüthrich and P. Stucki: An algorithmic comparison between square- and hexagonal-based grids, *CVGIP: Graphical Models and Image Proc.* **53** (1991) 324–339.

[60] M. Yamashita and N. Honda: Distance functions defined by variable neighborhood sequences, *Pattern Recognition* **17** (1984), 509–513.

[61] M. Yamashita and T. Ibaraki: Distances defined by neighborhood sequences, *Pattern Recognition* **19** (1986), 237–246.

In: Computational Mathematics
Editor: Peter G. Chareton, pp. 353-386
ISBN 978-1-60876-271-2

Chapter 12

A STREAM IN THE STUDY ON NORMALITY OF Σ-PRODUCTS

Yukinobu Yajima*
Department of Mathematics, Kanagawa University, Yokohama, 221-8686 Japan

Abstract

The concept of Σ-products was introduced and the normality has been investigated since 1959. This article is a survey for this study in a half century. There seems to be a certain stream in its long history. The purpose of this article is to show the vague stream as visible as possible. The contents consist of the following four chapters, each of which has several sections. We start from the beginning of the study and end with the most recent results which will appear in the near future.

Keywords: Σ-products, product, normal, paracompact, metric, collectionwise normal, countably paracompact, shrinking, countable tightness, p-space, (strong) Σ-space, semi-stratifiable, (strong) β-space, C-scattered, rectangular product, normal cover.

2010 Mathematics Subject Classification: Primary 54B10, 54D20; Secondary 54D15, 54E18.

I. Σ-Products and Infinite Products

This chapter begins with the Introduction and ends with an fundamental idea of the technique used through the many proofs in this article.

1 Introduction

The study of normality of Σ-products was begun by Corson [5] in 1959. The author gave a short survey [54] on this subject in 1987. This dealt with the advances of normality of Σ-products in about a quarter century. However, as it was published in a book with a small

*E-mail address: yajimy01@kanagawa-u.ac.jp

circulation, it is very difficult to be available nowadays. This article is not a simple update version, but a completely new one including the previous one.

When we look back the history of this study, there seems to be a certain *stream* in the advances of a half century, as if it naturally ran from a upper place to a lower one. Through this article, we shall try to make clear the certain stream. So we hope the reader to feel it. As is stated in the beginning, this article has the four chapters. We sometimes give the full proofs for some important results. The reasons are also stated around each of them.

In Chapter I, we introduce the concept of Σ-products and state the first result and problem from which the study of normality of Σ-products began. Afterward, we refer to a solution of this problem with the proof of our own style.

In the study of normality of Σ-products, the concept of countable tightness plays very important roles. In fact, the subsequent study is divided into the two cases by having countable tightness or not. Chapter II deals with Σ-products having countable tightness. We give a result showing the equivalence between normality and countable tightness of Σ-products. This was so beautiful that it seemed to make this study finish. However, it was only a beginning of the new advances. We end this chapter with a result which looks a final for the study of this direction.

In Chapter III, we discuss how to proceed the study on normality of Σ-products without countable tightness. We necessarily consider the relations between normality of Σ-products and the other properties such as collectionwise normality, countable paracompactness and the shrinking property.

In Chapter IV, we consider rectangularity of general products including infinite products. This concept originally came from finite products in dimension theory, but it was soon related with the investigation of certain subsets of infinite products. An interesting point is that the study of rectangularity of products is surprisingly resemble to that of normality of Σ-products. In fact, it have been advanced with quite similar manner. We end this chapter with the recent results which will appear in the near future.

Thus we deal with the study of normal Σ-products and rectangularity of products from the beginning to the present time. We believe that this study has been advanced into the direction to which it should be proceed as a natural stream during a half century. When you would finish to read this article, we would be so glad if you would feel a kind of stream in this study of such a long term. The names of an era are put on many results as the reader easily see the stream of times.

The notation and terminology follow the book [11] as a rule. It is desirable that the reader has the fundamental knowledge of General Topology (in particular, up to the "Chapter 5: Paracompact spaces" in this book).

Throughout this paper, all spaces are assumed to be *Hausdorff*. Let $\mathbb{N} = \{1,2,\cdots\}$ be the set of all natural numbers. Let ω denotes the first infinite cardinality, that is, $|\mathbb{N}| = \omega$. Otherwise, ω denotes the space $\mathbb{N} \cup \{0\}$ with the discrete topology. For a set Λ, we denote by $[\lambda]^{\leq\omega}$ and $[\lambda]^{<\omega}$ the sets of all countable and finite, respectively, subsets of Λ.

2 Products of Compact Factors

For making sure, let us begin with the most fundamental definitions.

Definition 1. A space X is *normal* if for any disjoint closed sets A and B in X there are disjoint open sets U and V such that $A \subset U$ and $B \subset V$.

For a collection $\{X_\lambda : \lambda \in \Lambda\}$ of spaces, the product $\prod_{\lambda\in\Lambda} X_\lambda$ has the topology generated by the base

$$\left\{ \prod_{\lambda\in\Lambda} U_\lambda : \text{each } U_\lambda \text{ is open in } X_\lambda \text{ and } U_\lambda = X_\lambda \text{ except finitely many } \lambda \right\}.$$

Next, we begin with the most classical theorem for products called the Tychonoff Theorem.

Theorem 2.1 (Tychonoff [49], 1935). *A product of compact spaces is compact.*

So there is no meaning to consider the normality of products of compact factors. On the other hand, there is one more important result.

Theorem 2.2 (Stone [45], 1948). *The product* ω^{ω_1} *is not normal.*

Since ω is the simplest space which is not compact, there seems to be no meaning to consider the normality of uncountable products of non-compact factors. In fact, the following puts the clincher on this conclusion.

Theorem 2.3 (Noble [36], 1971). *For a space X, the following are equivalent.*

(a) *X is compact.*

(b) *X^κ is normal for any cardinal κ.*

(c) *X^κ is normal for some uncountable cardinal $\kappa \geq w(X)$, where $w(X)$ denotes the weight of X.*

The proof found in [42, Corollary 6.6] seems to be easy to understand.

3 The Definition of Σ-Products

As stated in the previous section, there is no meaning to consider normality of uncountable products. However, there is a quite important normal dense subspace of uncountable products, introduced by Corson [5] in 1959.

Definition 2. Let $X = \prod_{\lambda\in\Lambda} X_\lambda$ be a product of spaces. Take a point $s = (s(\lambda)) \in X$. For each $x = (x(\lambda)) \in X$, let $\mathrm{Supp}(x) = \{\lambda \in \Lambda : x(\lambda) \neq s(\lambda)\}$, which is called the *support* of x. Then the subspace

$$\Sigma = \{x \in X : \mathrm{Supp}(x) \text{ is at most countable}\}$$

of X is called a *Σ-product* of the spaces $X_\lambda, \lambda \in \Lambda$. We say that such a Σ is a *Σ-product of ⋯ spaces* if each X_λ is a ⋯ space. Each X_λ is called a *factor* of Σ. We call $s \in \Sigma$ the *base point* of Σ, which is often omitted. A Σ-product Σ is *proper* if it has uncountably many factors, each of which contains at least two points. It is obvious that Σ is dense in X.

Proposition 3.1. *A proper Σ-product of compact spaces are countably compact, but not compact.*

This is easy to see. The proper Σ-product of 2-point spaces is contained in every proper Σ-product as a closed subspace. Since it contains a countably compact closed subspace which is not compact, it is non-sense to consider (generalized) paracompactness of Σ-products.

To be accurate, the concept of Σ-products should not be due to only Corson. In fact, Mazur [30] and Glicksberg [15] investigated continuous functions defined on Σ-products. Moreover, Corson himself dealt with such continuous functions in the same paper [5]. In this sense, they independently considered Σ-products each other. However, only Corson discussed normality of Σ-products and proved the following result.

Theorem 3.2 (Corson [5], 1959). *A Σ-product of complete metric spaces is normal.*

In this proof, Corson established a fundamental technique to prove normality of Σ-products. Moreover, he raised the following natural problem.

Problem 1 (Corson [5]). Can the "complete" in Theorem 3.2 be removed?

This problem was the start of the study on normality of Σ-products which has been an interesting subject during a half century. In this article, we focus only this subject.

4 Σ-Products of Metric Spaces

The Corson's problem (= Problem 1) had not been solved for a long time. This was independently solved by Gul'ko [16] and Rudin [43]. Here, we give a proof of the Gul'ko-Rudin's result which is a different style from the original one in [16].

The following is often used in the proofs for normality of general spaces.

Lemma 4.1 (well-known). *Let A and B be disjoint closed sets in a space X. If there is a σ-locally finite open cover $\mathcal{U}$ of X such that $\overline{U}$ misses A or B for each $U \in \mathcal{U}$, then there are disjoint open sets G and H in X such that $A \subset G$ and $B \subset H$.*

We use only the following quite fundamental result.

Theorem 4.2 (Stone [45], 1948). *Every metric space is paracompact.*

Now, we proceed the result which gave a great advance for the study of normality of Σ-products.

Theorem 4.3 (Gul'ko-Rudin [16, 43], 1977). *A Σ-product of metric spaces is normal.*

Before the proof of Theorem 4.3, we need some notations:

Let Σ be the Σ-product of spaces $X_\lambda, \lambda \in \Lambda$, with a base point $s \in \Sigma$. For each $R \in [\Lambda]^{\leq\omega}$, the product $\prod_{\lambda \in R} X_\lambda$, denoted by X_R, is called a *countable subproduct* of Σ.

Let Ξ be an index set of finite sequences. For $\xi = (\alpha_1, \cdots, \alpha_n) \in \Xi$, let $\xi|i = (\alpha_1, \cdots, \alpha_i)$ for each $i \leq n$. In particular, let $\xi_- = (\alpha_1, \cdots, \alpha_{n-1})$ and $\xi^\frown\alpha = (\alpha_1, \cdots, \alpha_n, \alpha)$.

Let Ξ be an index set. For each $R_\xi \in [\Lambda]^{\leq\omega}$, $\xi \in \Xi$, $X_{R_\xi} = \prod_{\lambda \in R_\xi} X_\lambda$ is a countable subproduct of Σ. The projections from Σ onto X_{R_ξ} and X_{R_η} onto X_{R_ξ}, where $R_\xi \subset R_\eta$, are denoted by p_{R_ξ} and $p_{R_\xi}^{R_\eta}$, respectively. Since these countable products and projections will appear quite frequently in the later proofs, for the simplicity, such a X_{R_ξ} is abbreviated by X_ξ, and such projections p_{R_ξ} and $p_{R_\xi}^{R_\eta}$ are abbreviated by p_ξ and p_ξ^η, respectively. For each $\Lambda' \subset \Lambda$, $p_{\Lambda'}$ denotes the projection from Σ onto $\Sigma_{\Lambda'}$, which denotes the Σ-products of $X_\lambda, \lambda \in \Lambda'$, with the base point $p_{\Lambda'}(s)$. For $R \in [\Lambda]^{\leq\omega}$, $V \subset \Sigma$ is *R-distinguished* in Σ if $p_R^{-1} p_R(V) = V$.

Proof of Theorem 4.3. Let Σ be the Σ-product of metric spaces $X_\lambda, \lambda \in \Lambda$, with some base point $s \in \Sigma$. For each $R \in [\Lambda]^{\leq\omega}, X_R$ has a metric d_R. Then, for each $Y \subset X_R$ and $\delta > 0$, $\operatorname{diam}_{X_R} Y < \delta$ means that $d_R(x, x') < \delta$ for any $x, x' \in Y$. Let A and B be any disjoint closed sets in Σ.

Now, for each $n \in \mathbb{N}$, we construct a collection $\mathcal{V}_n$ of open sets in Σ and an index set Ξ_n of n-tuple sequences such that, for each $\xi \in \Xi_n$, one can assign $V(\xi) \subset \Sigma, R_\xi \in [\Lambda]^{\leq\omega}$ and $a_\xi, b_\xi \in \Sigma$, satisfying the following conditions;

(a) $\mathcal{V}_n \cup \{V(\xi) : \xi \in \Xi_n\}$ is locally finite in Σ,

(b) for each $V \in \mathcal{V}_n$, $\overline{V} \cap A = \emptyset$ or $\overline{V} \cap B = \emptyset$,

(c) for each $\xi \in \Xi_n$,

  (1) $\xi_- \in \Xi_{n-1}$,
  (2) $V(\xi)$ is an R_{ξ_-}-distinguished open set in Σ with $V(\xi) \subset V(\xi_-)$,
  (3) $\operatorname{diam}_{X_{\xi|i}} p_{\xi|i}(V(\xi)) < 1/2^n$ for $i = 1, \cdots, n-1$,
  (4) $V(\xi) \subset (\bigcup \mathcal{V}_{n+1}) \cup \{V(\eta) : \eta \in \Xi_{n+1}$ with $\eta_- = \xi\}$,
  (5) $a_\xi \in \overline{V(\xi)} \cap A$ and $b_\xi \in \overline{V(\xi)} \cap B$,
  (6) $R_\xi = R_{\xi_-} \cup \operatorname{Supp}(a_\xi) \cup \operatorname{Supp}(b_\xi)$.

Let $\Xi_0 = \{\emptyset\}$ and $V(\emptyset) = \Sigma$. Pick any $\lambda_0 \in \Lambda$ and let $R_\emptyset = \{\lambda_0\}$. Assume that the construction above has been already performed for no greater than n.

Take any $\xi \in \Xi_n$ and fix it. Let $U(\xi) = p_\xi(V(\xi))$. For each $i \leq n$, since $X_{\xi|i}$ is metric, let $\mathcal{U}(\xi|i)$ be an open cover of $X_{\xi|i}$ with $\operatorname{diam}_{X_{\xi|i}} U < 1/2^{n+1}$ for each $U \in \mathcal{U}(\xi|i)$. Let

$$\mathcal{W}(\xi) = \{\bigcap_{i \leq n} (p_{\xi|i}^{\xi})^{-1}(U_i) \cap U(\xi) : U_i \in \mathcal{U}(\xi|i) \text{ and } i \leq n\}.$$

Since $\mathcal{W}(\xi)$ is an open cover of $U(\xi)$, it follows from Theorem 4.2 that there is a locally finite open refinement $\mathcal{U}(\xi) = \{U(\xi^\frown\alpha) : \alpha \in \Omega(\xi)\}$ of $\mathcal{W}(\xi)$. We let

$$\mathcal{V}(\xi) = \{p_\xi^{-1}(U) : U \in \mathcal{U}(\xi) \text{ such that } \overline{p_\xi^{-1}(U)} \text{ misses } A \text{ or } B\} \text{ and}$$

$$\Xi_{n+1}(\xi) = \{\xi^\frown\alpha : \alpha \in \Omega(\xi) \text{ such that } \overline{p_\xi^{-1}(U(\xi^\frown\alpha))} \text{ meets both } A \text{ and } B\}.$$

Take any $\eta = \xi^\frown\alpha \in \Xi(\xi)$. Let $V(\eta) = p_\xi^{-1}(U(\eta))$. Pick two points $a_\eta \in \overline{V(\eta)} \cap A$ and $b_\eta \in \overline{V(\eta)} \cap B$, and let $R_\eta = R_\xi \cup \mathrm{Supp}(a_\eta) \cup \mathrm{Supp}(b_\eta)$. Here, letting ξ run over Ξ_n, we let

$$\mathcal{V}_{n+1} = \bigcup\{\mathcal{V}(\xi) : \xi \in \Xi_n\} \text{ and } \Xi_{n+1} = \bigcup\{\Xi_{n+1}(\xi) : \xi \in \Xi_n\}.$$

Then it is easily verified that all of the above conditions are satisfied.

Let $\mathcal{V} = \bigcup_{n\in\mathbb{N}} \mathcal{V}_n$. It suffices from (a), (b) and Lemma 4.1 to show that $\mathcal{V}$ covers Σ. Assuming the contrary, pick a point $y \in \Sigma \setminus \bigcup \mathcal{V}$. By (4), we can inductively choose a sequence $\{\xi_n\}$ of indices such that $\xi_n \in \Xi_n$ with $(\xi_{n+1})_- = \xi_n$ and $y \in V(\xi_n)$ for each $n \in \mathbb{N}$. Let $R = \bigcup_{n\in\mathbb{N}} R_{\xi_n}$. Then note that $R \in [\Lambda]^{\leq\omega}$. Take the point $z \in \Sigma$ defined by $p_R(z) = p_R(y)$ and $p_{\Lambda\setminus R}(z) = p_{\Lambda\setminus R}(s)$. For simplicity, we abbreviate $R_{\xi_n}, p_{\xi_n}, p_{\xi_n}^{\xi_m}, a_{\xi_n}$ and b_{ξ_n} to R_n, p_n, p_n^m, a_n and b_n, respectively. We show that $z \in A \cap B$.

Let O be any basic open neighborhood of z in Σ. Since $\{R_{\xi_n}\}$ is an increasing sequence by (6), we can take $m \in \mathbb{N}$ such that $O = p_m(O) \times X_{R\setminus R_m} \times p_{\Lambda\setminus R}(O)$. By (3), we can take $k \geq m$ such that $p_m(y) = p_m(z) \in p_m(V(\xi_k)) \subset \overline{p_m(V(\xi_k))} \subset p_m(O)$. By (5), we have

$$\{p_k(a_k), p_k(b_k)\} \subset p_k(\overline{V(\xi_k)}) \subset (p_m^k)^{-1} p_m(\overline{V(\xi_k)}) \subset (p_m^k)^{-1} p_m(O) = p_k(O).$$

Since $p_{\Lambda\setminus R}(a_k) = p_{\Lambda\setminus R}(b_k) = p_{\Lambda\setminus R}(s) = p_{\Lambda\setminus R}(z) \in p_{\Lambda\setminus R}(O)$, we obtain $a_k \in A \cap O$ and $b_k \in B \cap O$. Since O is any open neighborhood of z, we conclude that $z \in \overline{A} \cap \overline{B} = A \cap B$, which contradicts $A \cap B = \emptyset$. Hence $\mathcal{V}$ covers Σ. □

The original proof of Gul'ko depends on the complicated metric property of countable products. We gave another proof in our survey [53]. However, it uses a complicated index set of square matrices. On the other hand, the above proof mainly uses the covering property of metric spaces with the index set of finite sequences. In this sense, we believe that our proof here is easier to understand than both of them. Moreover, it also suggests some further extensions.

II. Σ-Products with Countable Tightness

The concept of countable tightness is an essential property to investigate when Σ-products become normal. In fact, it has been used as a sufficient condition in so many results for normality of Σ-products.

5 Tightness and Products

Hereafter, κ denotes an infinite cardinal. We begin with the general concept of tightness by κ.

Definition 3. The *tightness* of a space X, denoted by $t(X)$, is the smallest cardinal κ such that for any $A \subset X$ and $x \in \overline{A}$, there is some $B \subset A$ such that $|B| \leq \kappa$ and $x \in \overline{B}$. In particular, we say that a space X has *countable tightness* if $t(X) \leq \omega$.

The following is obvious and it is often used.

Fact 5.1. *$t(X) \le \kappa$ implies $t(Y) \le \kappa$ for any $Y \subset X$.*

The following is found in [11, Exercise 1.7.3 (b)] without the proof.

Proposition 5.2. *For a space X, $t(X) \le \kappa$ if and only if, for any non-closed subset M in X, there is $N \subset M$ such that $|N| \le \kappa$ and $\overline{N} \setminus M \neq \emptyset$.*

Proof. The "if" part: Take any $A \subset X$. Let $A_\kappa = \bigcup\{\overline{B} : B \subset A$ with $|B| \le \kappa\}$. Since $A \subset A_\kappa \subset \overline{A}$, it suffices to show A_κ is closed in X. Assuming the contrary, by the assumption, there is $C \subset A_\kappa$ such that $|C| \le \kappa$ and $\overline{C} \setminus A_\kappa \neq \emptyset$. Let $C = \{x_\alpha : \alpha \in \kappa\}$. For each $\alpha \in \kappa$, there is $D_\alpha \subset A$ such that $|D_\alpha| \le \kappa$ and $x_\alpha \in \overline{D_\alpha}$. Let $D = \bigcup_{\alpha\in\kappa} D_\alpha$. Clearly, $D \subset A$ with $|D| \le \kappa$. So we have $C \subset \bigcup_{\alpha\in\kappa}\{\overline{D_\alpha} : \alpha \in \kappa\} \subset \overline{D} \subset A_\kappa$. Hence we obtain $\overline{C} \subset \overline{D} \subset A_\kappa$, which is a contradiction. The converse is obvious. □

It is obvious that every first countable space has countable tightness. By Proposition 5.2, more generally, every sequential space has countable tightness.

Now, we state two important results for tightness related to products.

Theorem 5.3 (Malyhin [29], 1972). *If X is a space and C is a compact space, then $t(X \times C) \le t(X) \cdot t(C)$ holds.*

Theorem 5.4 (Nogura [37], 1976). *A compact space X has tightness $\le \kappa$ if and only if $X \times \kappa^+$ is normal.*

Recall that a space X is called a *k-space* if for each $A \subset X$, A is closed in X whenever $A \cap K$ is closed in K for each compact subset K in X.

Proposition 5.5. *For a k-space X, the following are true.*

(1) *If $t(K) \le \kappa$ for each compact subset K in X, then $t(X) \le \kappa$ holds.*

(2) *If $X \times \kappa^+$ is normal, then $t(X) \le \kappa$ holds.*

Proof. (1): Let M be a non-closed subset in X. Then there is a compact subset K in X such that $M \cap K$ is non-closed in K. By Proposition 5.2, there is $N \subset M \cap K$ such that $|N| \le \kappa$ and $\overline{N} \setminus (M \cap K) \neq \emptyset$. Since K is closed in X, we have $N \subset M$ with $\overline{N} \setminus M \neq \emptyset$. By Proposition 5.2 again, we obtain $t(X) \le \kappa$.

(2): Let K be a compact subset in X. By the assumption, $K \times \kappa^+$ is normal. By Theorem 5.4, we have $t(K) \le \kappa$. By (1) above, $t(X) \le \kappa$ is assured. □

Let Σ be the Σ-product of spaces $X_\lambda, \lambda \in \Lambda$. Recall that $\prod_{\lambda\in R} X_\lambda$ is called a countable subproduct of Σ for each $R \in [\Lambda]^{\le\omega}$. Similarly, $\prod_{\lambda\in\theta} X_\lambda$ is called a *finite subproduct* of Σ for each $\theta \in [\Lambda]^{<\omega}$.

Proposition 5.6 ([28]). *Let Σ be a Σ-product of spaces. Then the following are equivalent.*

(a) *Σ has countable tightness.*

(b) *Each countable subproduct of Σ has countable tightness.*

(c) *Each finite subproduct of Σ has countable tightness.*

Proof. Let Σ be the Σ-product of spaces $X_\lambda, \lambda \in \Lambda$. Let $X = \prod_{\lambda \in \Lambda} X_\lambda$.

(a) $\Rightarrow$ (c): This is obvious from Fact 5.1.

(c) $\Rightarrow$ (b): Take any $R \in [\Lambda]^{\leq \omega}$. Let $A \subset X_R$ and $x \in \overline{A}$. For each $\theta \in [\Lambda]^{<\omega}$, since $\pi_\theta(x) \in \pi_\theta(\overline{A}) \subset \overline{\pi_\theta(A)}$, there is $B_\theta \subset A$ such that $|B_\theta| \leq \omega$ and $x \in \overline{\pi_\theta(B_\theta)}$. Let $C = \bigcup_{\theta \in [R]^{<\omega}} B_\theta$. Then we have $C \subset A$ with $|C| \leq \omega$ and $x \in \overline{C}$.

(b) $\Rightarrow$ (a): Let $M \subset \Sigma$ and $y \in \overline{M}$. Let $R_0 = \mathrm{Supp}(y)$ and $N_0 = \emptyset$. Suppose that $\{R_i : i \leq n\}$ and $\{N_i : i \leq n\}$ have been already defined. By $p_{R_n}(y) \in p_{R_n}(\overline{M}) \subset \overline{p_{R_n}(M)}$, there is $N_n \subset M$ such that $|N_n| \leq \omega$ and $p_{R_n}(y) \in \overline{p_{R_n}(N_n)}$. Let $R_{n+1} = R_n \cup \left(\bigcup\{\mathrm{Supp}(z) : z \in N_n\}\right)$. Then $|R_n| \leq \omega$ and $R_n \subset R_{n+1}$. Here let $N = \bigcup_{n \in \mathbb{N}} N_n$. Clearly, $N \subset M$ with $|N| \leq \omega$. It is easily verified that $y \in \overline{N}$. □

Corollary 5.7. *A Σ-product of first countable spaces has countable tightness.*

6 Σ-Products of Paracompact *p*-Spaces

After Corson's paper [5], it seems to be curious that no result for normality of Σ-products had appeared during more than 10 years. The following first smashed this sense of despair. Not only that, it also suggested the importance of countable tightness for Σ-products.

Theorem 6.1 (Kombarov [25], 1971). *Let Σ be a Σ-product of paracompact Čech-complete spaces. If Σ has countable tightness, then it is normal.*

We say that a class $\mathcal{C}$ of spaces is *countably productive* if for each countable $\{X_n : n \in \mathbb{N}\} \subset \mathcal{C}$, the product $\prod_{n \in \mathbb{N}} X_n$ belongs to $\mathcal{C}$.

The class of (paracompact) Čech-complete spaces is countably productive (see [11, Theorem 3.9.8]). As a class containing all metric and all Čech-complete spaces, the following concept was introduced.

Definition 4. A Tychonoff space X is called a *p-space* [1] if there is a sequence $\{\mathcal{U}_n\}$ of collections of open sets in βX such that each $\mathcal{U}_n$ covers X and $\bigcap_n \mathrm{St}(x, \mathcal{U}_n) \subset X$ for each $x \in X$, where βX denotes the Stone-Čech compactification of X.

Recall that a continuous map $f : X \to Y$ is a *perfect map* if it is a closed map such that $f^{-1}(y)$ is compact for each $y \in Y$.

Since the class of paracompact p-spaces coincides with that of paracompact M-spaces, the following is well-known.

Theorem 6.2 (Morita [33], 1964). *A Tychonoff space X is a paracompact p-space if and only if there are a metric space M and a perfect map f from X onto M.*

Proposition 6.3. *The following are true.*

(1) *The class of p-spaces is countably productive.*

(2) *The class of paracompact p-spaces is countably productive.*

(3) *Every p-space is a k-space.*

The proofs of (1) and (3) are stated in [14, Theorem 3.16 and 3.22], respectively. (2): By Theorem 6.2, let f_n be a perfect map from a paracompact p-space X_n onto a metric space M_n for each $n \in \mathbb{N}$, then $\prod_{n\in\mathbb{N}} f_n$ is a perfect map from $\prod_{n\in\mathbb{N}} X_n$ onto $\prod_{n\in\mathbb{N}} M_n$ (see [11, Theorem 3.7.9]).

Immediately after of the great advance by Theorem 4.3, the following quite nice result was proved. This was not only a generalization of Theorems 4.3 and 6.1, but also this showed that countable tightness is quite essential for normality of Σ-products. It was so beautiful that it seemed to mean an end of this kind of study.

Theorem 6.4 (Kombarov [27], 1978). *Let Σ be a proper Σ-product of paracompact p-spaces. Then the following are equivalent.*

(a) *Σ is normal.*

(b) *Σ is collectionwise normal.*

(c) *Σ has countable tightness.*

(d) *Each factor of Σ has countable tightness.*

The proof of (c) $\Rightarrow$ (a) is fully stated in [27]. Moreover, this part will be stated in a more generalized form later. The proof of (a) $\Rightarrow$ (b) is not stated here. (c) $\Rightarrow$ (d) is obvious. We give the proofs (a) $\Rightarrow$ (c) and (d) $\Rightarrow$ (c) in detail, according to his idea. Because they were written quite simply in [27].

Let $\Sigma 2^{\omega_1}$ be the Σ-product of ω_1 many copies of the discrete two-point space $\{0,1\}$. We denote by ω_1 the space $[0,\omega_1)$ with the usual order topology.

Fact 6.5. *The space ω_1 can be embedded as a closed subspace in $\Sigma 2^{\omega_1}$.*

In fact, the map $\varphi : \omega_1 \to \Sigma 2^{\omega_1}$ defined by $\varphi(\alpha)(\lambda) = 1$ if $\lambda < \alpha$ and $\varphi(\alpha)(\lambda) = 0$ if $\lambda \geq \alpha$ for each $\alpha \in \omega_1$ is a homeomorphism from ω_1 into $\Sigma 2^{\omega_1}$. Moreover, $\varphi(\omega_1)$ is closed in $\Sigma 2^{\omega_1}$.

Proposition 6.6. *Let Σ be a proper Σ-product, each finite subproduct of which is a k-space.*

(1) *If each factor of Σ has countable tightness, then Σ has countable tightness.*

(2) *If Σ is normal, then it has countable tightness.*

Proof. Let Σ be the proper Σ-product of spaces $X_\lambda, \lambda \in \Lambda$ such that each finite subproduct of Σ is a k-space.

(1): Take a $\theta \in [\Lambda]^{<\omega}$. It suffices from Proposition 5.6 to show $t(X_\theta) \leq \omega$. Take a compact subset K in X_θ. It suffices from Proposition 5.5 (1) to show $t(K) \leq \omega$. By Theorem 5.3, we have

$$t(K) \leq t(\prod_{\lambda\in\theta} \pi_\lambda(K)) \leq \max\{t(\pi_\lambda(K)) : \lambda \in \theta\} \leq \max\{t(X_\lambda) : \lambda \in \theta\} \leq \omega.$$

(2): Let X_θ be a finite subproduct of Σ. Since Σ is proper, it contains $X_\theta \times \Sigma 2^{\omega_1}$ as a closed set. Moreover, $X_\theta \times \omega_1$ is closed in Σ by Fact 6.5. So $X_\theta \times \omega_1$ is normal. Since X_θ is a k-space, it follows from Proposition 5.5 (2) that $t(X_\theta) \leq \omega$. By Proposition 5.6, we obtain $t(\Sigma) \leq \omega$. □

Thus, (a) ⇒ (c) and (d) ⇒ (c) follow from Propositions 6.3 and 6.6.

From Theorem 6.4, we immediately have

Corollary 6.7 (Kombarov [26], 1973). *If a Σ-product of compact spaces has countable tightness, then it is normal.*

7 Generalized Metric Spaces

For the generalizations of metric spaces, there is another direction different from p-spaces and M-spaces. This direction contains many important classes such as Lašnev spaces, M_1-spaces, stratifiable spaces (= M_3-spaces), σ-spaces, Σ-spaces and semi-stratifiable spaces.

Definition 5 (Some typical generalized metric spaces).

(1) A regular space X is called a *σ-space* [39] if it has a σ-locally finite net.

(2) A space X is called a *(strong) Σ-space* [34] if there are a σ-locally finite closed cover $\mathcal{F}$ of X and a closed cover $\mathcal{K}$ by countably compact (compact) sets in X such that, whenever $K \in \mathcal{K}$ and U is open in X with $K \subset U$, there is $F \in \mathcal{F}$ with $K \subset F \subset U$.

(3) A space X is called *semi-stratifiable* [6] if there is a function $g : X \times \mathbb{N} \to \mathrm{Top}(X)$, where $\mathrm{Top}(X)$ denotes the topology of X, satisfying

(i) $x \in g(x, n+1) \subset g(x, n)$ for each $x \in X$ and each $n \in \mathbb{N}$,

(ii) if $y \in \bigcap_{n\in\mathbb{N}} g(x_n, n)$, then $\{x_n\}$ converges to y.

Note that every paracompact Σ-space is a strong Σ-space and that each σ-space is a strong Σ-space and is semi-stratifiable. In particular, in the class of *paracompact* spaces, the following implications hold:

$$\begin{array}{ccccccc}
 & & \text{metric} \longrightarrow & \text{stratifiable} & \longrightarrow \sigma\text{-space} & \longrightarrow & \text{semi-stratifiable} \\
 & & \searrow & & \downarrow & & \downarrow \\
\check{\text{C}}\text{ech-complete} & \longrightarrow & & p\text{-space} & \longrightarrow \Sigma\text{-space} & \longrightarrow & \text{(strong) } \beta\text{-space}
\end{array}$$

Note that the three classes of σ-spaces, strong Σ-spaces and semi-stratifiable spaces are countably productive. Now, we refer to (strong) β-spaces, whose class is the widest one in the above generalized metric spaces.

Definition 6. A space X is called a *strong β-space* [58] (a *β-space* [19]) if there is a function $g : X \times \mathbb{N} \to \mathrm{Top}(X)$, satisfying

(i) $x \in g(x, n+1) \subset g(x, n)$ for each $x \in X$ and each $n \in \mathbb{N}$,

(ii) if $\bigcap_{n\in\mathbb{N}} g(x_n, n) \neq \emptyset$, then $\overline{\{x_n : n \in \mathbb{N}\}}$ is compact ($\{x_n\}$ has a cluster point).

Such a function g is called a *(strong) β-function for X*.

Although the original definition of strong β-spaces is somewhat different from the above, their equivalence is easily seen.

Proposition 7.1 ([58]). *The following are true.*

(1) *A paracompact β-space is a strong β-space.*

(2) *The class of strong β-spaces is preserved by perfect images and perfect preimages.*

(3) *The class of strong β-spaces is countably productive.*

Proof. As the above definition makes the proof of (3) much easier than the original, we only prove it here. Let $X = \prod_{i\in\mathbb{N}} X_i$, where each X_i is a strong β-space. Let g_i be a strong β-function for X_i for each $i \in \mathbb{N}$. Let $g(x,n) = \prod_{i\leq n} g_i(x(i),n) \times \prod_{i>n} X_i$ for each $x \in X$ and $n \in \mathbb{N}$. Let $y \in \bigcap_{n\in\mathbb{N}} g(x_n,n)$. Take any $k \in \mathbb{N}$. Since $y(k) \in \bigcap_{n\geq k} g_k(x_n(k),n)$, $K_k = \overline{\{x_n(k) : n \in \mathbb{N}\}}$ is compact. Then the compact subset $\prod_{k\in\mathbb{N}} K_k$ contains $\overline{\{x_n : n \in \mathbb{N}\}}$. Hence it is compact. □

The following is a result for the countable products of strong β-spaces, which is used in the proof of Σ-products of β-spaces.

Theorem 7.2 ([58], 2007). *Let X_i be a strong β-space for each $i \in \mathbb{N}$. If $\prod_{i\leq n} X_i$ is normal (paracompact) for each $n \in \mathbb{N}$, then $\prod_{i\in\mathbb{N}} X_i$ is normal (paracompact).*

The parenthetic part easily follows from Theorem 12.1 below.

8 Σ-Products of Generalized Metric Spaces

As is stated before, the Kombarov's result (= Theorem 6.4) is so beautiful that it looks a final form in the study of normality of Σ-products. However, the following means that it is not a final point but a passing point.

Theorem 8.1 ([50], 1984). *Let Σ be a Σ-product of paracompact Σ-spaces. If Σ has countable tightness, then it is normal.*

This was not only a generalization of (c) ⇒ (a) in Theorem 6.4, but also suggested a new direction of the study of normality of Σ-products. In fact, there are many interesting classes of generalized metric spaces such as stratifiable spaces, σ-spaces and semi-stratifiable spaces around Σ-spaces (as is seen the diagram in the previous section). Using one of them, another result was obtained as follows.

Theorem 8.2 ([53], 1987). *Let Σ be a Σ-product of semi-stratifiable spaces, each finite subproduct of which is paracompact. If Σ has countable tightness, then it is normal.*

The proofs of Theorems 8.1 and 8.2 are quite different. So it is natural to ask in [55, Question 4] whether these results can be unified in terms of β-spaces. This was affirmatively solved by the following. The proof is more resemble to that of Theorem 8.2 rather than Theorem 8.1. Since the full proof extends several papers, we give it for the reader's convenience. Here we also use the notations stated in the Section 4 without special mention.

Theorem 8.3 ([57], 2007). *Let Σ be a Σ-product of β-spaces, each finite subproduct of which is paracompact. If Σ has countable tightness, then it is normal.*

Proof. Let Σ be the Σ-product of spaces $X_\lambda, \lambda \in \Lambda$, with a base point $s \in \Sigma$. It follows from Proposition 7.1 and Theorem 7.2 that each countable subproduct of Σ is a paracompact, strong β-space with countable tightness. Let A and B be any disjoint closed sets in Σ. We may assume that $\overline{p_R(A)} \cap \overline{p_R(B)} \neq \emptyset$ for each $R \in [\Lambda]^{\leq\omega}$, since X_R is normal.

Now, for each $n \in \mathbb{N}$, we construct a collection $\mathcal{V}_n$ of open sets in Σ and an index set Ξ_n of n-tuple sequences such that for each $\xi \in \Xi_n$ one can assign $R_\xi \in [\Lambda]^{\leq\omega}, E(\xi) \subset \Sigma, H(\xi) \subset \Sigma, x_\xi \in X_{\xi_-}, A(\xi) \subset \Sigma, B(\xi) \subset \Sigma$ and a strong β-function g_ξ for X_ξ, satisfying the following conditions for each $n \in \mathbb{N}$:

(a) $\mathcal{V}_n = \bigcup\{\mathcal{V}(\mu) : \mu \in \Xi_{n-1}\}$ is locally finite in Σ.

(b) Each $V \in \mathcal{V}(\mu)$ is an R_μ-distinguished open set in Σ such that $\overline{V}$ misses A or B for each $\mu \in \Xi_{n-1}$.

(c) $\{H(\xi) : \xi \in \Xi_n\}$ is locally finite in Σ such that $\bigcup \mathcal{V}(\xi) \subset H(\xi) \subset H(\xi_-)$ for each $\xi \in \Xi_n$.

(d) For each $\xi \in \Xi_n$,

(1) $\xi_- \in \Xi_{n-1}$,

(2) $E(\xi)$ is a R_{ξ_-}-distinguished closed set in Σ and $H(\xi)$ is a R_{ξ_-}-distinguished open set in Σ such that $E(\xi) \subset H(\xi)$,

(3) $E(\xi) \subset \bigcup \mathcal{V}(\xi) \cup \left(\bigcup\{E(\nu) : \nu \in \Xi_{n+1} \text{ with } \nu_- = \xi\}\right)$,

(4) $g_\xi(x, k+1) \subset g_\xi(x,k)$ and $p^{\xi}_{\xi_-} \circ g_\xi(x,k) \subset g_\xi(p^{\xi}_{\xi_-}(x), k)$ for each $x \in X_\xi$ and each $k \in \mathbb{N}$,

(5) $p_{\xi_-}(E(\xi)) \subset g_{\xi_-}(x_\xi, n)$,

(6) $A(\xi)$ and $B(\xi)$ are countable subsets of A and B, respectively, such that $x_\xi \in \overline{p_{\xi_-}(A(\xi))} \cap \overline{p_{\xi_-}(B(\xi))}$,

(7) $R_\xi = R_{\xi_-} \cup \bigcup\{\mathrm{Supp}(c) : c \in A(\xi) \cup B(\xi)\}$.

Let $\Xi_0 = \{\emptyset\}$ and let $E(\emptyset) = H(\emptyset) = \Sigma$. Pick any $\lambda_0 \in \Lambda$ and let $R_\emptyset = \{\lambda_0\}$. Assume that the construction above has been already performed for no greater than n.

Take any $\xi \in \Xi_n$ and fix it. Let $\Phi = \overline{p_\xi(A)} \cap \overline{p_\xi(B)}$. Then Φ is a non-empty closed set in X_ξ. Note that $p_\xi(E(\xi))$ is a closed set and $p_\xi(H(\xi))$ is an open set in X_ξ such that $p_\xi(E(\xi)) \subset p_\xi(H(\xi))$. By the choice of Φ, there is a collection $\mathcal{W}$ of open sets in X_ξ such that

$$p_\xi(E(\xi)) \setminus \Phi \subset \bigcup \mathcal{W} \subset p_\xi(H(\xi)) \setminus \Phi$$

and, for each $W \in \mathcal{W}$, $\overline{W}$ misses $p_\xi(A)$ or $p_\xi(B)$. Since X_ξ is paracompact, there is a locally finite closed cover $\mathcal{F}$ of $p_\xi(E(\xi))$, which refines

$$\{W \cap p_\xi(E(\xi)) : W \in \mathcal{W}\} \cup \{g_\xi(x, n+1) \cap p_\xi(E(\xi)) : x \in p_\xi(E(\xi)) \cap \Phi\}.$$

Let $\mathcal{F}_+ = \{F \in \mathcal{F} : F \text{ is contained in some } W_F \in \mathcal{W}\}$. By the paracompactness of X_ξ again, there is a locally finite collection $\{U_F : F \in \mathcal{F}\}$ of open sets in X_ξ such that $F \subset U_F \subset p_\xi(H(\xi))$ for each $F \in \mathcal{F}$. Moreover, we may assume that $F \subset U_F \subset W_F \cap p_\xi(H(\xi))$ for each $F \in \mathcal{F}_+$. Let $\mathcal{V}(\xi) = \{p_\xi^{-1}(U_F) : F \in \mathcal{F}_+\}$. Then (a), (b) and (c) are satisfied. Let $\Xi(\xi)$ be an index set of $(n+1)$-tuple sequences such as $\mathcal{F} \setminus \mathcal{F}_+ = \{F_{\xi^\frown\alpha} : \xi^\frown\alpha \in \Xi(\xi)\}$. Pick any $\eta = \xi^\frown\alpha \in \Xi(\xi)$. Let $E(\eta) = p_\xi^{-1}(F_\eta)$ and $H(\eta) = p_\xi^{-1}(U_{F_\eta})$. By the choice of $\mathcal{F} \setminus \mathcal{F}_+$, we can find $x_\eta \in p_\xi(E(\xi)) \cap \Phi$ such that $F_\eta \subset g_\xi(x_\eta, n+1)$. Since X_ξ has countable tightness and $x_\eta \in \Phi$, we can choose some $A(\eta) \subset A$ and $B(\eta) \subset B$ satisfying (6). Define R_η as in (7). Here, letting ξ run over Ξ_n, we let

$$\mathcal{V}_{n+1} = \bigcup\{\mathcal{V}(\xi) : \xi \in \Xi_n\} \text{ and } \Xi_{n+1} = \bigcup\{\Xi(\xi) : \xi \in \Xi_n\}.$$

For each $\eta \in \Xi_{n+1}$, we can take a strong β-function for X_η satisfying (4), because X_η is a strong β-space and p_ξ^η is continuous. Then it is easily verified that (1)–(7) except (3) in (d) are satisfied. We only check about (3). Pick any $u \in E(\xi) \setminus \bigcup \mathcal{V}(\xi)$. Since $p_\xi(u) \in p_\xi(E(\xi)) = \bigcup \mathcal{F}$, there is $F_0 \in \mathcal{F}$ with $p_\xi(u) \in F_0$. Assuming $F_0 \in \mathcal{F}_+$, we have $u \in p_\xi^{-1}(F_0) \subset p_\xi^{-1}(U_{F_0}) \in \mathcal{V}(\xi)$, which is a contradiction. So F_0 must be in $\mathcal{F} \setminus \mathcal{F}_+$. Hence $u \in p_\xi^{-1}(F_\eta) = E(\eta)$ for some $\eta \in \Xi(\xi)$. Then $\eta \in \Xi_{n+1}$ with $\eta_- = \xi$. Thus we have inductively accomplished the desired construction above.

We let $\mathcal{V} = \bigcup_{n\in\mathbb{N}} \mathcal{V}_n$. By (a) and (b), $\mathcal{V}$ is a σ-locally finite collection of open sets in Σ such that $\overline{V}$ misses A or B. It suffices from Lemma 4.1 to show that $\mathcal{V}$ covers Σ. Assuming the contrary, there is some $y \in \Sigma \setminus \bigcup \mathcal{V}$. By $E(\emptyset) = \Sigma$ and (3), we can inductively choose a sequence $\{\xi_n\}$ of indices such that $\xi_n \in \Xi_n$ with $(\xi_{n+1})_- = \xi_n$ and $y \in E(\xi_n)$ for each $n \in \mathbb{N}$. Hereafter, we abbreviate $p_{\xi_n}, p_{\xi_n}^{\xi_k}, x_{\xi_n}$ and R_{ξ_n} to p_n, p_n^k, x_n and R_n, respectively.

Take any $n \in \mathbb{N}$. Let $C_{n,m} = \{p_n^k(x_{k+1}) : k \geq m\}$ for each $m \in \mathbb{N}$ with $m \geq n$. It follows from (4) that $p_n^k \circ g_{\xi_k}(x, j) \subset g_{\xi_n}(p_n^k(x), j)$ for each $x \in X_{\xi_k}$ and each $k, j \in \mathbb{N}$, where $k \geq n$. Using this, by (5), we have $p_n(y) \in \bigcap_{k\geq n} g_{\xi_n}(p_n^k(x_{k+1}), k+1)$. Since g_{ξ_n} is a strong β-function for X_{ξ_n}, each $\overline{C_{n,m}}$ is compact. Let $K_n = \bigcap_{m\geq n} \overline{C_{n,m}}$. Then K_n is a non-empty compact subset in X_{ξ_n} such that $p_n^{n+1}(K_{n+1}) \subset K_n$. Since $\{K_n, p_{n-1}^n\}$ is an inverse sequence of non-empty compact spaces, the inverse limit is not empty. That is, there is some $z_n \in K_n$ such that $p_n^{n+1}(z_{n+1}) = z_n$ for each $n \in \mathbb{N}$. Let $R = \bigcup_{n\in\mathbb{N}} R_n$. Since $|R| \leq \omega$, we can take the point $z \in \Sigma$ defined by $p_n(z) = z_n$ for each $n \in \mathbb{N}$ and $p_{\Lambda\setminus R}(z) = p_{\Lambda\setminus R}(s)$. We show that $z \in A \cap B$.

Let O be any basic open neighborhood of z in Σ. Since $\{R_n\}$ is an increasing sequence by (7), we can take $m \in \mathbb{N}$ such that

$$O = p_m(O) \times X_{R\setminus R_m} \times p_{\Lambda\setminus R}(O),$$

where recall $X_{R\setminus R_m} = \prod_{\lambda\in R\setminus R_m} X_\lambda$. Since $z_m \in \overline{C_{m,m}} \cap p_m(O)$, we can take $k > m$ with $p_m^k(x_{k+1}) \in p_m(O)$. Then we have

$$x_{k+1} \in (p_m^k)^{-1} p_m(O) = p_m(O) \times X_{R_k\setminus R_m} = p_k(O).$$

By (6), there are $a \in A(\xi_{k+1}) \subset A$ and $b \in B(\xi_{k+1}) \subset B$ such that $p_k(a)$ and $p_k(b)$ belong to $p_k(O)$. Then $p_{k+1}(a)$ and $p_{k+1}(b)$ are in $p_k(O) \times X_{R_{k+1}\setminus R_k} = p_{k+1}(O)$. Since $\mathrm{Supp}(a) \cup \mathrm{Supp}(b) \subset R_{k+1} \subset R$ by (7), we have $p_{\Lambda\setminus R}(a) = p_{\Lambda\setminus R}(b) = p_{\Lambda\setminus R}(s) = p_{\Lambda\setminus R}(z) \in p_{\Lambda\setminus R}(O)$. Hence we obtain $p_{\Lambda\setminus R_{k+1}}(a) = p_{\Lambda\setminus R_{k+1}}(b) \in p_{\Lambda\setminus R_{k+1}}(O)$. This implies that $a \in O \cap A$ and $b \in O \cap B$. Since O is any open neighborhood of z, we conclude that $z \in \overline{A} \cap \overline{B} = A \cap B = \emptyset$. This is a contradiction. Hence $\mathcal{V}$ covers Σ. □

We believe that Theorem 8.3 is a final result for normality of Σ-products with countable tightness. Because the class of strong β-spaces has been the widest interesting one in the classes of generalized metric spaces so far.

We say that a class $\mathcal{C}$ of spaces is *finitely productive* if, for each finite $\{X_i : i \leq n\} \subset \mathcal{C}$, the product $\prod_{i \leq n} X_i$ belongs to $\mathcal{C}$.

Remark 8.4. The class of paracompact Σ-spaces is countably productive (see [34]). On the other hand, the class of paracompact, semi-stratifiable spaces and that of paracompact β-spaces are not even finitely productive. In fact, it was shown in [3] that, under a certain set-theoretic assumption, there are two Lindelöf, semi-metric (= semi-stratifiable and first countable) spaces X and Y such that $X \times Y$ is not paracompact. Moreover, Ohta has pointed out that there are two β-spaces X and Y such that $X \times Y$ is not a β-space (see [59]).

From Remark 8.4, it is natural to ask the following problem.

Problem 2 ([58]). Are there two paracompact β-spaces (or semi-stratifiable spaces) X and Y such that $X \times Y$ is not paracompact in ZFC?

9 Σ-Products of Paracompact *C*-Scattered Spaces

The following is known as a generalization of locally compact spaces.

Definition 7. A space X is *C-scattered* [47] if each non-empty closed subset F in X has a point with a compact neighborhood in F. In particular, a space X is *scattered* if each non-empty (closed) subset A in X has an isolated point in A.

All of the classes of generalized metric spaces stated above are countably productive. On the other hand, the class of (paracompact) C-scattered spaces is finitely productive (see [47]), but it is not countably productive. In fact, the irrational space ω^ω is not C-scattered.

Nevertheless, we can obtain the following generalization of Corollary 6.7.

Theorem 9.1 (Tanaka-Yajima [46], 2002). *Let Σ is a Σ-product of paracompact C-scattered spaces. If Σ has countable tightness, it is normal*

The following is an immediate consequence of Theorem 9.1.

Corollary 9.2. *Let Σ is a Σ-product of paracompact scattered spaces. If Σ has countable tightness, it is normal.*

Telgársky introduced and studied topological games in [48], in which one of the most interesting classes is that of $\mathcal{DC}$-like spaces. This class contains all paracompact C-scattered spaces and all spaces with a σ-closure-preserving cover by compact sets. Subsequently, Theorem 9.1 was generalized in terms of $\mathcal{DC}$-like spaces as follows.

Theorem 9.3 (Hanaoka-Tanaka [18], 2002). *Let Σ be a Σ-product of paracompact $\mathcal{DC}$-like spaces. If Σ has countable tightness, it is normal*

10 Non-Normality of Σ-Products

A space X is *hereditarily separable* if any subspace of X is separable. The following implication is obvious:

separable metric $\longrightarrow$ hereditarily separable $\longrightarrow$ countable tightness

All of the results stated above are related to the normality of Σ-products of generalized metric factors or generalized compact factors. Hereditarily separable factor is neither of them. Using this factor, a result of another type was obtained.

Theorem 10.1 (Kombarov-Malyhin[28], 1973). *If Σ is a Σ-product, each countable subproduct of which is normal and hereditarily separable, then it is normal.*

It is natural to ask whether the "hereditarily separable" can be taken off from the above result. In fact, Morishita asked in his report [31]:

- If each countable subproduct of a Σ-product Σ is paracompact (or normal) and has countable tightness, is then Σ normal?

If this question would be solved affirmatively, the answer would be undoubtedly a dream result for the study of normality of Σ-products. However, it is negative as stated below.

Let us recall that $\Sigma 2^{\omega_1}$ denotes the Σ-product of ω_1 many copies of the discrete two-point space $\{0,1\}$. Similarly, $\Sigma\omega^{\omega_1}$ denotes the Σ-product of ω_1 many copies of $\omega = \{0,1,\cdots\}$.

To construct a non-normal Σ-product, it is a usual way to find a nice space X such that $X \times \omega_1$ is not normal. Then it follows from Fact 6.5 that the Σ-product $X \times \Sigma 2^{\omega_1}$ is not normal. Using this method, the above Morishita's problem was negatively solved by the following.

Example 10.2 (Daniel-Gruenhage [7], 1992). The following are true in ZFC.

(1) There is a normal, countably paracompact space X with countable tightness such that the Σ-product $\Sigma = X \times \Sigma 2^{\omega_1}$ is not normal.

(2) There is a first countable, paracompact, perfectly normal space X such that the Σ-product $\Sigma = X \times \Sigma\omega^{\omega_1}$ is not normal.

For (1), let G be the Fleissner's George in [13]. Then G is normal and countably paracompact, but not ω_1-collectionwise normal. Dow and Watson pointed out that a modification of G yields a normal, countably paracompact and non-ω_1-collectionwise normal space X such that it has countable tightness. It follows from [17, Theorem 2.3] that $X \times \omega_1$ is non-normal, hence so is the Σ-product $\Sigma = X \times \Sigma 2^{\omega_1}$. Since X is normal and countably paracompact, $X \times 2^{\omega}$ is normal. It follows from Theorem 5.3 that $X \times 2^{\omega}$ has countable tightness. So each countable subproduct of Σ is normal and has countable tightness. Hence the Σ-product $\Sigma = X \times \Sigma 2^{\omega_1}$ is a negative answer to the parenthetic part of the above Morishita's question.

For (2), let X be the space due to Todorčević. Then X is a left-separated, first countable and non-σ-discrete (see [7] for the detail of X). It was proved that the Σ-product $\Sigma = X \times \Sigma\omega^{\omega_1}$ is not normal. Moreover, X is perfectly normal and paracompact. Since X is a P-space in the sense of Morita, it follows from [33, Theorems 5.1 and 5.4] that the product of

X and a metric factor is perfectly normal and paracompact. Since $X \times \omega^\omega$ is first countable, perfectly normal and paracompact, so is each countable subproduct of Σ. Hence the Σ-product $\Sigma = X \times \Sigma\omega^{\omega_1}$ is a negative answer to the same question.

III. Σ-Products without Countable Tightness

The study of normality of Σ-products can be separated by having countable tightness or not. In this chapter, we discuss normality of Σ-products without countable tightness condition.

11 Collectionwise Normality of Σ-Products

All of the results for normality of Σ-products with countable tightness actually prove that the Σ-products are collectionwise normal. So it is natural to ask when a normal Σ-product of generalized metric spaces is collectionwise normal. Of course, we cannot assume the condition of countable tightness. The following was the first result answering this question after Kombarov's Theorem 6.4.

Theorem 11.1 ([52], 1984). *Let Σ be a Σ-product of paracompact Σ-spaces. Then Σ is collectionwise normal if and only if it is normal.*

This also generalizes (a) $\Leftrightarrow$ (b) in Theorem 6.4. Subsequently, an analogue was proved for Σ-products of semi-stratifiable spaces.

Theorem 11.2 ([53], 1987). *Let Σ be a Σ-product of semi-stratifiable spaces, each finite subproduct of which is paracompact. Then Σ is collectionwise normal if and only if it is normal.*

Seeing Theorem 8.3, it is natural to consider whether Theorems 11.1 and 11.2 can be unified in terms of β-spaces. That is, we raise

Problem 3. Let Σ be a Σ-product of β-spaces, each finite subproduct of which is paracompact. If Σ is normal, is it collectionwise normal?

We shall give a partial answer to the above in the remaining of this section. For the proof for collectionwise normality of spaces, we often use the following instead of Lemma 4.1.

Lemma 11.3 (well-known). *Let $\mathcal{D}$ be discrete collection of closed sets in a space X. If there is a σ-locally finite collection $\mathcal{U}$ of open sets in X such that $\bigcup \mathcal{D} \subset \bigcup \mathcal{U}$ and $\overline{U}$ meets at most one member of $\mathcal{D}$ for each $U \in \mathcal{U}$, then there is a disjoint collection $\{U_D : D \in \mathcal{D}\}$ of open sets in X such that $D \subset U_D$ for each $D \in \mathcal{D}$.*

The following is a partial answer to Problem 3. Since each open set in a semi-stratifiable space is an F_σ-set, this is also a slight generalization of Theorem 11.2. We give a full proof of this result because it is unpublished.

Theorem 11.4. *Let Σ be a Σ-product of β-spaces, each finite subproduct of which is paracompact and perfectly normal. Then Σ is collectionwise normal if and only if it is normal.*

Proof. Let Σ be the Σ-product of spaces $X_\lambda, \lambda \in \Lambda$, with a base point $s \in \Sigma$. First, recall the fact that if $\prod_{i \le n} X_i$ is paracompact and perfectly normal for each $n \in \mathbb{N}$, then so is $\prod_{i \in \mathbb{N}} X_i$ (see [39, Theorem 4.9]). By Proposition 7.1 and Theorem 7.2, each countable subproduct of Σ is paracompact, perfectly normal, strong β-space. Let $\mathcal{D}$ be a discrete collection of closed subsets in Σ.

Now, for each $n \in \mathbb{N}$, we construct a collection $\mathcal{V}_n$ of open sets in Σ and an index set Ξ_n of n-tuple sequences such that for each $\xi \in \Xi_n$ one can assign $R_\xi \in [\Lambda]^{\le\omega}, E(\xi) \subset \Sigma, H(\xi) \subset \Sigma, x_\xi \in \Sigma, D_\xi \in \mathcal{D}$ and a strong β-function g_ξ for X_ξ, satisfying the following conditions for each $n \in \mathbb{N}$:

(a) $\mathcal{V}_n$ is locally finite in Σ such that $\overline{V}$ meets at most one member of $\mathcal{D}$ for each $V \in \mathcal{V}_n$.

(b) $\{H(\xi) : \xi \in \Xi_n\}$ is locally finite in Σ such that $\bigcup \mathcal{V}(\xi) \subset H(\xi) \subset H(\xi_-)$ for each $\xi \in \Xi_n$.

(c) For each $\xi \in \Xi_n$,

(1) $\xi_- \in \Xi_{n-1}$,

(2) $E(\xi)$ is a R_{ξ_-}-distinguished closed set in Σ and $H(\xi)$ is a R_{ξ_-}-distinguished open set in Σ such that $E(\xi) \subset E(\xi_-) \cap H(\xi)$,

(3) $(\bigcup \mathcal{D}) \cap E(\xi) \subset \bigcup_{i \le n}(\bigcup \mathcal{V}_i) \cup \left(\bigcup\{E(\nu) : \nu \in \Xi_{n+1} \text{ with } \nu_- = \xi\}\right)$,

(4) $p^{\xi}_{\xi_-} \circ g_\xi(t,k) \subset g_\xi(p^{\xi}_{\xi_-}(t),k)$ for each $t \in X_\xi$ and each $k \in \mathbb{N}$,

(5) $x_\xi \in D_\xi \cap E(\xi_-) \setminus \bigcup_{i \le n-1}(\bigcup \mathcal{V}_i)$,

(6) $p_{\xi_-}(E(\xi)) \subset g_{\xi_-}(p_{\xi_-}(x_\xi), n)$,

(7) $R_\xi = R_{\xi_-} \cup \mathrm{Supp}(x_\xi)$,

(8) $D_\xi \cap E(\xi) \subset \bigcup \mathcal{V}_n$.

Let $\Xi_0 = \{\emptyset\}$ and let $E(\emptyset) = H(\emptyset) = \Sigma$. Pick any $\lambda_0 \in \Lambda$ and let $R_\emptyset = \{\lambda_0\}$. Assume that the construction above has been already performed for no greater than n.

Take any $\xi \in \Xi_n$ and fix it. Note that $p_\xi(E(\xi))$ is closed in X_ξ. Let

$$\mathcal{W} = \{g_\xi(p_\xi(x), n+1) \cap p_\xi(E(\xi)) : x \in \left(\bigcup \mathcal{D}\right) \cap E(\xi) \setminus \bigcup_{i \le n}\left(\bigcup \mathcal{V}_i\right)\}.$$

Since $\bigcup \mathcal{W}$ is an F_δ-set in X_ξ which is paracompact, there is a σ-locally finite collection $\mathcal{F}$ of closed subsets in X_ξ such that $\bigcup \mathcal{F} = \bigcup \mathcal{W}$ and each $F \in \mathcal{F}$ is contained in some $W_F \in \mathcal{W}$. Since X_ξ is paracompact again, there is a σ-locally finite collection $\{U_F : F \in \mathcal{F}\}$ of open sets in X_ξ such that $F \subset U_F \subset W_F \cap p_\xi(H(\xi))$ for each $F \in \mathcal{F}$. Let $\Xi(\xi)$ be an index set of $(n+1)$-tuple sequences such as $\mathcal{F} = \{F_{\xi^\frown\alpha} : \xi^\frown\alpha \in \Xi(\xi)\}$.

Pick any $\eta = \xi^\frown\alpha \in \Xi(\xi)$. Let $E(\eta) = p_\xi^{-1}(F_\eta)$ and $H(\eta) = p_\xi^{-1}(U_{F_\eta})$. By the choice of $\mathcal{F}$, we can find $x_\eta \in (\bigcup \mathcal{D}) \cap E(\xi) \setminus \bigcup_{i \le n}(\bigcup \mathcal{V}_i)$ such that $p_\xi(E(\eta)) = F_\eta \subset g_\xi(p_\xi(x_\eta), n+1) \cap p_\xi(E(\xi))$. Take a $D_\eta \in \mathcal{D}$ with $x_\eta \in D_\eta$, and let $R_\eta = R_\xi \cup \mathrm{Supp}(x_\eta)$. Since Σ is normal, there is an open set V_η in Σ such that

$$D_\eta \cap E(\eta) \subset V_\eta \subset \overline{V_\eta} \subset H(\eta) \setminus \bigcup\{D \in \mathcal{D} : D \ne D_\eta\}.$$

Here, letting ξ run over Ξ_n, we let $\Xi_{n+1} = \bigcup\{\Xi(\xi) : \xi \in \Xi_n\}$. Moreover, let $\mathcal{V}_{n+1} = \{V_\eta : \eta \in \Xi_{n+1}\}$. For each $\eta \in \Xi_{n+1}$, we take a strong β-function g_η for X_η satisfying (4). Then it is easily verified that all of the above conditions except (3) are satisfied. By the choice of $\mathcal{W}$, since $p_\xi\big((\bigcup \mathcal{D}) \cap E(\xi) \setminus \bigcup_{i \le n}(\bigcup \mathcal{V}_i)\big) \subset \bigcup \mathcal{W} = \bigcup \mathcal{F}$, we have

$$\Big(\bigcup \mathcal{D}\Big) \cap E(\xi) \setminus \bigcup_{i \le n}\Big(\bigcup \mathcal{V}_i\Big) \subset p_\xi^{-1}\Big(\bigcup \mathcal{F}\Big) = \bigcup\{p_\xi^{-1}(F_\eta) : \eta \in \Xi(\xi)\}$$
$$= \bigcup\{E(\eta) : \eta \in \Xi_{n+1} \text{ with } \eta_- = \xi\}.$$

Thus (3) is satisfied. We have inductively accomplished the desired construction above.

We let $\mathcal{V} = \bigcup_{n \in \mathbb{N}} \mathcal{V}_n$. By (a), $\mathcal{V}$ is a σ-locally finite collection of open sets in Σ such that $\overline{V}$ meets at most one member of $\mathcal{D}$ for each $V \in \mathcal{V}$. It suffices from Lemma 11.3 to show that $\mathcal{V}$ covers $\bigcup \mathcal{D}$. Assuming the contrary, there is some $y \in \bigcup \mathcal{D} \setminus \bigcup \mathcal{V}$. By $E(\emptyset) = \Sigma$ and (3), we can inductively choose a sequence $\{\xi_n\}$ of indices such that $\xi_n \in \Xi_n$ with $(\xi_{n+1})_- = \xi_n$ and $y \in E(\xi_n)$ for each $n \in \mathbb{N}$. We similarly abbreviate $p_{\xi_n}, p_{\xi_n}^{\xi_k}, x_{\xi_n}$ and R_{ξ_n} to p_n, p_n^k, x_n and R_n, respectively.

Take any $n \in \mathbb{N}$. Let $C_{n,m} = \{p_n(x_k) : k \ge m\}$ for each $m \in \mathbb{N}$ with $m \ge n$. It follows from (4) and (6) that for each $k \ge m\,(\ge n)$,

$$p_n(y) \in p_n^{k-1} \circ p_{k-1}(E(\xi_k)) \subset p_n^{k-1} \circ g_{\xi_{k-1}}(p_{k-1}(x_k), k) \subset g_{\xi_n}(p_n(x_k), k).$$

By the choice of g_{ξ_n}, $\overline{C_{n,m}}$ is compact for each $m \ge n$. Let $K_n = \bigcap_{m \ge n} \overline{C_{n,m}}$. Then K_n is a non-empty compact subset in X_{ξ_n} such that $p_n^{n+1}(K_{n+1}) \subset K_n$. So there is some $z_n \in K_n$ such that $p_n^{n+1}(z_{n+1}) = z_n$ for each $n \in \mathbb{N}$. Let $R = \bigcup_{n \in \mathbb{N}} R_n$. We take the point $z \in \Sigma$ defined by $p_n(z) = z_n$ for each $n \in \mathbb{N}$ and $p_{\Lambda \setminus R}(z) = p_{\Lambda \setminus R}(s)$. We show that every open neighborhood of z meets infinitely many members of $\mathcal{D}$.

Let O be any basic open neighborhood of z in Σ. By (7), we can take $m \in \mathbb{N}$ such that $O = p_m(O) \times X_{R \setminus R_m} \times p_{\Lambda \setminus R}(O)$. Since $z_m \in \overline{C_{m,\ell}} \cap p_m(O)$ for each $\ell \ge m$, we can inductively choose a sequence $\{k_i\}$ of natural numbers such that $m < k_1 < k_2 < \cdots$ and $\{p_m(x_{k_1}), p_m(x_{k_2}), \cdots\} \subset p_m(O)$. Then we have

$$p_R(x_{k_i}) = \big(p_R(x_{k_i}), p_{R \setminus R_m}(x_{k_i})\big) \in p_m(O) \times X_{R \setminus R_m} = p_R(O).$$

Since $\mathrm{Supp}(x_{k_i}) \subset R$ by (7), we have $p_{\Lambda \setminus R}(x_{k_i}) = p_{\Lambda \setminus R}(s) = p_{\Lambda \setminus R}(z) \in p_{\Lambda \setminus R}(O)$. Hence, by (5), we obtain $x_{k_i} \in O \cap D_{\xi_{k_i}}$ for each $i \in \mathbb{N}$. Let $j < i$. By (8), we have $D_{\xi_{k_j}} \cap E(\xi_{k_j}) \subset \bigcup \mathcal{V}_{k_j}$. By (2) and (5), we have

$$x_{k_i} \in D_{\xi_{k_i}} \cap E(\xi_{k_i - 1}) \setminus \bigcup_{\ell < k_i}\Big(\bigcup \mathcal{V}_\ell\Big) \subset D_{\xi_{k_i}} \cap E(\xi_{k_j}) \setminus \bigcup \mathcal{V}_{k_j}.$$

This means that $D_{\xi_i} \ne D_{\xi_j}$ if $i \ne j$. Hence any open neighborhood O of z meets infinitely many members of $\mathcal{D}$. This contradicts the discreteness of $\mathcal{D}$. Hence $\mathcal{V}$ covers $\bigcup \mathcal{D}$. □

Moreover, we raise the following similar problem, which is also still open.

Problem 4. Let Σ be a Σ-product of paracompact C-scattered (scattered) spaces. If Σ is normal, is it collectionwise normal?

Remark 11.5. As well as the above, if we additionally assume that each finite subproduct of Σ is perfect normal, Problem 4 can be solved affirmatively.

12 Countable Paracompactness of Σ-Products

All proper Σ-products do not have a covering property which ω_1 does not have, since they contains ω_1 as a closed subspace. However, ω_1 is countably compact, hence it is countably paracompact. Moreover, normality and countable paracompactness are resemble in a sense. So we consider the relation between normality and countable paracompactness in Σ-products.

We begin with the equivalence of normality and countable paracompactness in countable products.

Theorem 12.1 (Zenor-Nagami [61, 35], 1971–72). *Let $X = \prod_{n\in\mathbb{N}} X_n$ be a countable product such that $\prod_{i\leq n} X_i$ is normal (paracompact) for each $n \in \mathbb{N}$. Then X is normal if and only if it is countably paracompact (paracompact).*

This result is also found in [42, theorem 6.1] with the full proof. Using this, we see that normality is always stronger than countable paracompactness for Σ-products.

Proposition 12.2. *If a proper Σ-product is normal, then it is countably paracompact.*

Proof. Let Σ be the proper Σ-product of spaces $X_\lambda, \lambda \in \Lambda$ with a base point $s \in \Sigma$. We can take an infinite countable subset $\Gamma = \{\lambda_1.\lambda_2, \cdots\}$ of Λ. Let Σ_0 be the Σ-product of $X_\lambda, \lambda \in \Lambda \setminus \Gamma$ with the base point $p_{\Lambda\setminus\Gamma}(s)$. Then Σ_0 is proper. Since Σ is normal, so is $\Sigma_0 \times \prod_{i\leq n} X_{\lambda_i}$ for each $n \in \mathbb{N}$. By Theorem 12.1, the normality of Σ implies again that $\Sigma = \Sigma_0 \times \prod_{i\in\mathbb{N}} X_{\lambda_i}$ is countably paracompact. □

Conversely, it is natural to ask when a Σ-product is normal if it is countably paracompact. A key point to consider this question, countable paracompactness has a certain separation property as follows.

A subset Z in a space X is called a *zero-set* (or a *functionally closed*) if there is a continuous function $f : X \to [0,1]$ such that $Z = \{x \in X : f(x) = 0\}$.

Proposition 12.3 ([20]). *Let X be a countably paracompact space. If Z is a zero-set and F is a closed set in X such that $Z \cap F = \emptyset$, then there are two disjoint open sets U and V in X such that $Z \subset U$ and $F \subset V$.*

Using this, the following nice result was proved.

Theorem 12.4 (Yang [60], 1994). *Let Σ be a Σ-product of paracompact σ-spaces. Then Σ is countably paracompact if and only if it is normal.*

Since σ-spaces are semi-stratifiable, this was soon generalized.

Theorem 12.5 (Kemoto-Yajima [23], 1994). *Let Σ be a Σ-product of semi-stratifiable spaces, each finite subproduct of which is paracompact. Then Σ is countably paracompact if and only if it is normal.*

It seems to be difficult to generalize Theorem 12.4 by replacing σ-spaces with Σ-spaces, using a simple modification of the proof.

13 The Shrinking Property of Σ-Products

A space X is said to be *shrinking* (or has *the shrinking property*) if for every open cover $\mathcal{U} = \{U_\alpha : \alpha \in \Omega\}$ of X, there is a closed cover $\mathcal{F} = \{F_\alpha : \alpha \in \Omega\}$ of X with the same index such that $F_\alpha \subset U_\alpha$ for each $\alpha \in \Omega$. Such a closed cover $\mathcal{F}$ is called a *shrinking* of $\mathcal{U}$. Note that the shrinking property implies normality and countable paracompactness.

Recall that the closed subspace ω_1 in a proper Σ-product has the shrinking property as a covering property. The shrinking property of Σ-products was first proved as a strengthening of Gul'ko-Rudin's result (= Theorem 4.3).

Theorem 13.1 (Rudin [44], 1983). *A Σ-product of metric spaces is shrinking.*

Subsequently, it was generalized as follows.

Theorem 13.2 (Donne [9], 1985). *A Σ-product of paracompact p-spaces is shrinking if and only if it is normal.*

Observe from Theorem 6.4 that such a Σ-product still belongs to the category of countable tightness, because it is normal.

A space X is said to be *subshrinking* if if for every open cover $\{U_\alpha : \alpha \in \Omega\}$ of X, there is a closed cover $\{F_{\alpha,n} : \alpha \in \Omega \text{ and } n \in \mathbb{N}\}$ of X such that $F_{\alpha,n} \subset U_\alpha$ for each $\alpha \in \Omega$ and each $n \in \mathbb{N}$.

Proposition 13.3. *A space is normal and subshrinking if and only if it is shrinking.*

Note that a normal and subshrinking space is countably paracompact. So the proof is easily seen. It is found in [2, Lemma 3.1].

A space X is *subnormal* if for any disjoint closed sets A and B in X there are disjoint G_δ-sets U and V such that $A \subset U$ and $B \subset V$. Obviously, every subshrinking space is subnormal.

The following means that the shrinking property of Σ-products mainly depends on the subnormality rather than the normality.

Theorem 13.4 ([56], 1993). *A Σ-product of strong Σ-spaces is subshrinking if and only if it is subnormal.*

Finally, we obtain an extension of Theorem 13.2 as an immediate consequence of Theorem 13.4 and Proposition 13.3.

Corollary 13.5 ([55], 1989). *A Σ-product of strong Σ-spaces is shrinking if and only if it is normal.*

This means that the shrinking property of Σ-products is neither related to paracompactness nor countable tightness of finite subproducts.

14 Non-Normality of Σ-Products, Revisited

Let C be a compact space with tightness $> \omega$ (for example, $C = \omega_1 + 1$). Then it follows from Theorem 6.4 that $C \times \Sigma 2^{\omega_1}$ is a non-normal Σ-product of compact spaces. So there are many non-normal Σ-products of compact spaces.

A space is called an M_1-*space* if it has a σ-closure preserving base. Recall that every M_1-space is a paracompact σ-space. Ohta pointed out the following simple examples.

Example 14.1 (Ohta, see [50]). The following are true.

(1) There is a normal Σ-product $\Sigma = X \times \Sigma 2^{\omega_1}$ with X being an M_1-spaces which has no countable tightness.

(2) There is a non-normal Σ-product $\Sigma = X \times \Sigma 2^{\omega_1}$ with X being an M_1-spaces.

For (1), since $\Sigma 2^{\omega_1}$ is normal but not paracompact, it follows from [38, theorem 1] that there is a paracompact space X such that $X \times \Sigma 2^{\omega_1}$ is normal. From the proof of it, since the space X is a σ-space with only one non-isolated point, X is an M_1-space. Since it is easily seen that X has no countable tightness, we have $t(\Sigma) > \omega$. This example (1) means that we cannot hope a beautiful result such as Kombarov's Theorem 6.4 for Σ-products of generalized metric spaces.

For (2), by Fact 6.5, this immediately follows from the result that there is an M_1-space X such that $X \times \omega_1$ is not normal, which was due to Chiba [4]. This example (2) means that a Σ-product of generalized metric spaces needs a certain condition such as countable tightness for its normality.

A space is called a *Lašnev space* if it is a closed image of a metric space. The following implications hold:

$$\text{metric} \longrightarrow \text{Lašnev} \longrightarrow M_1 \longrightarrow \text{stratifiable} \longrightarrow \text{paracompact } \sigma\text{-space}$$

For an infinite cardinal κ, $S(\kappa)$ denotes the space obtained from the disjoint union of κ many convergent sequences by identifying all the limit points to a single point. The space $S(\kappa)$ is called the *sequential fan with* κ *many spines*. Obviously, $S(\kappa)$ is a Lašnev space.

Kodama asked whether every Σ-product of Lašnev spaces is normal. The following is an answer to this under some set-theoretic assumption.

Example 14.2 (Eda et al.[10], 1995). The product $S(\omega_2) \times S(\omega_2) \times \omega_1$ is not normal in some model of set theory.

Note that $S(\omega_2) \times S(\omega_2) \times \omega_1$ is a closed subspace of the Σ-product $\Sigma = S(\omega_2)^2 \times \Sigma 2^{\omega_1}$ of Lašnev spaces. Hence it is consistent that there is a non-normal Σ-product of Lašnev spaces. As a consequence, Kodama's problem turns out the following.

Problem 5 (Kodama). Is there a non-normal Σ-product of Lašnev spaces in ZFC?

This problem has not been solved yet.

Finally, we refer to a problem of Σ-products concerning the base point. The base point of a Σ-product has been omitted to describe so far. Because all of the results stated above do not depend on the choice of the base point in Σ-products. However, it is not known whether we can generally say so.

Problem 6 (van Douwen, see [9]). Can normality of a Σ-product depend on the choice of the base point?

Remark 14.3. Corson [5, p.786] pointed out that there are two Σ-products consisting of the same collection of spaces with the different base points such that they are not homeomorphic. On the other hand, recall that σ-products are dense subspaces of Σ-products. Replacing Σ-products with σ-products, it was shown in [22] that countable paracompactness of σ-products depends on the choice of the base point.

IV. Rectangular Products

The concept of rectangular product came from the product theorem in dimension theory. First, it was defined for finite products. Later, it was defined for general products in the natural way. This concept also has some interesting aspects for a covering property of products. Here, we deal with rectangularity of infinite products, since we can see some common features with normality of Σ-products.

15 Rectangular Products and Covering Dimension

A subset U in a space X is called a *cozero-set* (or a *functionally open*) if there is a continuous function $f : X \to [0,1]$ such that $U = \{x \in X : f(x) > 0\}$. That is, a cozero-set is the complement of a zero-set (defined to Proposition 12.3). A space X has the *covering dimension* $\leq n$, denoted by $\dim X \leq n$, if every finite cozero cover of X has a finite cozero refinement of order $\leq n$ (see [11] for the detail).

Let $X = \prod_{i \leq n} X_i$ be a finite product. A subset of the form $\prod_{i \leq n} A_i$ is called a *rectangle* in X. A rectangle $\prod_{i \leq n} U_i$ in X is called a *cozero rectangle* if each U_i is a cozero-set in X_i. An open cover $\mathcal{U}$ of X is said to be *rectangular* (*cozero*) if each member of $\mathcal{U}$ is a (cozero) rectangle in X.

Pasynkov [40] introduced the following concept.

Definition 8. A finite product $X = \prod_{i \leq n} X_i$ is said to be *rectangular* if every finite cozero cover of X has a σ-locally finite cozero rectangular refinement.

Using this concept, he proved the following remarkable result.

Theorem 15.1 (Pasynkov [40], 1975). *If a finite product* $X = \prod_{i \leq n} X_i$ *of Tychonoff spaces is rectangular, then* $\dim X \leq \sum_{i \leq n} \dim X_i$ *holds.*

As is listed in [40], many typical products are rectangular. Hence this unifies many product theorems in dimension theory which had been proved before. Here we state one of them which will be used later.

Proposition 15.2 ([34], see [57]). *A finite product of paracompact* Σ*-spaces is rectangular.*

However, the following Pasynkov's problem has not been solved yet.

Problem 7 ([40]). Let $X \times Y$ be a paracompact product.

(1) Is $X \times Y$ rectangular?

(2) Does $\dim(X \times Y) \leq \dim X + \dim Y$ hold?

Next, we naturally define a general product version of rectangularity. Let $X = \prod_{\lambda\in\Lambda} X_\lambda$ be the product of spaces $X_\lambda, \lambda \in \Lambda$. For each $\theta \in [\Lambda]^{<\omega}$, $X_\theta = \prod_{\lambda\in\theta} X_\lambda$ is called a *finite subproduct* of X, and let π_θ denote the projection from X onto X_θ. In particular, $\pi_{\{\lambda\}}$ is denoted by π_λ for each $\lambda \in \Lambda$. For each $\theta, \theta' \in [\Lambda]^{<\omega}$ with $\theta \subset \theta'$, $\pi^{\theta'}_\theta$ denotes the projection from $X_{\theta'}$ onto X_θ.

Now, let $\theta \in [\Lambda]^{<\omega}$. A subset of the form $\bigcap_{\lambda\in\theta} \pi_\lambda^{-1}(U_\lambda)$ in X is called a *rectangle*, where $U_\lambda \subset X_\lambda$ for each $\lambda \in \theta$. Then $\bigcap_{\lambda\in\theta} \pi_\lambda^{-1}(U_\lambda)$ is called a *cozero* (*open*) *rectangle* in X if each U_λ is a cozero-set (an open set) in X_λ. A cover $\mathcal{G}$ of the product $X = \prod_{\lambda\in\Lambda} X_\lambda$ is said to be *rectangular* if each member of $\mathcal{G}$ is a rectangle in X. So $\mathcal{G}$ is called *rectangular cozero* (*open*) if each member of $\mathcal{G}$ is a cozero (open) rectangle in X.

A subset of the form $\pi_\theta^{-1}(U)$ in X is called a *cylinder*, where $U \subset X_\theta$. Then $\pi_\theta^{-1}(U)$ is called a *cozero* (*open*) *cylinder* in X if U is cozero (open) in X_θ. A cover $\mathcal{G}$ of X is said to be *cylindrical* if each member of $\mathcal{G}$ is a cylinder in X. So $\mathcal{G}$ is called *cylindrical cozero* (*open*) if each member of $\mathcal{G}$ is a cozero (open) cylinder in X. A subset V in X is said to be θ-*distinguished* in X if $\pi_\theta^{-1}\pi_\theta(V) = V$. So each rectangle or each cylinder in X is θ-distinguished in X for some $\theta \in [\Lambda]^{<w}$.

The rectangularity and cylindricality of general products are defined by the same way as the case of finite products.

Definition 9. A product $X = \prod_{\lambda\in\Lambda} X_\lambda$ is said to be *rectangular* (*cylindrical*) if every finite cozero cover of X has a σ-locally finite rectangular (cylindrical) cozero refinement.

The concept of cylindricality is also useful for the infinite product theorem in dimension theory as follows.

Theorem 15.3 (Pasynkov-Yajima [41, 51], 1984–85). *Let* $X = \prod_{\lambda\in\Lambda} X_\lambda$ *be a cylindrical product. Then* $\dim X \leq \sup\{\dim \prod_{\lambda\in\theta} X_\lambda : \theta \in [\Lambda]^{<\omega}\}$ *holds.*

In this survey, we do not deal with the relations between rectangularity (or cylindricality) and covering dimension more than this. Rectangularity and cylindricality of products have rather an interesting aspect for considering certain subsets of infinite products. This is seen by the following easy equivalence.

Proposition 15.4. *For a product* $X = \prod_{\lambda\in\Lambda} X_\lambda$*, the following are equivalent.*

(a) *X is rectangular (cylindrical).*

(b) *Every binary cozero cover of X has a σ-locally finite rectangular (cylindrical) cozero refinement.*

(c) *Every cozero set in X is the union of a σ-locally finite collection by cozero rectangles (cylinders).*

Proof. (c) $\Rightarrow$ (a) $\Rightarrow$ (b) are obvious. We show (b) $\Rightarrow$ (c). Let G be a cozero-set in X. Then there is a sequence $\{Z_n\}$ of zero-sets in X such that $G = \bigcup_{n\in\mathbb{N}} Z_n$ and $Z_n \subset Z_{n+1}$ for each $n \in \mathbb{N}$. By the assumption, the binary cozero cover $\{G, X \setminus Z_n\}$ of X has a σ-locally finite rectangular cozero refinement $\mathcal{U}_n$. Let $\mathcal{V}_n = \{U \in \mathcal{U}_n : U \subset G\}$ for each $n \in \mathbb{N}$. Since $Z_n \subset \bigcup \mathcal{V}_n \subset G$, $\bigcup_{n\in\mathbb{N}} \mathcal{V}_n$ is a σ-locally finite (in X) collection by cozero rectangles with $G = \bigcup \mathcal{V}$. □

Moreover, the following shows a relation between rectangularity and cylindricality of products.

Proposition 15.5. *Let $X = \prod_{\lambda\in\theta} X_\lambda$ be a cylindrical product. If each finite subproduct of X is rectangular, then X is rectangular.*

As the proof is easy, it is left to the reader.

16 Products of Metric Spaces

A typical rectangular products was given in order to consider the structure of open F_σ-sets in infinite products. By Proposition 15.4, it coincides with to consider the rectangularity of products. The first one is the following.

Theorem 16.1 (Klebanov [24], 1980). *A product of metric spaces is rectangular.*

However, the proof is quite similar to that of Gul'ko [16] for Σ-products. To make sure this, we try to give a proof of Theorem 16.1 according to our proof of Theorem 4.3.

Let $X = \prod_{\lambda\in\Lambda} X_\lambda$ be the product of spaces $X_\lambda, \lambda \in \Lambda$. For each $R \in [\Lambda]^{\leq\omega}$, the product $\prod_{\lambda\in R} X_\lambda$, denoted by X_R, is called a *countable subproduct* of X.

Let Ξ be an index set of finite sequences. For $\xi = (\alpha_1, \cdots, \alpha_n) \in \Xi$, we similarly let $\xi|i = (\alpha_1, \cdots, \alpha_i)$ for each $i \leq n$ and let $\xi_- = (\alpha_1, \cdots, \alpha_{n-1})$ and $\xi^\frown\alpha = (\alpha_1, \cdots, \alpha_n, \alpha)$.

For each $R_\xi \in [\Lambda]^{\leq\omega}$, $\xi \in \Xi$, $X_{R_\xi} = \prod_{\lambda\in R_\xi} X_\lambda$ is a countable subproduct of X. The projections from X onto X_{R_ξ} and from X_{R_η} onto X_{R_ξ}, where $R_\xi \subset R_\eta$, are denoted by π_{R_ξ} and $\pi^{R_\eta}_{R_\xi}$, respectively. For the simplicity, such a X_{R_ξ} is abbreviated by X_ξ, and such projections π_{R_ξ} and $\pi^{R_\eta}_{R_\xi}$ are abbreviated by π_ξ and π^η_ξ, respectively. For each $\Lambda' \subset \Lambda$, $\pi_{\Lambda'}$ denotes the projection from X onto $X_{\Lambda'} = \prod_{\lambda\in\Lambda'} X_\lambda$.

Proof of Theorem 16.1. Let $X = \prod_{\lambda\in\Lambda} X_\lambda$ be the product of metric spaces $X_\lambda, \lambda \in \Lambda$. Note that each open rectangle in X is a cozero rectangle. Let G be any cozero-set in X.

Now, for each $n \in \mathbb{N}$, we construct a collection $\mathcal{V}_n$ of open rectangles in Σ and an index set Ξ_n of n-tuple sequences such that, for each $\xi \in \Xi_n$, one can assign $V(\xi) \subset X, R_\xi \in [\Lambda]^{\leq\omega}$ and $a_\xi \in X \setminus G$, satisfying the following conditions;

(a) $\mathcal{V}_n \cup \{V(\xi) : \xi \in \Xi_n\}$ is locally finite in Σ,

(b) each $V \in \mathcal{V}_n$ is an open rectangle in X contained in G,

(c) for each $\xi \in \Xi_n$,

 (1) $\xi_- \in \Xi_{n-1}$,

(2) $V(\xi)$ is an R_{ξ_-}-distinguished open set in Σ with $V(\xi) \subset V(\xi_-)$,

(3) $\mathrm{diam}_{X_{\xi|i}} \pi_{\xi|i}(V(\xi)) < 1/2^n$ for $i = 1, \cdots, n-1$, where $\mathrm{diam}_{X_\xi} Y$ denotes as in the proof of Theorem 4.3,

(4) $V(\xi) \subset (\bigcup \mathcal{V}_{n+1}) \cup \{V(\eta) : \eta \in \Xi_{n+1}$ with $\eta_- = \xi\}$,

(5) $a_\xi \in V(\xi) \setminus G$,

(6) $\pi_\xi^{-1}\pi_\xi(a_\xi) \cap G = \emptyset$ and $R_{\xi_-} \subset R_\xi$.

Assume that the construction above has been already performed for no greater than n. Take any $\xi \in \Xi_n$ and fix it. Let $U(\xi) = \pi_\xi(V(\xi))$. For each $i \le n$, since $X_{\xi|i}$ is metric, let $\mathcal{U}(\xi|i)$ be an open cover of $X_{\xi|i}$ with $\mathrm{diam}_{X_\xi} U < 1/2^{n+1}$ for each $U \in \mathcal{U}(\xi|i)$. Let

$$\mathcal{W}(\xi) = \{\bigcap_{i \le n} (\pi^{\xi}_{\xi|i})^{-1}(U_i) \cap U(\xi) : U_i \in \mathcal{U}(\xi|i) \text{ and } i \le n\}.$$

Then there is a locally finite open refinement $\mathcal{U}(\xi) = \{U(\xi^\frown\alpha) : \alpha \in \Omega(\xi)\}$ of $\mathcal{W}(\xi)$. We let

$$\mathcal{V}(\xi) = \{p_\xi^{-1}(U) : U \in \mathcal{U}(\xi) \text{ with } p_\xi^{-1}(U) \subset G\} \text{ and}$$
$$\Xi_{n+1}(\xi) = \{\xi^\frown\alpha : \alpha \in \Omega(\xi) \text{ such that } p_\xi^{-1}(U(\xi^\frown\alpha)) \not\subset G\}.$$

Take any $\eta = \xi^\frown\alpha \in \Xi(\xi)$. Let $V(\eta) = \pi_\xi^{-1}(U(\eta))$. Pick a points $a_\eta \in V(\eta) \setminus G$. Since G is an F_σ-set, there is $Q(\eta) \in [\Lambda]^{\le\omega}$ such that $\pi_{Q(\eta)}^{-1}\pi_{Q(\eta)}(a_\eta) \cap G = \emptyset$. Let $R_\eta = R_\xi \cup Q(a_\eta)$. Here, letting ξ run over Ξ_n, we let

$$\mathcal{V}_{n+1} = \bigcup\{\mathcal{V}_{n+1}(\xi) : \xi \in \Xi_n\} \text{ and } \Xi_{n+1} = \bigcup\{\Xi_{n+1}(\xi) : \xi \in \Xi_n\}.$$

Then it is easily verified that all of the above conditions are satisfied.

Let $\mathcal{V} = \bigcup_{n \in \mathbb{N}} \mathcal{V}_n$. By (a) and (b), $\mathcal{V}$ is a σ-locally finite (in X) collection of cozero rectangles with $\bigcup \mathcal{V} \subset G$. We show that $\mathcal{V}$ covers G. Assume that there is a point $y \in G \setminus \bigcup \mathcal{V}$. By (4), we can inductively choose a sequence $\{\xi_n\}$ of indices such that $\xi_n \in \Xi_n$ with $(\xi_{n+1})_- = \xi_n$ and $y \in V(\xi_n)$ for each $n \in \mathbb{N}$. Let $R = \bigcup_{n \in \mathbb{N}} R_{\xi_n} \in [\Lambda]^{\le\omega}$. For simplicity, we abbreviate $\pi_{\xi_n}, \pi^{\xi_m}_{\xi_n}$ and a_{ξ_n} to π_n, π^m_n and a_n, respectively. Let O be some basic open neighborhood of y contained in G. Since $\{R_{\xi_n}\}$ is an increasing sequence, we can take some $m \in \mathbb{N}$ such that $O = \pi_m(O) \times X_{R \setminus R_{\xi_m}} \times \pi_{\Lambda \setminus R}(O)$. By (3), we can take $k \ge m$ such that $\pi_m(y) \in \pi_m(V(\xi_k)) \subset \pi_m(O)$. Take the point $z \in X$ defined by $\pi_R(z) = \pi_R(a_k)$ and $\pi_{\Lambda \setminus R}(z) = \pi_{\Lambda \setminus R}(y)$. By (5), we have

$$\pi_k(z) = \pi_k(a_k) \in \pi_k(V(\xi_k)) \subset (\pi^k_m)^{-1}\pi_m(V(\xi_k)) \subset (\pi^k_m)^{-1}\pi_m(O) = \pi_k(O).$$

Since $\pi_{\Lambda \setminus R}(z) = \pi_{\Lambda \setminus R}(y) \in \pi_{\Lambda \setminus R}(O)$, we obtain $z \in O \subset G$. However, by (6), we have $z \in \pi_R^{-1}\pi_R(a_k) \subset \pi_k^{-1}\pi_k(a_k) \subset X \setminus G$. This is a contradiction. Hence we conclude that $\bigcup \mathcal{V} = G$. By Proposition 15.4, X is rectangular. □

The similarity of the two proofs of Theorems 4.3 and 16.1 suggests that there is a closed relation between normality of Σ-products and rectangularity of products.

17 Products of Generalized Metric Spaces

Let $X = \prod_{\lambda\in\Lambda} X_\lambda$ be the product of spaces $X_\lambda, \lambda \in \Lambda$. For each $\theta \in [\Lambda]^{<\omega}$, the product $\prod_{\lambda\in\theta} X_\lambda$, denoted by X_θ, is called a *finite subproduct* of X.

Let Ξ be an index set (of finite sequences). For each $\theta_\xi \in [\Lambda]^{<\omega}$, $X_{\theta_\xi} = \prod_{\lambda\in\theta_\xi} X_\lambda$ is a finite subproduct of X. The projections π_{θ_ξ} and $\pi_{\theta_\xi}^{\theta_\eta}$ are defined in the similar way as the above. For the simplicity, X_{θ_ξ} is abbreviated by X_ξ, and π_{θ_ξ} and $\pi_{\theta_\xi}^{\theta_\eta}$ are abbreviated by π_ξ and π_ξ^η, respectively.

Subsequently, the Klevanov's result above was extended by the following. This is just the same case of normality of Σ-products. The Filippov's paper [12] seems to be very difficult to be available and it was written by Russian. So, for the reader's convenience, we give a proof of it according to his idea.

Theorem 17.1 (Filippov [12], 1983). *A product of paracompact p-spaces is rectangular.*

Proof. Let $X = \prod_{\lambda\in\Lambda} X_\lambda$ be the product of paracompact p-spaces $X_\lambda, \lambda \in \Lambda$. Since every paracompact p-space is a strong Σ-space, it follows from Proposition 15.2 that each finite subproduct of X is rectangular. So it suffices from Proposition 15.5 to prove that X is cylindrical.

Let A and B be any disjoint zero-sets in X. There is a continuous function $g : X \to [0,1]$ such that $g \restriction A \equiv 0$ and $g \restriction B \equiv 1$. Take an ε with $0 < \varepsilon < 1$ and fix it.

For each $\lambda \in \Lambda$, there are a metric space Y_λ and a perfect map f_λ from X_λ onto Y_λ (by Theorem 6.2). Let $Y = \prod_{\lambda\in\Lambda} Y_\lambda$ and $f = \prod_{\lambda\in\Lambda} f_\lambda$. For each $\theta \in [\Lambda]^{<\omega}$, let $X_\theta = \prod_{\lambda\in\theta} X_\lambda$ and $Y_\theta = \prod_{\lambda\in\theta} Y_\lambda$, and let π_θ and p_θ denote the projections from X onto X_θ and from Y onto Y_θ, respectively.

Note that f is a perfect map from X onto Y (see [11, Theorem 3.7.9]). For each $y \in Y$, there is some $\varphi(y) \in [\Lambda]^{<\omega}$ such that $x, x' \in X$ with $x(\lambda) = x'(\lambda) \in f_\lambda^{-1}(y(\lambda))$ for each $\lambda \in \varphi(y)$ implies $|g(x) - g(x')| < \varepsilon/3$.

Now, for each $n \in \mathbb{N}$, we will construct an index sets Δ_n of n-tuple sequences and a subset Ξ_n of Δ_n such that each $\xi \in \Delta_n$ assigns $V(\xi) \subset Y$ and each $\xi \in \Xi_n$ assigns $\theta_\xi \in [\Lambda]^{<\omega}$ and $a_\xi, b_\xi \in X$, satisfying the following conditions:

(a) $\{V(\xi) : \xi \in \Delta_n\}$ is locally finite in Y.

(b) For each $\xi \in \Delta_n$,

  (1) $\xi_- \in \Xi_{n-1}$,

  (2) $V(\xi)$ is a θ_{ξ_-}-distinguished open cylinder in Y,

  (3) $\operatorname{diam}_{X_\rho} p_\rho(V(\xi)) < 1/2^n$ for each $\rho \subset \theta_{\xi_-}$.

(c) $\Xi_n := \{\xi \in \Delta_n$: There are two points $a_\xi, b_\xi \in f^{-1}(V(\xi))$ satisfying $\pi_{\theta_{\xi_-}}(a_\xi) = \pi_{\theta_{\xi_-}}(b_\xi)$ and $|g(a_\xi) - g(b_\xi)| \geq \varepsilon\}$,

(d) For each $\xi \in \Xi_n$,

  (4) $V(\xi) = \bigcup\{V(\nu) : \nu \in \Delta_{n+1}$ with $\nu_- = \xi\}$,

(5) $\theta_\xi = \theta_{\xi_-} \cup \varphi(f(a_\xi)) \cup \varphi(f(b_\xi))$, where $\varphi(y) \in [\Lambda]^{<\omega}$ is stated as above.

For each $\xi \in \Xi_n, n \in \mathbb{N}$, the finite products X_{θ_ξ} and Y_{θ_ξ} are abbreviated by X_ξ and Y_ξ, respectively, and the projections π_{θ_ξ} and p_{θ_ξ} are abbreviated by π_ξ and p_ξ, respectively.

Let $\Delta_0 = \Xi_0 = \{\emptyset\}$ and let $V(\emptyset) = Y$. Pick any $\lambda_0 \in \Lambda$ and let $\theta_\emptyset = \{\lambda_0\}$. Assume that the construction above has been already performed for no greater than n.

Take any $\xi \in \Xi_n$ and fix it. For each $y \in Y_\xi$, there is an open neighborhood N_y in Y_ξ such that $\mathrm{diam}_{X_\rho} p^\xi_\rho(N_y) < 1/2^{n+1}$ for each $\rho \subset \theta_\xi$, where p^ξ_ρ denotes the projection from Y_ξ onto Y_ρ. There is a locally finite open refinement $\{W(\alpha) : \alpha \in \Omega(\xi)\}$ of $\{N_y : y \in Y_\xi\}$. Let

$$\Delta_{n+1}(\xi) = \{\xi^\frown\alpha : \alpha \in \Omega(\xi) \text{ with } p_\xi^{-1}(W(\alpha)) \cap V(\xi) \neq \emptyset\}.$$

For each $\eta = \xi^\frown\alpha \in \Delta_{n+1}(\xi)$, let $V(\eta) = V(\xi^\frown\alpha) = p_\xi^{-1}(W(\alpha)) \cap V(\xi)$. Letting ξ range over Ξ_n, we put $\Delta_{n+1} = \bigcup\{\Delta_{n+1}(\xi) : \xi \in \Xi_n\}$. Then it is verified that the conditions (a), (1)–(3) in (b) and (4) in (d) are satisfied. We define $\Xi_{n+1} \subset \Delta_{n+1}$ such as in (c). For each $\eta \in \Xi_{n+1}$ with $\eta_- = \xi$, since there are $a_\eta, b_\eta \in f^{-1}(V(\eta))$ satisfying (c), we put $\theta_\eta = \theta_\xi \cup \varphi(f(a_\eta)) \cup \varphi(f(b_\eta)) \in [\Lambda]^{<\omega}$. So (5) in (d) are satisfied. Our construction has been accomplished.

For each $n \in \mathbb{N}$, we put $\mathcal{V} = \{V(\xi) : \xi \in \Delta_n \setminus \Xi_n \text{ and } n \in \mathbb{N}\}$.

Claim 1. $\mathcal{V}$ covers Y.

Proof. Assume that there is a point $z \in Y \setminus \bigcup \mathcal{V}$. By (1) and (4), we can inductively choose a sequence $\{\xi_n\}$ such that $\xi_n \in \Xi_n$ with $(\xi_{n+1})_- = \xi_n$ and $z \in V(\xi_n)$ for each $n \in \mathbb{N}$. Let $R = \bigcup_{n\in\mathbb{N}} \theta_{\xi_n}$. Pick a point $u \in f^{-1}(z)$. Take the two points $a_n, b_n \in X$ defined by $\pi_R(a_n) = \pi_R(a_{\xi_n}), \pi_R(b_n) = \pi_R(b_{\xi_n})$ and $\pi_{\Lambda\setminus R}(a_n) = \pi_{\Lambda\setminus R}(b_n) = \pi_{\Lambda\setminus R}(u)$ for each $n \in \mathbb{N}$.

Subclaim 1. $\{f(a_n)\}$ and $\{f(b_n)\}$ converge to z.

Proof. Let O be any basic open neighborhood of z in Y. By (3) and (5), there is $m \in \mathbb{N}$ such that $p_R(O)$ is $\theta_{\xi_{m-1}}$-distinguished in Y_R and $p_{\xi_{m-1}}(z) \in p_{\xi_{m-1}}(V(\xi_m)) \subset p_{\xi_{m-1}}(O)$. So we have $p_R(z) \in p_R(V(\xi_m)) \subset p_R(O)$. Then it follows from (4) that, for each $n \geq m$,

$$\begin{aligned} f(a_n) = (p_R(f(a_{\xi_n})), p_{\Lambda\setminus R}(z)) &\in p_R(V(\xi_n)) \times p_{\Lambda\setminus R}(O) \\ &\subset p_R(V(\xi_m)) \times p_{\Lambda\setminus R}(O) \subset p_R(O) \times p_{\Lambda\setminus R}(O) = O. \end{aligned}$$

Hence $f(a_n) \to z$ $(n \to \infty)$. Similarly, we have $f(b_n) \to z$ $(n \to \infty)$.

Since f is a perfect map, the $\{a_n\}$ has a cluster point $v \in f^{-1}(z)$.

Subclaim 2. Every open neighborhood of v in X contains some a_k and b_k.

Proof. Let G be any basic open neighborhood of v in X. Take $\ell \in \mathbb{N}$ such that $\pi_{\xi_{\ell-1}}(G) \times X_{R\setminus\theta_{\xi_{\ell-1}}} \times \pi_{\Lambda\setminus R}(G) = G$. Choose $k \geq \ell$ with $a_k \in G$. By (c), we have $\pi_{\xi_{k-1}}(b_k) = \pi_{\xi_{k-1}}(b_{\xi_k}) = \pi_{\xi_{k-1}}(a_{\xi_k}) = \pi_{\xi_{k-1}}(a_k) \in \pi_{\xi_{k-1}}(G)$. Note that $\pi_{\Lambda\setminus R}(v) = \pi_{\Lambda\setminus R}(u)$. So we have $\pi_{\Lambda\setminus R}(b_k) = \pi_{\Lambda\setminus R}(a_k) = \pi_{\Lambda\setminus R}(v) \in \pi_{\Lambda\setminus R}(G)$. Hence we obtain $a_k, b_k \in G$.

Subclaim 3. $|g(a_n) - g(b_n)| \geq \varepsilon/3$ for each $n \in \mathbb{N}$.

Proof. Take any $n \in \mathbb{N}$. By (c), observe $|g(a_{\xi_n}) - g(b_{\xi_n})| \geq \varepsilon$. By (5) and $\pi_R(a_{\xi_n}) = \pi_R(a_n)$, we have $\pi_{\varphi(f(a_{\xi_n}))}(a_{\xi_n}) = \pi_{\varphi(f(a_{\xi_n}))}(a_n)$. It follows from the choice of $\varphi(f(a_{\xi_n}))$ that

$|g(a_{\xi_n}) - g(a_n)| < \varepsilon/3$. Similarly, we have $|g(b_{\xi_n}) - g(b_n)| < \varepsilon/3$. Hence these inequalities assure Subclaim 3.

Since g is continuous at v in X, it follows from Subclaim 2 that we can take $k \in \mathbb{N}$ such that $|g(a_k) - g(v)| < \varepsilon/6$ and $|g(b_k) - g(v)| < \varepsilon/6$. So we obtain $|g(a_k) - g(b_k)| < \varepsilon/3$, which contradicts Subclaim 3. □

Now, take any $\xi \in \bigcup_{n\in\mathbb{N}}(\Delta_n \setminus \Xi_n)$. Pick an $x_\xi \in f^{-1}(V(\xi))$ and fix it. Let $U^*(\xi) = \pi_{\xi_-}(f^{-1}(V(\xi))) \times X_{\theta_\xi \setminus \theta_{\xi_-}}$. We let

$$U_0^*(\xi) = \{x' \in U^*(\xi) : g(x', \pi_{\Lambda \setminus \theta_\xi}(x_\xi)) < 2/3\},$$
$$U_1^*(\xi) = \{x' \in U^*(\xi) : g(x', \pi_{\Lambda \setminus \theta_\xi}(x_\xi)) > 1/3\},$$

Moreover, we put $U_j(\xi) = \pi_\xi^{-1}(U_j^*(\xi))$ for $j = 0, 1$. Then $U_0(\xi) \cup U_1(\xi) = f^{-1}(V(\xi))$. Since $U_j^*(\xi)$ is a cozero-set in X_ξ, $U_j(\xi)$ is a cozero cylinder in X for $j = 0, 1$. Here, letting ξ range, we put

$$\mathcal{U} = \{U_j(\xi) : \xi \in \Delta_n \setminus \Xi_n, n \in \mathbb{N} \text{ and } j = 0, 1\}.$$

By Claim 1 and (a), $\mathcal{U}$ is a σ-locally finite cover of X by cozero cylinders.

Take any $\xi \in \bigcup_{n\in\mathbb{N}}(\Delta_n \setminus \Xi_n)$ again and put $\varepsilon = 1/3$. In order to complete the proof, from Proposition 15.4, it remains to show the following claim.

Claim 2. $U_0(\xi) \cap B = \emptyset$ and $U_1(\xi) \cap A = \emptyset$.

Proof. Let $\xi \in \Delta_n \setminus \Xi_n$. Pick any $x \in U_0(\xi)$. Take the point $x^* \in X$ defined by $\pi_\xi(x^*) = \pi_\xi(x)$ and $\pi_{\Lambda\setminus\theta_\xi}(x^*) = \pi_{\Lambda\setminus\theta_\xi}(x_\xi)$. By $\pi_\xi(x^*) = \pi_\xi(x) \in U_0^*(\xi)$, we have $x, x^* \in U_0(\xi) \subset f^{-1}(V(\xi))$. Since $\pi_{\xi_-}(x) = \pi_{\xi_-}(x^*)$ and $\xi \notin \Xi_n$, it follows from (c) that $|g(x) - g(x^*)| < 1/3$. By $\pi_\xi(x^*) \in U_0^*(\xi)$, we have $g(x^*) = g(\pi_\xi(x^*), \pi_{\Lambda\setminus\theta_\xi}(x_\xi)) < 2/3$. This implies $g(x) < 1$, that is, $x \notin B$. Similarly, $x \in U_1(\xi)$ implies $x \notin A$. □

Recently, the above result was extended as follows.

Theorem 17.2 ([57], 2007). *For each $\lambda \in \Lambda$, let f_λ be a perfect map from X_λ onto Y_λ. If each finite subproduct of $Y = \prod_{\lambda\in\Lambda} Y_\lambda$ is a paracompact β-space with countable tightness, then $X = \prod_{\lambda\in\Lambda} X_\lambda$ is cylindrical.*

This proof is very long. However, reading the proof of Theorem 17.1, it is easier to understand. Theorem 17.2 immediately yields

Corollary 17.3. *Let $X = \prod_{\lambda\in\Lambda} X_\lambda$ be a product of β-spaces. If each finite subproduct of X is paracompact and has countable tightness, then X is cylindrical.*

Moreover, Propositions 15.2 and 15.5 and Corollary 17.3 yield

Corollary 17.4 ([51], 1984). *Let $X = \prod_{\lambda\in\Lambda} X_\lambda$ be a product of paracompact Σ-spaces. If each finite subproduct of X has countable tightness, then X is rectangular.*

Comparing the two results of Kombarov and Filippov (= Theorems 6.4 and 17.1), it seems that the latter has no relation to countable tightness. However, seeing Theorem 17.2 which is a generalization of it, we have to consider that the Filippov's result is still under the influence of countable tightness.

18 Normal Covers of Products

As well as the case of normality of Σ-products, it is natural to consider that the study of rectangular products should proceed to the next stage of products without the influence of countable tightness. For that, we take notice the fact that every σ-locally finite rectangular cozero cover of a product is a normal cover.

Let us recall the concept of normal covers. As is well-known, it is very important because of the following which yields Theorem 4.2 as a corollary.

Theorem 18.1 (Stone [45], 1948). *A space X is paracompact if and only if every open cover of X is normal.*

Normal covers are usually defined in terms of a sequence of star-refinements (see [11]). However, we only have to know the following characterization of normal covers instead of the definition.

Theorem 18.2 (Stone-Michael-Morita, 1948-64). *Let X be a space and O an open cover of X. Then the following are equivalent.*

(a) *O is normal.*

(b) *O has a σ-locally finite cozero refinement.*

(c) *O has a σ-discrete cozero refinement.*

(d) *O has a locally finite cozero refinement.*

(e) *O has a locally finite and σ-discrete cozero refinement.*

(f) *O has a locally finite and σ-discrete cozero refinement with a shrinking by zero-sets.*

This was arranged in the Morita's papers [32, 33] except (f). This was also stated in [21, Theorems 1.2 and 1.4] including (f).

The rectangularity of products is stated in terms of normal covers as follows.

Proposition 18.3. *A product $\prod_{\lambda\in\Lambda} X_\lambda$ is rectangular if and only if every normal cover of it has a σ-locally finite rectangular cozero refinement.*

In fact, the "only if" part is easily seen from Theorem 18.2 (f) and the converse is obvious. So, not only that we do not assume countable tightness, but also we do not assume rectangularity of products. This naturally lead us to the following problem.

Problem 8. For a (non-rectangular) product $X = \prod_{\lambda\in\Lambda} X_\lambda$, when does a rectangular (or cylindrical) normal cover of X has a σ-locally finite rectangular cozero refinement?

Considering this problem, we have been obtained some characterizations of normal covers of products of generalized metric spaces.

Theorem 18.4 ([59], 200?). *Let $X = \prod_{\lambda\in\Lambda} X_\lambda$ be a product of β-spaces, each finite subproduct of which is paracompact. For a cylindrical cozero cover O of X, the following are equivalent.*

(a) *$\mathcal{O}$ is normal.*

(b) *$\mathcal{O}$ has a locally finite open refinement.*

(c) *$\mathcal{O}$ has a locally finite closed refinement.*

(d) *$\mathcal{O}$ has a locally finite cylindrical cozero refinement.*

(e) *$\mathcal{O}$ has a locally finite and σ-discrete cylindrical cozero refinement.*

Remark 18.5. There is a product $X = \prod_{\lambda\in\omega_1} X_\lambda$ of β-spaces, each finite subproduct of which is paracompact and rectangular, but X is not cylindrical. Hence the cylindricality of $\mathcal{O}$ in Theorem 18.4 cannot be taken off.

Moreover, we can obtain the following two results.

Theorem 18.6 ([59], 200?). *Let $X = \prod_{\lambda\in\Lambda} X_\lambda$ be a product of semi-stratifiable spaces, each finite subproduct of which is paracompact. For a cylindrical open cover $\mathcal{O}$ of X, the following are equivalent.*

(a) *$\mathcal{O}$ is normal.*

(b) *$\mathcal{O}$ has a σ-locally finite cylindrical cozero refinement.*

(c) *$\mathcal{O}$ has a σ-discrete cylindrical cozero refinement.*

Theorem 18.7 ([59], 200?). *Let $X = \prod_{\lambda\in\Lambda} X_\lambda$ be a product of paracompact Σ-spaces. For a rectangular open cover $\mathcal{O}$ of X, the following are equivalent.*

(a) *$\mathcal{O}$ is normal.*

(b) *$\mathcal{O}$ has a locally finite open refinement.*

(c) *$\mathcal{O}$ has a locally finite rectangular cozero refinement.*

(d) *$\mathcal{O}$ has a locally finite and σ-discrete rectangular open refinement with a rectangular shrinking.*

(e) *$\mathcal{O}$ has a locally finite and σ-discrete rectangular cozero refinement.*

Observe that there are delicate differences between these three results above. Unfortunately, we cannot find an example of non-rectangular products such as non-normal Σ-products in Example 14.1 (2). So we remain it as a problem.

Problem 9 ([59]). Is there a non-rectangular product of paracompact σ-spaces?

Remark 18.8. Rcall that many non-normal Σ-products are one of two types $X \times \Sigma 2^{\omega_1}$ or $X \times \Sigma\omega^{\omega_1}$. Since every product with a compact factor is rectangular, $X \times 2^{\omega_1}$ is always rectangular. So, to solve Problem 9, it might be useful to consider whether there is a paracompact σ-space X such that $X \times \omega^{\omega_1}$ is not rectangular.

References

[1] A. V. Arhangel'skiĭ, *On a class of spaces containing all metric and all locally bicompact spaces*, Soviet Math. Dokl. **4** (1963), 1051–1055.

[2] A. Bešlagić, *Normality in products*, Topology Appl. **22** (1986), 71-82.

[3] D. K. Burke and S. W. Davis, *Subset of* ω^{ω} *and generalized metric spaces*, Pacific J. Math. **110** (1984), 273–281.

[4] K. Chiba, *On products of normal spaces*, Rep. Fac. Sci. Shizuoka univ. **9** (1974), 1–11.

[5] H. H. Corson, *Normality in subsets of product spaces*, Amer. J. Math. **81** (1959), 785–796.

[6] G. D. Creed, *Concerning semi-stratifiable spaces*, Pacific J. Math. **32** (1970),@47–54.

[7] T. Daniel and G. Gruenhage, *Some non-normal Σ-products*, Topology and Appl. **43** (1992), 19–25.

[8] A. L. Donne, *Shrinking property in Σ-products of paracompact p-spaces*, Topology and Appl. **19** (1985), 95–101.

[9] E. K. van Douwen, *Problem section*, Topology Proc. **7** (1982), 386.

[10] K. Eda, G. Gruenhage, P. Koszmider, K. Tamano and S. Todorčević, *Sequential fans in topology*, Topology and Appl. **67** (1995) 189–220.

[11] R. Engelking, *General Topology*. Herdermann Verlag, Berlin (1989).

[12] V. V. Filippov, *Normally situated subspaces*, Trudy Mat. Inst. Steklov. **154** (1983) 239–251.

[13] W. G. Fleissner, *A normal collectionwise Hausdorff not collectionwise normal*, Topology and Appl. **6** (1976) 57–64.

[14] G. Gruenhage, *Generalized metric spaces*, Handbook of Set-theoretic Topology (K. Kunen and J.E. Vaughan, eds), North-Holland, Amsterdam 423–501 (1984).

[15] I. Glicksberg, *Stone-Čech compactifications of products*, Trans. Amer. Math. Soc. **90** (1959), 369–382.

[16] S. P. Gul'ko, *On the properties of subsets of Σ-products*, Soviet Math. Dokl. **18** (1977), 1438–1442.

[17] G. Gruenhage, T. Nogura, S. Purisch, *Normality of* $X \times \omega_1$, Topology and Appl. **39** (1991) 263–275.

[18] J. Hanaoka and H. Tanaka, *Σ-products of paracompact* $\mathcal{DC}$*-like spaces*, Topology Proc. **26** (2001/2002), 199-212.

[19] R. E. Hodel, *More spaces and wΔ-spaces*, Pacific J. Math. **38** (1971), 641–652.

[20] T. Hoshina, *Products of normal spaces with Lašnev spaces*, Fund. Math. **124** (1984), 143–153.

[21] T. Hoshina, *Extensions of mappings II*, Topics in General Topology (K. Morita and J. Nagata, Eds.), North-Holland (1989), 41-80.

[22] N. Kemoto and P. J. Szeptycki, *Countable paracompactness of σ-products*, Topology and Appl. **149** (2005), 259–271.

[23] N. Kemoto and Y. Yajima, *Remarks on normality of Σ-products*, Topology Proc. **19** (1994), 161–168.

[24] B. S. Klebanov, *On subspaces of products of metric spaces*, Soviet Math. Dokl. **21** (1980), 303–306.

[25] A. P. Kombarov, *On Σ-products of topological spaces*, Soviet Math. Dokl. **13** (1971), 1101–1104.

[26] A. P. Kombarov, *On the normality of Σ_m-products*, Soviet Math. Dokl. **14** (1973), 1050–1054.

[27] A. P. Kombarov, *On tightness and normality of Σ-products*, Soviet Math. Dokl. **19** (1978), 403–407.

[28] A. P. Kombarov and V. I. Malyhin, *On Σ-products*, Soviet. Math. Dokl. **14** (1973), 1780–1783.

[29] V. I. Malyhin, *On the tightness and the Souslin number of* $\exp X$ *and in a product of spaces*, Soviet Math. Dokl. **13** (1972), 496–499.

[30] S. Mazur, *On continuous mappings on Cartesian products*, Fund. Math. **39** (1952), 229-238.

[31] M. Morishita, *On Σ-product*, General Topology Summer Seminary Note (1975).

[32] K. Morita, *Paracompactness and product spaces*, Fund. Math. **50** (1962), 223-236.

[33] K. Morita, *Products of normal spaces with metric spaces*, Math. Ann. **154** (1964), 365-382.

[34] K. Nagami, *Σ-spaces*, Fund. Math. **65** (1969), 169–192.

[35] K. Nagami, *Countable paracompactness of inverse limit and products*, Fund. Math. **73** (1972), 261–270.

[36] N. Noble, *Products with closed projections, II*, Trans. Amer. Math. Soc. **160** (1971), 169–183.

[37] T. Nogura, *Tightness of compact Hausdorff spaces and normality of product spaces*, J. Math. Soc. Japan **28** (1976), 360–362.

[38] H. Ohta, *On normal, non-rectangular products*, Quart. J. Math. **32** (1981), 339–344.

[39] A. Okuyama, *Some generalizations of metric spaces, their metrization theorems and product spaces*, Sci. Rep. Tokyo Kyouiku Daigaku Sect. A **9** (1967), 236–254.

[40] B. A. Pasynkov, *On the dimension of rectangular products*, Soviet Math. Dokl. **16** (1975), 344-347.

[41] B. A. Pasynkov, *On dimension theory*, Aspects of Topology, London Math. Soc. Lecture Note Ser. **93** Cambridge Univ. Press, Cambridge 227–250 (1985).

[42] T. C. Przymusiński, *Products of normal spaces*, Handbook of Set-theoretic Topology (K. Kunen and J.E. Vaughan, eds), North-Holland, Amsterdam 781–826 (1984).

[43] M. E. Rudin, *Book review*, Bull. Amer. Math. Soc. **84** (1977), 271–272.

[44] M. E. Rudin, *The shrinking property*, Canad. Math. Bull. **26** (1983), 385–388.

[45] A. H. Stone, *paracompactness and product spaces*, Bull. Amer. Math. Soc. **54** (1948), 977–982.

[46] H. Tanaka and Y. Yajima, *Σ-products of paracompact C-scattered spaces*, Topology and Appl. **124** (2002), 39–46.

[47] R. Telgársky, *C-scattered and paracompact spaces*, Fund. Math. **73** (1971), 59–74.

[48] R. Telgársky, *Spaces defined by topological games*, Fund. Math. **88** (1975), 193–223.

[49] A. Tychonoff, *Ein Fixpunktsatz*, Math. Ann. **111** (1935), 767–776.

[50] Y. Yajima, *On Σ-products of Σ-spaces*, Fund. Math. **123** (1984), 29–37.

[51] Y. Yajima, *On the dimension of limits of inverse systems*, Proc. Amer. Math. **91** (1984), 461–466.

[52] Y. Yajima, *The normality of Σ-products and the perfect κ-normality of Cartesian products*, J. Math. Soc. Japan **36** (1984), 689–699.

[53] Y. Yajima, *On Σ-products of semi-stratifiable spaces*, Topology and Appl. **25** (1987), 1–11.

[54] Y. Yajima, *Some advances in the study of normality of Σ-products*, Topology and Computer Sciences (S. Suzuki ed.), Kinokuniya Company Ltd., (1987), 315–330.

[55] Y. Yajima, *The shrinking property of Σ-products*, Tsukuba J. Math. **13** (1989), 83–98.

[56] Y. Yajima, *Subnormality of* $X \times \kappa$ *and Σ-products*, Topology and Appl. **54** (1993), 111–122.

[57] Y. Yajima, *Normal covers of infinite products and normality of Σ-products*, Topology and Appl. **154** (2007), 103–114.

[58] Y. Yajima, *Strong β-spaces and their countable products*, Houston J. Math. **33** (2007), 531–540.

[59] Y. Yajima, *Normal covers of various products*, preprint.

[60] L. Yang, *Countable paracompactness of Σ-products*, Proc. Amer. Math. Soc. **122** (1994), 949–956.

[61] P. Zenor, *Countable paracompactness in product spaces*, Proc. Amer. Math. Soc. **30** (1971), 199–201.

In: Computational Mathematics
Editor: Peter G. Chareton, pp. 387-403

ISBN 978-1-60876-271-2

Chapter 13

THE COMPLETION OF FUZZY METRIC SPACES AND OF OTHER RELATED STRUCTURES

Salvador Romaguera*
Instituto de Matemática Pura y Aplicada,
Universidad Politécnica de Valencia, 46022 Valencia, Spain

Abstract

The problem of the completion of fuzzy metric spaces is one that has received major attention by researchers in the area of fuzzy topology. In fact, several authors have studied such a problem for different notions of metric fuzziness. In this article we discuss the fuzzy metric completion for fuzzy metric spaces in the sense of Kramosil and Michalek. This notion of metric fuzziness has an evident appeal due to its close relationship with probabilistic metric spaces. In particular, it is well known that Sherwood's construction of the completion of a Menger space endowed with a continuous t-norm, can be extended in a natural way to the fuzzy setting to deduce that every fuzzy metric space (in the sense of Kramosil and Michalek) has a completion which is unique up to isometry. We apply this construction to the study of the problem of the completion of strong fuzzy metric spaces, non-Archimedean fuzzy metric spaces, fuzzy metric groups and intuitionistic fuzzy metric spaces. The completion of fuzzy metric spaces (in the sense of George and Veeramani) is also discussed.

MSC (2000): 54A40, 54E35, 54E50, 54E70, 54H11.

Keywords: Fuzzy metric space, isometry, completion, strong fuzzy metric space, non-Archimedean fuzzy metric space, fuzzy metric group, intuitionistic fuzzy metric space.

1 Introduction and Preliminaries

In [19], Kramosil and Michalek introduced their celebrated notion of a fuzzy metric space and established, among other results, its relationship with the notion of a probabilistic

*E-mail address: sromague@mat.upv.es

metric space. Later on, George and Veeramani [9] presented and discussed an attractive variant of Kramosil-Michalek's notion. In the last years several authors have contributed to the development of the theory of theses spaces and their applications (see [3, 10, 12, 13, 14, 17, 21, 22, 23, 24, 26, 28, etc]).

In this paper we discuss the problem of the completion for fuzzy metric spaces in the sense of Kramosil and Michalek and for other related structures.

The paper is organized as follows. In Section 2 we recall the construction of the completion of a fuzzy metric space from the well-known construction of the completion of a Menger space endowed with a continuous t-norm, as given by Sherwood in [34]. Moreover, and following [13, 14], we also observe that the situation is quite different in studying the completion of fuzzy metric spaces in the sense of George and Veeramani. In Section 3 we apply the construction presented in Section 2 to deduce that the completion of any strong fuzzy metric space is strong; then we derive, as an special case, that the completion of any non-Archimedean fuzzy metric space is non-Archimedean. Section 4 is devoted to analyze the completion of fuzzy metric groups. Finally, in Section 5 we prove that the completion of any intuitionistic fuzzy metric space, in the sense of [1], is an intuitionistic fuzzy metric space.

Our basic reference for General Topology is [8].

2 The Completion of Fuzzy Metric Spaces

We start this section by recalling some pertinent concepts related to Menger spaces.

Following [31] by a triangular norm (briefly, a t-norm) we mean a binary operation $* : [0,1] \times [0,1] \to [0,1]$ which satisfies the following conditions:

(i) $*$ is associative and commutative;

(ii) $a * 1 = a$ for every $a \in [0,1]$;

(iiii) $a * b \leq c * d$ whenever $a \leq c$ and $b \leq d$, and $a,b,c,d \in [0,1]$.

If, in addition $*$ is continuous, then it is called a continuous t-norm.

Three paradigmatic examples of continuous t-norms are $\wedge$, Prod and $*_L$ (the Lukasiewicz t-norm), which are defined by $a \wedge b = \min\{a,b\}$, $a\text{Prod}b = a \cdot b$ and $a *_L b = \max\{a + b - 1, 0\}$, respectively.

Note that $*_L \leq \text{Prod} \leq \wedge$. Actually, it is well known and easy to see from condition (iii) above, that $* \leq \wedge$ for every t-norm $*$.

A distribution function [32] is a function $F : \mathbb{R} \to [0,1]$ such that:

(i) F is nondecreasing (i.e., $F(s) \leq F(t)$ whenever $s \leq t$);

(ii) F is left continuous;

(iii) $\lim_{t \to -\infty} F(t) = 0$ and $\lim_{t \to +\infty} F(t) = 1$.

We denote by Δ the set of distribution functions, and by Δ^+ the subset of Δ consisting of those distribution functions F such that $F(0) = 0$.

A probabilistic metric space [32] is a pair $(X, \mathcal{F})$ such that X is a (nonempty) set and $\mathcal{F}$ is a mapping from $X \times X$ into Δ^+ such that for all $x,y,z \in X$:

(i) $\mathcal{F}(x,y)(t) = 1$ for all $t > 0$ if and only if $x = y$;

(ii) $\mathcal{F}(x,y) = \mathcal{F}(y,x)$;
(iii) If $\mathcal{F}(x,y)(t) = 1$ and $\mathcal{F}(y,z)(s) = 1$, then $\mathcal{F}(x,z)(t+s) = 1$.

As usual, we shall write F_{xy} instead of $\mathcal{F}(x,y)$ if no confusion arises.

A sequence $(x_n)_n$ in a probabilistic metric space $(X,\mathcal{F})$ is said to be a Cauchy sequence if for each $r \in (0,1)$ and each $t > 0$ there exists $n_0 \in \mathbb{N}$ such that $F_{x_n x_m}(t) > 1 - r$ for all $n,m \geq n_0$.

$(X,\mathcal{F})$ is called complete if for each Cauchy sequence $(x_n)_n$ there exists $x \in X$ satisfying the following condition: for each $r \in (0,1)$ and each $t > 0$ there exists $n_0 \in \mathbb{N}$ such that $F_{xx_m}(t) > 1 - r$ for all $n \geq n_0$; i.e., if for each $t > 0$, $\lim_n F_{xx_n}(t) = 1$.

A Menger space is a triple $(X,\mathcal{F},*)$ such that $(X,\mathcal{F})$ is a probabilistic metric space and $*$ is a t-norm such that for all $x,y,z \in$ and $t,s \geq 0$:

(iv) $F_{xy}(t+s) \geq F_{xz}(t) * F_{zy}(s)$.

Since every Menger space can be considered as a probabilistic metric space, the notions of a Cauchy sequence in Menger spaces and of a complete Menger space are defined in the obvious manner.

In fact, each Menger space $(X,\mathcal{F},*)$ induces a topology $\tau_{\mathcal{F}}$ on X defined as follows:

$$\tau_{\mathcal{F}} = \{A \subseteq X : \text{for each } x \in A \text{ there exist } r \in (0,1), t > 0, \text{ such that } B_{\mathcal{F}}(x,r,t) \subseteq A\},$$

where $B_{\mathcal{F}}(x,r,t) = \{y \in X : F_{xy}(t) > 1 - r\}$ for all $x \in X$, $r \in (0,1)$, $t > 0$.

Furthermore, if the t-norm $*$ is continuous, then $(X,\tau_{\mathcal{F}})$ is a metrizable topological space because the countable collection

$$\{\{(x,y) \in X \times X : F_{xy}(1/n) > 1 - 1/n\} : n \in \mathbb{N}\},$$

is a base for a uniformity on X such that its induced topology coincides with $\tau_{\mathcal{F}}$ (see [32]).

Note that, in particular, a sequence $(x_n)_n$ in $(X,\mathcal{F},*)$ converges to $x \in X$ with respect to $\tau_{\mathcal{F}}$ if and only if for each $t > 0$, $\lim_n F_{xx_n}(t) = 1$.

A mapping f from a Menger space $(X,\mathcal{F},*)$ to a Menger space $(Y,\mathcal{G},\star)$ is called an isometry if for each $x,y \in X$:

$$G_{f(x)f(y)} = F_{xy}.$$

It is clear that any isometry is one-to-one.

Two Menger spaces $(X,\mathcal{F},*)$ and $(Y,\mathcal{G},\star)$ are called isometric if there is an isometry from $(X,\mathcal{F},*)$ onto $(Y,\mathcal{G},\star)$.

A complete Menger space is said to be a completion of a given Menger space $(X,\mathcal{F},*)$ if it has a dense subspace isometric to $(X,\mathcal{F},*)$.

Sherwood proved in [34] that every Menger space $(X,\mathcal{F},*)$ such that $*$ is continuous has a completion which is unique up to isometry (a different construction of the completion, based on the theory of fixed point, was obtained by Sempi in [33]).

Sherwood's construction is strongly based on the properties of Lévy's metric L on Δ, which is defined as follows [11, 20]:

$$L(F,G) = \inf\{h > 0 : F(t-h) - h \leq G(t) \leq F(t+h) + h, \text{ for all } t > 0\},$$

whenever $F, G \in \Delta$.

It is well known that (Δ, L) is a complete metric space. Since Δ^+ is a closed subset of Δ it follows that (Δ^+, L) is also a complete metric space.

Then, and following Sherwood [34], denote by $\sim$ the binary relation defined on the set S of all Cauchy sequences in $(X, \mathcal{F}, *)$ by

$$(x_n)_n \sim (y_n)_n \Longleftrightarrow \lim_n F_{x_n y_n}(t) = 1 \quad \text{for all } t > 0,$$

whenever $(x_n)_n, (y_n)_n \in S$.

Thus $\sim$ is an equivalence relation on S. Let $\widetilde{X_{\mathcal{F}}}$ be the quotient $S/\sim$. The elements of $\widetilde{X_{\mathcal{F}}}$ will be denoted by $[(x_n)_n]$, where $(x_n)_n \in S$.

For each pair $(x_n)_n, (y_n)_n \in S$, it follows that $(F_{x_n y_n})_n$ is a Cauchy sequence in the complete metric space (Δ^+, L), so this sequence converges to an element of Δ^+ which we denote by $\lim_n^L F_{x_n y_n}$.

Furthermore, for each $(x'_n)_n \in [(x_n)_n]$, and each $(y'_n)_n \in [(y_n)_n]$, one has that

$$\lim_n^L F_{x_n y_n} = \lim_n^L F_{x'_n y'_n}.$$

Consequently, we can define a function $\widetilde{\mathcal{F}} : \widetilde{X_{\mathcal{F}}} \times \widetilde{X_{\mathcal{F}}} \to [0,1]$, by

$$\widetilde{\mathcal{F}}([(x_n)_n], [(y_n)_n])(0) = 0,$$

and

$$\widetilde{\mathcal{F}}([(x_n)_n], [(y_n)_n])(t) = \lim_n^L F_{x_n y_n}(t)$$

whenever $t > 0$.

Then $(\widetilde{X_{\mathcal{F}}}, \widetilde{\mathcal{F}}, *)$ is a complete Menger space, and the mapping

$$i_{\mathcal{F}} : X \to \widetilde{X_{\mathcal{F}}}$$

given by

$$i_{\mathcal{F}}(x) = [(x, x, \ldots)]$$

whenever $x \in X$, is an isometry between $(X, \mathcal{F}, *)$ and a dense subspace of $(\widetilde{X_{\mathcal{F}}}, \widetilde{\mathcal{F}}, *)$.

Moreover, if $(Y, \mathcal{G}, \star)$ is a complete Menger space having a dense subspace isometric to $(X, \mathcal{F}, *)$, then $(Y, \mathcal{G}, \star)$ is isometric to $(\widetilde{X_{\mathcal{F}}}, \widetilde{\mathcal{F}}, *)$.

The (complete) Menger space $(\widetilde{X_{\mathcal{F}}}, \widetilde{\mathcal{F}}, *)$ is called the completion of $(X, \mathcal{F}, *)$.

Next we shall apply Sherwood's construction to directly construct the completion of any fuzzy metric space in the sense of Kramosil and Michalek.

Definition 1 ([19]). A fuzzy metric space is a triple $(X, M, *)$ such that X is a (nonempty) set, $*$ is a continuous t-norm, and M is a fuzzy set on $X \times X \times [0, +\infty)$ such that for all $x, y, z \in X$:

(i) $M(x, y, 0) = 0$;
(ii) $M(x, y, t) = 1$ for all $t > 0$ if and only if $x = y$;
(iii) $M(x, y, t) = M(y, x, t)$ for all $t > 0$;
(iv) $M(x, y, t+s) \geq M(x, z, t) * M(z, y, s)$ for all $t, s \geq 0$;
(v) $M(x, y, \_) : [0, +\infty) \to [0, 1]$ is left continuous;
(vi) $\lim_{t \to +\infty} M(x, y, t) = 1$.

Then the pair $(M, *)$ is called a fuzzy metric on X.

Kramosil and Michalek also observed in [19] that there exists a natural "equivalence" between the class of Menger spaces with continuous t-norm and the class of fuzzy metric spaces.

Indeed, if $(X, \mathcal{F}, *)$ is a Menger space such that $*$ is a continuous t-norm, then we define $M_{\mathcal{F}} : X \times X \times [0, +\infty) \to [0, 1]$ by

$$M_{\mathcal{F}}(x, y, t) = F_{xy}(t),$$

for all $t \geq 0$, and thus $(X, M_{\mathcal{F}}, *)$ is a fuzzy metric space.

Conversely, if $(X, M, *)$ is a fuzzy metric space, then we define $\mathcal{F}_M : X \times X \to \Delta^+$ by

$$\mathcal{F}_M(x, y)(t) = M(x, y, t),$$

for all $t > 0$, and $\mathcal{F}_M(x, y)(t) = 0$ for all $t \leq 0$.

Thus $(X, \mathcal{F}_M, *)$ is a Menger space, and we shall write M_{xy} instead of $\mathcal{F}_M(x, y)$ if no confusion arises.

Consequently, each fuzzy metric space $(X, M, *)$ induces a topology τ_M on X which has as a base the family of open balls

$$\{B_M(x, r, t) : x \in X,\ 0 < r < 1,\ t > 0\},$$

where $B_M(x, r, t) = \{y \in X : M(x, y, t) > 1 - r\}$ for all $x \in X$, $r \in (0, 1), t > 0$.

Actually (X, τ_M) is a metrizable topological space because the countable collection

$$\{\{(x, y) \in X \times X : M(x, y, 1/n) > 1 - 1/n\} : n \in \mathbb{N}\},$$

is a base for a uniformity on X such that its induced topology coincides with τ_M [12].

Conversely, if (X, d) is a metric space, then for each continuous t-norm $*$, the pair $(M_d, *)$ is a fuzzy metric on X such that τ_{M_d} coincides with the topology τ_d induced by d, where for each $x, y \in X$,

$$M_d(x, y, 0) = 0,$$

and

$$M_d(x, y, t) = \frac{t}{t + d(x, y)},$$

for all $t > 0$.

In particular, the fuzzy metric (M_d, Prod) is called the standard fuzzy metric of (X,d) (see, for instance, [9]).

Therefore, we have that a topological space (X,τ) admits a fuzzy metric $(M,*)$ on X such that $\tau = \tau_M$ if and only if it is metrizable.

Moreover, the notions given above for Menger spaces can be translated *mutatis mutandis* to the fuzzy metric setting as follows.

A fuzzy metric space $(X,M,*)$ is complete [9, 35] provided that each Cauchy sequence in X is convergent with respect to τ_M, where a sequence $(x_n)_n$ in X is said to be a Cauchy sequence if for each $r \in (0,1)$ and each $t > 0$ there is $n_0 \in \mathbb{N}$ such that $M(x_n, x_m, t) > 1 - r$ for all $n, m \geq n_0$.

A mapping f from a fuzzy metric space $(X,M,*)$ to a fuzzy metric space $(Y,N,\star)$ is called an isometry if for each $x, y \in X$ and each $t > 0$, $N(f(x), f(y), t)) = M(x,y,t)$. It is clear that every isometry is a one-to-one mapping.

Two fuzzy metric spaces $(X,M,*)$ and $(Y,N,\star)$ are called isometric if there is an isometry from X onto Y.

A complete fuzzy metric space is said to be a completion of a given fuzzy metric space $(X,M,*)$ if it has a dense subspace isometric to $(X,M,*)$.

Then, we can immediately adapt Sherwood's construction to the fuzzy metric context as follows.

Let $(X,M,*)$ be a fuzzy metric space. Denote by $\sim$ the binary relation defined on the set S of all Cauchy sequences in $(X,M,*)$ by

$$(x_n)_n \sim (y_n)_n \Longleftrightarrow \lim_n M(x_n, y_n, t) = 1 \quad \text{for all } t > 0,$$

whenever $(x_n)_n, (y_n)_n \in S$.

Then $\sim$ is an equivalence relation on S. Let $\widetilde{X_M}$ be the quotient $S/\sim$. The elements of $\widetilde{X_M}$ will be denoted by $[(x_n)_n]$, where $(x_n)_n \in S$.

For each pair $(x_n)_n, (y_n)_n \in S$, we have that $(M_{x_n y_n})_n$ is a Cauchy sequence in the complete metric space (Δ^+, L), so this sequence converges to an element of Δ^+ which is denoted by $\lim_n^L M_{x_n y_n}$.

Furthermore, for each $(x'_n)_n \in [(x_n)_n]$, and each $(y'_n)_n \in [(y_n)_n]$, one has that

$$\lim_n^L M_{x_n y_n} = \lim_n^L M_{x'_n y'_n}.$$

Consequently, we can define a function $\widetilde{M} : \widetilde{X_M} \times \widetilde{X_M} \times [0,\infty) \to [0,1]$, by

$$\widetilde{M}([(x_n)_n], [(y_n)_n], 0) = 0,$$

and

$$\widetilde{M}([(x_n)_n], [(y_n)_n], t) = \lim_n^L M_{x_n y_n}(t)$$

whenever $t > 0$.

Thus $(\widetilde{X_M}, \widetilde{M}, *)$ is a complete fuzzy metric space, and the mapping

$$i_M : X \to \widetilde{X_M}$$

given by

$$i_M(x) = [(x, x, \ldots)]$$

whenever $x \in X$, is an isometry between $(X, M, *)$ and a dense subspace of $(\widetilde{X_M}, \widetilde{M}, *)$.

Moreover, if $(Y, M_Y, *_Y)$ is a complete fuzzy metric space having a dense subspace isometric to $(X, M, *)$, then $(Y, M_Y, *_Y)$ is isometric to $(\widetilde{X_M}, \widetilde{M}, *)$.

Therefore we have the following.

Theorem 1. *Every fuzzy metric space has a completion which is unique up to isometry.*

The (complete) fuzzy metric space $(\widetilde{X_M}, \widetilde{M}, *)$ is called the completion of $(X, M, *)$.

In the rest of this section we will consider the completion of fuzzy metric spaces in the sense of George and Veeramani [9, 10].

Thus, we shall define a GV-fuzzy metric space as a triple $(X, M, *)$ such that X is a (nonempty) set, $*$ is a continuous t-norm, and M is a fuzzy set on $X \times X \times (0, +\infty)$ such that for all $x, y, z \in X$ and $t, s > 0$:

(i) $M(x, y, t) = 1$ if and only if $x = y$;
(ii) $M(x, y, t) = M(y, x, t)$;
(iii) $M(x, y, t + s) \geq M(x, z, t) * M(z, y, s)$;
(iv) $M(x, y, \_) : (0, +\infty) \to (0, 1]$ is continuous.

Then the pair $(M, *)$ will be called a GV-fuzzy metric on X.

Clearly, a GV-fuzzy metric space $(X, M, *)$ such that $\lim_{t \to +\infty} M(x, y, t) = 1$ for all $x, y \in X$, can be considered a fuzzy metric space (in Kramosil and Michalek's sense), putting $M(x, y, 0) = 0$ for all $x, y \in X$.

Therefore, the notions of a Cauchy sequence, completeness and completion for GV-fuzzy metric spaces are defined as for fuzzy metric spaces in the sense of Definition 1.

Note also that the fuzzy metric $(M_d, *)$ induced by any metric space (X, d) is actually a GV-fuzzy metric on X.

The next example presented in [13] shows that there exist GV-fuzzy metric spaces which do not admit a GV-fuzzy metric completion.

Example 1. Let $(x_n)_{n \geq 3}$ and $(y_n)_{n \geq 3}$ be two sequences of distinct points such that $A \cap B = \emptyset$, where $A = \{x_n : n \geq 3\}$ and $B = \{y_n : n \geq 3\}$. Put $X = A \cup B$ and define a fuzzy set M on $X \times X \times (0, +\infty)$ by:

$$M(x_n, x_m, t) = M(y_n, y_m, t) = 1 - [1/(n \wedge m) - 1/(n \vee m)],$$

and

$$M(x_n, y_m, t) = M(y_m, x_n, t) = 1/n + 1/m,$$

for all $n, m \geq 3$, $t > 0$.

Then, it was shown in [13] that $(M, *_L)$ is a GV-fuzzy metric on X such that both $(x_n)_{n \geq 3}$ and $(y_n)_{n \geq 3}$ are Cauchy sequences in $(X, M, *_L)$. Suppose that $(Y, N, \star)$ is a complete GV-fuzzy metric space that has a τ_N-dense subspace isometric to $(X, M, *_L)$. Then $(Y, N, \star)$ is isometric to $(\widetilde{X_M}, \widetilde{M}, *_L)$ by Theorem 1. But an easy computation shows that

$$\widetilde{M}([(x_n)_n], [(y_n)_n], t) = 0,$$

for all $t > 0$. Hence $N([(x_n)_n],[(y_n)_n],t) = 0$ for all $t > 0$, which contradicts our assumption that $(Y,N,\star)$ is a GV-fuzzy metric space.

In the light of the above example it suggests, in a natural way, the problem of characterizing completable GV-fuzzy metric spaces, i.e., those GV-fuzzy metric spaces that admit a fuzzy metric completion which is a GV-fuzzy metric space; such a completion if exists is called a GV-fuzzy metric completion. This problem is solved in [14] as follows.

Theorem 2. *A GV-fuzzy metric space $(X,M,*)$ is completable if and only if for each pair $(a_n)_n, (b_n)_n$, of Cauchy sequences in X, the assignment*

$$t \mapsto \lim_n M(a_n,b_n,t)$$

is a continuous function on $(0,+\infty)$ with values in $(0,1]$.

Furthermore, if a GV-fuzzy metric space is completable, then its GV-fuzzy metric completion is unique up to isometry.

3 The Completion of Strong Fuzzy Metric Spaces and of Non-Archimedean Fuzzy Metric Spaces

An important class of fuzzy metric spaces are the so-called non-Archimedean fuzzy metric spaces.

Following Hadzic and Pap [18], by a non-Archimedean fuzzy metric space we mean a fuzzy metric space $(X,M,*)$ such that

(iv') $M(x,y,t) \geq M(x,z,t) * M(z,y,t)$ for all $x,y,z \in X$ and $t,s \geq 0$.

It is clear that every non-Archimedean fuzzy metric space is a fuzzy metric space.

Since a topological space (X,τ) is metrizable if and only if there is a fuzzy metric $(M,*)$ on X such that $\tau = \tau_M$, one could expect that this fact also holds for the non-Archimedean case. However, the following result shows that actually the topology induced by a fuzzy metric space can be also induced by a non-Archimedean fuzzy metric.

Proposition 1. *Let $(X,M,*)$ be a fuzzy metric space. Then there exists a non-Archimedean fuzzy metric (N,Prod) on X such that $\tau_M = \tau_N$.*

Proof. Let d be a metric on X such that $\tau_d = \tau_M$. Then it is easy to see that the standard fuzzy metric (M_d,Prod) is actually a non-Archimedean fuzzy metric on X. The fact that $\tau_d = \tau_{M_d}$ concludes the proof. □

In [30], Sapena defined a non-Archimedean fuzzy metric space as a triple $(X,M,*)$ such that X is a (nonempty) set, $*$ is a continuous t-norm and M is a fuzzy set on $X \times X \times [0,+\infty)$ satisfying conditions (i), (ii), (iii), (v) and (vi) in Definition 1, and, for each $x,y,z \in X$ and $t,s \geq 0$:

(iv'') $M(x,y,t) \geq \min\{M(x,z,t), M(z,y,t)\}$.

He also proved that a topological space (X,τ) is non-Archimedeanly metrizable if and only if there is a non-Archimedean fuzzy metric $(M,*)$ on X such that $\tau = \tau_M$.

Due to this fact, we will use in the rest of this section, the term non-Archimedean fuzzy metric space for non-Archimedean fuzzy metric spaces in Sapena's sense, whereas non-Archimedean fuzzy metric spaces in Hadzic-Pap's sense will be called strong fuzzy metric spaces. Obviously, every non-Archimedean fuzzy metric space is a strong fuzzy metric space, and the converse does not hold in general.

Theorem 3. *The completion of a strong fuzzy metric space is a strong fuzzy metric space.*

Proof. Let $(x_n)_n$, $(y_n)_n$ and $(z_n)_n$ be Cauchy sequences in a strong fuzzy metric space $(X, M, *)$. Put

$$F = \lim_n^L M_{x_n y_n}, \quad G = \lim_n^L M_{x_n z_n}, \quad \text{and} \quad H = \lim_n^L M_{z_n y_n}.$$

Then, it will suffices to show that $G(t) * H(t) \leq F(t)$ for all $t > 0$.

To this end, let $t > 0$. If $G(t) = 0$ or $H(t) = 0$, then $G(t) * H(t) = 0 \leq F(t)$. Hence, we shall suppose that $G(t) > 0$ and $H(t) > 0$.

Choose an arbitrary $\varepsilon > 0$ such that $t > 2\varepsilon$, $G(t) > 2\varepsilon$ and $H(t) > 2\varepsilon$. Since G and H are left continuous there exists $\delta \in (0, \varepsilon)$ such that $G(t - 2\delta) + \varepsilon > G(t)$ and $H(t - 2\delta) + \varepsilon > H(t)$.

Therefore, there exists $n_\varepsilon \in \mathbb{N}$ such that for each $n \geq n_\varepsilon$,

$$L(F, M_{x_n y_n}) < \delta, \quad L(G, M_{x_n z_n}) < \delta, \quad \text{and} \quad L(H, M_{z_n y_n}) < \delta.$$

Consequently, for each $n \geq n_\varepsilon$ there exists $h_n \in (0, \delta)$ such that

$$F(t - \delta - h_n) - h_n \leq M_{x_n y_n}(t - \delta) \leq F(t - \delta + h_n) + h_n,$$

$$G(t - \delta - h_n) - h_n \leq M_{x_n z_n}(t - \delta) \leq G(t - \delta + h_n) + h_n,$$

and

$$H(t - \delta - h_n) - h_n \leq M_{z_n y_n}(t - \delta) \leq H(t - \delta + h_n) + h_n.$$

(Note that $t - \delta - h_n > t - 2\varepsilon > 0$.)

Hence

$$\begin{aligned} 0 &< G(t) - 2\varepsilon < G(t - 2\delta) - h_n \leq G(t - \delta - h_n) - h_n \\ &\leq M_{x_n z_n}(t - \delta), \end{aligned}$$

and

$$\begin{aligned} 0 &< H(t) - 2\varepsilon < H(t - 2\delta) - h_n \leq H(t - \delta - h_n) - h_n \\ &\leq M_{z_n y_n}(t - \delta), \end{aligned}$$

for all $n \geq n_\varepsilon$. Thus

$$\begin{aligned} (G(t) - 2\varepsilon) * (H(t) - 2\varepsilon) &\leq M(x_n, z_n, t - \delta) * M(z_n, y_n, t - \delta) \\ &\leq M(x_n, y_n, t - \delta) \\ &\leq F(t - \delta + h_n) + h_n < F(t) + \varepsilon. \end{aligned}$$

Since ε is arbitrary, it follows from continuity of $*$ that

$$G(t) * H(t) \leq F(t).$$

We conclude that $(\widetilde{X_M}, \widetilde{M}, *)$ is a strong fuzzy metric space. □

If in the preceding theorem we put $* = \wedge$, we obtain the following

Theorem 4. *The completion of a non-Archimedean fuzzy metric space is a non-Archimedean fuzzy metric space.*

4 The Completion of Fuzzy Metric Groups

In [27] it was introduced a notion of fuzzy metric group which permits us to obtain satisfactory extensions of classical theorems on metric groups to the fuzzy framework. Here, we shall give a suitable modification of this notion which is more appropriate to our context.

Let us recall that a topological group is a triple $(X,\cdot,\tau)$ such that $(X,\cdot)$ is a group and τ is a topology on X such that $\cdot$ is continuous from $(X\times X,\tau\times\tau)$ into (X,τ).

Definition 2. A fuzzy metric $(M,*)$ on a group $(X,\cdot)$ is called left subinvariant if for each $x,y,z\in X$ and each $t>0$, one has

$$M(zx,zy,t)\geq M(x,y,t).$$

Similarly, $(M,*)$ is said to be right subinvariant if for each $x,y,z\in X$ and each $t>0$, one has

$$M(xz,yz,t)\geq M(x,y,t).$$

Definition 3. A fuzzy metric group is a 4-tuple $(X,\cdot,M,*)$ such that $(X,\cdot)$ is a group and $(M,*)$ is a left and right subinvariant fuzzy metric on X.

The following result is an immediate consequence of the above definitions.

Proposition 2. *If $(X,\cdot,M,*)$ is a fuzzy metric group, then $(X,\cdot,\tau_M)$ is a topological group.*

Proof. Let $(x_n)_n$ and $(y_n)_n$ be sequences in X that converge to $x\in X$ and $y\in X$, respectively. Fix $t>0$, and let $r\in(0,1)$. Then, there are $r'\in(0,1)$, with $(1-r')*(1-r')>1-r$, and $n_0\in\mathbb{N}$ such that $M(x,x_n,t/2)>1-r'$ and $M(y,y_n,t/2)>1-r'$ for all $n\geq n_0$. Hence

$$\begin{aligned}M(xy,x_ny_n,t) &\geq M(xy,x_ny,t/2)*M(x_ny,x_ny_n,t/2)\\ &\geq M(x,x_n,t/2)*M(y,y_n,t/2)\\ &\geq (1-r')*(1-r')>1-r,\end{aligned}$$

for all $n\geq n_0$. Consequently $\lim_n M(xy,x_ny_n,t)=1$, and hence $(X,\cdot,\tau_M)$ is a topological group.

Theorem 5. *The completion of a fuzzy metric group is a fuzzy metric group.*

Proof. Let $(X,\cdot,M,*)$ be a fuzzy metric group and let $(\widetilde{X_M},\widetilde{M},*)$ be the completion of $(X,M,*)$. For each pair of Cauchy sequences $(x_n)_n$ and $(y_n)_n$ define

$$[(x_n)_n]\circ[(y_n)_n]=[(x_ny_n)_n].$$

It is routine to check that $(\widetilde{X},\circ)$ is a group. Furthermore $(\widetilde{M},*)$ is left subinvariant. In fact, if $(x_n)_n$, $(y_n)_n$ and $(z_n)_n$ are Cauchy sequences in $(X,M,*)$, we have

$$\widetilde{M}([(z_n)_n] \circ [(x_n)_n], [(z_n)_n] \circ [(y_n)_n], t) = \lim_n^L M(z_n x_n, z_n y_n, t)$$

$$\geq \lim_n^L M(x_n, y_n, t) = \widetilde{M}([(x_n)_n], [(y_n)_n], t),$$

for all $t > 0$. Therefore $(\widetilde{M}, *)$ is left subinvariant. Similarly, we show that it is right subinvariant. □

Definition 4 ([27]). A fuzzy metric $(M, *)$ on a group $(X, \cdot)$ is called left invariant if for each $x, y, z \in X$ and each $t > 0$, one has

$$M(zx, zy, t) = M(x, y, t).$$

Similarly, $(M, *)$ is said to be right invariant if for each $x, y, z \in X$ and each $t > 0$, one has

$$M(xz, yz, t) = M(x, y, t).$$

The following interesting fact may be found in [27].

Proposition 3. *Every first countable T_0 topological group admits a compatible left invariant fuzzy metric.*

Theorem 6. *Let $(X, \cdot)$ be a group and let $(M, *)$ be a left (right) invariant fuzzy metric on X. Then $(\widetilde{M}, *)$ is a left (right) invariant fuzzy metric on $(\widetilde{X}, \circ)$.*

Proof. Suppose that $(M, *)$ is left invariant and let $(x_n)_n$ and $(y_n)_n$ be Cauchy sequences in $(X, M, *)$. Then

$$\widetilde{M}([(z_n)_n] \circ [(x_n)_n], [(z_n)_n] \circ [(y_n)_n], t) = \lim_n^L M(z_n x_n, z_n y_n, t)$$
$$= \lim_n^L M(x_n, y_n, t) = \widetilde{M}([(x_n)_n], [(y_n)_n], t).$$

Hence $(\widetilde{M}, *)$ is left invariant.

Similarly, we prove that $(\widetilde{M}, *)$ is right invariant whenever $(M, *)$ is right invariant. □

5 The Completion of Intuitionistic Fuzzy Metric Spaces

Motivated, in part, by the potential applicability of fuzzy topology to quantum particle physics (see [4, 5, 6, 7, etc]), and based on the idea of an inuitionistic fuzzy set in the sense of Atanassov [2], Park [25] and Alaca, Turkoglu and Yildiz [1], have introduced and studied intuitionistic versions of the notion of a fuzzy metric space in the sense of George and Veeramani, and Kramosil and Michalek, respectively.

However, Gregori, Romaguera and Veeramani showed in [16] that the main metric and topological properties of Park's approach can be easily derived from the corresponding ones for fuzzy metric spaces. A similar situation occurs for intuitionistic fuzzy metric spaces in the sense of [1] as it is observed in [29].

In this section we discuss the problem of the completion of intuitionistic fuzzy metric spaces in the sense of [1]. We prove that every intuitionistic fuzzy metric space has a completion which is unique up to isometry. Our approach is based on the well-known construction of the completion of a fuzzy metric space in the sense of Kramosil and Michalek as given in Section 2.

Following [31] by a triangular conorm (briefly, a t-conorm) we mean a binary operation $\Diamond : [0,1] \times [0,1] \rightarrow [0,1]$ which satisfies the following conditions:

(i) $\Diamond$ is associative and commutative;
(ii) $a\Diamond 0 = a$ for every $a \in [0,1]$;
(iiii) $a\Diamond b \leq c\Diamond d$ whenever $a \leq c$ and $b \leq d$, and $a,b,c,d \in [0,1]$.

If, in addition $\Diamond$ is continuous, then it is called a continuos t-conorm.

It is well known that if $*$ is a continuous t-norm (respectively, a continuous t-conorm), then $*'$ is a continuous t-conorm (respectively, a continuous t-norm), where $a *' b = 1 - [(1-a)*(1-b)]$ for all $a,b \in [0,1]$.

The following is a slight modification of the notion of an intuitionistic fuzzy metric space introduced by Alaca, Turkoglou and Yidilz in [1]

Definition 5. An intuitionistic fuzzy metric space is a 5-tuple $(X,M,N,*,\Diamond)$ such that X is a (nonempty) set, $*$ is a continuous t-norm, $\Diamond$ is a continuous t-conorm and M,N are fuzzy sets on $X \times X \times [0,+\infty)$ satisfying the following conditions, for all $x,y,z \in X$:

(i) $M(x,y,t)+N(x,y,t) \leq 1$;
(ii) $M(x,y,0) = 0$;
(iii) $M(x,y,t) = 1$ for all $t > 0$ if and only if $x = y$;
(iv) $M(x,y,t) = M(y,x,t)$ for all $t > 0$;
(v) $M(x,y,t+s) \geq M(x,z,t) * M(z,y,s)$ for all $t,s \geq 0$;
(vi) $M(x,y,\_) : [0,+\infty) \rightarrow [0,1]$ is left continuous;
(vii) $\lim_{t\rightarrow+\infty} M(x,y,t) = 1$;
(viii) $N(x,y,0) = 1$;
(ix) $N(x,y,t) = 0$ for all $t > 0$ if and only if $x = y$;
(x) $N(x,y,t) = N(y,x,t)$ for all $t > 0$;
(xi) $N(x,y,t+s) \leq N(x,z,t)\Diamond N(z,y,s)$ for all $t,s \geq 0$;
(xii) $N(x,y,\_) : [0,+\infty) \rightarrow [0,1]$ is left continuous.

Remark 1. Note that if $(X,M,N,*,\Diamond)$ is an intuitionistic fuzzy metric space, then both $(X,M,*)$ and $(X,1-N,\Diamond')$ are fuzzy metric spaces. Conversely, if $(X,M,*)$ is a fuzzy metric space, then $(X,M,1-M,*,*')$ is an intuitionistic fuzzy metric space.

In order to define a suitable topology on an intuitionistic fuzzy metric space $(M,N,*,\Diamond)$ it seems natural to consider "balls" $B(x,r,t)$ defined, similarly to [25], by:

$$B(x,r,t) = \{y \in X : M(x,y,t) > 1-r, N(x,y,t) < r\}$$

for all $x \in X$, $r \in (0,1)$ and $t > 0$. Then, one can prove, as in [25], that the family of sets of the form $\{B(x,r,t) : x \in X, r \in (0,1), t > 0\}$ is a base for a topology $\tau_{(M,N)}$ on X.

However, condition (i) in Definition 5, permits us to easily prove the following crucial facts (compare [16]), which provide the key to construct the theory of intuitionistic fuzzy metric almost directly from the well-established theory of fuzzy metric spaces.

Proposition 4. *Let $(X,M,N,*,\Diamond)$ be an intuitionistic fuzzy metric space. Then:*
(a) $B(x,r,t) = B_M(x,r,t)$ *for all* $x \in X$, $r \in (0,1)$, $t > 0$;
(b) $\tau_{(M,N)} = \tau_M$ *and* $\tau_{1-N} \subseteq \tau_M$.

A sequence $(x_n)_n$ in an intuitionistic fuzzy metric space $(X,M,N,*,\Diamond)$ is said to be Cauchy if for each $r \in (0,1)$ and each $t > 0$ there is $n_0 \in \mathbb{N}$ such that $M(x_n,x_m,t) > 1-r$ and $N(x_n,x_m,t) < r$ whenever $n,m \geq n_0$ (compare [25]). An intuitionistic fuzzy metric space $(X,M,N,*,\Diamond$ is called complete if every Cauchy sequence is convergent with respect to $\tau_{(M,N)}$.

Then, we obtain the following immediate but useful consequence of Proposition 4.

Proposition 5. *Let $(X,M,N,*,\Diamond)$ be an intuitionistic fuzzy metric space. Then:*
(a) *A sequence in X is a Cauchy sequence in $(X,M,N,*,\Diamond)$ if and only if it is a Cauchy sequence in $(X,M,*)$. Hence every Cauchy sequence in $(X,M,*)$ is a Cauchy sequence in $(X,1-N,\Diamond')$.*
(b) *$(X,M,N,*,\Diamond)$ is complete if and only if $(X,M,*)$ is complete.*

Similarly to [1], a mapping f from an inutitionistic fuzzy metric space $(X,M,N,*,\Diamond)$ to an intuitionistic fuzzy metric space $(Y,M_Y,N_Y,*_Y,\Diamond_Y)$ is called an isometry if for each $x,y \in X$ and each $t > 0$,

$$M_Y(f(x),f(y),t)) = M(x,y,t) \quad \text{and} \quad N_Y(f(x),f(y),t)) = N(x,y,t).$$

It is clear that every isometry is a one-to-one mapping.

Two intuitionistic fuzzy metric spaces $(X,M,N,*,\Diamond)$ and $(Y,M_Y,N_Y,*_Y,\Diamond_Y)$ are called isometric if there is an isometry from X onto Y.

A complete intuitionistic fuzzy metric space is said to be a completion of a given intuitionistic fuzzy metric space $(X,M,N,*,\Diamond)$ if it has a dense subspace isometric to $(X,M,N,*,\Diamond)$.

In the following we shall proceed to construct the completion of an intuitionistic fuzzy metric space.

Let $(X,M,N,*,\Diamond)$ be an intuitionistic fuzzy metric space. Consider the fuzzy metric space $(X,M,*)$, and let $(\widetilde{X_M},\widetilde{M},*)$ be its completion as constructed above in Section 2.

Recall that, in particular, the mapping $i_M : X \to \widetilde{X_M}$ is an isometry between $(X,M,*)$ and a dense subspace of $(\widetilde{X_M},\widetilde{M},*)$.

On the other hand, since, by Remark 1, $(X,1-N,\Diamond')$ is a fuzzy metric space, it follows that for each pair $(x_n)_n,(y_n)_n$, of Cauchy sequences in $(X,1-N,\Diamond')$, there exists $\lim_n^L(1-N)_{x_ny_n}$.

Therefore, if $(x_n)_n,(y_n)_n$, are Cauchy sequences in $(X,M,*)$, then there exists $\lim_n^L(1-N)_{x_ny_n}$, because, by Proposition 5 (a), $(x_n)_n$ and $(y_n)_n$ are Cauchy sequences in the fuzzy metric space $(X,1-N,\Diamond')$.

Furthermore, if $(x'_n)_n,(y'_n)_n$, are Cauchy sequences in $(X,M,*)$ with $(x_n)_n \sim (x'_n)_n$ and $(y_n) \sim (y'_n)_n$, it follows from condition $M+N \leq 1$, that

$$\lim_n(1-N)(x_n,y,t) = \lim_n(1-N)(x'_n,y'_n,t) = 1 \quad \text{for all } t > 0.$$

Consequently

$$\lim_n^L (1-N)_{x_n y_n} = \lim_n^L (1-N)_{x'_n y'_n},$$

and thus, one can define a function $\widetilde{1-N} : \widetilde{X_M} \times \widetilde{X_M} \times [0,\infty) \to [0,1]$, by

$$\widetilde{1-N}([(x_n)_n],[(y_n)_n],0) = 0,$$

and

$$\widetilde{1-N}([(x_n)_n],[(y_n)_n],t) = \lim_n^L (1-N)_{x_n y_n}(t),$$

whenever $t > 0$.

As in [34], it is proved that $(\widetilde{X_M}, \widetilde{1-N}, \lozenge')$ is a fuzzy metric space.
Moreover, by condition (i) in Definition 5, we have

$$M_{x_n y_n}(t) \leq 1 - N_{x_n y_n}(t),$$

for each pair of Cauchy sequences $(x_n)_n, (y_n)_n$, each $n \in \mathbb{N}$ and each $t > 0$. Therefore

$$\lim_n^L M_{x_n y_n}(t) \leq \lim_n^L (1 - N_{x_n y_n})(t),$$

for each $t > 0$, and thus $\widetilde{M} \leq \widetilde{1-N}$.

We conclude that $(\widetilde{X_M}, \widetilde{M}, 1-(\widetilde{1-N}), *, \lozenge)$ is an intuitionistic fuzzy metric space, which is complete by Proposition 5 (b).

Furthermore, the mapping i_M defined above satisfies for each $x, y \in X$ and $t > 0$,

$$\widetilde{M}(i_M(x), i_M(y), t) = M(x,y,t),$$

and, also,

$$(1-(\widetilde{1-N}))(i_M(x), i_M(y), t) = N(x,y,t).$$

Thus i_M is an isometry between $(X,M,N,*,\lozenge)$ and the subspace $i_M(X)$ of $(\widetilde{X_M}, \widetilde{M}, 1-(\widetilde{1-N}), *, \lozenge)$, which is dense with respect to τ_M and hence with respect to $\tau_{(M,N)}$ by Proposition 4 (b).

We have proved that $(\widetilde{X_M}, \widetilde{M}, 1-(\widetilde{1-N}), *, \lozenge)$ is a completion of $(X,M,N,*,\lozenge)$.

Finally, suppose that $(Y, M_Y, N_Y, *_Y, \lozenge_Y)$ is any completion of $(X,M,N,*,\lozenge)$. Then, there is an isometry j from $(X,M,N,*,\lozenge)$ to $(Y, M_Y, N_Y, *_Y, \lozenge_Y)$. On the other hand, since, by Propositions 5 (b) and 4 (b), the fuzzy metric space $(Y, M_Y, *_Y)$ is a completion of $(X,M,*)$, there is a unique isometry F from $(\widetilde{X_M}, \widetilde{M}, *)$ onto $(Y, M_Y, *_Y)$ such that $F(i_M) = j$. Taking into account that every Cauchy sequence in $(X,M,N,*,\lozenge)$ is a Cauchy sequence in $(X, 1-N, \lozenge')$ and that $\tau_{1-(\widetilde{1-N})} \subseteq \tau_{\widetilde{M}}$ and $\tau_{N_Y} \subseteq \tau_{M_Y}$, we deduce from standard arguments (see for instance the proof of [15, Proposition 4.5]) that

$$N_Y(F(\widetilde{x}), F(\widetilde{y}), t) = (1-(\widetilde{1-N}))(\widetilde{x}, \widetilde{y}, t),$$

whenever $\widetilde{x}, \widetilde{y} \in \widetilde{X}$ and $t > 0$. Therefore F is an isometry from $(\widetilde{X_M}, \widetilde{M}, 1-(\widetilde{1-N}), *, \lozenge)$ onto $(Y, M_Y, N_Y, *_Y, \lozenge_Y)$.

Thus, we have proved the following.

Theorem 7. *Each intuitionistic fuzzy metric space has a completion which is unique up to isometry.*

Acknowledgments

The author thanks the support of the Spanish Ministry of Education and Science, and FEDER, under grant MTM2006-14925-C02-01, and of the Generaliat Valenciana, under grant ACOMP2009/005.

References

[1] C. Alaca, D. Turkoglu, C. Yildiz, Fixed points in intuitionistic fuzzy metric spaces, *Chaos, Solitons & Fractals* **29** (2006), 1073-1078.

[2] K. Atanassov, Intuitionistic fuzzy sets, *Fuzzy Sets and Systems* **20** (1986), 87-96.

[3] Y.J. Cho, M. Grabiec, V. Radu, *On Nonsymmetric Topological and Probabilistic Sructures*, Nova Sci. Publ., New York, 2006.

[4] M.S. El Naschie, On the verification of heterotic strings theory and $\varepsilon^{(\infty)}$ theory, *Chaos, Solitons & Fractals* **11** (2000), 2397-2408.

[5] M.S. El Naschie, The two-slit experiment as the foundation of E-infinity of high energy physics, *Chaos, Solitons & Fractals* **25** (2005), 509-514.

[6] M.S. El Naschie, On the cohomology and instantons number in E-infinity Cantorian spacetime, *Chaos, Solitons & Fractals* **26** (2005), 13-17.

[7] M.S. El Naschie, On a class of fuzzy Kähler-like manifolds, *Chaos, Solitons & Fractals* 26 (2005), 257-261.

[8] R. Engelking, *General Topology,* PWN-Polish Sci. Publ., Warsaw, 1977.

[9] A. George, P. Veeramani, On some results in fuzzy metric spaces, *Fuzzy Sets and Systems* **64** (1994), 395-399.

[10] A. George, P. Veeramani, On some results of analysis of fuzzy metric spaces, *Fuzzy Sets and Systems* **90** (1997), 365-368.

[11] B.V. Gnedenko, A.N. Kolmogorov, *Limit Distributions for Sums of Independent Random Variables*, Cambridge, Addison-Wesley, 1954.

[12] V. Gregori, S. Romaguera, Some properties of fuzzy metric spaces, *Fuzzy Sets and Systems* **115** (2000), 485-489.

[13] V. Gregori, S. Romaguera, On completion of fuzzy metric spaces, *Fuzzy Sets and Systems* **130** (2002), 399-404.

[14] V. Gregori, S. Romaguera, Characterizing completable fuzzy metric spaces, *Fuzzy Sets and Systems* **144** (2004), 411-420.

[15] V. Gregori, S. Romaguera, Fuzzy quasi-metric spaces, Appl. Gen. Topology **5** (2004), 129-136.

[16] V. Gregori, S. Romaguera, P. Veeramani, A note on intuitionistic fuzzy metric spaces, *Chaos, Solitons & Fractals* **28** (2006), 902-905.

[17] J. Gutiérrez-García, M. A. De Prada Vicente, S. Romaguera, Completeness of Hutton [0,1]-quasi-uniform spaces, *Fuzzy Sets and Systems* **158** (2007), 1791-1802.

[18] O. Hadzic, E. Pap, *Fixed Point Theory in Probabilistic Metric Spaces* , Kluwer Acad. Publ., Dordrecht, 2001.

[19] I. Kramosil, J. Michalek, Fuzzy metrics and statistical metric spaces, *Kybernetika* **11** (1975), 326-334.

[20] M. Loève, *Probability Theory*, New York, Van Nostrand, 1963.

[21] D. Mihet, A Banach contraction theorem in fuzzy metric spaces, *Fuzzy Sets and Systems* **144** (2004), 431-439.

[22] D. Mihet, Fuzzy ψ-contractive mappings in non-Archimedean fuzzy metric spaces, *Fuzzy Sets and Systems* **159** (2008), 739-744.

[23] S. Morillas, V. Gregori, G. Peris-Fajarnés, A. Sapena, New adaptative vector filter using fuzzy metrics, *Journal of Electronic Imaging* **16** (2007), 1-15

[24] S. Morillas, V. Gregori, G. Peris-Fajarnés, A. Sapena, Local self-adaptative impulsive noise filter using fuzzy metrics, *Signal Processing* **88** (2008), 390-398

[25] J.H. Park, Intuitionistic fuzzy metric spaces, *Chaos, Solitons & Fractals* 22 (2004), 1039-1046.

[26] J. Rodríguez-López, S. Romaguera, The Hausdorff fuzzy metric on compact sets, *Fuzzy Sets and Systems* **147** (2004), 273-283.

[27] S. Romaguera, M. Sanchis, On fuzzy metric groups, *Fuzzy Sets and Systems* **124** (2001), 109-115.

[28] S. Romaguera, A. Sapena, P. Tirado, The Banach fixed point theorem in fuzzy quasi-metric spaces with application to the domain of words, *Topology Appl.* **154** (2007), 2196-2203.

[29] S. Romaguera, P. Tirado, On fixed point theorems in intuitionistic fuzzy metric spaces, *Internat. J. Nonlinear Sci. Numer. Simul.* **8** (2007), 233-238.

[30] A. Sapena, A contribution to the study of fuzzy metric spaces, *Appl. Gen. Topology* **2** (2001), 63-76.

[31] B. Schweizer, A. Sklar, Statistical metric spaces, *Pacific J. Math.* 10 (1960), 314-334.

[32] B. Schweizer, A. Sklar, *Probabilistic Metric Spaces*, North-Holland, Amsterdam, 1983.

[33] C. Sempi, Hausdorff distance and the completion of probabilistic metric spaces, *Boll. U.M.I. (7)* **6-B** (1992), 317-327.

[34] H. Sherwood, On the completion of probabilistic metric spaces, *Z. Wahrsch. verw. Geb.* **6** (1966), 62-64.

[35] R. Vasuki, P. Veeramani, Fixed point theorems and Cauchy sequences in fuzzy metric spaces, *Fuzzy Sets and Systems* **135** (2003), 409-413.

In: Computational Mathematics
Editor: Peter G. Chareton, pp. 405-434
ISBN 978-1-60876-271-2

Chapter 14

Homotopies and the Instability of Economic Equilibria

Debora Di Caprio[a],[*] and Francisco J. Santos-Arteaga[b]
[a] School of Economics and Management, Free University of Bozen-Bolzano,
Piazza Università 1, 39100 Bolzano, Italy
[b] GRINEI, Universidad Complutense de Madrid,
Campus de Somosaguas, 28223 Pozuelo, Spain

Abstract

Homotopy theory has been sporadically applied to economic theory mainly in order to simplify the aggregation of preferences among agents (decision makers) in social choice and to design stable algorithms in computable general equilibrium models. These applications, while dealing with relevant issues, do not consider explicitly the influence of information asymmetries on the behaviour of agents, which constitutes a leading argument in current economic theory. The present paper aims to fill the existing gap and illustrate the main consequences derived from applying homotopy theory to an economic system where agents are asymmetrically informed. Indeed, we show that, when information asymmetries among agents are explicitly considered, a homotopic approach can be used as a destabilizing device in economic equilibrium theory. We use homotopy techniques to illustrate how the information sets determining the choices of agents can be modified to induce any a priori assigned economic equilibrium. More precisely, we investigate the conditions under which a homotopy can be defined such that a predetermined choice is imposed on an economic agent. In this way, choices and, consequently, equilibria are proved to be perfectly manipulable when such conditions apply. Besides its already important economic applications, our model displays immediate extensions in management and decision theory.

1 Introduction

General economic equilibrium theory illustrates how rational agents, defined by complete and transitive preference orders, are able to optimally exchange goods and agree on their

[*]E-mail address: DDiCaprio@unibz.it, dicaper@mathstat.yorku.ca

price without the help of a central coordination mechanism. The theory, while formally impeccable, imposes several simplifying assumptions, the most restrictive of all being the public information character of all relevant variables, see [2] for a classic reference.

The dispersion of information among different sources constitutes an important phenomenon whose study was originally formalized by Stigler ([25]) in economic theoretical terms. The author states that no agent is able to know all the prices that various sellers quote at any given time, and that such price dispersion reflects the ignorance existing in the market. At the same time, and even though prices are directly observable, other characteristics defining the overall quality of goods require consumption to be observed.

The transmission of information between an informed sender and an uninformed but rational decision maker and the subsequent strategic analysis derived from the unverifiability of the information being transmitted was first introduced by Crawford and Sobel ([6]) within the economic literature. In their model, an unilaterally informed agent, who observes privately a signal realization regarding a one-dimensional variable, i.e. defined by a unique characteristic, sends a noisy message to a decision maker, who takes an action determining the wealth of them both. Their model extended to a purely strategic environment that of Akerlof ([1]), who exposed the inability of the decentralized market mechanism to solve the inefficiencies resulting from information asymmetries between informed senders and uninformed decision makers.[1] In this regard, Dulleck and Kerschbamer ([14]) provide sufficient conditions for the existence of a mechanism that induces the full revelation of all transmitted information when a unique unknown characteristic is considered. This result has been partially generalized to settings where either multiple one-dimensional real variables or a single multi-dimensional real variable are considered, see [5] and [3], respectively. Clearly, if such a mechanism cannot be correctly defined, the decisions taken, or choices made, by a given decision maker can be easily manipulated by the information sender.[2]

The economic theoretical literature described requires the common knowledge of all utility and probability functions on the set of variables, or goods, before being able to define an optimal full revelation mechanism. That is, decision makers are endowed with a preference order, represented by a continuous and real utility function, and subjectively defined probability density functions on the space of variables, or goods. Besides, all functions are assumed to be publicly known.[3] The probability density functions represent the beliefs of the decision maker regarding the probability of finding a given good within the set of all available ones.

The main purpose of the current chapter is to highlight how the problem of strategic information transmission and choice manipulation hides complex scenarios that remain overlooked due to the simplifying assumptions imposed by the cited economic literature. In order to do so, we generalize the space of analysis, usually restricted to the reals, to

[1]Economic decision makers are generally assumed to be female, and we will refer to them as such throughout the chapter. On the other hand, information senders are assumed to be male.

[2]In general, the idea of manipulation is usually relegated to social choice theory and the design of voting mechanisms, where the preference profiles of voters are exogenously given and publicly known by all involved decision makers, see [26].

[3]In order to complete the strategic economic environment, a set of density functions on the information being transmitted should be assumed. These density functions would depend, at the same time, on the set of variable realizations observed by the information sender. We do not consider such a requirement, since all transmitted information will be assumed verifiable by the decision maker.

a generic product of abstract spaces. This generalization allows for the use of multiple multi-dimensional (not necessarily real) variables.

We assume that the only information available to the decision maker is encoded by the sender in a *multifunction*. After receiving the multifunction, the decision maker *naturally decomposes* the available information dimension by dimension using coordinate functions on the set of variables. These multifunctions are mechanisms that force the decision maker to choose according to the preferences induced by their encoded information. It must be remarked that the preferences induced in this way will not generally coincide with those defined in a complete information environment, where all variables and their characteristics are known to decision makers.

The information provided by the sender is completely verifiable by the decision maker. That is, the information sender is not allowed to lie, leading to a much less restrictive environment than that considered by the cited economic literature. However, the sender is able to display the information subsets he finds more convenient. Indeed, we show that the common knowledge assumption regarding utilities and probability functions allows for the manipulation of the decision maker's preferences before the strategic information transmission process modelled in [6], [3], and [5] takes place. Thus, we show how it is possible to manipulate the choices made by decision makers even if all transmitted information is assumed to be verifiable.

In addition, we provide sufficient conditions for the utility functions induced by the information sender on the decision maker to be continuous. In this way, the main requirement[4] imposed by the cited economic literature to guarantee the existence of a full revelation mechanism would be satisfied, even though, as we show through the chapter, the mechanism remains manipulable.

Finally, we use homotopy techniques to illustrate how the information sets encoded by the sender in the multifunctions determining the choices of a decision maker can be smoothly modified to induce *any* a priori assigned equilibrium. In other words, we investigate the conditions under which a homotopy can be defined such that the choices made by a decision maker can be directed in a smooth (i.e. continuous) way towards *any* predetermined variable (i.e. good). Therefore, choices and, consequently, economic equilibria are proved to be perfectly manipulable when such conditions apply.

The chapter has been structured in a self-contained way so as to be accessible to a vast audience, ranging from economic theorists to set theoretical and algebraic topologists. However, some basic knowledge of economic theory and general topology is required from the reader.

2 Preliminaries and Notations

2.1 Basic Topological Notions and Results

Let (X, τ_X) and (Y, τ_Y) be two topological spaces. We will refer to the standard continuity definition: a function $f : X \to Y$ is said to be **continuous** if for each open subset V of Y, the set $f^{-1}(V)$ is an open subset of X.

[4]The other main condition, not necessary in our setting, is the compactness of the domain of the utility functions.

Let us remark that, in the topological sense, the continuity of a function depends not only on the function itself, but also on the topologies specified for its domain and range. If necessary, and in order to emphasize this fact, we will write that f is **continuous with respect to τ_X and τ_Y**.

Regarding the methods used to construct continuous functions, we recall the main and most general ones (see Theorem 18.2, Theorem 18.3 and Theorem 18.4 in [24]).

Theorem 2.1. *Let (X,τ_X), (Y,τ_Y) and (Z,τ_Z) be topological spaces.*

(I) *(Constant function) Every function $f : X \to Y$ defined by $x \to y_0$, where y_0 is fixed in Y, is continuous.*

(II) *(Composite) If $f : X \to Y$ and $g : Y \to Z$ are continuous, then $g \circ f : X \to Z$ is continuous.*

(III) *(Restricting the domain) If $f : X \to Y$ is continuous and A is a subspace of X, then the restriction $f \restriction_A : A \to Y$ is continuous.*

(IV) *(Local formulation of continuity) A function $f : X \to Y$ is continuous if $X = \bigcup_\alpha U_\alpha$, where, for each α, U_α is open in X and $f \restriction_{U_\alpha}$ is continuous.*

(V) *(The pasting lemma) Suppose that*

- *$X = A \cup B$, where A and B are closed in X,*
- *$f : A \to Y$ and $g : B \to Y$ are continuous,*
- *for every $x \in A \cap B$, $f(x) = g(x)$.*

Then, the function $h : X \to Y$ defined by setting $h(x) = f(x)$ if $x \in A$ and $h(x) = g(x)$ if $x \in B$, is continuous.

(VI) *(Maps into products) Endow $X \times Y$ with the product topology τ_p and let $f : Z \to X \times Y$ be defined by:*

$$f(z) = (f_1(z), f_2(z))$$

where $f_1 : Z \to X$ and $f_2 : Z \to Y$. The **product function** *f is continuous (with respect to τ_Z and τ_p) if and only if both* **coordinate functions** *f_1 and f_2 are continuous (with respect to τ_Z and, respectively, τ_X and τ_Y).*

Part (VI) of Theorem 2.1 can be generalized to the product of any indexed family of topological spaces.

Theorem 2.2 (Theorem 19.6 in [24]). *Let $\prod_{\alpha \in J} X_\alpha$ be a product of topological spaces endowed with the product topology and Y be a topological space. Let $f : Y \to \prod_{\alpha \in J} X_\alpha$ be given by the equation:*

$$f(y) = (f_\alpha(y))_{\alpha \in J},$$

where $f_\alpha : Y \to X_\alpha$ for each α. Then, the function f is continuous if and only if each function f_α is continuous.

The Cartesian product of n nonempty sets $X_1, \dots, X_n$ will be denoted by $\prod_{i \le n} X_i$, where $i \le n$ is a short for $i \in \{1, 2, \dots, n\}$. Henceforth, all Cartesian products are to be considered non-trivial (that is, $n \ge 2$). Given $f : Y \to \prod_{i \le n} X_i$, the i-th coordinate function of f is the function $f_i : Y \to X_i$ such that $f_i(y)$ is the i-th component of the n-tuple $f(y)$.

The pasting lemma, Theorem 2.1(V), can be used to prove several interesting results, such as the following.

Theorem 2.3 (Exercise 18.9 in [24]). *Let X and Y be two topological spaces. Let $f : X \to Y$ be a function and $\mathcal{A} = \{A_\alpha\}_{\alpha \in \Lambda}$ be a collection of subsets of X such that:*

- $X = \bigcup_{\alpha \in \Lambda} A_\alpha$,
- *for every α, $f \restriction_{A_\alpha}$ is continuous.*

The following hold true.

(*a*) *If the family $\mathcal{A}$ is finite and consists of closed subsets, then f is continuous.*

(*b*) *If the family $\mathcal{A}$ is locally finite and consists of closed subsets, then f is continuous.*

Recall that an indexed family $\mathcal{A} = \{A_\alpha\}_{\alpha \in \Lambda}$ of subsets of a topological space X is said to be **locally finite** if

$$\forall x \in X,\ \exists U \text{ neighbourhood of } x \text{ such that } \{\alpha \in \Lambda : A_\alpha \cap U \neq \emptyset\} \text{ is finite.}$$

Every finite family of closed subsets is locally finite; thus, in Theorem 2.3, part (*b*) is actually a generalization of part (*a*).

So far we have shown how to construct a continuous function from a space into another when their topologies have been assigned. However, it is possible to look at the problem of obtaining continuous functions from a different point of view. Given two sets X and Y and a function $f : X \to Y$, which topologies can be assigned on X and Y such that f is continuous?

When either the domain X or the range Y has already been endowed with a topology, requiring the continuity of f allows to define a topology on either the range or the domain, respectively. We refer to such topologies as induced by the function f. In particular, if Y is a topological space, the topology induced on the domain X such that $f : X \to Y$ is continuous is known as the **weak topology** on X induced by f.

More in general, a topology can be determined by a family of functions. In the literature, the definition of weak topology is often given for Hausdorff[5] spaces (see Section 1.2 in [4]), but it is indeed valid for every topological space.

[5] A topological space X is said to be **Hausdorff**, or T_2, if for each pair x, y of distinct points of X, there exist disjoint open sets containing x and y, respectively. This is known as the *Hausdorff separation axiom*.

Every Hausdorff space is also T_1 (all one-point sets are closed). Among the basic results, we recall in particular that

(*a*) every subspace of a Hausdorff space (endowed with the subspace topology) is still a Hausdorff space;

(*b*) every product of Hausdorff spaces (endowed with the product topology) is still a Hausdorff space;

(*c*) every metric topology, and hence, any n-dimensional real space endowed with the standard Euclidean topology, satisfies the Hausdorff axiom.

Definition 2.4. Let $\{X_\alpha : \alpha \in \Lambda\}$ be a family of topological spaces and let X be a nonempty set. Suppose that $\mathcal{F} = \{f_\alpha : \alpha \in \Lambda\}$ is a family of functions, where for every α, $f_\alpha : X \to X_\alpha$. The **weak topology on X determined by $\mathcal{F}$** is the topology having as a subbase all sets of the form $f_\alpha^{-1}(V_\alpha)$, where V_α is open in X_α.

It is easy to check that the weak topology on a set X induced by the family of functions $\mathcal{F}$ is the weakest topology with respect to which each of the functions in $\mathcal{F}$ is continuous.

If each space X_α is Hausdorff, then the weak topology induced by $\mathcal{F}$ is Hausdorff if and only if $\mathcal{F}$ satisfies the following property:

$$\mathcal{F} \textit{ separates points}: \quad x_1 \neq x_2 \implies \exists \alpha \text{ such that } f_\alpha(x_1) \neq f_\alpha(x_2).$$

There are many examples of weak topologies commonly used in the literature. We would like to highlight the following two cases.

Example 2.5. Let $\{X_\alpha : \alpha \in \Lambda\}$ be a family of topological spaces. The *product topology* on $\prod_{\alpha\in\Lambda} X_\alpha$ is the weak topology defined by the family of projection maps $\{p_\alpha : \alpha \in \Lambda\}$ where $p_\beta : \prod_{\alpha\in\Lambda} X_\alpha \to X_\beta$ is defined by $p_\beta((x_\alpha)_{\alpha\in\Lambda}) = x_\beta$.

Example 2.6. Let X be a topological space and A be a nonempty subset of X. The *subspace topology* (also known as *relative topology*) on A is the weak topology induced by a single map, namely, the inclusion map $i_A : A \to X$ defined by $i_A(a) = a$.

2.2 Preferences and Utilities

Let X be a nonempty set. A **preference relation** on X is a binary relation $R \subseteq X \times X$ satisfying:

reflexivity: $\forall x \in X,\ (x,x) \in R$;

completeness: $\forall x,y \in X,\ (x,y) \in R \ \vee\ (y,x) \in R$;

transitivity: $\forall x,y,z \in X,\ (x,y) \in R \ \wedge\ (y,z) \in R \ \Rightarrow\ (x,z) \in R$.

Preference relations are usually denoted by the symbol $\succsim$. We will write $x \succsim y$ in place of $(x,y) \in \succsim$ and read: *x is preferred or indifferent to y.*

The strict preference and the indifference relations associated to a preference relation $\succsim$ are defined as follows:

$$x \succ y \overset{def}{\Longleftrightarrow} x \succsim y \ \wedge\ y \not\succsim x,$$

$$x \sim y \overset{def}{\Longleftrightarrow} x \succsim y \ \wedge\ y \succsim x.$$

We read $x \succ y$ as *x is preferred to y*, while $x \sim y$ is read *x is indifferent to y*.

From the definition it is clear that preference relations are complete preorders. Also, preference relations that are complete and transitive are usually called *rational*. Hence, all the preference relations in this paper are rational.

In particular, the symbols $\geq$ and $>$ will denote the standard partial and linear order on the reals, respectively.

A **utility function** representing $\succsim$ is any function $u : X \to \mathbb{R}$ such that:

$$\forall x, y \in X,\ x \succsim y \iff u(x) \geq u(y).$$

It is known that any (rational) preference relation $\succsim$ on a nonempty set X can be represented by a utility function (not necessarily continuous) if and only if it is *perfectly separable*, that is, if there exists a countable subset V of X such that for all $x \succ y$ there exists $z \in V$ with $x \succsim z \succsim y$ (see [27]).

A preference relation $\succsim$ on a Cartesian product $\prod_{i \leq n} X_i$ is called **additive** (see [28]) if it is representable by an additive utility function, that is, if there exist $u : \prod_{i \leq n} X_i \to \mathbb{R}$ and $u_i : X_i \to \mathbb{R}$, where $i \leq n$, such that:

$(A.1)$ $\forall (x_1, \ldots, x_n) \in \prod_{i \leq n} X_i,\ u(x_1, \ldots, x_n) = u_1(x_1) + \cdots + u_n(x_n);$

$(A.2)$ $\forall (x_1, \ldots, x_n), (y_1, \ldots, y_n) \in \prod_{i \leq n} X_i,$
$(x_1, \ldots, x_n) \succsim (y_1, \ldots, y_n) \Leftrightarrow u(x_1, \ldots, x_n) \geq u(y_1, \ldots, y_n).$

If $u : \prod_{i \leq n} X_i \to \mathbb{R}$ is an additive utility function, then for every nonempty set Y and every function $f : Y \to \prod_{i \leq n} X_i$, we have $(u \circ f) = \sum_{i \leq n} (u_i \circ f_i)$, where f_i is the i-th coordinate function of f (see Subsection 2.1). Clearly, $(u \circ f)$ satisfies an additive-like property. Thus, abusing notation but in order to be formally consistent, **we introduce the following extension of the notion of additivity** for preference relations defined on a generic nonempty set.

Definition 2.7. Let $\prod_{i \leq n} X_i$ be the Cartesian product of n nonempty sets endowed with a preference relation $\succsim$. Given a nonempty set Y and a function $f : Y \to \prod_{i \leq n} X_i$, a preference relation can be defined on Y as follows:

$$\forall y_1, y_2 \in Y,\ y_1 \succsim_f y_2 \overset{def}{\iff} f(y_1) \succsim f(y_2).$$

The preference relation $\succsim_f$ will be called the **f-relation induced by** $\succsim$.

Definition 2.8. Let $\prod_{i \leq n} X_i$ be the Cartesian product of n nonempty sets endowed with a preference relation $\succsim$. Let Y be a nonempty set and fix $f : Y \to \prod_{i \leq n} X_i$. The f-relation $\succsim_f$ will be called **additive** if the inducing relation $\succsim$ is additive on $\prod_{i \leq n} X_i$.

2.3 Topology versus Preferences

Let (X, τ) be a topological space. A **preference relation** $\succsim$ on X is called **continuous with respect to** τ if for every $x \in X$, $[x, \to) = \{t \in X : t \succsim x\}$ and $(\leftarrow, x] = \{t \in X : x \succsim t\}$ are closed subsets of (X, τ).

Equivalently, $\succsim$ is continuous with respect to τ if τ is finer than the order topology induced by $\succsim$, see [27].

For the sake of completeness, recall that the **order topology** induced by a complete preorder $\geq$ on a set X, denoted by $\tau_{\geq}$, is the topology having as a subbase all subsets $(x, \to) = \{t \in X : t > x\}$ and $(\leftarrow, x) = \{t \in X : x > t\}$, where $x \in X$. All the intervals of the form (a, b), where $a, b \in X$ and $a < b$, are open subsets of $(X, \tau_{\geq})$, while all $\geq$-rays of the form $(\leftarrow, x]$ and $[x, \rightarrow)$, are closed subsets of $(X, \tau_{\geq})$. In particular, for every $x \in X$, the singleton $\{x\}$ is a closed subset of X. As a result, the order topology induced by $\geq$ is the smallest topology on X with respect to which $\geq$ is continuous.

Unless further assumptions are considered (for example, that the order topology is separable and connected, see Corollary 3.2 in [27]), the fact that a preference relation $\succsim$ is continuous with respect to a topology τ does not necessary imply that the preference relation is representable by a continuous utility function.

Consider, for instance, the lexicographic order $>_{Lex}$ on $\mathbb{R}^2$. This order is continuous with respect to the order topology $\tau_{>_{Lex}}$, but it is well-known that it cannot be represented by a continuous utility function (see Example 3.C.1 in [18]).

Among the seminal works on the problem of the existence of continuously representable preference relations, we refer the reader to Debreu ([7], [8]). Mas-Colell et al. ([18]) state the corresponding results when $X \subseteq \mathbb{R}^L_+$, with $L \in \mathbb{N}$.

2.4 Multifunctions

Given two nonempty sets X and Y, a **multifunction** (or **set-valued map**) from X to Y is a map assigning to each element of X a (possibly empty) subset of Y. We will write $T : X \Rightarrow Y$ to indicate the fact that T is a multifunction (see Chapter 6 in [4]). Of course, any function, or single-valued map, is a multifunction: we will say that a multifunction is *proper* if it is not single-valued.

The set $gr(T) = \{(x,y) \in X \times Y : y \in T(x)\}$ is the *graph* of T, while the domain, $Dom(T)$, and the range, $Range(T)$, of T are the sets $\{x \in X : T(x) \neq \emptyset\}$ and $\{y \in Y : \exists x \in X \text{ with } y \in T(x)\}$, respectively. In particular, $Dom(T) = X$ means that T takes only nonempty values.

Historically speaking, the multifunction tool is related to the problem of existence of continuous selections. A selection for a multifunction is a function whose graph is contained in that of the multifunction. The existence of a selection for an arbitrary multifunction is merely equivalent to the axiom of choice. The existence of well-behaved selections for well-behaved multifunctions is a much more interesting and difficult question. In particular, the existence of continuous selections is strictly connected to the possibility of extending continuous functions.

We refer the interested reader to the seminal papers of Michael ([20], [21], [22], [23]) and the subsequent literature. A brief literature review and some extensions can be found in [12]. The current chapter provides an innovative interpretation of the idea of multifunction in economic theoretical terms. In particular, this notion will be used to model the strategic display of information between rational agents within an economic equilibrium setting. We believe that this set-theoretical tool, barely used by economic theorists, may be successfully applied in decision theory.

2.5 Homotopies

Let X and Y be two topological spaces. Two continuous functions $f : X \rightarrow Y$ and $g : X \rightarrow Y$ are called **homotopic** if there exists a continuous map $F : X \times [0,1] \rightarrow Y$ such that

$$F(x,0) = f(x) \text{ and } F(x,1) = g(x)$$

for every $x \in X$. The map F is said to be a **homotopy** between f and g.

We may think of a homotopy as a continuous one-parameter family of maps from X into Y. More precisely, for every $t \in [0,1]$, define $\varphi_t : X \to Y$ as follows:

$$\varphi_t(x) = F(x,t), \text{ for every } x \in X$$

If we think of the parameter t as representing time, then, the homotopy F represents a *continuous deformation* of the function f into the function g. At time $t = 0$, we have the function $f = \varphi_0$, but, as t varies, the function φ_t varies *continuously* so that at time $t = 1$, we are left with the function $g = \varphi_1$.

An equivalent way of interpreting homotopies between functions is the following. Consider the set $C(X,Y)$ of all continuous functions from X to Y. Endow $C(X,Y)$ with the compact-open topology (see [16]). Then, any two elements $f, g \in C(X,Y)$ are homotopic if and only if there exists an arc in the function space $C(X,Y)$ joining f and g. Indeed, the map $t \to \varphi_t$ is a path from f to g (see [19]).

In general, a **path**, or **arc**, in a topological space is a continuous function from some closed real interval $[a,b]$ into the space itself. The images of the end points of the interval are called the *end points* of the path, and the path is said to *join* its end points. It is technically simpler to demand all paths to be functions defined on the fixed interval $[0,1]$.

The following example is particularly interesting (see Section 51 in [24]).

Example 2.9. Let f and g be two continuous functions from a space X into $\mathbb{R}^2$. The map $F : X \times [0,1] \to \mathbb{R}^2$ defined by

$$F(x,t) = (1-t)f(x) + tg(x)$$

for every $x \in X$, is a homotopy between f and g. Figure 2.1 represents a deformation of f in g when they are both paths in $\mathbb{R}^2$.

The homotopy of the previous example is well-known as the **straight line homotopy**, since the point $f(x)$ is deformed in the point $g(x)$ through the straight line segment joining both points. This property can be generalized as follows.

Proposition 2.10. *Let X be a topological space and Y be a convex subspace of $\mathbb{R}^n$. Then, any two continuous functions from X into Y are homotopic. (The straight line homotopy continuously deforms one into the other.)*

For any other topological concepts and/or standard results not explicitly stated in this section, the reader may refer to [16], [19] and [24].

3 Main Assumptions and Properties

Henceforth, we will let $\mathcal{G}$ denote **the set of all goods**, or commodities, and fix $n \geq 2$. Moreover, for every $i \leq n$, X_i will represent the set of all possible variants for the i-th characteristic or attribute of any commodity in $\mathcal{G}$, while X will stand for the Cartesian product $\prod_{i \leq n} X_i$.

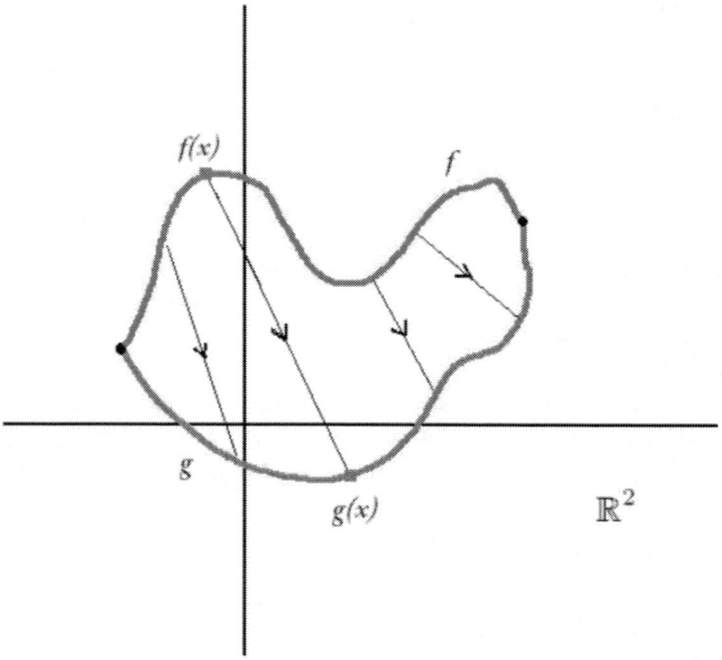

Figure 2.1

Thus, an element $x_i^G \in X_i$ specifies the i-th characteristic of a given good $G \in \mathcal{G}$, while an n-tuple $(x_1^G, \ldots, x_n^G)$ lists all its characteristics.

Definition 3.1. For every $i \leq n$, X_i will be called the **i-th characteristic factor**. The Cartesian product $X = \prod_{i \leq n} X_i$ will be referred to as the **characteristic space**.

The preference relation on X depends on the preference relations defined on the characteristic factors according to the following assumptions, which will hold through the paper.

Assumption 1. For every $i \leq n$, let $\succsim_i$ be a preference relation on X_i and u_i be a bounded (above and below) utility function representing $\succsim_i$.

Henceforth, let $u : X \to \mathbb{R}$ be defined by:

$$\forall x = (x_1, \ldots, x_n) \in X, \ u(x) = \sum_{i \leq n} u_i(x_i).$$

Since each u_i is an increasing real function, the sum function u is increasing and it induces a preference relation $\succsim_u$ on X, defined as follows:

$$\forall x, y \in X, \ x \succsim_u y \overset{def}{\Longleftrightarrow} u(x) \geq u(y).$$

Assumption 2. Endow X with the preference relation $\succsim_u$.

The following two results can be easily obtained.

Proposition 3.2. *The preference relation* $\succsim_u$ *is additive on* X.

Proof. By definition of u. □

Proposition 3.3. *For every* $f : \mathcal{G} \to X$*, the* f*-relation* $\succsim_f$*, induced by* $\succsim_u$*, is additive on* $\mathcal{G}$.

Proof. By Definition 2.8. □

Henceforth, let $\varphi : \mathcal{G} \to X$ be defined by $\varphi(G) = (x_1^G, x_2^G, \ldots, x_n^G)$, for every $G \in \mathcal{G}$.
Note that X may contain tuples of characteristics that do not necessarily describe any existing good. Therefore, φ is injective, but not necessarily bijective. Without loss of generality, we will work under the assumption that $X = \varphi(\mathcal{G})$, that is:

Assumption 3. φ is bijective.

Clearly, by Assumption 3, every G in $\mathcal{G}$ corresponds to exactly one n-tuple of X.
Moreover, by means of the map φ, the relation $\succsim_u$ induces the preference relation $\succsim_\varphi$ on $\mathcal{G}$ (see Definition 2.7) which is additive (by Proposition 3.3).

Assumption 4. Endow $\mathcal{G}$ with the additive φ-relation $\succsim_\varphi$.

We also assume the decision maker to be endowed with a subjective probability (density) function over each characteristic factor X_i. Abusing notation, each X_i can be considered a random variable.

Assumption 5. For every $i \leq n$, $\mu_i : X_i \to [0,1]$ is a non-atomic probability density function if X_i is absolutely continuous, and a non-degenerate probability function if X_i is discrete.[6]

The probability densities $\mu_1, \ldots, \mu_n$ must be interpreted as the subjective "beliefs" of the decision maker. For $i \leq n$, $\mu_i(Y_i)$ is the subjective probability that a randomly observed good from $\mathcal{G}$ displays an element $x_i \in Y_i \subseteq X_i$ as its i-th characteristic.[7]

Finally, following the standard economic theory of choice under uncertainty (see [18]), we assume that every decision maker assigns to each unknown i-th characteristic $x_i \in X_i$ the i-th certainty equivalent value induced by her subjective probability density μ_i.

Definition 3.4. Let $i \leq n$. The **certainty equivalent of** μ_i **and** u_i, denoted by c_i, is a characteristic in X_i that the decision maker is indifferent to accept in place of the expected one to be obtained through (μ_i, u_i).

In other words, for every $i \leq n$, c_i is an element of X_i whose utility $u_i(c_i)$ equals the expected value of u_i. Hence, $c_i \in u_i^{-1}\left(\int_{X_i} u_i(x_i)\mu_i(x_i)dx_i\right)$, if X_i is absolutely continuous, and $c_i \in u_i^{-1}\left(\sum_{x_i \in X_i} u_i(x_i)\mu_i(x_i)\right)$, if X_i is discrete.

The existence of the i-th certainty equivalent characteristic defined by the decision maker in X_i is trivially equivalent to $u_i^{-1}\left(\int_{X_i} u_i(x_i)d\mu_i(x_i)\right)$, or $u_i^{-1}\left(\sum_{x_i \in X_i} u_i(x_i)\mu_i(x_i)\right)$, being a nonempty set. It is not difficult to provide examples of pairs (μ_i, u_i) on the set X_i

[6]We do not consider atomic probability density functions or degenerate probability functions, since they do not necessarily induce risk on the choices made by the decision maker.

[7]Note that the probability densities $\mu_1, \ldots, \mu_n$ can be assumed either independent or correlated, without this fact affecting our results.

such that c_i does not exist. In these cases, the decision maker can fix an element of X_i whose utility provides the subjectively closest approximation to the expected value (that is, $\int_{X_i} u_i(x_i)\mu_i(x_i)dx_i$, or $\sum_{x_i \in X_i} u_i(x_i)\mu_i(x_i)$); see [10] and [11]. Clearly, any approximation process generates a bias on the choice of the decision maker. However, as it will become evident below, our results remain unaffected by this fact. Hence, without loss of generality, we will work under the following assumption.

Assumption 6. For every $i \leq n$, c_i exists.

The use of certainty equivalent values implies that if the known characteristic delivers a higher (lower) utility than the corresponding subjective certainty equivalent value, the decision maker prefers the good defined by the former (latter) one.

4 Info-multifunctions and Induced Preferences

Consider a multifunction $T : \mathcal{G} \Rightarrow \{1,2,\ldots,n\}$. Such a multifunction associates to each good G a (possibly empty) finite set of indices. These indices can be interpreted as those corresponding to the known characteristics of the good G. Following this interpretation, a multifunction T becomes a mechanism describing which information and from which good is made available to the decision maker by the information sender.

In a manipulation oriented setting, it is also natural to assume that the information sender would disclose information so as to direct the choice of the decision maker towards a predetermined subset of goods in $Dom(T)$. In order to do so, the information sender cannot allow for the whole information or no information at all to be displayed. Thus, he cannot use as a mechanism the *global info-multifunction* T^* defined by:

$$T^*(G) = \{1,2,\ldots,n\}, \text{ whenever } G \in \mathcal{G},$$

or the *empty valued info-multifunction* $T_\emptyset$ defined by:

$$T_\emptyset(G) = \emptyset, \text{ whenever } G \in \mathcal{G}.$$

Clearly, $Dom(T^*) = \mathcal{G}$. However, requiring $Dom(T) = \mathcal{G}$ for a multifunction $T : \mathcal{G} \Rightarrow \{1,2,\ldots,n\}$ does not necessarily imply that $T = T^*$. Examples of multifunctions $T \neq T^*$ such that $Dom(T) = \mathcal{G}$ can be easily given: consider, for instance, $T : \mathcal{G} \Rightarrow \{1,2,\ldots,n\}$ defined by $T(G) = \{1\}$, whenever $G \in \mathcal{G}$.

This idea yields the following definition.

Definition 4.1. A multifunction $T : \mathcal{G} \Rightarrow \{1,2,\ldots,n\}$ is an **information multifunction**, or **info-multifunction**, if it is different from both T^* and $T_\emptyset$, and $Dom(T) \neq \mathcal{G}$. We will denote the set of all info-multifunctions by $\mathcal{M}(\mathcal{G},n)$.

Note that it makes intuitive sense to assume that the goods belonging to $Dom(T)$ provide the decision maker with a higher utility than those in the complementary set. That is,

$$\forall G \in Dom(T), \quad \sum_{i \in T(G)} u_i(x_i^G) > \sum_{i \in T(G)} u_i(c_i).$$

Indeed, the main results presented through the chapter would not be modified if this condition is imposed on the analysis. In the current more general setting, however, we are also allowing the information sender to disclose information on a given set of goods so as to deliver a lower utility to the decision maker than the corresponding sum of certainty equivalent values, i.e. $\sum_{i\in T(G)} u_i(x_i^G) < \sum_{i\in T(G)} u_i(c_i)$. Clearly, the goods belonging to such a set would be excluded from the choice of the decision maker, but not from the domain of the info-multifunction.

Furthermore, according to our interpretation, any set of information is induced by a multifunction in $\mathcal{M}(\mathcal{G},n)$, and vice versa. Assigning a multifunction, however, is more general than assigning an information set, since the multifunction does not specify the value of each of the known characteristics. The value of each of the known characteristics remains specified by means of the info-map associated to the given info-multifunction.

Definition 4.2. Let $T \in \mathcal{M}(\mathcal{G},n)$. For every $i \leq n$, let $\psi_i^T : \mathcal{G} \to X_i$ be defined by

$$\psi_i^T(G) = \begin{cases} x_i^G & \text{if } i \in T(G), \\ c_i & \text{otherwise,} \end{cases}$$

with c_i being the i-th certainty equivalent of μ_i and u_i. The product function $\prod_{i\leq n} \psi_i^T : \mathcal{G} \to X$ defined by

$$\left(\prod_{i\leq n} \psi_i^T\right)(G) = (\psi_1^T(G), \ldots, \psi_n^T(G)),$$

where $G \in \mathcal{G}$, and denoted by ψ^T, is the **info-map determined by** T.

Given $T \in \mathcal{M}(\mathcal{G},n)$, the info-map ψ^T allows to describe each good as an n-tuple where all unknown characteristics are substituted by their corresponding certainty equivalent values. Clearly, info-maps are not necessarily bijective.[8]

Figure 4.1 represents the graph of an info-multifunction $T \in \mathcal{M}(\mathcal{G},n)$, in the particular case when $\mathcal{G} = \{G_1, G_2, G_3, G_4, G_5\}$ and $n = 5$. The corresponding info-map ψ^T is also described in the lower part of the figure.

Given $T \in \mathcal{M}(\mathcal{G},n)$, the decision maker is endowed with an incomplete information set, that may force her to change her original preference relation, $\succsim_\varphi$. Indeed, in place of $\succsim_\varphi$, the decision maker will base her choice on the ψ^T-relation induced by $\succsim_u$; see Definition 2.7.

More precisely:

$$\forall G, H \in \mathcal{G},\ G \succsim_{\psi^T} H \overset{def}{\Longleftrightarrow} \psi^T(G) \succsim_u \psi^T(H).$$

Thus, by the definitions of u and ψ^T,

$$\forall G, H \in \mathcal{G},\ G \succsim_{\psi^T} H \Longleftrightarrow u(\psi^T(G)) \geq u(\psi^T(H))$$

$$\Longleftrightarrow \sum_{i\leq n} u_i(\psi_i^T(G)) \geq \sum_{i\leq n} u_i(\psi_i^T(H)).$$

Proposition 3.3 immediately yields the following.

[8]If we were to consider them, the info-map ψ^{T^*} would be equal to the identification map φ, while the info-map $\psi^{T_\emptyset}$ would be the constant function defined by $\psi^{T_\emptyset}(G) = (c_1, \ldots, c_n)$, whenever $G \in \mathcal{G}$.

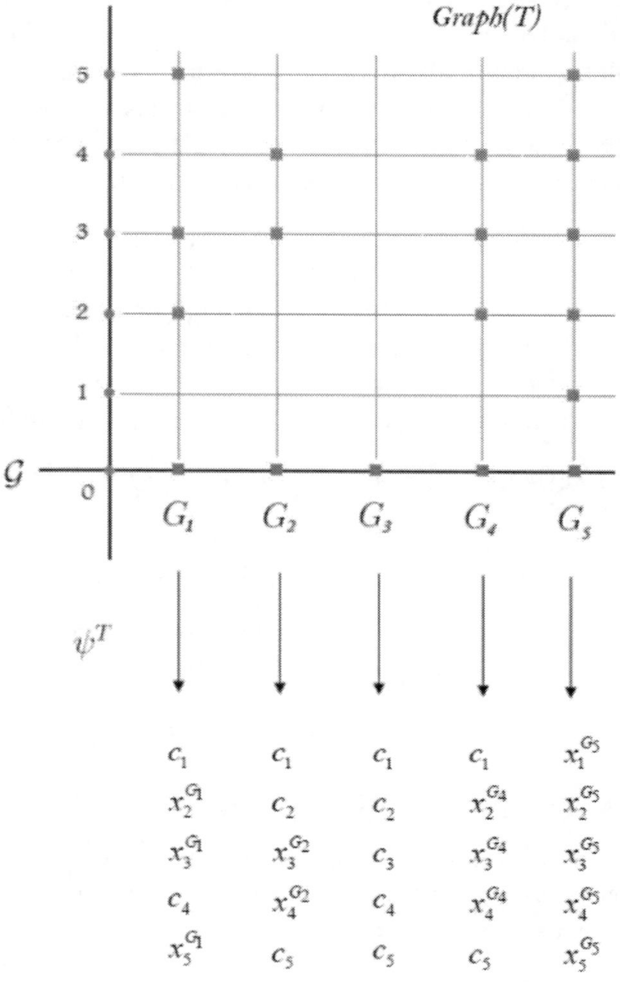

Figure 4.1

Proposition 4.3. *For every $T \in \mathcal{M}(\mathcal{G}, n)$, the preference relation $\succsim_{\psi^T}$ is additive on $\mathcal{G}$ and represented by $u \circ \psi^T$.*

Note that $\succsim_{\psi^T}$ is in general different from $\succsim_{\varphi}$. Therefore, different preference relations can be induced depending on the information set presented to the decision maker. This implies that knowing the original preference relation of a decision maker, $\succsim_{\varphi}$, allows for displaying information sets in such a way so as to manipulate her final choice.

More precisely, an information sender who knows u_i, for $i \leq n$ (equivalently, $\succsim_i$ for $i \leq n$), as well as μ_i, for $i \leq n$, can manipulate the choice made by the decision maker. As already mentioned (see Section 1), this assumption is in line with the common knowledge of utilities and beliefs used by the cited economic literature to define full revelation mechanisms.

5 Continuity Conditions for Induced Preferences

We shall study now sufficient conditions for a generic info-multifunction to generate continuously representable preference relations.

More precisely, we shall provide conditions such that the induced preference relation $\precsim_{\psi^T}$ is representable by a continuous utility function when restricted to $Dom(T)$.[9]

In order to reinforce our initial assumptions, replace Assumption 1 with the following.

Assumption 1(Bis). For every $i \leq n$, let τ_i and $\precsim_i$ be a connected topology and a preference relation on X_i such that there exists a continuous utility function $u_i : (X_i, \tau_i) \to \mathbb{R}$ representing $\precsim_i$.

Besides, we need to add the following assumptions to those introduced in Section 3. The reals $\mathbb{R}$ are, of course, endowed with the standard Euclidean topology.

Assumption 7. Let X be endowed with the product topology τ_p and assume the utility function $u = u_1 + u_2 + \cdots + u_n : (X, \tau_p) \to \mathbb{R}$ to be continuous.

Assumption 8. Let $\mathcal{G}$ be endowed with the weak topology τ_φ induced by the function $\varphi : \mathcal{G} \to (X, \tau_p)$, that is, the coarsest topology on $\mathcal{G}$ with respect to which φ is continuous (see Definition 2.4).

We leave it to the reader to check that, by Assumption 7, the preference relation $\precsim_u$ must be continuous with respect to the topology τ_p (see Subsection 2.3).

At the same time, the preference relation $\precsim_\varphi$, in the way it has been defined (see Definition 2.7 and Assumptions 3 and 4), is also forced to be continuous with respect to the topology τ_φ.

Nevertheless, neither the order topology induced by $\precsim_u$ needs to coincide with τ_p, nor the order topology induced by $\precsim_\varphi$ needs to coincide with τ_φ.

For a discussion concerning the conditions sufficient for Assumption 1(bis) and Assumption 7 to be verified we refer the reader to [7], [8], [9].

The new set of assumptions leads to the following results.

Proposition 5.1. *For every $i \leq n$, the function $\varphi_i : \mathcal{G} \to X_i$ defined by $\varphi_i(G) = x_i^G$ is continuous.*

Proof. Note that $\varphi = \prod_{i \leq n} \varphi_i$. Since, φ is continuous (Assumption 8), each φ_i is continuous by Theorem 2.2. □

Proposition 5.2. *The preference relation $\precsim_\varphi$ is additive and continuously representable on $\mathcal{G}$.*

Proof. By Assumptions 7 and 8, $u \circ \varphi$ is a continuous function from $\mathcal{G}$, endowed with τ_φ, to the reals (endowed with the standard topology). Clearly, $u \circ \varphi$ represents $\precsim_\varphi$ and its additivity follows from Assumption 4. □

[9]The problem of getting continuous utility functions representing preference relations determined by specific information sets in a more general topological setting is studied in [9].

Let $T \in \mathcal{M}(\mathcal{G},n)$. If $Dom(T)$ is discrete and finite, it is straightforward to show that the restriction of ψ^T to $Dom(T)$, denoted by $\psi^T\restriction_{Dom(T)}$, is continuous. Hence, by Proposition 4.3, $u \circ \left(\psi^T\restriction_{Dom(T)}\right)$ is an additive continuous utility function representing $\succsim_{\psi^T}$ over $Dom(T)$.

We shall show that the restriction $\psi^T\restriction_{Dom(T)}$ can be proved to be continuous even if we relax the finiteness assumption on the set $Dom(T)$.

For every $G \in Dom(T)$, let

$$\mathcal{D}_G = \{H \in Dom(T) : T(H) = T(G)\}$$

Since for every $G \in Dom(T)$, $T(G) \subseteq \{1,2,\ldots,n\}$, it is clear that there exist at most 2^n sets of the form $\mathcal{D}_G$. Thus, whatever is the cardinality (finite or infinite) of $Dom(T)$, there exists a finite subset $\{G_1,\ldots,G_{h(T)}\}$ of $Dom(T)$ of cardinality at most 2^n with the property that:

$$\forall G \in Dom(T),\ \exists j \in \{1,\ldots,h(T)\} \text{ such that } \mathcal{D}_G = \mathcal{D}_{G_j}.$$

Moreover, it is easy to check that

$$Dom(T) = \bigcup_{j=1}^{h(T)} \mathcal{D}_{G_j}$$

and

$$\forall i,j \in \{1,\ldots,h(T)\}, \text{ with } i \neq j,\ \mathcal{D}_{G_i} \cap \mathcal{D}_{G_j} = \emptyset.$$

These last remarks yield the following result.

Proposition 5.3. *For every $T \in \mathcal{M}(\mathcal{G},n)$ there exist a set of indexes $\Lambda(T)$ of cardinality at most 2^n and a unique family $\Delta(T) = \{\mathcal{D}_\alpha : \alpha \in \Lambda(T)\}$ of pairwise disjoint subsets of $\mathcal{G}$ such that*

(1) $Dom(T) = \bigcup_{\alpha \in \Lambda(T)} \mathcal{D}_\alpha$;

(2) $\forall \alpha \in \Lambda(T)$,

$$G,H \in \mathcal{D}_\alpha \Longleftrightarrow T(G) = T(H)$$

Proposition 5.3 allows us to associate to each info-multifunction T a unique finite family of pairwise disjoint subsets of $\mathcal{G}$ whose union equals $Dom(T)$.

Definition 5.4. Let $T \in \mathcal{M}(\mathcal{G},n)$. The index set $\Lambda(T)$ and the family $\Delta(T)$ will be called the **index set associated with** T and the **family associated with** T, respectively.

Lemma 5.5. *Let $T \in \mathcal{M}(\mathcal{G},n)$ and $\Delta(T) = \{\mathcal{D}_\alpha : \alpha \in \Lambda(T)\}$ be the family associated with T. Then, for every $\alpha \in \Lambda(T)$, $\psi^T\restriction_{\mathcal{D}_\alpha}$ is continuous.*

Proof. Fix $\alpha \in \Lambda(T)$ and note that $\psi^T\restriction_{\mathcal{D}_\alpha} = \prod_{i \leq n} \psi_i^T\restriction_{\mathcal{D}_\alpha}$.

If $i \in T(G)$, with $G \in \mathcal{D}_\alpha$, then $\psi_i^T\restriction_{\mathcal{D}_\alpha} = \varphi_i\restriction_{\mathcal{D}_\alpha}$. If $i \notin T(G)$, with $G \in \mathcal{D}_\alpha$, then $\psi_i^T\restriction_{\mathcal{D}_\alpha}$ is the constant function $H \in \mathcal{D}_\alpha \to c_i \in X_i$. Thus, $\psi^T\restriction_{\mathcal{D}_\alpha}$ is the product of continuous functions; hence, it is continuous. □

Figure 5.1 provides a graphical representation of the proof of Lemma 5.5 with respect to the single i-th characteristic factor space X_i. Far from being a general case, but in order to gain some intuition, the family associated with the info-multifunction T has been represented by connected intervals. Comparing the graph of φ_i (in red) with the graph of ψ_i^T (in green) it is clear that each restriction $\psi_i^T \upharpoonright_{\mathcal{D}_j}$, where $i \leq n$ and $j \in \{1, \ldots, h(T)\}$, is continuous.

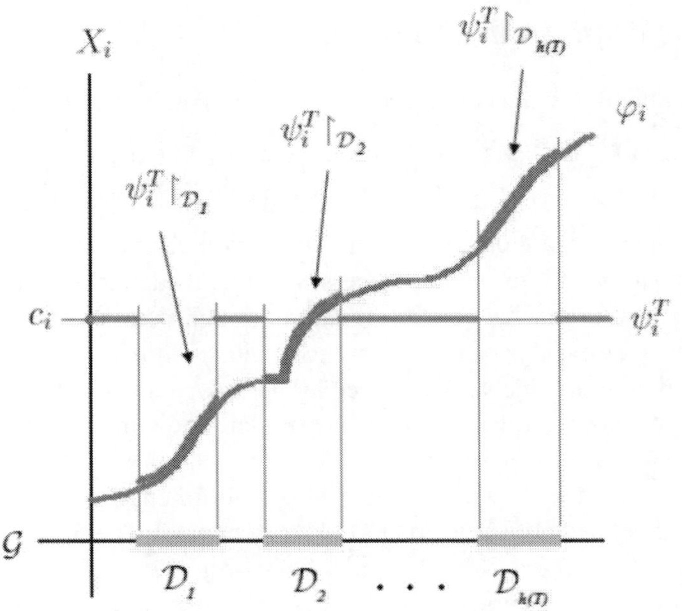

Figure 5.1

Lemma 5.6. *Let $T \in \mathcal{M}(\mathcal{G}, n)$ and $\Delta(T) = \{\mathcal{D}_\alpha : \alpha \in \Lambda(T)\}$ be the family associated with T. If, for every $\alpha \in \Lambda(T)$:*

(*a*) *$\mathcal{D}_\alpha$ is a closed subset of $\mathcal{G}$;*

(*b*) *$\psi^T \upharpoonright_{\mathcal{D}_\alpha}$ is continuous;*

then, the map $\psi^T \upharpoonright_{Dom(T)}$ is continuous.

Proof. Note that $Dom(T)$ is the disjoint union of finitely many sets of the form $\mathcal{D}_\alpha$. Apply Theorem 2.3. □

Note that condition (a) of Lemma 5.6. can hold true for an info-multifunction T thanks to the requirement $Dom(T) \neq \mathcal{G}$. Allowing for $Dom(T) = \mathcal{G}$ would imply that $\mathcal{G}$ is disconnected, contradicting Assumptions 1(Bis), 7 and 8.

To simplify notation, **we will write that an info-multifunction T satisfies property** ($\star$) to indicate the fact that the family associated with T, $\Delta(T) = \{\mathcal{D}_\alpha \in \Lambda(T)\}$, satisfies the following property.

$(\star)$ $\forall \alpha \in \Lambda(T)$, $\mathcal{D}_\alpha$ is a closed subset of $\mathcal{G}$.

Theorem 5.7. *Let $T \in \mathcal{M}(\mathcal{G},n)$. If T satisfies property $(\star)$, then the additive preference relation $\succsim_{\psi^T}$ is continuously representable on Dom (T).*

Proof. $u \circ \psi^T$ represents $\succsim_{\psi^T}$ on $\mathcal{G}$ and, by Lemmas 5.5 and 5.6, $u \circ \psi^T \restriction_{Dom(T)}$ is continuous. □

6 Defining Homotopic Preferences

We will consider the following partial order on the set of all info-multifunctions on $\mathcal{G}$.[10]

Definition 6.1. Let $T_1, T_2 \in \mathcal{M}(\mathcal{G},n)$. We say that T_1 is a **shrinking** of T_2, or that T_2 is a **dilatation** of T_1, if $gr(T_1) \subseteq gr(T_2)$ (equivalently, $\forall G \in \mathcal{G}$, $T_1(G) \subseteq T_2(G)$).

The shrinking and dilatation concepts defined above can be respectively associated to the loss and acquisition of information processes. While the latter concept is reasonable, as decision makers tend to receive new information through time, the former one may seem counterintuitive. However, it can be theoretically justified using the concept of limited memory. This notion is widely used in economic theory to model environments where decision makers display a limited capacity to assimilate and remember information, see for example [13].

In Section 5, we have shown that every info-multifunction T satisfying property $(\star)$ determines a well defined preference relation $\succsim_{\psi^T}$, which, by Theorem 5.7, is continuously representable on the set $Dom(T)$ by the utility function $u \circ \psi^T$.

As remarked after Definition 4.1, we are assuming that any info-multifunction T displays information so as to direct the choice of the decision maker towards a given set of goods, namely $Dom(T)$. Therefore, we will refer to the set $Dom(T)$ as the **choice set** induced by T.

Intuitively speaking, increasing (or decreasing) the information available to the decision maker through a dilatation (or a shrinking) of T after a certain interval of time may lead to a different preference relation and consequently to a different choice.

What we show with the use of basic homotopic results is that continuously representable preference relations determined by dilatations or shrinkings of a given info-multifunction can be continuously transformed one into another. This property implies that *an informed sender is able to manipulate the choice of the decision maker through time and in a smooth way.*

Let T be given at time $t = 0$ and suppose that after a (continuous) time interval, the sender releases further information encoded in a dilation of T denoted by T^d. By Proposition 2.10, if both $u \circ \psi^T$ and $u \circ \psi^{T^d}$ are continuous on a certain subset $\mathcal{H}$ of $\mathcal{G}$, then there exists a homotopy $F^d : \mathcal{H} \times [0,1] \to \mathbb{R}$ continuously transforming $u \circ \psi^T$ in $u \circ \psi^{T^d}$.

Similarly, if the sender decides to reduce the initial information encoded in T using a shrinking T^s such that $u \circ \psi^T$ and $u \circ \psi^{T^s}$ are both continuous on a given subset $\mathcal{H}$ of $\mathcal{G}$, then there exists an homotopy $F^s : \mathcal{H} \times [0,1] \to \mathbb{R}$ continuously transforming $u \circ \psi^T$ in $u \circ \psi^{T^s}$.

[10]Definition 6.1 is an interpretation in information terms of the standard definition of submultifunction/supermultifunction. See for example Section 6.1 in [4].

In this respect, note that, by Theorem 5.7, if both the info-multifunction T and its dilatation T^d satisfy property $(\star)$, then both $u \circ \psi^T$ and $u \circ \psi^{T^d}$ are continuous but not necessarily on the same domain. In fact, $u \circ \psi^T$ is continuous when restricted to $Dom(T)$, but it does not need to be continuous on $Dom(T^d)$, possibly larger than $Dom(T)$.

Similarly, if both the info-multifunction T and its shrinking T^s satisfy property $(\star)$, then both $u \circ \psi^T$ and $u \circ \psi^{T^s}$ are continuous but not necessarily on the same domain. In fact, $u \circ \psi^{T^s}$ is continuous when restricted to $Dom(T^s)$, but it does not need to be continuous on $Dom(T)$, possibly larger than $Dom(T^s)$.

Thus, it may not exist a homotopy between $u \circ \psi^T$ and $u \circ \psi^{T^d}$ (resp. between $u \circ \psi^T$ and $u \circ \psi^{T^s}$). A necessary condition for such a homotopy to exist is that $Dom(T^d) = Dom(T)$ (resp. $Dom(T^s) = Dom(T)$).

In other words, in order to be able to apply Proposition 2.10 and to define a homotopy between the initially induced preference relation $\succsim_{\psi^T}$ and the new one $\succsim_{\psi^{T^d}}$ (resp. $\succsim_{\psi^{T^s}}$) resulting from dilatating (resp. shrinking) the information transmitted, both these preference relations must be defined and continuously representable on the same subset of $\mathcal{G}$, namely $Dom(T)$.

To simplify the presentation of the results and avoid any misunderstanding, we extend the homotopic relationship to the induced preference relations.

Definition 6.2. Let $T_1, T_2 \in \mathcal{M}(\mathcal{G}, n)$ and $\mathcal{H} \subseteq \mathcal{G}$. The preference relation $\succsim_{\psi^{T_1}}$ is said to be **homotopic** to the preference relation $\succsim_{\psi^{T_2}}$ on $\mathcal{H}$ if the utility functions $u \circ \psi^{T_1}$ and $u \circ \psi^{T_2}$ are continuous and homotopic on $\mathcal{H}$.

The previous remarks together with Definition 6.2 justify the introduction of the following notion of *compatibility*.

Definition 6.3. Let $T \in \mathcal{M}(\mathcal{G}, n)$. A dilatation T^d (resp. shrinking T^s) of T is called **compatible** if $Dom(T^d) = Dom(T)$ (resp. $Dom(T^s) = Dom(T)$).

Note that examples of info-multifunctions that do not admit any compatible dilatation or shrinking can be easily constructed.

Example 6.4. Fix $G^\star$ in $\mathcal{G}$. Consider the info-multifunction $T \in \mathcal{M}(\mathcal{G}, n)$ defined as follows:

$$T(G^\star) = \{1, \dots, n\} \text{ and } T(G) = \emptyset, \ \forall G \neq G^\star.$$

Clearly, $Dom(T) = \{G^\star\}$. Moreover, a dilatation T^d such that $Dom(T^d) = \{G^\star\}$ cannot be defined. In fact, $T^d(G^\star)$ cannot be larger than $\{1, \dots, n\}$.

Figure 6.1 represents the graph of the info-multifunction defined in Example 6.4 in the particular case when $n = 5$.

Example 6.5. Fix $G^\star$ in $\mathcal{G}$. Consider the info-multifunction $T \in \mathcal{M}(\mathcal{G}, n)$ defined as follows:

$$T(G^\star) = \{1\} \text{ and } T(G) = \emptyset, \ \forall G \neq G^\star.$$

Clearly, $Dom(T) = \{G^\star\}$. At the same time, the only possible way to shrink T is associating the empty set to $G^\star$. The resulting multifunction is $T^\emptyset$, already excluded from the set $\mathcal{M}(\mathcal{G}, n)$.

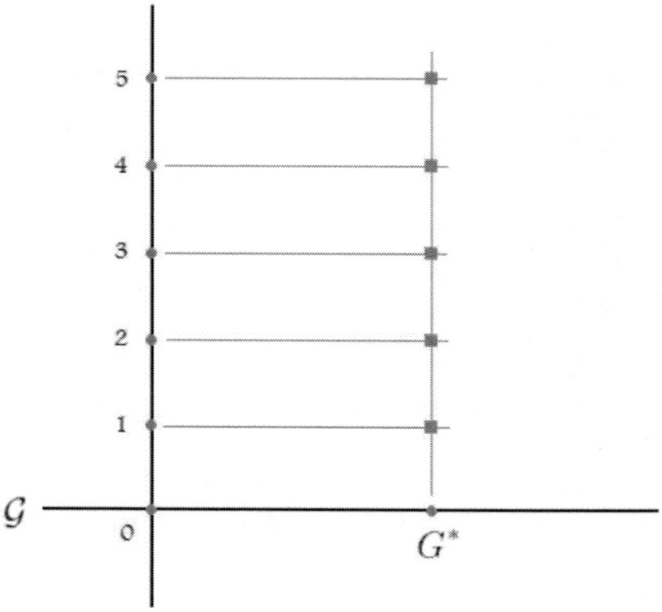

Figure 6.1

Figure 6.2 represents the graph of the info-multifunction defined in Example 6.5 in the particular case when $n = 5$.

Proposition 6.6. *Let $T \in \mathcal{M}(\mathcal{G}, n)$. The following are equivalent:*

(a) *T admits a compatible dilatation;*

(b) *$\exists G \in Dom(T)$ such that $|T(G)| < n$.*

Proposition 6.7. *Let $T \in \mathcal{M}(\mathcal{G}, n)$. The following are equivalent:*

(a) *T admits a compatible shrinking;*

(b) *$\exists G \in Dom(T)$ such that $|T(G)| \geq 2$.*

Theorem 6.8. *Let $T \in \mathcal{M}(\mathcal{G}, n)$ satisfy property $(\star)$.*

(a) *For every compatible dilatation T^d satisfying property $(\star)$, $\succsim_{\psi^{T^d}}$ is homotopic to $\succsim_{\psi^T}$ on $Dom(T)$.*

(b) *For every compatible shrinking T^s satisfying property $(\star)$, $\succsim_{\psi^{T^s}}$ is homotopic to $\succsim_{\psi^T}$ on $Dom(T)$.*

Proof. (a). By Theorem 5.7, $u \circ \psi^T \restriction_{Dom(T)}$ and $u \circ \psi^{T^d} \restriction_{Dom(T^d)}$ are both continuous. Since $Dom(T) = Dom(T^d)$, by Proposition 2.10, there exists a homotopy $F : Dom(T) \times [0,1] \to \mathbb{R}$ continuously transforming $u \circ \psi^T \restriction_{Dom(T)}$ into $u \circ \psi^{T^d} \restriction_{Dom(T^d)}$.

(b). Similar to the proof of (a). □

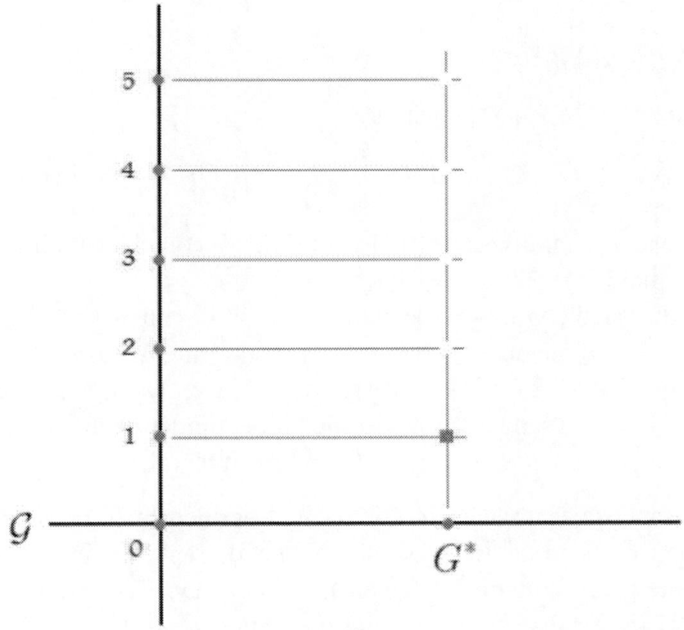

Figure 6.2

7 Compatible Dilatations and Shrinkings preserving Property($\star$)

Let $T \in \mathcal{M}(\mathcal{G},n)$ be defined so as to satisfy condition (b) of Proposition 6.6 and property $(\star)$. Then,

$(b.1)$ there exist $\alpha^\star \in \Lambda(T)$ and an index j such that the j-th characteristic is still unknown for all goods belonging to $\mathcal{D}_{\alpha^\star}$.

Furthermore, the associated family $\Delta(T) = \{\mathcal{D}_\alpha : \alpha \in \Lambda(T)\}$ consists of closed subsets of $\mathcal{G}$. The easiest way of obtaining a compatible dilatation preserving property $(\star)$ is the following.

Let $T^{d(\alpha^\star,j)}$ be defined by

$$T^{d(\alpha^\star,j)}(G) = \begin{cases} T(G) \cup \{j\} & \text{if } G \in \mathcal{D}_{\alpha^\star}, \\ T(G) & \text{otherwise.} \end{cases}$$

It is easy to check that $Dom(T^{d(\alpha^\star,j)}) = Dom(T)$ and that one of the following holds:

Case 1.

- $\Lambda(T^{d(\alpha^\star,j)}) = \Lambda(T)$
- for every $\alpha \in \Lambda(T) \backslash \{\alpha^\star\}$,

$$G \in \mathcal{D}_{\alpha^\star} \text{ and } H \in \mathcal{D}_\alpha \implies T(G) \cup \{j\} \neq T(H).$$

Case 2.

- $\Lambda(T^{d(\alpha^\star,j)}) = \Lambda(T)\setminus\{\alpha^\star\}$
- there exist $\alpha' \in \Lambda(T)\setminus\{\alpha^\star\}$ such that

$$G \in \mathcal{D}_{\alpha^\star} \text{ and } H \in \mathcal{D}_{\alpha'} \implies T(G)\cup\{j\} = T(H).$$

In the first case, the family associated with $T^{d(\alpha^\star,j)}$ coincides with the one already associated with T, that is, $\Delta(T^{d(\alpha^\star,j)}) = \Delta(T)$.

In the second case, the family associated with $T^{d(\alpha^\star,j)}$ consists of all the sets $\mathcal{D}_\alpha$, where $\alpha \in \Lambda(T)\setminus\{\alpha^\star,\alpha'\}$, and the union set $\mathcal{D}_{\alpha^*} \cup \mathcal{D}_{\alpha'}$, which are all closed subsets of $\mathcal{G}$.

Consequently, $T^{d(\alpha^\star,j)}$ satisfies property $(\star)$. Hence, Theorem 6.8 can be applied to guarantee the information sender the possibility of manipulating the preferences of the decision maker in a continuous way during a fixed time interval.

Example 7.1. Suppose that each good in $\mathcal{G}$ is described by $n = 5$ characteristics. Let $T \in \mathcal{M}(\mathcal{G},5)$ be the multifunction whose domain and graph are represented in Figure 7.1 by the red and the green horizontal segments, respectively. The family associated with T consists of the disjoint subsets $\mathcal{D}_1$, $\mathcal{D}_2$ and $\mathcal{D}_3$. Suppose that each $\mathcal{D}_i$ is closed, so that property $(\star)$ holds.

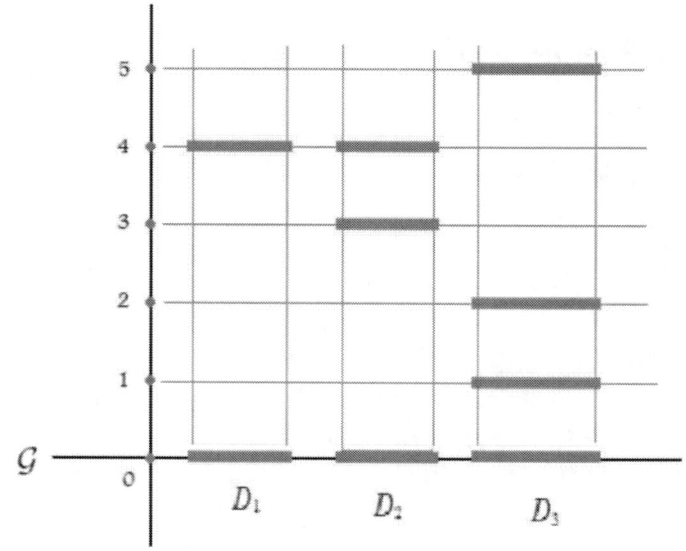

Figure 7.1

Clearly T satisfies condition (b) of Proposition 6.6, and, hence, condition $(b.I)$. Among the possible compatible dilatations, consider $T^{d(1,1)}$, whose graph is represented in Figure 7.2.

The dilatation $T^{d(1,1)}$ belongs to Case 1 above. Similarly, the dilatation $T^{d(1,3)}$, whose graph is illustrated in Figure 7.3, is also compatible but belongs to Case 2. Both dilatations preserve property $(\star)$.

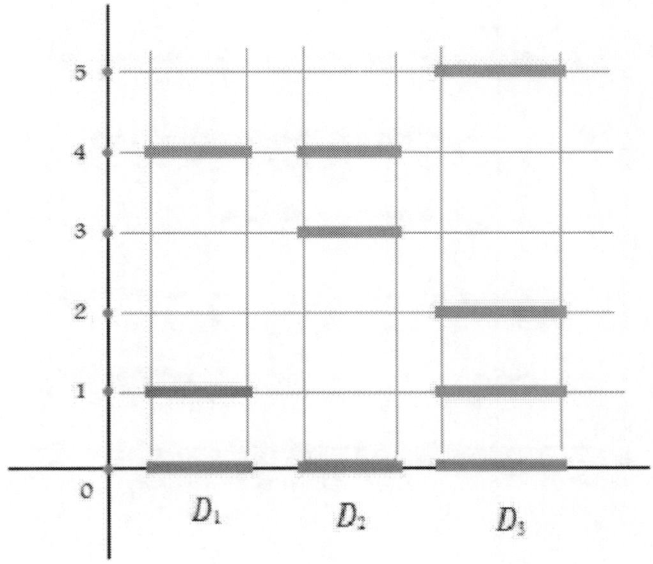

Figure 7.2

The dual of condition $(b.I)$ used above allows us to define compatible shrinkings that preserve property $(\star)$.

Assume that $T \in \mathcal{M}(\mathcal{G}, n)$ satisfies condition (b) of Proposition 6.7 and property $(\star)$. Thus,

$(b.II)$ there exist $\alpha^\star \in \Lambda(T)$ and an index j such that the j-th characteristic is already known for all goods belonging to $\mathcal{D}_{\alpha^\star}$,

and the associated family $\Delta(T) = \{\mathcal{D}_\alpha : \alpha \in \Lambda(T)\}$ consists of closed subsets of $\mathcal{G}$.

Let $T^{s(\alpha^\star, j)}$ be defined by

$$T^{s(\alpha^\star,j)}(G) = \begin{cases} T(G)\backslash\{j\} & \text{if } G \in \mathcal{D}_{\alpha^\star}, \\ T(G) & \text{otherwise.} \end{cases}$$

It is easy to check that $Dom(T^{s(\alpha^\star,j)}) = Dom(T)$ and that one of the following holds:

Case 1.

- $\Lambda(T^{s(\alpha^\star,j)}) = \Lambda(T)$
- for every $\alpha \in \Lambda(T)\backslash\{\alpha^\star\}$,

$$G \in \mathcal{D}_{\alpha^\star} \text{ and } H \in \mathcal{D}_\alpha \implies T(G)\backslash\{j\} \neq T(H).$$

Case 2.

- $\Lambda(T^{s(\alpha^\star,j)}) = \Lambda(T)\backslash\{\alpha^\star\}$

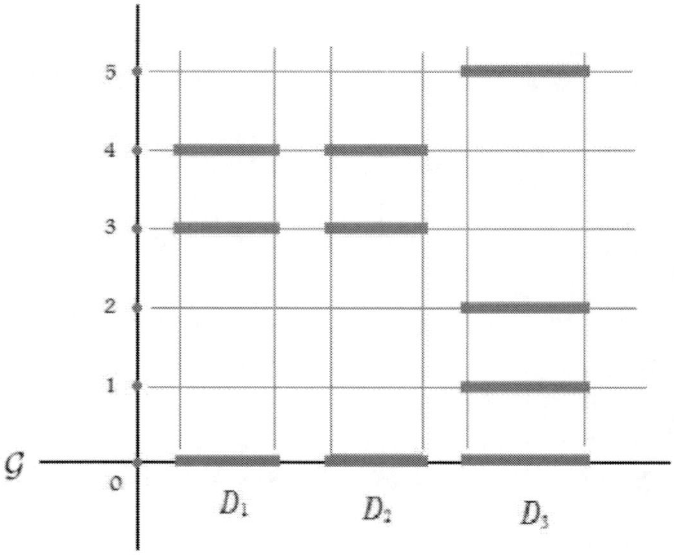

Figure 7.3

- there exist $\alpha' \in \Lambda(T)\setminus\{\alpha^\star\}$ such that

$$G \in \mathcal{D}_{\alpha^\star} \text{ and } H \in \mathcal{D}_{\alpha'} \implies T(G)\setminus\{j\} = T(H).$$

In the first case, the family associated with $T^{s(\alpha^\star,j)}$ coincides with the one already associated with T, that is, $\Delta(T^{s(\alpha^\star,j)}) = \Delta(T)$.

In the second case, the family associated with $T^{s(\alpha^\star,j)}$ consists of all the sets $\mathcal{D}_\alpha$, where $\alpha \in \Lambda(T)\setminus\{\alpha^\star,\alpha'\}$, and the union set $\mathcal{D}_{\alpha^*} \cup \mathcal{D}_{\alpha'}$, which are all closed subsets of $\mathcal{G}$.

Consequently, $T^{s(\alpha^\star,j)}$ satisfies property $(\star)$.

Example 7.2. Suppose that each good in $\mathcal{G}$ is described by $n = 5$ characteristics and let $T \in \mathcal{M}(\mathcal{G},5)$ be the multifunction of Example 7.1.

Note that T also satisfies condition (b) of Proposition 6.7, and, hence, condition $(b.II)$. Consider the following two compatible shrinkings: $T^{s(3,2)}$ and $T^{s(2,3)}$. The first shrinking, represented in Figure 7.4(a), belongs to Case 1, while the second, illustrated in Figure 7.4(b), belongs to Case 2. Both shrinkings preserve property $(\star)$.

The idea behind the procedure that we have followed to define compatible dilatations and shrinkings preserving property $(\star)$ can be generalized so as to define a class of suitable dilatations and shrinkings for any given info-multifunction.

7.1 Uniform Dilatations

Let $T \in \mathcal{M}(\mathcal{G},n)$ be defined so as to satisfy condition (b) of Proposition 6.6 and let $\Delta(T) = \{\mathcal{D}_\alpha : \alpha \in \Lambda(T)\}$ be the family associated with T.

For every $\alpha \in \Lambda(T)$, let $J_\alpha \subset \{1,\ldots,n\}$ be the set of the indexes corresponding to the characteristics which are still unknown for all the goods in $\mathcal{D}_\alpha$. Clearly, $J_\alpha = \emptyset$ if $T(G) = \{1,\ldots,n\}$ for $G \in \mathcal{D}_\alpha$.

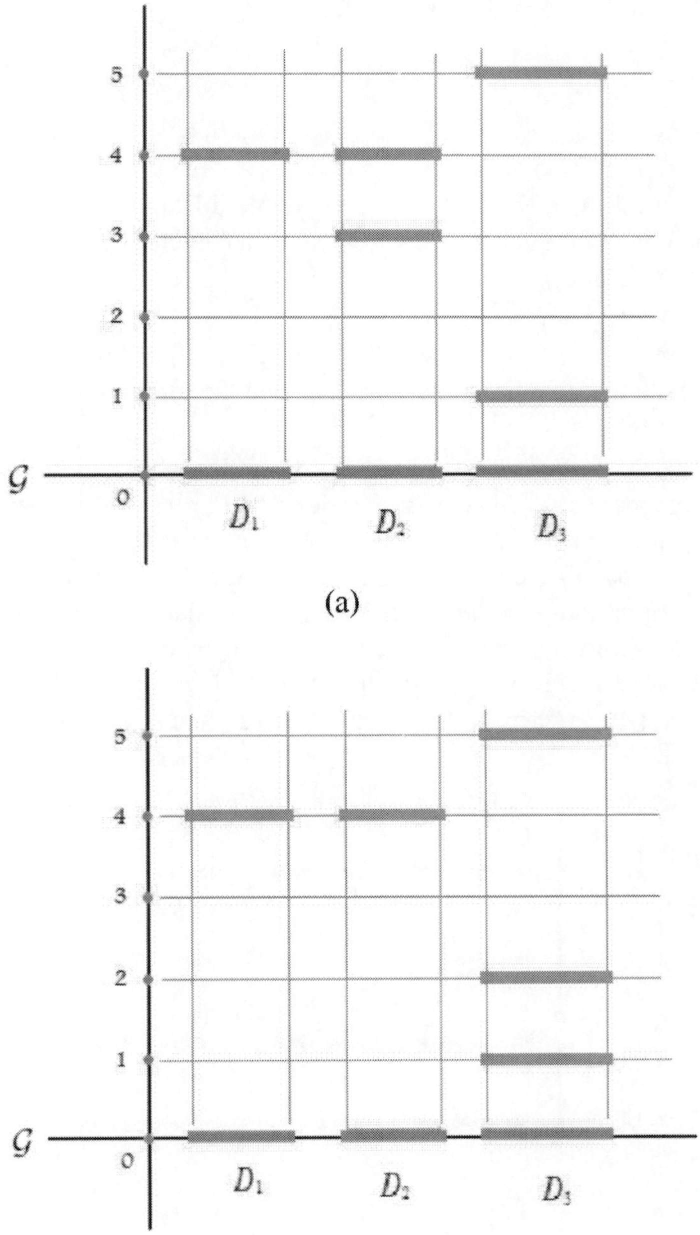

(b)

Figure 7.4

Now, let $\Gamma(T) = \{\alpha \in \Lambda(T) : J_\alpha \neq \emptyset\}$. Note that $\Gamma(T) \neq \emptyset$ since condition (b) of Proposition 6.6 holds.

For every $\Theta \subseteq \Gamma(T)$ and for every family $\{I_\alpha : \alpha \in \Theta\}$, where for every $\alpha \in \Theta$, $I_\alpha \subseteq J_\alpha$, consider the info-multifunction $T^{d(\Theta, I_{\alpha\in\Theta})}$ defined by:

$$T^{d(\Theta, I_{\alpha\in\Theta})}(G) = \begin{cases} T(G) \cup I_\alpha & \text{if } \alpha \in \Theta \text{ and } G \in \mathcal{D}_\alpha, \\ T(G) & \text{otherwise.} \end{cases}$$

It is easy to check that $T^{d(\Theta, I_{\alpha\in\Theta})}$ is a compatible dilatation of T. Given its uniform character on each one of the sets of the family associated with T, we can classify this type of dilatation as "uniform".

Definition 7.3. Let $T \in \mathcal{M}(\mathcal{G}, n)$ be such that condition (b) of Proposition 6.6 is satisfied. An info-multifunction of the form $T^{d(\Theta, I_{\alpha\in\Theta})}$ is called the **uniform dilatation determined by Θ and $\{I_\alpha : \alpha \in \Theta\}$.**

Example 7.4. Suppose that each good in $\mathcal{G}$ is described by $n = 5$ characteristics and let $T \in \mathcal{M}(\mathcal{G}, 5)$ be the multifunction of Example 7.1. We can compatibly dilatate the information encoded in T by expanding the information known for the goods in $\mathcal{D}_1$, $\mathcal{D}_2$ and $\mathcal{D}_3$ as follows. Add the information on the 2-nd, 3-rd and 5-th characteristics for all the goods in $\mathcal{D}_1$, on the 5-th characteristic for all the goods in $\mathcal{D}_2$, and on the 3-rd and 4-th characteristics for all the goods in $\mathcal{D}_3$. This leads to the dilatation $T^{d(\Theta, I_{\alpha\in\Theta})}$ determined by $\Theta = \{1, 2, 3\}$ and the family $\{I_1, I_2, I_3\}$, where $I_1 = \{2, 3, 5\}$, $I_2 = \{5\}$ and $I_3 = \{3, 4\}$. See Figure 7.5.

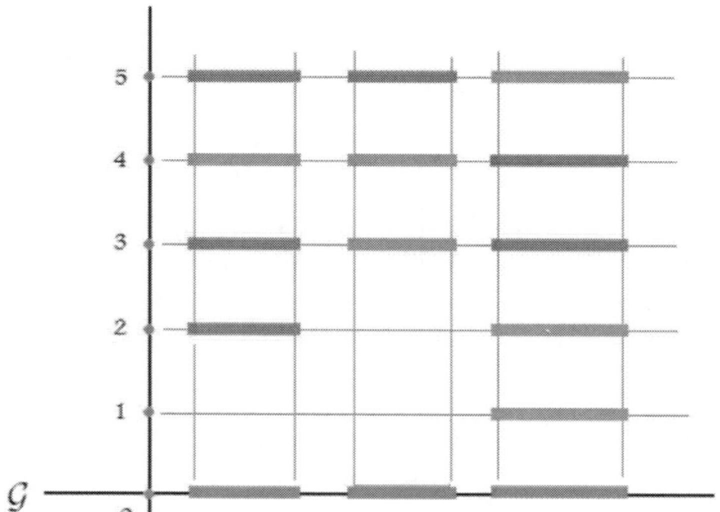

Figure 7.5

7.2 Uniform Shrinkings

Let $T \in \mathcal{M}(\mathcal{G},n)$ satisfy condition (b) of Proposition 6.7 and let $\Delta(T) = \{\mathcal{D}_\alpha : \alpha \in \Lambda(T)\}$ be the family associated with T.

For every $\alpha \in \Lambda(T)$, let $J_\alpha \subseteq \{1,\ldots,n\}$ be the set of the indexes of the characteristics which are already known for all the goods in $\mathcal{D}_\alpha$. Clearly, by definition of associated family, $J_\alpha \neq \emptyset$, for all α.

Let $\Gamma(T) = \{\alpha \in \Lambda(T) : |J_\alpha| \geq 2\}$. Note that $\Gamma(T) \neq \emptyset$ since condition (b) of Proposition 6.7 holds.

For every $\Theta \subseteq \Gamma(T)$ and for every family $\{I_\alpha : \alpha \in \Theta\}$, where for every $\alpha \in \Theta$, I_α is a proper subset of J_α, consider the info-multifunction $T^{s(\Theta,I_{\alpha\in\Theta})}$ defined by:

$$T^{s(\Theta,I_{\alpha\in\Theta})}(G) = \begin{cases} T(G)\backslash I_\alpha & \text{if } \alpha \in \Theta \text{ and } G \in \mathcal{D}_\alpha, \\ T(G) & \text{otherwise.} \end{cases}$$

Clearly, $T^{s(\Theta,I_{\alpha\in\Theta})}$ is a compatible "uniform" shrinking of T.

Definition 7.5. Let $T \in \mathcal{M}(\mathcal{G},n)$ satisfy condition (b) of Proposition 6.7. An info-multifunction of the form $T^{s(\Theta,I_{\alpha\in\Theta})}$ is called the **uniform shrinking determined by Θ and $\{I_\alpha : \alpha \in \Theta\}$**.

Example 7.6. Suppose that each good in $\mathcal{G}$ is described by $n = 5$ characteristics and let $T \in \mathcal{M}(\mathcal{G},5)$ be the multifunction of Example 7.1. We can compatibly shrink the information encoded in T by restricting the information known for the goods in $\mathcal{D}_2$ and $\mathcal{D}_3$ in the following way. Cancel the information on the 4-th characteristic for all the goods in $\mathcal{D}_2$, and on the 1-st and 2-nd characteristics for all the goods in $\mathcal{D}_3$. We obtain the shrinking $T^{s(\Theta,I_{\alpha\in\Theta})}$ determined by $\Theta = \{2,3\}$ and the family $\{I_2, I_3\}$, where $I_2 = \{4\}$ and $I_3 = \{1,2\}$. See Figure 7.6.

Note that no compatible shrinking can be obtained by cancelling information on the goods in $\mathcal{D}_1$.

7.3 Uniformity and Homotopic Preferences

The following result states that the property $(\star)$ is preserved by any uniform dilatation or shrinking of a given info-multifunction. The proof is analogous to those given at the beginning of this section for $T^{d(\alpha^*,j)}$ and $T^{s(\alpha^*,j)}$, and is, therefore, left to the reader.

Proposition 7.7. *Let $T \in \mathcal{M}(\mathcal{G},n)$.*

(a) *If T satisfies condition (b) of Proposition 6.6 and property $(\star)$, then every uniform dilatation of T is compatible and preserves property $(\star)$.*

(b) *If T satisfies condition (b) of Proposition 6.7 and property $(\star)$, then every uniform shrinking of T is compatible and preserves property $(\star)$.*

Finally, Proposition 7.7 combined with Theorem 6.8, allow us to conclude that an information sender can manipulate in a continuous way and through time the preferences of a decision maker via uniform dilatations or shrinkings of an initially assigned info-multifunction.

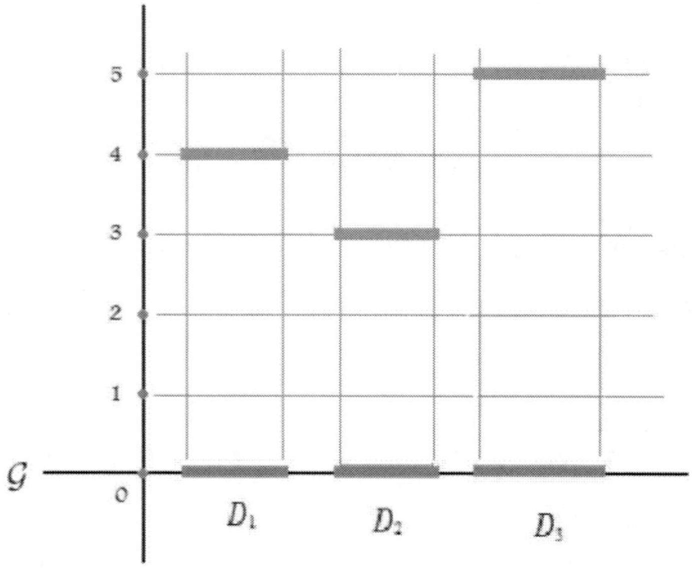

Figure 7.6

Theorem 7.8. *Let $T \in \mathcal{M}(\mathcal{G}, n)$.*

(a) *If T satisfies condition (b) of Proposition 6.6 and property $(\star)$, then for every uniform dilatation D of T, $\succsim_{\psi^D}$ is homotopic to $\succsim_{\psi^T}$ on $Dom(T)$.*

(b) *If T satisfies condition (b) of Proposition 6.7 and property $(\star)$, then for every uniform shrinking S of T, $\succsim_{\psi^S}$ is homotopic to $\succsim_{\psi^T}$ on $Dom(T)$.*

Proof. By Proposition 7.7 and Theorem 6.8. □

8 Conclusion

Homotopy theory has been sporadically applied to economic theory mainly in order to simplify the aggregation of preferences among decision makers in social choice (see [17]), and to design stable algorithms in computable general equilibrium models (see [15]). These applications, while dealing with relevant issues, do not consider explicitly the influence of information asymmetries on the behaviour of decision makers, which constitutes a leading argument in current economic theory.

The present chapter has shown how homotopy theory can be used to design destabilizing mechanisms in economic equilibrium theory when information asymmetries among decision makers are explicitly considered. Even though the homotopy techniques employed may not be complex in themselves, it is the design of a theoretical economic decision scenario allowing for their application what deserves particular attention. Indeed, once the asymmetric information structure is built, homotopy theory can be used to illustrate how the information sets determining the choices of decision makers are modifiable so as to induce any a priori assigned economic equilibrium.

Finally, the current paper has aimed at extending the limits of economic theory in dealing with complex theoretical structures not explicitly defined by economists. At the same time, it has introduced new applications of homotopy theory in economics, that had not yet been considered.

References

[1] G. Akerlof, The Market for "Lemons": Quality Uncertainty and the Market Mechanism, *The Quarterly Journal of Economics*, **84** (1970), 488–500.

[2] K. Arrow and F. Hahn, *General Competitive Analysis*, North-Holland, Amsterdam,1991.

[3] M. Battaglini, Multiple Referrals and Multidimensional Cheap Talk, *Econometrica*, **70** (2002), 1379–1401.

[4] G. Beer, *Topologies on Closed and Closed Convex Sets*, Kluwer Academic Publishers, 1993.

[5] A. Chakraborty and R. Harbaugh, Comparative Cheap Talk, *Journal of Economic Theory*, **132** (2007), 70–94.

[6] V. Crawford and J. Sobel, Strategic Information Transmission, *Econometrica*, **50** (1982), 1431– 1451.

[7] G. Debreu, Representation of a preference ordering by a numerical function (1954). In: G. Debreu, *Mathematical economics: Twenty papers of Gerard Debreu*, Cambridge University Press, 1983.

[8] G. Debreu, Topological methods in cardinal utility theory (1960). In: G. Debreu, *Mathematical economics: Twenty papers of Gerard Debreu*, Cambridge University Press, 1983.

[9] D. Di Caprio and F. Santos-Arteaga, Continuous Lexicographic Choice Through Incomplete Information, *Working Paper no. 52*, Free University of Bozen-Bolzano, 2007.

[10] D. Di Caprio and F. Santos-Arteaga, Rationally Induced Choice Errors: Error Multifunctions and Generalized Certainty Equivalents, *Working Paper no. 53*, Free University of Bozen-Bolzano, 2007.

[11] D. Di Caprio and F. Santos-Arteaga, Error-Induced Certainty Equivalents: A Set-Theoretical Approach to Choice under Risk, *International Journal of Contemporary Mathematical Sciences*, **3** (2008), 1121–1131.

[12] D. Di Caprio and S. Watson, Continuous Selections and Purely Topological Convex Structures, *Topology Proceedings*, **29** (2005), 75– 103.

[13] J. Dow, Search Decisions with Limited Memory, *Review of Economic Studies*, **58** (1991), 1– 14.

[14] U. Dulleck and R. Kerschbamer, On Doctors, Mechanics, and Computer Specialists: The Economics of Credence Goods, *Journal of Economic Literature*, **XLIV** (2006), 5–42.

[15] B.C. Eaves and K. Schmedders, General Equilibrium Models and Homotopy Methods, *Journal of Economic Dynamics & Control*, **23** (1999), 1249–1279.

[16] R. Engelking, *General Topology*, Heldermann Verlag, Berlin, 1989.

[17] L. Lauwers, Topological Social Choice, *Mathematical Social Sciences*, **40** (2000), 1–39.

[18] A. Mas-Colell, M.D. Whinston and J.R. Green, *Microeconomic Theory*, Oxford University Press, New York, 1995.

[19] W.S. Massey, *A Basic Course in Algebraic Topology*, Springer-Verlag, New York, 1991.

[20] E. Michael, Continuous selections I, *Ann. Math.*, **63** (1956), 361–382.

[21] E. Michael, Continuous selections II, *Ann. Math.*, **64** (1956), 562–580.

[22] E. Michael, Continuous selections III, *Ann. Math.*, **65** (1957), 375–390.

[23] E. Michael, Convex Structures and Continuous Selections, *Can. J. Math.*, **11** (1959), 556–575.

[24] J.R. Munkres, *Topology*, Prentice Hall, Inc, 2000.

[25] G. Stigler, The Economics of Information, *Journal of Political Economy*, **LXIX** (1961), 213– 225.

[26] A. Taylor, *Social Choice and the Mathematics of Manipulation* , Cambridge University Press, 2005.

[27] P. Wakker, Continuity of preference relations for separable topologies, *International Economic Review*, **29** (1988), 105–110.

[28] P. Wakker, *Additive Representations of Preferences, A New Foundation of Decision Analysis*, Dordrecht, Kluwer Academic Publishers, 1989.

In: Computational Mathematics
Editor: Peter G. Chareton, pp. 435-437
ISBN: 978-1-60876-271-2

Chapter 15

MINGGEN CUI (1942-) BIOGRAPHICAL ARTICLE

Fazhan Geng[1] and Wei Jiang[1]
Department of Mathematics, Harbin Institute of Technology, Weihai,
Shandong 264209, P.R.China[1]

In the field of reproducing kernel theory, Cui is an outstanding pioneer. He is highly gifted and regarded as a scientist. He often writes for prestigious journals and became a professor after he became a mathematics teacher in junior middle school for twelve years.

Cui was born in 1942 in Yanbian Korean municipality, Jilin Province. His parents are authentic farmers. Influenced by the industry and simplicity of his parents, Cui is diligent to do anything. As a child, Cui was an avid reader in all areas of study and his skills in study certainly were apparent----he entered Senior Middle Schools.

Cui entered Heilongjiang University in 1962, and received his B.S. in 1966. After graduation, he became a mathematics teacher in junior middle school. In this period, he studied ruby laser with his colleagues, so he was redeployed to Changchun Mitsumoto College and still studied optical following Daheng Wang who is the main designer of Chairman Mao's crystal coffin. But three months later, he discovered that he is still interested in mathematics. Fortunately, the cultural revolution was just over and he successfully passed his postgraduate exam of Harbin Institute of Technology the first year. In this University, he obtained his M.S. in computational mathematics in 1981 and Ph.D. in fundamental mathematics in 1986. During the study, he solved two world difficult problems in Hermite interpolation.

As a graduate student, Cui mainly researched the approximation theory in reproducing kernel space and numerical methods since 1980s. He was able to continue his research at the Harbin Institute of Technology when he became an assistant professor after obtaining his Doctorate. Cui proved that $W_2^1[a,b]$ is a reproducing kernel Hilbert space and gave concrete expression of reproducing kernel [1]. After that, he initiated the research in the field of reproducing kernel in China.

Cui's work on reproducing kernel made important contributions to the field of interpolation problem. By using the character of reproducing kernel space, Cui obtained the best expression of interpolation in $W_2^1[a,b]$ [2] and discussed surface interpolation in

$W_2^1[a,b]$ [3]. There are many two-dimensional problems eager to solve both in theory and practice. So Cui established the reproducing kernel space $W_2^1[D]$ in two dimensional rectangular domain $D=[a,b]\times[c,d]\subset R^2$, which establishes the foundation of studying multivariate interpolation. He constructed multivariate interpolation formula and an iterative algorithm for multivariate interpolation formula in $W_2^1[D]$. Cui researched not only the reproducing kernel theory, but also the application of reproducing kernel theorem to solve linear operator equations. Working with Zhongxing Deng, Cui explored the expression of solution of differential equation with determinate conditions [4]. Furthermore, they gave analytical expression of exact solution to ill-posed operator equation of first kind. At the same time, Cui carried out research on reproducing kernel in infinity domain and showed exact solution of integral-differential equations in $W_2^2[0,+\infty)$ [5].

As more and more scientists paid attention to nonlinear problems, Cui also turned the focal point studied to the numerical method for solving nonlinear operator equations in reproducing kernel space. He firstly studied factorization method and characteristic value method for the approximation solution of nonlinear operator equation formed by $AuBu+Cu=f$ in $W_2^1[a,b]$ [6], where $A,B,C: W_2^1[a,b]\to W_2^1[a,b]$ are bounded linear operators. Using reproducing kernel theory, he presented methods for solving singular boundary value problems, singularly perturbed boundary value problems, systems of boundary value problems and partial differential equation [7-15]

Throughout his 30-year career at the Harbin Institute of Technology, Cui was active as a professor on campus and was well known. As a young scientist, he was awarded the Special Contribution to aviation and space from National Aeronautics and Space Administration in 1991. In 2006, he was appointed as evaluation expert of National Science and Technology Award, which is the highest quality award in China. He had been asked many times to give lectures in Japan and Korea and had been invited to deliver a forty-five minute talk at the Fifth World Congress of Nonlinear Analysts (2008) on the area of reproducing kernel space.

Cui was an active member of many professional organizations. He was a board chairperson of Heilongjiang Institute of Computational Mathematics Institute, a member of the Standing Council of China Institute of Computational Mathematics.

Cui is an extraordinary teacher, many of his students like him. He was granted the first prize of first teaching in Harbin Institute of Technology.

References

[1] M.G. Cui, Z.X. Deng, Boying *Wu,Numerical functional method in reproducing kernel space,* Harbin Institute of Technology Press, 1988.

[2] M.G.Cui, Z.X.Deng. *On the Best Operator of Interpolation.* Math.Numerical Sinica. 8(2) (1986)207-218.

[3] M.G.Cui. *Two-Dimensional Reproducing Kernel and Surface Interpolation.* J.Comp.Math. 4(2)(1986)177-181.

[4] M.G.Cui, Z.X.Deng. *Solutions to the Definite Solution Problem of Differential Equation in Space*. Advances in Math. 17(2)(1998)327-329.

[5] Y.H. Li, Minggen Cui, *The representation of exact solution og the first inter-differential equation in the reproducing kernel space,* Mathematics of Computation 21(2)(1999)189-198.

[6] C.L.Li, M.G.Cui. *How to Solve the Equation AuBu+Cu=f.* Appl. Math. Comput.19(2002)285-302.

[7] M.G. Cui, F.Z. Geng, *Solving singular two-point boundary value problem in reproducing kernel space,* Journal of Computational and Applied Mathematics 205(2007)6-15.

[8] F.Z. Geng, M.G. Cui, *Solving singular nonlinear second-order periodic boundary value problems in the reproducing kernel space*, Applied Mathematics and Computation 192 (2007)389-398.

[9] F.Z. Geng, M.G. Cui, *Solving a nonlinear system of second order boundary value problems,* Journal of Mathematical Analysis and Applications 327(2007)1167-1181.

[10] M.G. Cui, F.Z. Geng, *A computational method for solving one-dimensional variable-coefficient Burgers equation*, Applied Mathematics and Computation 188(2007)1389-1401.

[11] M.G. Cui, Y.Z. Lin, A new method of solving the coefficient inverse problem of differential equation, Science in China Series A 50(4)(2007)561-572.

[12] Y.Z. Lin, M.G. Cui, Y. Zheng, *Representation of the exact solution for infinite system of linear equations,* Applied Mathematics and Computation 168(2005)636-650.

[13] Y.Z. Lin, M.G. Cui, Lihong Yang, *Representation of the exact solution for a kind of nonlinear partial differential eqations,* Applied Mathematics Letters 19(2006)808-813.

[14] M.G. Cui, C. Chen, *The exact solution of nonlinear age-structured population model, Nonlinear Analysis*: Real World Applications 8(2007)1096-1112.

[15] C.L. Li, M.G. Cui, *The exact solution for solving a class nonlinear operator equations in the reproducing kernel space*, Applied Mathematics and Computation 143 (2003) 393-399.

INDEX

A

B

C

D

E

F

G

H

I

K

L

M

N

O

P

W